Biogeography and Geological Evolution of SE Asia

BIOGEOGRAPHY AND GEOLOGICAL EVOLUTION OF SE ASIA

edited by
Robert Hall and Jeremy D. Holloway

BACKHUYS PUBLISHERS
LEIDEN
1998

Cover

Centre: part of the global marine gravity anomaly map from satellite altimetry. The surface of the ocean bulges outward and inward mimicking the topography of the ocean floor. The bumps, too small to be seen, can be measured by a radar altimeter aboard a satellite. The sea surface height measurements are then converted to variations in the pull of gravity (gravity anomaly). From the marine gravity anomaly map from satellite altimetry by D. T. Sandwell and W. H. F. Smith, copyright David T. Sandwell 1995.

Photographs copyright Robert Hall, Jeremy D. Holloway, Moyra Wilson. Clockwise from top left: orchid, Sulawesi; tree fern, Java; above the tree line, Seram; orang-utan, Sabah; foraminiferal limestone, Sulawesi; crocodile, Singapore; jungle perch, northern Australia; islands with fringing reefs off South Sulawesi.

ISBN 90-73348-97-8

© Backhuys Publishers, Leiden, The Netherlands, 1998

All rights reserved. Nothing from this publication may be reproduced, stored in a computerized system or published in any form or in any manner, including electronic, mechanical, reprographic or photographic, without prior written permission from the publishers, Backhuys Publishers, P.O. Box 321, 2300 AH Leiden, The Netherlands

Printed in The Netherlands.

Contents

SE Asian geology and biogeography: an introduction
by J. D. Holloway and R. Hall 1

PALAEOZOIC AND MESOZOIC GEOLOGY AND BIOGEOGRAPHY

Palaeozoic and Mesozoic geological evolution of the SE Asian region: multidisciplinary constraints and implications for biogeography
by I. Metcalfe 25

Biogeography and palaeogeography of the Sibumasu terrane in the Ordovician: a review
by R. A. Fortey and L. R. M. Cocks 43

Permian marine biogeography of SE Asia
by G. R. Shi and N. W. Archbold 57

Upper Palaeozoic floras of SE Asia
by J. F. Rigby 73

The biogeographical significance of the Mesozoic vertebrates from Thailand
by E. Buffetaut and V. Suteethorn 83

The distribution of *Apsilochorema* Ulmer, 1907: biogeographic evidence for the Mesozoic accretion of a Gondwana microcontinent to Laurasia
by W. Mey 91

CENOZOIC TO RECENT GEOLOGY AND BIOGEOGRAPHY

The plate tectonics of Cenozoic SE Asia and the distribution of land and sea
by R. Hall 99

Biogeographic implications from the Tertiary palaeogeographic evolution of Sulawesi and Borneo
by S. J. Moss and M. E. J. Wilson 133

Implications of paucity of corals in the Paleogene of SE Asia: plate tectonics or Centre of Origin?
by M. E. J. Wilson and B. R. Rosen 165

Genetic structure of marine organisms and SE Asian biogeography
by J. A. H. Benzie 197

Palynological evidence for Tertiary plant dispersals in the SE Asian region in relation to plate tectonics and climate
by R. J. Morley 211

Noteworthy disjunctive patterns of Malesian mosses
by B. C. Tan 235

Patterns of distribution of Malesian vascular plants
by W. J. Baker, M. J. E. Coode, J. Dransfield, S. Dransfield, M. M. Harley,
P. Hoffmann and R. J. Johns 243

Historical biogeography of *Spatholobus* (Leguminosae-Papilionoideae) and allies in SE Asia
by J. W. A. Ridder-Numan 259

Biogeography of *Aporosa* (Euphorbiaceae): testing a phylogenetic hypothesis
using geology and distribution patterns
    by A. M. Schot 279

Geological signal and dispersal noise in two contrasting insect groups in the
Indo-Australian tropics: R-mode analysis of pattern in Lepidoptera and cicadas
    by J. D. Holloway 291

Halmahera and Seram: different histories, but similar butterfly faunas
    by R. de Jong 315

Assembling New Guinea: 40 million years of island arc accretion as indicated
by the distributions of aquatic Heteroptera (Insecta)
    by D. A. Polhemus and J. T. Polhemus 327

Marine water striders (Heteroptera, Gerromorpha) of the Indo-Pacific:
cladistic biogeography and Cenozoic palaeogeography
    by N. M. Andersen 341

Biogeography of Sulawesi grasshoppers, genus *Chitaura*, using DNA sequence data
    by R. K. Butlin, C. Walton, K. A. Monk and J. R. Bridle 355

Terrestrial birds of the Indo-Pacific
    by B. Michaux 361

Pre-glacial Bornean primate impoverishment and Wallace's line
    by D. Brandon-Jones 393

Glossary of terms 405

INDEX 411

SE Asian geology and biogeography: an introduction

Jeremy D. Holloway
Department of Entomology, The Natural History Museum, Cromwell Road, London, SW7 5BD, UK

Robert Hall
SE Asia Research Group, Department of Geology, Royal Holloway University of London, Egham, Surrey TW20 0EX, UK

Introduction

This volume originated in discussion between the editors at the Geological Society of London during a conference on the *Tectonic Evolution of SE Asia* in December 1994, the results of which have been published (Hall and Blundell, 1996). We agreed that much might be gained in trying to bring geologists and biogeographers together to discuss the development of biogeographic patterns in the SE Asian and Australasian tropics in relation to the tectonic history of the area and planned a further meeting. This meeting, *Biogeography and Geological Evolution of SE Asia*, organised by the editors and B. R. Rosen, took place in March 1996 over two days at The Natural History Museum. Many of the presentations on that occasion have been developed into chapters for this book. It also draws on an earlier meeting on *Biogeography and Biodiversity of Wallacea and Adjacent Areas*, organised by N. M. Andersen and R. I. Vane-Wright as part of the 13th meeting of the Willi Hennig Society in Copenhagen in 1994, and a later one on *Biodiversity and Biogeography of South-East Asia and South-West Pacific* organised by J. P. Duffels and J. D. Holloway for the International Congress of Entomology in Florence in August 1996. A further convergence of geologists and biogeographers with an interest in the region is planned for the year 2000 in Leiden, *Biogeography of SE Asia 2000*, when it is hoped that some of the seeds germinating during this sequence of meetings will bear further fruit.

The first major occasion at which geologists, palaeontologists and biogeographers came together was in December 1971, for a symposium entitled *Organisms and Continents through Time*, organised by the Geological Society, the Palaeontological Association and the Systematics Association (Hughes, 1973). Two of the authors in this volume also contributed on that occasion (Cocks and McKerrow, 1973; Holloway, 1973). It followed directly from the major resurgence of interest in continental drift arising through discoveries in palaeomagnetism, seafloor spreading and plate tectonics in the mid to late sixties, reviewed in the volume by Smith *et al.* (1973), who also provided a set of palaeogeographic maps for the major continental areas that biogeographers referred to frequently thereafter. Most of the chapters in Hughes (1973) were palaeontological, with only a handful of neontological contributions whereas the balance, or lack of it, is reversed in this volume. Palaeontology, even if not explicitly identified as such, has always been intimately involved in efforts by geologists to elucidate Earth history. In contrast, data from the distributions of modern organisms, for example terrestrial invertebrates, which mostly lack a diverse and continuous fossil record, are much less frequently brought into consideration. However, these organisms may represent a major untapped source of information on the extent and distribution of land areas in the past. One aim of the conference and this book was to assess this possibility.

By the early seventies, there was some consensus on the position of the major continents through Phanerozoic time (Fig.1), and a number

Era	Period	Epoch	Stage	Age
CENOZOIC	Quaternary	Holocene		0.1
		Pleistocene		1.6
	Tertiary — Neogene	Pliocene	Piacenzian	
			Zanclian	5
		Miocene	Messinian	
			Tortonian	
			Serravallian	
			Langhian	
			Burdigalian	
			Aquitanian	23
	Tertiary — Paleogene	Oligocene	Chattian	
			Rupelian	35
		Eocene	Priabonian	
			Bartonian	
			Lutetian	
			Ypresian	57
		Paleocene	Thanetian	
			Danian	65
MESOZOIC	Cretaceous	Late — Senonian	Maastrichtian	
			Campanian	
			Santonian	
			Coniacian	
		Gallic	Turonian	
			Cenomanian	97
			Albian	
			Aptian	
		Early	Barremian	
		Neocomian	Hauterivian	
			Valanginian	
			Berriasian	146
	Jurassic	Late	Tithonian	
			Kimmeridgian	
			Oxfordian	157
		Middle	Callovian	
			Bathonian	
			Bajocian	
			Aalenian	178
		Early	Toarcian	
			Pliensbachian	
			Sinemurian	
			Hettangian	208
	Triassic	Late	Rhaetian	
			Norian	
			Carnian	235
		Middle	Ladinian	
			Anisian	241
		Early	Spathian	
			Nammalian	
			Griesbachian	245

Era	Period	Epoch	Stage	Age
PALAEOZOIC	Permian	Late	Tatarian — Changxingian	
			Tatarian — Wujiapinian	
			Kazanian — Midian	
			Kazanian — Murgabian	
			Kubergandian	256
		Early	Kungurian	
			Artinskian	
			Sakmarian	
			Asselian	290
	Carboniferous — Pennsylvanian	Late	Gzelian	
			Kasimovian	
			Moscovian	
			Bashkirian	322
	Carboniferous — Mississippian	Early	Serpukhovian	
			Visean	
			Tournaisian	363
	Devonian	Late		377
		Middle		386
		Early		409
	Silurian	Late		424
		Early		439
	Ordovician	Late	Ashgill	
			Caradoc	443
		Middle	Llandeilo	
			Llanvirn	476
		Early	Arenig	
			Tremadoc	510
	Cambrian	Late		517
		Middle		536
		Early		570
PRECAMBRIAN		PROTEROZOIC		2500
		ARCHAEAN		

Fig.1. The geological timescale as used by authors in this book. Stage names for the Cambrian to Carboniferous periods are omitted. Note that Tatarian and Kazanian are Permian stages, not epochs. Ages are in Ma and are those of Harland *et al.* (1990).

of attempts had been made to relate biotic distribution patterns to this, such as by Raven and Axelrod (1974), but there was little information on the history of complex areas such as the Indonesian, Pacific or Caribbean archipelagos, a factor noted in the contribution by Holloway (1973). This contribution would now be criticised as too 'process-oriented' with its call for comprehensive, testable hypotheses for the ecological and evolutionary development of biotas to facilitate interpretation of patterns of distribution, coupled with a more ecologically oriented approach to biological survey to accumulate data on biotas in such complex areas. However, such hypotheses will be essential to any approach involving modelling of biogeographic pattern generation over past archipelagic geographies, a point we will return to later, and which was brought up at the discussion in the 1996 meeting by Roger Butlin, who commented on an early draft of this chapter.

The development of biogeographic methodology

The eighties and early nineties saw a major re-examination of biogeographic goals, philosophy and methodology, stemming from the increasing acceptance of cladistic analysis in biosystematics, the methodology being seen as the best means of generating phylogenetic hypotheses for the groups studied. Cladistic methodology and its applications are elucidated by Kitching *et al.* (1998). Cladistic methods, in conjunction with various interpretations of the writings of Leon Croizat (*e.g.,* Croizat, 1958, 1964), led to the development of two main schools of biogeographic thought, cladistic biogeography and panbiogeography. The former focuses on the relationships of areas indicated by the phylogenetic relationships of organisms endemic to these areas, and the latter on coincidence of patterns (generalised tracks) across areas, intersection of different patterns, and their relationship to the major ocean basins. Holloway (1998*; * marks chapters in this volume – these are not listed in the references) discusses this dichotomy of philosophy in relation to Q-mode and R-mode methods of analysis and reiterates the suggestion by Holloway and Jardine (1968) that such approaches are complementary rather than conflicting. Implicit in Q-mode approaches, leading to trees of biotas representing areas, is the assumption that biotas are hierarchical (Rosen, 1988b); the R-mode approach offers a partial escape from this.

The geological reader wishing to explore this debate further will find references to these topics in many of the chapters that follow, but some 'landmark' publications that give a review of these philosophies and the arguments surrounding them are, for cladistic biogeography, Nelson and Platnick (1981), Nelson and Rosen (1981), Humphries and Parenti (1986) and Humphries *et al.* (1988), and for panbiogeography, Craw (1988) and Matthews (1989). There have also been a number of volumes published in the past two decades that focus on particular areas of the world but also contain applications of these various methods. Of particular relevance to the focus of this volume are Whitmore (1981, 1987), for the Indo-Australian archipelago, Keast and Miller (1996), for the Pacific, and Ladiges *et al.* (1991) and Matthews (1989) for their southern connections. There are also two recent compendia on the similar complex archipelagic situation in the Caribbean (Leibherr, 1988; Woods, 1989).

Biogeography and geology: symbiosis or parasitism?

In most of the volumes cited above there is a chapter or chapters contributed by geologists reviewing geological knowledge of the area under study in an attempt to 'set the scene' for the more biological chapters that make up the bulk of each book. Rarely, if ever, are there biogeographic contributions to geological volumes or symposia except for palaeontological ones. Neontological evidence in the absence of any fossil record is commonly seen by geologists engaged in the task of reconstructing Earth history as peripheral to the major bodies of evidence from, for example, stratigraphy, palaeontology, palaeomagnetism and the many other geological sub-disciplines listed by Metcalfe (1998*). Attendance at our meeting in 1996 was weighted to the biogeographical side. Therefore, one of the questions we address in this chapter is what benefits can each discipline draw from a symbiosis, or is the relationship largely parasitic from the geological standpoint?

The relationship between the two disciplines is currently unequal, with biogeographers either seeking to draw geologists into their circles or attempting themselves to generate syntheses of the complex and often conflicting geological literature available at the time (*e.g.,* Holloway, 1979; Boer, 1995). Because of the complexity and volume of the geological literature, biogeographers often appear to cling to ideas which have been rejected by almost all geologists, such as expanding Earth hypotheses, hypothetical land bridges, and submerged continents, or palaeogeographies which have been replaced or discredited by the accumulation of new data. In this context, it is worth noting that in the early 1970s the oceans were still almost unexplored, allowing considerable room for speculation about global tectonics. Much of the last 200 Ma of the Earth's tectonic history is now known to be recorded in the ocean floor. Since the first meeting of geologists and biogeographers almost thirty years ago, marine geophysical cruises and ocean drilling campaigns have mapped and clearly established the age of the ocean floor, and in the last ten years satellite mapping has added immense new detail of the ocean floor structure, enabling rapid and accurate tectonic reconstructions. Plate tectonics has become firmly established (see Hallam (1994) for a good discussion of alternative explanations and their value, and an incisive dismissal of the expanding Earth hypothesis), dating of major global events is vastly improved, and there is an understanding of the connection between plate tectonics and sea level change. As we move into the 21st century the links between tectonics and climate, and between tectonics and ocean and atmospheric circulation, are the challenges being addressed, investigated and modelled.

There are a number of obstacles to a closer collaboration. Geologists have been reluctant, or attach low importance, to identifying the areas of the globe that were above sea level at any time. Hall (1998*) identifies some of the reasons for this. Although the absence of palaeogeographic maps may not be too serious for marine biogeographers, it is critical for those studying the terrestrial biota. However, the distribution of land and sea at any point in time, though reflecting geology, may not lead to biogeographic pattern that reproduces the underlying geology or its tectonic divisions, particularly in an oceanic archipelagic context. Terrestrial and marine patterns may also be influenced to a greater or lesser degree by climate. Indeed, climate, particularly variation in solar energy input (and on land, evapo-transpiration) is increasingly seen as the major factor controlling the development of high levels of biological diversity at low latitudes (*e.g.*, Currie, 1991; Wilson and Rosen, 1998*). Hence strong biogeographic signals of relationships of areas could be misleading in terms of geology and tectonics.

This also leads to problems with current biogeographical methodology which generates diagrams of area relationships that are essentially dichotomous, suggestive of progressive fragmentation of a previously continuous land area. In instances where this has actually occurred, such as the break-up of Gondwanaland, there is no methodological conflict. But the complex episodes of archipelagic evolution, divergence and coalescence that have characterised the relationship of eastern Asia and Australasian Gondwanaland through long periods of geological time are of a much more reticulate nature, involving in some cases of coalescence a sort of inverted dichotomy through time. Hallam (1994) briefly reviews the history of the vicariant and dispersalist schools of biogeographers, and the geological response to the extremes of the argument. To a geologist the dismissal of the role of dispersal seems bizarre, and judging from the geological history of SE Asia (Metcalfe, 1998*; Hall, 1998*) it seems highly improbable that any understanding of biogeographic pattern can be achieved without considering both vicariance and dispersal. Certainly, the Cenozoic development of the region is characterised more by amalgamation than fragmentation. Thus a tree of areas derived from biological evidence could reasonably be interpreted as: the fragmentation of a previously continuous land area such as Gondwanaland, or the Pleistocene division of Sundaland through marine transgression (Ruedi, 1996); the approach and accretion of terranes (and thus dispersal to) onto a larger land mass (*e.g.*, the assembly of Borneo producing patterns discussed by Ridder-Numan, 1998*); or the slow dispersal of organisms, with speciation, through an archipelago with stable geography (*e.g.*, possibly some of the Oriental allopatric groups discussed by Holloway, 1998*). Biological (phylogenetic) parallels of convergence are few but the hypothesis of hybridisation of two plant lineages in New Guinea by Schot (1998*) may be one. Objective comparison of biogeographic information with geological information, both of which are incomplete and extensively based on hypothesis, remains a major challenge despite recent methodological advances on both sides.

How many patterns?

Analysis of biogeographic pattern itself is imperfect, and the cladistic biogeographic goal of a fully resolved solution of area relationships for the world (*e.g.*, Humphries *et al.*, 1988) may now be an impediment to further progress, in that there is unlikely to be a unique solution (Holloway, 1998*; Polhemus and Polhemus, 1998*), but more probably a multiplicity of patterns. Attempts to reduce these to a lowest common denominator of area relationships may be akin to those of a child with its first paintbox trying to mix more and more exciting colours but always ending up with a sludgy brown. This can be observed as loss of resolution as more and more raw area cladograms for taxa are combined, leading to ever greater numbers of alternative trees in the results of an analysis and an increasingly comb-like consensus tree. A preliminary method of pattern (or 'colour') recognition, might help define pattern compatibility and ensure that analyses of pattern groups led to sets of general area relationship hypotheses with enhanced rather than obscured resolution.

Definition of areas of endemism

Platnick (1991) suggested areas of endemism could "be defined by the congruent distributional limits of two or more species", and recommended that biogeographers should prefer to initiate their studies with those taxa that are maximally endemic. He was far from convinced of the value of incorporating areas in an analysis defined on a geographical or geological (*e.g.*, terranes) basis that do not conform to a minimal endemism definition. Small terranes may harbour no endemism in the groups under study. This problem is discussed by Ridder-Numan (1998*).

Platnick also warned against analysis based on large geographic areas on the presumption that these areas represent natural groupings of areas of minimal endemism, a presumption he suggests may be false in most cases. This can be a counsel of perfection, as most groups include a wide range of types of distribution, from narrowly endemic species to very widespread ones. Add to that another recommendation of cladistic biogeography, that one should explore commonality of pattern across as wide a range of higher taxa as possible, then problems in the analysis presented by widespread, partially overlapping and paralogous taxa become legion. Lepidoptera generally have much larger areas of endemism than cicadas (Holloway, 1998*), and Andersen (1998*) selected even larger areas of endemism in his analysis of marine water striders. Perhaps it is possible to apply some nesting criterion to such patterns, *e.g.*, the small areas of endemism within Borneo in one genus and those within peninsular Malaysia in another genus (Ridder-Numan, 1998*), appear to add detail to the pattern derived on the basis of four genera rather than conflicting with it.

Paralogy

The pattern of paralogy – redundancy or repetition of information – in biogeographic data is addressed by Andersen (1998*), following the thoughtful analysis of the topic by Nelson and Ladiges (1996). A somewhat different 'tree-pruning' approach to the problem of redundancy and widespread taxa, to recognise 'syntaxa' or basic information units within a tree, was offered by Kluge (1988), and applied to longhorn beetle groups in Australia and New Guinea by Wang *et al.* (1996).

A feature of both paralogy-free subtree analysis and coding for parsimony analysis of trees reduced to syntaxa is that they appear to be biased towards the more terminal structure of trees. In the example given by Andersen (1998*) in his Fig.2, the only paralogy-free subtree that can be derived by the algorithm has 'Aust' in a distal position, yet analysis of this tree by means of three-taxon statements, and, indeed, by subjective inspection, suggests an alternative equally paralogous tree with 'Aust' in a more basal position. The consensus trees for all water strider data generated by Andersen (his Fig.7) also have 'Aust' in this basal position. Coding trees through three-taxon statements appears to offer the best means of capturing information on more basal positioning of areas when they appear more than once in a cladogram. Lability of positioning of certain areas relative to a more consistent relationship of others in sets of trees resulting from biogeographic parsimony analyses might indicate either that these areas are composites, incorporating terranes with biotas of different origins, or that they have had more than one phase of interaction with other areas through time.

The phenomenon of paralogy itself may be worthy of study in its own right, as it may be indicative of a period of increase in land area, particularly when isolation has prevented biotic enrichment by dispersal from elsewhere. The development of the Hawaiian biota presents an extreme example of this (Wagner and Funk, 1995), but extensive cladogenesis can be observed within the current areas of Sulawesi and New Guinea in many groups of organisms, particularly the less dispersive ones (Holloway, 1991). Any attempt to model biogeographic pattern generation based on hypotheses of geological evolution in the area should try to include such cladogenesis, perhaps using the situation in Hawaii as a yardstick.

Phenetic biogeographic methods

Before considering prospects for modelling of biogeographic pattern in a geological context, a brief reprise of phenetic Q-mode biogeography is apposite for, although out of vogue, such analyses are still being published where cladistic treatments of organisms are not widely avail-

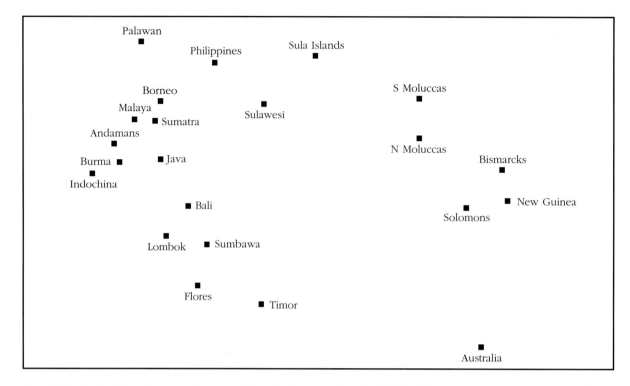

Fig.2. Butterfly faunistic distance scaling 'map' from Holloway and Jardine (1968). This map represents the best two-dimensional summary in respect of distortion to the rank-order of distance measures between all pairs of N taxa that would be arrayed by the distance measures in N-1 dimensions.

able. Phenetic methods derive trees of relationships for the items being classified by means of various clustering procedures applied to arrays of similarity or dissimilarity measures between each pair of items. The raw data are usually in the form of tables indicating the presence/absence (or some other value such as number of species for higher taxa) for a set of taxa across a set of areas. These tables can be analysed in the Q-mode (areas in respect of similarity of taxa present) or in the R-mode (taxa in respect of their representation across the areas). How and Kitchener (1997) applied these methods in the Q-mode to data on Indonesian snakes to show that Weber's Line of Faunal Balance represented the major distributional discontinuity in the archipelago at both specific and generic levels, as it did for butterflies and birds in the analysis by Holloway and Jardine (1968). They also identified strong endemism within the Lesser Sunda Islands.

Shi and Archbold (1998*) apply an essentially phenetic approach to the biogeographic affinities of the Cimmerian continent though time in relation to Gondwanan and Cathaysian biotas. The temporal changes from one source biota to the other are similar to the spatial changes noted by Holloway and Jardine (1968) in proportions of Oriental and Australian components in the faunas of the Lesser Sunda Islands as one passes from west to east. The use of phenetic versus cladistic methods in palaeobiogeography is reviewed by Rosen (1988b).

One advantage that phenetic methods have over cladistic ones is that the derivation of a distance measure between all pairs of areas in terms of faunal or floral similarity (or more accurately for a distance measure, dissimilarity) facilitates the application of various scaling or ordination measures to generate 'biotic maps' of the areas that may allow some sort of interpretation in terms of past geography (*e.g.*, Fig.2, showing the 'butterfly map' of Holloway and Jardine). The analyses of Holloway and Jardine (1968) are discussed further by Holloway (1998*), but classic examples of this approach involve the projection of areas onto a sphere in terms of distances between them based on shared modern genera, for conifers by Sneath (1967) and for freshwater crustaceans by Sneath and McKenzie (1973). Sheppard (1998) uses similar methodology in an analysis of modern

Indian Ocean coral faunas, and identifies the central role of the Chagos archipelago in their biogeography.

If we take the geological reconstructions of Hall (1998*) as converging closely on what actually happened, then the development of the modern biotas of most areas east of Sundaland and north of Australia will have owed a lot to the random processes of dispersal through time as submarine features became emergent and available for colonisation, and it might be as appropriate to apply phenetic methods as cladistic ones in any analysis of the 'biotas' generated by modelling experiments. Indeed, the comparison could be highly informative, indicating which methods are better able to recapture the geological input to the model.

Pattern versus process: prospects for modelling

At the 1996 meeting the topic of modelling was discussed: whether it would be possible to simulate mathematically such biological processes as dispersal, speciation and vicariance over various scenarios of past geography in an attempt to generate patterns of distribution from defined hypotheses for comparison with those seen in modern biotas.

The methodological debates of the past decades have sought to separate the analysis of pattern from considerations of the processes that may have generated those patterns. But the intense efforts to avoid circularity of argument may have led to methodology that excludes certain processes, particularly those involving dispersal. It may therefore be timely to return to process questions and a more pragmatically deductive approach as recommended by de Jong (1998*) and implicit in the general analyses of evolution in the Hawaiian biota assembled by Wagner and Funk (1995), without losing the essential rigour of cladistic biogeographic methodology. Rosen (1988a) explored in detail the contrasts between inductive and deductive approaches in biogeography.

It is beyond the scope of this introductory chapter to set out a clear research agenda for a modelling programme, but we list here some essential ingredients. The series of papers brought together by Myers and Giller (1988) forms a good basis for further reading; some we refer to individually in the paragraphs following.

The geographical input should include both fixed and mobile options (including increasing and diminishing the size of areas, and fragmentation and amalgamation of areas). There should be some strategy for exploring the effects of sea-level changes. There will need to be an evaluation of a relief component to assess differences between the biogeography of montane versus lowland biotas. R. Butlin (pers. comm., 1998) suggests that the geographical input could be seen in two ways: as a framework for assessing hypotheses of biological process; or as providing hypotheses to be tested using data on biological patterns, particularly with regard to the question of emergent land. These alternative approaches have some potential for reciprocal illumination.

Butlin also notes that the biological starting point will require careful consideration. Decisions will need to be made on the number of ancestral taxa, their distributions, and the variation allowed in their properties such as speciation rate and powers of dispersal. A related aspect to consider in the starting point of any simulation is the degree of occupancy of any land area in relation to potential occupancy, from the colonisation of virgin territory to saturation point (B. R. Rosen, pers. comm., 1998).

There is a relationship between island area and the size of a biota. The parameters of this relationship are well documented even if our understanding of it is still imperfect. The equilibrium theory of MacArthur and Wilson (1967) suggests that the relative isolation of an area should be included. Williamson (1988) has reviewed ideas on the relationship between distance, area and number of species in island biogeography.

The exponential model for dispersal between areas of MacArthur and Wilson (1967) offers a good starting point for modelling exchange between areas, as it has been supported by data where the age of an island is known and the potential sources of its biota, with different areas and different distances, are clear cut (Holloway, 1996). The model incorporates both distance and area components for exchange of biotas. The scaling methods referred to in the previous section relate to this type of model.

The model also requires some estimate of mean dispersal power of potential propagules. Hall (1998*) suggested that patterns in poorly dispersive groups are more likely to reflect past geography than those from more dispersive ones, yet the former are unlikely to reach the remoter archipelagos of his mid-Cenozoic reconstruction. R. Butlin (pers. comm., 1998) suggests that some kind of fitting approach might be more powerful than simulation, for example, using iterative optimisation to find the maximum likelihood values of a small number of dispersal parameters for any given geological scenario and current biological pattern (distribution and phylogeny). This type of approach could also be applied to all aspects of biological process, specifying each with as small a number of parameters as possible.

It might be necessary to model varying dispersal rates, holding these both constant and allowing them to decline differentially in relation to area and isolation. There is considerable evidence that loss of vagility occurs regularly in island biotas for both plants and animals (*e.g.,* Carlquist, 1974). Wilson (1961) suggested that observed patterns in the Melanesian ant fauna might best be explained in terms of a taxon cycle, with initial dispersal of marginal or ephemeral habitat species, followed by adaptation to more stable forest habitats, with loss of vagility. Some species may then reinvade marginal habitats and reacquire dispersive powers according to the theory, but it might prove too complex to model such reversals. Certainly the moth faunas of oceanic islands today are mostly drawn from lineages with high proportions of mobile, ephemeral habitat, r-selected species (Holloway and Nielsen, 1998). One experiment in any modelling programme should be to model process hypotheses across the Polynesian archipelagos as they are today to test predictions against actual patterns, and, when these are modelled accurately, apply the same rules across the oceanic areas in the Cenozoic reconstructions of Hall (1998*). This topic is discussed further below.

The next major factor to model is the speciation process (Barton, 1988; Grant, 1998), both between and within areas. Between areas the exponential model could again be used with a time factor to determine when two island populations with a common ancestral origin diverged sufficiently to become distinct, despite still exchanging genetic material by dispersal. The relationship of the populations could be symmetric (both derived initially from an outside source) or asymmetric (one gives rise to the other through dispersal). Some consideration of genetic drift and the founder effect will be necessary in both cases. McCall *et al.* (1996) analysed data for island-endemic bird species in relation to a model of peripheral isolated speciation and a null model, and concluded there were significant positive associations of endemism with continental species numbers per family and with density of neighbouring islands.

The frequency of speciation within an area could be related to its holding capacity in terms of number of species and the extent to which the area has 'filled up' to that level.

The development of areas of endemism within a large island and in relation to geomorphology and local geology (*e.g.,* Polhemus and Polhemus, 1998*), and the fate of previously separate terrane biotas after docking with another area are additional facets to be considered for biotic enrichment in an archipelago. Climate and other ecological factors also influence the development of heterogeneity in island organisms, and it is often difficult to identify from all these factors the causes of observed phenetic and genetic heterogeneity (Thorpe and Malhotra, 1998).

Speciation processes in models could be calibrated with reference to pattern in the Hawaiian archipelago to ensure they generate approximately compatible results under a similar scenario of island generation over a hot-spot and its subsequent disappearance through subsidence (Wagner and Funk, 1995). For the alternative fitting approach suggested by R. Butlin (pers. comm., 1998), it may be possible to assess if the parameters needed to fit a distribution pattern to a geological scenario are reasonable.

An extinction factor (Marshall, 1988) would also have to be included, perhaps based on the equilibrium model of MacArthur and Wilson (1967), but rules would have to be considered for probability of extinction of any species on an island, perhaps based on longevity on the island, or in relation to the position of a given species in some sort of randomly varying abundance ranking for all species, with vulnerability to extinction being related to rarity. Current concepts of metapopulation dynamics in relation to islands are likely to be applicable (*e.g.,* Gilpin and Hanski, 1991), as is the related concept of rarity (Gaston, 1994), although Gaston concluded that there is no general theory of rarity.

Even with all these factors included in a model, many ecological factors are still omitted, particularly the influence of climate, including the directional effects it may have on dispersal, as in the references cited by Holloway (1998*), and its role in the development of centres of biological richness. Climatic modelling itself is proving a complex and challenging task.

Modelling for marine biogeography would have to take into account the prevalence of a planktonic dispersive larval stage for a high proportion of shallow water benthic organisms and therefore should consider hypotheses of ocean current systems at different times in the past, as indicated by Hall (1998*). Benzie (1998*) demonstrates the importance of ocean currents in the development of diversity and taxonomic structure in marine organisms, and Jokiel and Martinelli (1992) model this with a computer simulation. There are some precedents for modelling transport potential for terrestrial organisms by surface currents (references cited by Holloway, 1998*).

Despite the potential interest of numerical modelling, geologists will remain reluctant to use a biogeographic pattern with no fossil record to test a geological hypothesis because the factors above reinforce our earlier point that a pattern can be produced in so many ways. The problem with modelling patterns is that there will always be a way of achieving a match because of the host of variables. The most interesting insights will probably come from models that achieve good matches with few variables. There are interesting parallels here to the plasticine indentor models of India-Asia collision and the development of SE Asia which are still the subject of controversy amongst geologists (see Hall, 1998*). Does the great similarity between the models and the apparent structure of SE Asia mean that the plasticine models have validity? This debate continues.

Pacific past, Pacific present

It is hoped that the 40-10 Ma reconstructions of Hall (1998*) will help focus future research on biogeographic pattern in Wallacea and the Melanesian archipelagos and may provide the stimulus for further examination of data already to hand. For example, if one regards detailed patterns of endemism in groups such as cicadas (Boer, 1995), water bugs (Polhemus and Polhemus, 1998*) or plants (*e.g.,* Turner, 1995) as at least partial snapshots of past geography, it might be possible to seek a 'best match' with geological units at a particular time, either overall for a higher group (tribal level) or clade by clade, as attempted by Michaux (1998*) for birds. It might then be possible to make predictions that could be tested by further research, including fieldwork.

For example, the northern versus southern patterns noted by Holloway (1998*) could have arisen through dispersal over the geography of 40-30 Ma. The Cosmopsaltriaria cicadas, however, have a number of interesting features within their general 'northern' nature. The first is the sister-relationship indicated for the Bird's Head and the Solomons by the genera *Rhadinopyga* and *Inflatopyga* (Duffels, 1997). Boer (1995) reviewed geological evidence for a more easterly position for the Bird's Head, adjacent to the Solomons, that would provide an explanation for this sister-relationship. Geologically, this is unlikely within the time period considered, and several other patterns referred to in this book, such as that of *Myrmephytum* (Baker *et al.,* 1998*) and one *Aporosa* lineage (Schot, 1998*), are consistent with a westerly position for the Bird's Head. An alternative most parsimonious hypothesis might be that dispersal occurred between the proto-North Moluccas (including Waigeo) and the Solomons prior to the formation of the South Caroline arc (perhaps with ancestral *Diceropyga* already associated with it) between them. The Bird's Head, as seen in the review of the cicada patterns by Holloway (1998*), is otherwise involved in patterns with the two tribes of the unrelated 'southern' group, with the Prasiini to the west and with the Chlorocystini to the east, patterns consistent with its position in the reconstructions, though the chlorocystine genus *Aedeastria* tracks the western components of the South Caroline arc and those of proto-North Moluccas as closely as it does the Bird's Head. The presence of *Rhadinopyga* in the Bird's Head could therefore have arisen much later, say in the period from 15-10 Ma as the proto-North Moluccas and northern New Guinea terranes moved westwards into its vicinity. The biogeography of the northern Moluccas is also discussed by de Jong (1998*).

This hypothesis places certain constraints on, and makes predictions for, the timing of various cladogenetic events within the Cosmopsaltriaria. The *Rhadinopyga/Inflatopyga* split would have occurred at about 35-30 Ma. The arrival and divergence of the *Cosmopsaltria* ancestor in the 'southern' component of New Guinea would have preceded this, and indeed the establishment of the remaining Cosmopsaltriaria genera through the Melanesian arc to Fiji, including invasion of the South Caroline arc. The split between the Sulawesi and Melanesian genera would have been even earlier. The Cosmopsaltriaria phylogeny is mapped onto a 35 Ma reconstruction in Fig.3, with the ancestral ranges of the tribes Prasiini and Chlorocystini also suggested: see Boer and Duffels (1996a: Fig.11) for illustration of their alternative hypothesis. Cladogenetic events within *Rhadinopyga*, particularly within the Bird's Head itself, would

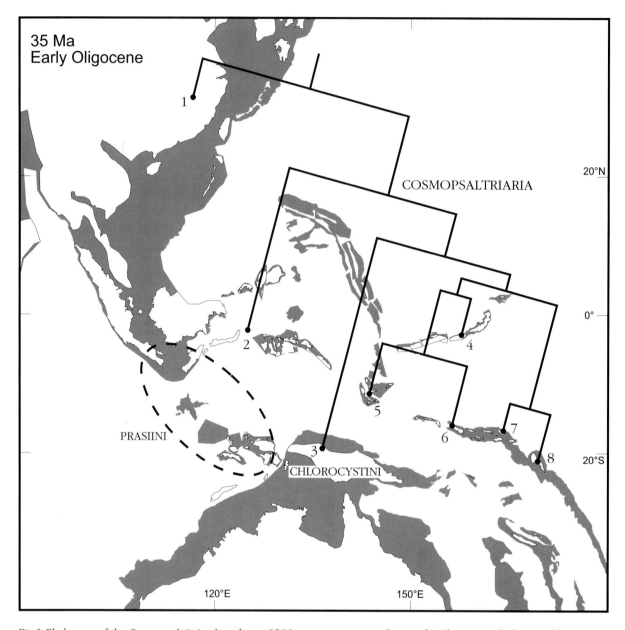

Fig.3. Phylogeny of the Cosmopsaltriaria plotted on a 35 Ma reconstruction as discussed in the text, with the possible distribution of the Prasiini and the Chlorocystini at that time also shown. 1 = *Meimuna*, 2 = *Dilobopyga* and *Brachylobopyga*, 3 = *Cosmopsaltria*, 4 = *Diceropyga*, 5 = *Rhadinopyga*, 6 = *Inflatopyga*, 7 = *Moana*, 8 = *Aceropyga*.

probably be younger than 10 Ma. Establishment of such dates in the absence of a fossil record may only be feasible through reference to 'molecular clock' data, a contentious issue that is discussed in the next section.

Another feature of the biogeography of the Cosmopsaltriaria is their abundance in Sulawesi and absence from the Philippines, a feature at odds both with the reconstructions and with the involvement of the two areas in other 'northern' patterns where the Philippines predominate, often with the exclusion of Sulawesi. This leads to the possibilities that the Cosmopsaltriaria have become extinct in the Philippines, or remain to be discovered there.

Implicit in any hypothesis of development of northern patterns over the terranes and areas of the 40-30 Ma West Pacific margins is that land areas throughout the region were sufficiently numerous and contiguous to permit this. Similar arguments apply to the components of the 'southern' patterns over Sulawesi and the Banda

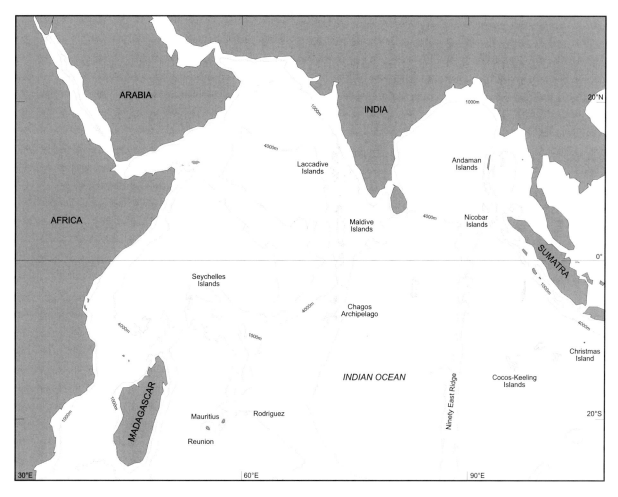

Fig.4. Present-day geography of the Indian Ocean showing the location of the Chagos and Cocos-Keeling groups of islands.

arcs, *e.g.,* the Prasiini cicadas. It may therefore be informative to consider briefly biogeographic patterns in the more remote islands and archipelagos of the Indian and Pacific Oceans today. An understanding of this may be necessary to indicate when discrepancies between geological reconstructions and general biogeographic patterns may require us to re-examine the former rather than the methodological basis for deriving the latter. Examples are drawn mainly from the Lepidoptera.

Indian Ocean: Chagos and Cocos-Keeling groups, and Christmas Island

Both the Chagos and Cocos-Keeling groups of islands consist of coral atolls (Fig.4). The biota of the first, in a central and highly isolated position, though closest to the Indian subregion, consists predominantly of extremely widespread, mobile taxa offering little indication of precise biogeographical origin except a general Oriental tropical nature. Barnett *et al.* (1998) have analysed the moth fauna, and Sheppard (1998) has provided a marine perspective. The moth fauna of the Cocos-Keeling Islands (Holloway, 1983a) is of a similar size and character. The islands are closer to Sundaland (1000 km) and Australia (2000 km) than is the Chagos archipelago, but more distant from the Indian subregion. Surprisingly, the affinities of moths which are not generally distributed Indo-Australian taxa, are with Australia rather than Sundaland. Both archipelagos are of such low relief that it is likely that they were largely submerged during periods of high sea-level since the last glaciation (Barnett *et al.*, 1998). Christmas Island is much closer to Java and currently more lofty than the other islands, although because it is

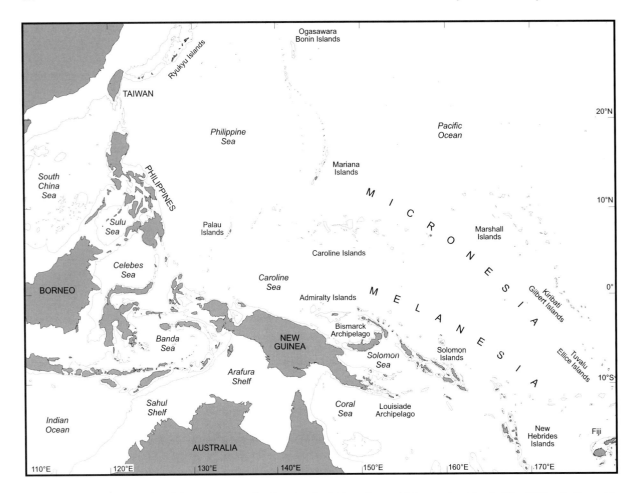

Fig.5. Present-day geography of the western Pacific showing the Melanesian and Micronesian islands. The 1000 m bathymetric contour shown further emphasises the relative continuity of the Melanesian island arcs to the east and southeast of New Guinea, in contrast to the relative isolation of the Micronesian, Marianas and Bonin islands.

situated on the Indian plate, its past position is likely to have been more remote from Sundaland without any change in its distance from Australia. Species richness and endemism are much higher in both flora and fauna than on the Chagos and Cocos-Keeling Islands, affinities being both Oriental and Australasian (Holloway, 1983a). Ackery and Vane-Wright (1984) noted a much stronger Australian character to the Christmas Island butterflies. The floristics of Indian Ocean islands have been reviewed by Renvoize (1979).

Thus it is entirely feasible that the 'southern' type of biogeographic pattern exemplified by the Prasiini/Chlorocystini group of cicadas could have developed on any land emergent at or near the margin of the Indian plate between Australia, the Bird's Head and Sundaland, but it would possibly have been restricted to more mobile organisms.

Pacific Ocean: Micronesia and Polynesia

The geography of the Micronesian island arcs today (Fig.5) bears a superficial resemblance to the 30 Ma reconstruction (Fig.6) of Hall (1998*), with the Marianas and Ogasawaras (Bonins) in a roughly north-south orientation, meeting the Carolines, Marshalls, and Kiribati and Tuvalu chains in the vicinity of Palau. One could loosely equate the Marianas and Bonins to the east Philippines and other fragments, and the Carolines and other archipelagos to the South Caroline and Melanesian arcs of 30 Ma. The junction at Palau is equivalent to that near the proto-Halmahera of 30 Ma.

Floristically this region has a weak, but distinct character of its own (Balgooy *et al.*, 1996; Holloway, 1979). Munroe (1996) indicated that the modest moth faunas of the topographically higher Micronesian islands show moderate en-

SE Asian geology and biogeography

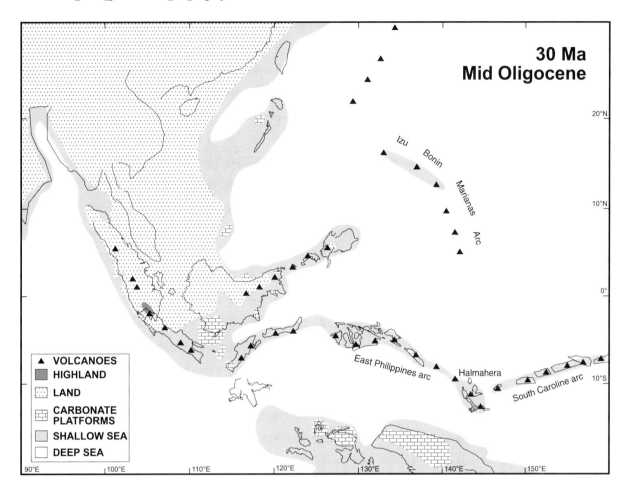

Fig.6. Postulated distribution of land and sea in SE Asia at 30 Ma from Hall (1998*). The volcanic islands of the Izu-Bonin-Marianas arc would have resembled the present-day Bonin and Marianas islands in their isolation, in contrast to the relatively continuous South Caroline, Halmahera and East Philippine arcs which probably resembled the present-day Melanesian arcs.

demism, and distribution patterns are largely independent of those of Polynesia. The atolls generally have very small faunas. The Ogasawaras, as for the flora, have a subtropical east Asian character, including some endemic genera, whereas the other groups have mainly Papuan affinities with some association with the Philippines. Palau has the richest fauna, perhaps due to its proximity to source areas in the Papuan subregion (Boer and Duffels, 1996b). Six cicada species are present in the Caroline Islands, with mainly Australasian affinities.

In the more remote archipelagos of Polynesia, particularly those such as the Marquesas and Hawaii with high volcanic islands, there have been radiations within a few genera of microlepidoptera, with species-rich, monophyletic assemblages in each archipelago (Munroe, 1996). This is less evident in the macrolepidoptera (Holloway, 1983b), but localised patterns of endemism are still observed, with groups of related species found over a number of archipelagos, *e.g.*, in the geometrid genus *Cleora* and the noctuid genera *Callopistria* and *Hydrillodes*. At its western end the *Callopistria* pattern is one of those like the cicada genus *Moana* and the *Baeturia bloetei* group (Duffels, 1988) that extends from the Solomons (and sometimes Vanuatu and Rotuma) to Samoa, missing out the main Fiji group, and possibly, therefore, this pattern was established after the opening of the North Fiji basin in the last 10 Ma.

Consideration of current Pacific island biogeography in relation to that of the 30 Ma reconstruction must also take into account the relative lack of potential source areas between Sundaland and the northern margin of Australia comparable to the Indonesian archipelago today. In the past these, for organisms of Oriental affinity, may well have been limited to the Philippines and Sulawesi sources suggested by Morley (1998*) for the flora.

Fossil evidence

The use of fossil data in biogeography has proved as much of a focus for debate as has the use of data on the distributions of living organisms, a debate reviewed by Rosen (1988b). The reluctance of some biogeographers to consider palaeontological data is something that geologists find difficult to understand. As Rosen pointed out, the further back in time one seeks to elucidate Earth history from the distribution of organisms, the more important fossil evidence becomes. Some of the difficulties are similar: problems of sampling, missing data and of definition of assemblages of species, though with an added time dimension of stratigraphy; problems of analysis, with both phenetic and cladistic methodologies applied (Rosen described the cladistic approach he developed, PAE: parsimony analysis of endemicity); problems of interpretation, differing for marine versus land biotas; problems of assessing the influence of climate versus geography on changes in past biotas (fossils are used as much as ecological indicators as palaeogeographic ones); problems of interpreting past biodiversity in an historical context (originations and extinctions).

All or many of these issues are illustrated in the more strictly palaeobiogeographical chapters of this book. Shi and Archbold (1998*) attempt to unravel the influences of climate and geography on the biota of the moving Cimmerian continent in the Permian, as does Rigby for the floras. Rigby's (1998*) paper shows that palaeobiogeographic patterns could be used to test geological hypotheses, and he plots the distributions on a reconstruction of Carey (1996) based on an expanding Earth hypothesis, and on that of Metcalfe (1996) based on plate tectonics. Both appear consistent with the distributions of Permian Pantropical and Gondwanaland floras so in this case the floras do not help falsify one or the other hypothesis. However, the map of Carey's illustrates another pitfall for biogeographers dependent on geologists for their base maps. The hypothesis of the expanding Earth has little credibility (see Hallam, 1994), but the reconstruction has less, simply because a very large number of the fragments reassembled to fill the map did not exist in Permian or Triassic times. The map is little more than a jigsaw rather than a reconstruction.

Buffetaut and Suteethorn (1998*) use fossil evidence from terrestrial faunas to explore the interaction after docking of terranes in SE Asia: Fortey and Cocks (1998*) use a similar approach to demonstrate continental integrity with evidence from Ordovician faunas. These two contributions, and the 'rules' tabulated by Rosen (1988b) for statistical interpretation of assemblages in palaeobiogeography, highlight an interesting divergence of approach from that of some biogeographers working with modern distribution patterns. The latter, particularly pan-biogeographers (*e.g.*, Heads, 1989), postulate a high degree of terrane fidelity by biota after docking and during subsequent tectonic movements. Such biogeographers would no doubt argue that, whilst homogeneity of terrestrial fossil biota across a terrane suture might indicate that terranes were behaving as one landmass, a biotic discontinuity across the suture would not necessarily indicate that the two terranes were not united. Some degree of terrane fidelity is suggested in this volume for modern groups, such as the cicadas already mentioned, some plant taxa (Ridder-Numan, 1998*) and water bugs (Polhemus and Polhemus, 1998*), but it might be dangerous to extrapolate this to a generality. This may be the case with the cicadas of New Guinea because cicada dispersal was slow and the age of arrival of terranes is very young (possibly even younger than suggested by Polhemus and Polhemus (1998*), cf. Hall, 1998*). We also note that some of the distributions in New Guinea reflect not specifically the terrane but the rock types within it. The terrane fidelity of water bugs identified in New Guinea may therefore have a simple explanation, since some rock types are missing in adjacent terranes, thus preventing dispersal. On the other hand, one would need to be certain that the pattern is not a reflection of speciation, with adaptation to rock-controlled features such as water chemistry or soil type.

Dating: fossils and the molecular clock

A topic that has seen intense debate, again reviewed by Rosen (1988b), is the contribution fossil evidence can make to deciphering biogeographic pattern in relation to Earth history in modern organisms, particularly with regard to dating cladogenetic events. The dating of the establishment of biogeographic patterns, particularly in groups today lacking any fossil record, is the topic that is usually foremost in the minds of geologists when working with biogeographers: dating, absolute or relative, is intrinsic to the geological approach to Earth history.

Fossil data always exist in some time context through stratigraphy, and indeed often form the basis for fine detail in dating. Their contribution to dating and providing positive evidence of a taxon in a given area at a given time for modern groups of organisms can be valuable but needs to be approached cautiously. Rosen (1988b) has reviewed problems of incorporating fossils in phylogenetic analyses of modern taxa, and recommended the approach of 'stratocladistics', comparing the results of analyses based on different geological horizons using fossils, and including the Recent as an independent horizon, using living organisms. It may not be possible to resolve situations where the results at different horizons appear incompatible, as strictly no pattern can falsify any other. However, the comparisons can have some inductive value if the patterns share common elements (Rosen, 1988b). Rosen suggested that neobiogeography can only say less and less about the past, although with rigour, whereas paleobiogeography can say more, if less rigorously.

If fossil data are too rare for this sort of stratocladistic approach, they can nevertheless provide helpful adjuncts to constrain interpretations of modern patterns: they can falsify hypotheses of primitive absence of a taxon from an area; they can indicate the presence of a taxon in an area at a particular time horizon, but not necessarily persistence above or below that time horizon even if the taxon is currently present in the area; they can also provide some dating constraints on the appearance of certain character-states and combinations of characters in groups of organisms. Wilson and Rosen (1998*) show the importance of examining the fossil record, reinforcing the point that identical present-day biogeographic patterns can be produced in different ways, and showing that only the fossil record can distinguish between them.

Unfortunately, the fossil record is scanty or non-existent for many plant groups and most terrestrial invertebrates, potentially the major sources of finely detailed (in terms of areas of endemism) neobiogeographic pattern. Boer and Duffels (1996b) reviewed the fossil record for the cicada groups they studied which is restricted to the Middle Miocene to Upper Pleistocene of Japan. Only one record, a *Meimuna* species from the Middle Miocene, was relevant to their studies and the fossil species is a close relative of a modern *Meimuna* occurring in Japan. They suggested *Meimuna* is a sister-group to the Cosmopsaltriaria (Fig.3; and see review in Holloway, 1998*) and currently distributed widely in mainland East Asia, including Japan. Boer and Duffels concluded that the genus or its ancestors occurred in East Asia at the time when the West Pacific island arc collided with East Asia, and Asian biota could therefore disperse into it. However, the Middle Miocene is considerably younger than the Early Oligocene age suggested for development of patterns within the Cosmopsaltriaria earlier in this chapter, based on most parsimonious matching of the patterns with the reconstructions of Hall (1998*). By the Middle Miocene the Bird's Head and North Moluccan terranes were well separated from the Solomons, posing problems for the apparent sister-relationship of these areas in the *Rhadinopyga/Inflatopyga* dichotomy. This suggests that a *Meimuna* ancestor was present in east Asia long before the Middle Miocene.

Andersen (1998*) reviews fossil data for marine water striders. A 20-30 Ma record of *Halovelia* in Dominican amber may indicate a wider distribution than at present, much as does the fossil record for the mangrove palm, *Nypa*, reviewed by Baker *et al.* (1998*). Andersen suggests that the role of extinction in shaping modern distribution patterns is likely to be significant.

The only other potential means of dating the development of patterns is through molecular clock data, as applied by Butlin *et al.* (1998*) and Walton *et al.* (1997), and discussed by Benzie (1998*). The concept of a molecular clock is one that is not likely to go away despite trenchant criticism from cladists, *e.g.*, as reviewed by Lynch (1988), Eggleton and Vane-Wright (1994), and debated in 1994 in the pages of *Cladistics* (references cited by Holloway, 1998*). Lynch reviewed debate over biogeographic patterns in South America that had been related to the expansion and contraction of Pleistocene refugia by some authors. Molecular clock evidence suggested the patterns could have developed much earlier (Oligocene to Miocene) in some of the groups sharing the patterns. Conversely, congruent patterns shown by a number of groups for areas in the Mediterranean, and related to events in the Oligocene by one set of authors, were reproduced in molecular data for another group that indicated a much more recent derivation of the pattern, as reviewed by Holloway and Nielsen (1998).

Neutral theories of molecular evolution are central to the concept of a molecular clock, but this foundation is undermined by evidence that

CROCIDURA

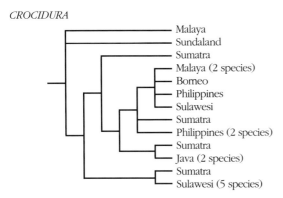

Fig. 7. Area cladogram for shrews of the genus *Crocidura*, after Ruedi (1996: Fig.5).

different taxonomic groups and traits have different rates of molecular evolution (although this does not contradict neutral theories themselves). Molecular data may enable us to date pattern, provided that the data satisfy certain constraints: genetic distance data conform globally or locally to ultrametric structure (Jardine *et al.*, 1969), and some calibration for the group concerned is available. Even without calibration, some relative dating of events can be deduced, enabling the ordering of speciation events in separate branches of a phylogeny (R. Butlin, pers. comm., 1998).

The study by Butlin *et al.* (1998*) and Walton *et al.* (1997) on *Chitaura* grasshoppers in Sulawesi details the methodology used and the sort of problems that can be encountered. They suggest that within-island diversification began by at least 7-14 Ma, suggested to be consistent with presence of the common ancestor in the component of Sulawesi that was separated from Borneo, *i.e.*, the initial separation in their tree between the Sulawesi group and a Sundanian representative for Java was due to vicariance rather than dispersal. However, the palaeogeographic maps by Moss and Wilson (1998*) suggest this geological event was very much earlier, and that there was a considerable water-gap between Borneo and Sulawesi from 34 Ma onwards. The Neogene age is more easily interpreted as the result of emergence of Sulawesi following collision of microcontinental fragments derived from the east, allowing migration into the area from Java, across the eastern end of the Sunda shelf.

The only other study for the region, using DNA/DNA hybridization, is that by Ruedas and Kirsch (1997) on *Maxomys*, a genus of murid rodents. This includes a review of the value of calibration based on a *Mus-Rattus* divergence of 10 or 12.2 Ma. Most of their samples were from Bornean populations of taxa but they did include representatives of three taxa from Queensland, one a species of *Rattus* and the others species from two related genera restricted to tropical Australasia, *Uromys* and *Melomys*. The Australian and Asian taxa separated at or before 12 Ma, indicating that it was possible for at least some non-volant animals to reach Australia from Asia at that time. Most of the diversification of the other Asian taxa was indicated to have occurred since 5 Ma.

There are two other recent molecular studies worth reviewing here, as they involve the Philippines and Sulawesi. Both were enzyme electrophoretic studies. Peterson and Heaney (1993) measured intraspecific differentiation in two species of Philippines fruit bat, and found it was positively related to degree of geographic isolation during Pleistocene periods of low sea level and to general patterns of mammalian faunal similarity for the Philippines as a whole. The two species were stated to have significantly different ecology and powers of dispersion.

Ruedi (1996) studied 20 species of the widespread Old World shrew genus *Crocidura* that are distributed over Sundaland, the Philippines and Sulawesi. The greatest richness is found in Africa where there are about 150 species. There is much sympatry in a Sundanian complex of species that Ruedi suggested developed through both vicariance and dispersal during periods of Pleistocene sea-level fluctuation. The dispersal episodes led to evolution of species complexes in the Philippines and one species in Sulawesi (Fig.7). The Sundanian complex is sister to another clade in which a Sumatran species is sister to a monophyletic complex of five species in Sulawesi. Earlier branches in the Sundanian complex are of species in Sumatra and Java, with the former acting as a source area/refugium in the interpretation suggested by Ruedi for this pattern. Hence there are broad parallels with the situation in gibbons and leaf monkeys described by Brandon-Jones (1998*). The macaques also have a complex of species in Sulawesi, referred to by Brandon-Jones and by Butlin *et al.* (1998*). Molecular analysis of these primate groups would provide a valuable comparison with the shrew data and might help elucidate some of the taxonomic problems still outstanding for the primates. No indication is given by Ruedi of whether there are *Crocidura* on the Mentawai Islands, nor does he suggest a date for the arrival of the ancestors of the Sulawesi complex, though he proposed it was earlier than that of the Sulawesi member of the Sundanian complex, both by dispersal. Ruedi does not suggest anything other than relative ages for development of these patterns, including that they are recent compared to most of the patterns for the group in Africa or the Western Palaearctic.

Molecular dating has also been attempted in plant groups whose ancestors are thought to have been separated during the break-up of Gondwanaland (Martin and Dowd, 1991). The biological dates suggested for taxa separated by African/Australian and Australian/New Zealand vicariance conform closely with those in the geophysical literature (130 Ma and 80 Ma). Division between northern and southern primitive angiosperm families (Magnoliaceae v. Winteraceae, Calycanthaceae v. Idiospermataceae) was dated at 162 Ma (Upper Jurassic). This would accord with vicariance involving the West Burma and Woyla terranes in the reconstructions of Metcalfe (1998*), but this is not supported for angiosperms generally by the fossil evidence reviewed by Morley (1998*). Martin and Dowd also suggested a Triassic (244 Ma) age for the angiosperm lineage. Association of a phylogenetic split with a well-dated vicariance event may thus also increase confidence in calibration of molecular dating.

Molecular dating, despite its shortcomings, may therefore offer an additional means of elucidating the contribution of the biotas of fragments of Gondwanaland, particularly that of India, to the development of the biological richness seen in tropical Asia, including the western part of the Indo-Australian archipelago, as discussed in the next section. Molecular analyses may also be helpful in organisms such as mosses (Tan, 1998*) in which both anagenesis and cladogenesis appear to have been extremely conservative, such that pattern shown at a high taxonomic level in seed plant groups is reproduced at the level of individual moss species. Benzie (1998*) uses molecular analyses to reveal taxonomic structure and cryptic species within apparently widespread species identified on gross morphology. They give a much more precise picture of gene flow patterns within the Pacific that may be critical for biogeographic interpretation.

Detection of old Gondwanan biotic elements in Asia

Morley (1998*) considers that palynological evidence points to a major contribution from the pre-collision angiosperm flora of India to that now present in the Oriental tropics, and also that India acted as a stepping-stone during its drift phase, enabling African taxa to reach SE Asia after contact. Very few, if any, of these pre-collision biotic elements show any 'terrane fidelity' today and, if they lack a fossil record, it may be very difficult to identify their modern counterparts, as seen in early attempts to do this by Holloway (1969, 1974) and Mani (1974). Indeed, as a result of climatic changes in the area, such groups may not currently be represented in India, but may show strong centres of richness in SE Asia and Sundaland, as suggested by Morley (1998*) for the dipterocarps, durians and some palm taxa such as the Calamoidea (Baker *et al.*, 1998*), and perhaps indicated by some of the disjunctions noted by Baker *et al.* (1998*).

Given this geographical fluidity of major elements of whole biotas, there may be little hope of recapturing geological signal from their modern distributions, unless these are viewed in conjunction with fossil evidence to establish some hypothesis of the ancestral range of the group. The situation in the caddis genus *Apsilochorema* described by Mey (1998*) may therefore be exceptional, and even that has been modified by subsequent dispersals. The *Apsilochorema* relationships are characterised by, firstly, a bipolar split at the family level, with considerable diversity at a generic level in the southern family, Hydrobiosidae, that permits identification of Oriental clades within one of these genera as potential components of the biotas of two separate Gondwanan terranes that have moved north to dock with Asia.

Groups in the Lepidoptera that may reflect at least some of this pattern include the troidine swallowtails (Parsons, 1996) and the uraniine uraniids (Lees and Smith, 1991). Both the troidines and uraniines have specialist host-plant requirements (in Aristolochiaceae and Euphorbiaceae respectively), and the cited studies attempted to identify similar biogeographic relationships amongst the plants, with only partial success. The area relationships in each show little congruence, and the Indo-Australian components in the uraniine tree are allopatric arrays of species that extend throughout the Indo-Australian tropics. Simplified area cladograms for these groups are shown in Fig.8.

Two other potentially Gondwanan moth groups (Fig.8) also show little congruence with the two just described. The Castniidae (Miller, 1989) have a Sundanian group as sister to an Australian/South American pair. The former is a

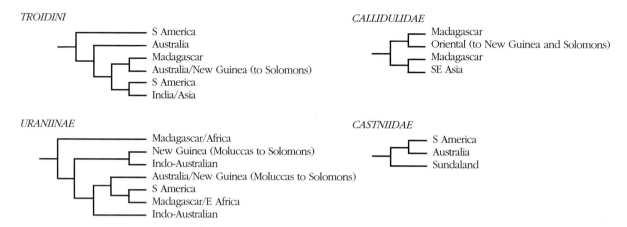

Fig.8. Area cladograms for some Gondwanan Lepidoptera groups as described in the text. Bracketed areas support lower diversity of the group concerned.

potential Burma/Woyla terrane traveller, but its species (five in two genera) are extremely rare (perhaps 10-20 specimens only in world collections), and extinction of the group elsewhere cannot be ruled out, *e.g.*, in India. Nothing is known of the biology of the Sundanian species, but the South American species feed as larvae internally on monocotyledons, including palms, so the Sundanian taxa could well have tracked the centre of palm diversity to Sundaland from India as indicated above.

The family Callidulidae (Minet, 1989) consists of two pairs of sister-taxa showing a Madagascar/Oriental split. One of the Oriental groups extends east to the Solomons but is absent from Australia and more diverse at a generic level in the Oriental Region. Minet suggested the family was a Gondwanan group with a biogeography shaped by events in the Indian Ocean.

Some insect herbivore groups feed exclusively on particular plant groups, and, where the latter have an obvious Gondwanan history and an extensive palynological record, it might be fruitful to investigate the phylogenies of both groups. Attempts to do this in the past have met with some, albeit limited success (*e.g.*, Humphries *et al.* (1986) and the studies cited above). In the geometrid moths there are a number of groups that could be suitable.

The robust Oenochrominae are predominantly Australian (Scoble and Edwards, 1989) and are associated with the Gondwanan plant family Proteaceae for larval hosts. There is one Oriental (though extending weakly to New Guinea) genus, *Sarcinodes*, where the few larval host records are from *Helicia* (Sommerer, 1995), one of two mainly Oriental proteaceous genera (Weston and Crisp, 1996). Weston and Crisp showed that different classes of evidence support the hypothesis of a Gondwanan vicariance history for the Proteaceae, with minimal dispersal over marine barriers. *Helicia* today ranges from Sri Lanka to Japan, the Bismarck Islands and eastern Australia, and is perhaps most diverse in New Guinea, whereas *Sarcinodes* has its greatest diversity in SE Asia and Sundaland.

There are two diverse geometrid groups that appear to be associated with the southern conifers, particularly Podocarpaceae, but with one also recorded from Araucariaceae. Host records are sparse for both, and therefore this specificity requires further investigation. However, very few moth genera feed on both gymnosperms and angiosperms, and those tend to be highly polyphagous in character (Holloway and Hebert, 1979). One conifer-feeding group is the ennomine genus *Milionia* (Holloway, 1993) and the other is the larentiine genus *Tympanota* (Dugdale, 1988). Both genera are most diverse in New Guinea and tend to be montane. *Milionia* has a small number of species in mainland Asia and Sundaland, some of which may prove to represent an old Indian Gondwanan lineage of the genus, but the phylogeny has yet to be investigated. Several *Tympanota* species extend west into Sundaland along with the plant taxa discussed by Morley (1998*), several other moth groups (Holloway, 1997) and the third Oriental lineage of *Apsilochorema* described by Mey (1998*). A possible sister-genus is *Episteira*, also recorded from *Podocarpus*, but with greater Oriental diversity and a few species in Africa (Holloway, 1997). *Episteira* may therefore be an Indian Gondwanan vicariant of *Tympanota*.

The influence of climate

Perhaps biogeographers need to work downwards from gross climatic considerations, given that these, driven by solar energy input, have a predominant influence on development of biological richness, particularly within the tropics (B. R. Rosen, pers. comm., 1998). Tectonics, eustasy, and more localised influences add finer detail to the gross pattern controlled by climate, and therefore it is important to assess this influence first.

Pleistocene climatic changes and related fluctuations in sea-level are seen as a major factor in the development of biogeographic pattern in Sundaland, certainly amongst the mammals in the hypotheses for shrew (Ruedi, 1997) and primate (Brandon-Jones, 1998*) evolution referred to above. Morley (1998*) discusses changes in climate throughout the Tertiary and concludes they had a major influence on the development of biogeographic pattern in the SE Asian flora, as have Rosen (1984) and Wilson and Rosen (1998*) for corals, although the latter authors argue that plate tectonics had a more important role in development of the Indo-West Pacific centre of coral diversity.

It was suggested earlier that hypotheses of strict terrane fidelity by biotas is not supported by fossil evidence, and Holloway (1998*) cites evidence from New Zealand that plants introduced there by Europeans can reproduce in response to climate the 'parallel arcs' patterns shown by indigenous plants (see also Michaux, 1998*). Indeed, biotas themselves may be largely transient phenomena with components associating, disassociating and reassociating in different combinations in response to climatic change, and capable, in continental areas at least, of translocation over massive distances in the process (*e.g.*, Coope, 1975; Huntley and Webb, 1989). In the absence of physical barriers to spread, such as water gaps, mountain ranges, deserts, etc., the range of a species can be seen as an optimum response surface that will track the combination of ecological factors under which that species is most successful, *i.e.*, a dynamic biogeography (Hengeveld, 1990, 1997).

The potential influence of climate on development of biogeographic patterns has been emphasised in several contributions to this book (*e.g.*, Shi and Archbold, Morley, Butlin *et al.*, Brandon-Jones). When terranes come together, any subsequent segregation of their biotas may have an ecological, rather than geological, basis, such as the plant lineages in New Guinea discussed by Schot (1998*). The water-bugs may be somewhat exceptional in this respect (Polhemus and Polhemus, 1998*), although see our comments above.

In an archipelagic setting, biotas are to a large extent captive to their terranes, escaping only by episodes of dispersal over water. At continental margins, such as on the Sunda and Sahul shelves, changes in climate and sea-level may have led to patterns that reflect both the migration of ecological zones and the disruptive influence of marine transgression. A clear appreciation of any ecological restrictions on species in a biogeographic study will be essential in such circumstances, such as the rain forest requirements of the primates investigated by Brandon-Jones (1998*).

An Oriental species complex of emperor moths, the *Antheraea frithi* group, illustrates the importance of data on ecological constraints as it contains a range of species more or less specific to forest types from mangrove to montane (Nässig *et al.*, 1996) and a small monophyletic group of five species endemic to Sulawesi (Holloway *et al.*, 1996). This Sulawesi group is found across a range of hill to montane rainforest types. A complex of species including the mainland Asian *frithi*, an imperfectly resolved group of taxa in Sundaland such as *gschwandneri* in Sumatra, *celebensis* in Sulawesi and *rumphii* in the South Moluccas (Seram, Buru, Ambon), is characteristic of lowland rain forest, though there are also montane forms in Sumatra currently referred to *gschwandneri*, and others located in the Mentawai Islands. This rain forest group may therefore show some parallels with the primate patterns. There is a montane species, *alleni*, endemic to Borneo, that also occurs in wet heath forest on river terraces in the lowlands. Two further species occur in Borneo, restricted to (*brunei*) or closely associated with (*moultoni*) mangrove forest. *A. moultoni* is known only from Brunei and Sarawak, but may be related to *A. billitonensis* from Belitung Island situated on the 'watershed' of the Sunda shelf between Sumatra and Borneo. *A. brunei* was described from Brunei but has since been taken in Palawan, the coast of Sabah, Belitung Island, and one possible specimen from Sumatra (Tapanuli on the west coast, opposite Nias). During full exposure of the Sunda shelf during the glaciations such species might be expected to range across the coastline of Sundaland in the South China Sea between Brunei and South Vietnam and then to track this coastline back as sea-level rose subsequently, with populations becoming stranded along current coastlines and remnant islands. Mangrove forests could also have extended far inland along the major rivers of low sea-level Sundaland as they do along some rivers today (Whitmore, 1975).

Morley (1998*) reviews the increasing evidence for periods with a much greater extent of seasonally dry climate in the Indo-Australian tropics, particularly during the Quaternary, when taxa characteristic of such habitats might have spread widely. Resumption of moist conditions could have led to their extinction through much of the archipelago, perhaps even with the development of highly disjunct patterns of a spurious Gondwanan nature, with taxa represented in the savannahs of Africa, India and Australia, such as in the noctuid moth genus *Tathorhynchus* (Hayes, 1980).

Conclusions

In the absence of widespread terrane fidelity in terrestrial biotas, with evidence from fossils that undermines any concept of strict biotic integrity, and with no means of dating pattern with confidence, with or without fossil evidence, geologists might be sceptical of what biogeography might offer them to help elucidate the highly complex histories of archipelagic areas such as the Indo-Australian and Pacific tropics in relation to their mainland confines. They might suggest that the goal of many (but by no means all) historical biogeographers to elucidate Earth history was largely misplaced, apart from offering some scope for fine-tuning geological hypotheses, such as the degree of separation of Sulawesi from Borneo in the early Cenozoic (Moss and Wilson, 1998*; Hall, 1998*), and its interaction with Java and the Lesser Sunda Islands, or to help resolve conflict between opposing ones, perhaps as in the case of the Bird's Head. They will continue to treat neobiogeographic evidence as a last resort for elucidating Earth history, when geophysical, structural, stratigraphic and palaeontological evidence is inconclusive or ambiguous.

Geologists would argue that they are offering increasingly reliable tectonic templates, albeit with unknowns about which parts constituted dry land at any time, against which biogeographic pattern can be set to elucidate life history. Is it therefore time for biogeographers, with these templates and the array of methods of phylogenetic and pattern analysis at their disposal, to renew their interest in biological process? The contribution by Wilson and Rosen (1998*), exploring the development of coral diversity in the area within the constraints of current geological hypotheses, provides an example of this. This may be anathema to many historical biogeographers, but to others it may offer an exciting prospect. It may also be the best means of their retaining the interest and collaboration of their geological colleagues.

An example of such a pragmatic approach may be seen in the collection of papers on Hawaiian biogeography in Wagner and Funk (1995). One particular feature of this Hawaiian collection is that it contains about 20 modern cladistic treatments of groups from a wide range of organisms that have speciated within the Hawaiian archipelago. Our sample of such treatments for groups that might be of value in elucidating the biogeography of the Indo-Australian archipelago is still pitifully small. Future generations of both biogeographers and geologists may well criticise today's biogeographic cohort for spending too little time on basic taxonomic spadework and too much time on theoretical abstractions.

We hope that readers will find the diversity of papers in this book stimulating and useful. Many of the references cited have been seminal in the development of biogeographic debate in recent decades. But we would argue that, as the geological picture comes more sharply into focus, the objectives of historical biogeography may require redefinition. We still believe that geologists and biogeographers can usefully learn from one another and that the answer to the question posed at the beginning of this chapter is that the two disciplines can benefit from a symbiotic relationship, to which we hope this volume will contribute.

Acknowledgements

The meeting in March 1996 was sponsored by the Geological Society of London Stratigraphic Commission, the Linnean Society, The Natural History Museum, and the University of London SE Asia Research Group. The Royal Society contributed financial support to the meeting as part of UK participation in three International Geological Correlation Projects: *IGCP306 Stratigraphic Correlation in SE Asia*, IGCP321 *Gondwana Dispersion and Accretion*, and IGCP355 *Neogene Evolution of Pacific Ocean Gateways*. Brian Rosen and Diane Cameron made major contributions at numerous stages to the meeting and to this publication for which we thank them. The contributions to the book were reviewed by Nils Andersen, Peter Ashton, Peter Barnard, Arnold de Boer, Clive Burrett, Roger Butlin, Josephine Camus, Fabio Cassola, David Chivers, Robin Cocks, Mark Coode, Edward Dickinson, John Dransfield, Hans Duffels, Sue Evans, Richard Fortey, William Foster, Neville Haile, Tony Hallam, Neil Harbury, Kevin Hill, Chris Humphries, Charles Hutchison, Ken Johnson, Michelle Kelly-Borges, Ian Metcalfe, Angela Milner, Bob Morley, Dan Polhemus, Brian Rosen, Randall T Schuh, Guang Shi, Charles Sibley, Jan van Tol, Dick Vane-Wright, Robert H Wagner, Alice Wells, Peter Weston, Anthony Whitten and Tim Whitmore, to all of whom we offer our thanks. Ian Metcalfe and Ian Kitching provided helpful comments and assistance with the glossary.

References

Ackery, P. R. and Vane-Wright, R. I. 1984. Milkweed Butterflies. British Museum (Natural History), London.

Balgooy, M. M. J. van, Hovenkamp, P. H. and Welzen, P. C. van. 1996. Phytogeography of the Pacific – floristic and historical distribution patterns in plants. *In* The Origin and Evolution of Pacific Island Biotas, New Guinea to Eastern Polynesia: patterns and processes. pp. 191-213. Edited by A. Keast and S. E. Miller. SPB Academic Publishing, Amsterdam.

Barnett, L. K., Emms, C. W. and Holloway, J. D. 1998. The moths of the Chagos Archipelago with notes on their biogeography. Journal of Natural History, in press.

Barton, N. H. 1988. Speciation. *In* Analytical Biogeography: an integrated approach to the study of animal and plant distributions. pp. 219-254. Edited by A. A. Myers and P. S. Giller. Chapman and Hall, London.

Boer, A. J. de 1995. Islands and cicadas adrift in the west-Pacific. Biogeographic patterns related to plate tectonics. Tijdschrift voor Entomologie 138: 169-241.

Boer, A. J. de and Duffels, J. P. 1996a. Historical biogeography of the cicadas of Wallacea, New Guinea and the West Pacific: a geotectonic exploration. Palaeogeography, Palaeoclimatology, Palaeoecology 124: 153-177.

Boer, A. J. de and Duffels, J. P. 1996b. Biogeography of Indo-Pacific cicadas east of Wallace's Line. *In* The Origin and Evolution of Pacific Island Biotas, New Guinea to Eastern Polynesia: Patterns and Processes. pp. 297-330. Edited by A. Keast, and S. E. Miller. Academic Publishing, Amsterdam.

Carey, S. W. 1996. Earth, Universe, Cosmos. Geology Department, University of Tasmania, 231 pp.

Carlquist, S. 1974. Island Biology. Columbia University Press, New York.

Cocks, L. R. M. and McKerrow, W. S. 1973. Brachiopod distributions and faunal provinces in the Silurian and Lower Devonian. *In* Organisms and Continents through Time: a Symposium. Edited by N. F. Hughes. Special Papers in Palaeontology 12: 291-304.

Coope, G. R. 1975. Climatic fluctuations in north-west Europe since the last Interglacial indicated by fossil assemblages of Coleoptera. *In* Ice Ages: Ancient and Modern. Edited by A. E. Wright and F. Moseley. Geological Journal, Special Issue 6: 153-158.

Craw, R. 1988. Panbiogeography: method and synthesis in biogeography. *In* Analytical Biogeography: an integrated approach to the study of animal and plant distributions. pp. 405-435. Edited by A. A. Myers and P. S. Giller. Chapman and Hall, London.

Croizat, L. 1958. Panbiogeography. The author, Caracas.

Croizat, L. 1964. Space, Time and Form: the Biological Synthesis. The author, Caracas.

Currie, D. J. 1991. Energy and large-scale patterns of animal- and plant-species richness. American Naturalist 137: 227-49.

Duffels, J. P. 1988. The cicadas of the Fiji, Samoa and Tonga islands, their taxonomy and biogeography (Homoptera, Cicadoidea). Entomonograph, 10.

Duffels, J. P. 1997. *Inflatopyga*, a new cicada genus (Homoptera: Cicadoidea: Cicadidae) endemic to the Solomon Islands. Invertebrate Taxonomy 11: 549-568.

Dugdale, J. S. 1988. Australian Trichopterygini (Lepidoptera: Geometridae) with descriptions of eight new taxa. Australian Journal of Zoology 28: 301-340.

Eggleton, P. and Vane-Wright, R. I. 1994. Some principles of phylogenetics and their implications for comparative biology. *In* Phylogenetics and Ecology. Edited by P. Eggleton and R. I. Vane-Wright. Linnean Society Symposium Series 17: 345-366. Academic Press, London.

Gaston, K. J. 1994. Rarity. Population and Community Biology Series 13. Chapman Hall, London.

Gilpin, M. E. and Hanski, I. (Editors) 1991. Metapopulation Dynamics. Academic Press, London.

Grant, P. R. (Editor) 1998. Evolution on Islands. Oxford University Press, 334 pp.

Hallam, A. 1994. An Outline of Phanerozoic Biogeography. Oxford University Press, 246 pp.

Hall, R. and Blundell, D. J. (Editors) 1996. Tectonic evolution of Southeast Asia. Geological Society of London Special Publication 106: 566 pp.

Harland, W. B., Armstrong, R. L., Cox, A. V., Craig, L. E., Smith, A. G. and Smith, D. G. 1990. A Geologic Time Scale 1989. Cambridge University Press, 263 pp.

Hayes, A. H. 1980. A revision of the pantropical genus *Tathorhynchus* Hampson (Lep. : Noctuidae, Ophiderinae). Proceedings of the British Entomological and Natural History Society 13: 25-30.

Heads, M. 1989. Integrating earth and life sciences in New Zealand natural history: the parallel arcs model. New Zealand Journal of Zoology 16: 549-585.

Hengeveld, R. 1990. Dynamic Biogeography. Cambridge University Press.

Hengeveld, R. 1997. Impact of biogeography on a population-biological paradigm shift. Journal of Biogeography 24: 541-547.

Holloway, J. D. 1969. A numerical investigation of the biogeography of the butterfly fauna of India, and its relation to continental drift. Biological Journal of the Linnean Society 1: 373-385.

Holloway, J. D. 1973. Problems with small islands. Organisms and Continents through Time: a Symposium. Edited by N. F. Hughes. Special Papers in Palaeontology 12: 107-112.

Holloway, J. D. 1974. The biogeography of Indian butterflies. *In* Ecology and Biogeography in India. Edited by M. S. Mani. Monographiae biologicae 23: 473-499. W. Junk, The Hague.

Holloway, J. D. 1979. A Survey of the Lepidoptera, Biogeography and Ecology of New Caledonia. Series Entomologica 15. W. Junk, The Hague.

Holloway, J. D. 1983a. On the Lepidoptera of the Cocos-Keeling Islands in the Indian Ocean, with a review of the *Nagia linteola* complex (Noctuidae). Entomologia Generalis 8: 99-110.

Holloway, J. D. 1983b. The biogeography of the macrolepidoptera of south-eastern Polynesia. GeoJournal 7: 517-525.

Holloway, J. D. 1993. The Moths of Borneo: family Geometridae, subfamily Ennominae. Malayan Nature Journal 47: 1–309.

Holloway, J. D. 1996. The Lepidoptera of Norfolk Island, actual and potential, their origins and dynamics. In The Origin and Evolution of Pacific Island Biotas, New Guinea to Eastern Polynesia: Patterns and Processes. pp. 123-151. Edited by A. Keast and S. E. Miller. SPB Academic Publishing, Amsterdam.

Holloway, J. D. 1997. The Moths of Borneo: family Geometridae, subfamilies Sterrhinae and Larentiinae. Malayan Nature Journal 51: 1-242.

Holloway, J. D. and Hebert, P. D. N. 1979. Ecological and taxonomic trends in macrolepidopteran host plant selection. Biological Journal of the Linnean Society 11: 229-251.

Holloway, J. D. and Jardine, N. 1968. Two approaches to zoogeography: a study based on the distributions of butterflies, birds and bats in the Indo-Australian area. Proceedings of the Linnean Society of London 179: 153-188.

Holloway, J. D., Naumann, S. and Nässig, W. A. 1996. The *Antheraea* Hübner (Lepidoptera: Saturniidae) of Sulawesi, with descriptions of new species. Part 2: The species of the *frithi*-group. Nachrichten entomologischen Vereins Apollo, N. F. 17: 225-258.

Holloway, J. D. and Nielsen, E. S. 1998. Biogeography of the Lepidoptera. In Lepidoptera, Volume 1. Edited by N. P. Kristensen. Handbuch der Zoologie 35. Walter de Gruyter, Berlin, in press.

How, R. A. and Kitchener, D. J. 1997. Biogeography of Indonesian snakes. Journal of Biogeography 24: 725-735.

Hughes, N. F. (Editor). 1973. Organisms and Continents through Time: a Symposium. Special Papers in Palaeontology 12: vi + 334 pp.

Humphries, C. J., Cox, J. M. and Nielsen, E. S. 1986. *Nothofagus* and its parasites: a cladistic approach to coevolution. In Coevolution and Systematics. pp. 55-76. Edited by A. R. Stone and D. L. Duckworth. Clarendon Press, Oxford.

Humphries, C. J., Ladiges, P. Y., Roos, M and Zandee, M. 1988. Cladistic biogeography. In Analytical Biogeography: an integrated approach to the study of animal and plant distributions. pp. 372-404. Edited by A. A. Myers and P. S. Giller. Chapman and Hall, London.

Humphries, C. J. and Parenti, L. R. 1986. Cladistic Biogeography. Oxford Monographs in Biogeography 2. Clarendon Press, Oxford.

Huntley, B. and Webb, T. III. 1989. Migration: species response to climatic variations caused by changes in the earth's orbit. Journal of Biogeography 16: 5-19.

Jardine, N., van Rijsbergen, C. J. and Jardine, C. J. 1969. Evolutionary rates and the inference of evolutionary tree forms. Nature, London 224: 195.

Jokiel, P. and Martinelli, F. J. 1992. The vortex model of coral reef biogeography. Journal of Biogeography 19: 449-458.

Keast, A. and Miller, S. E. (Editors) 1996. The Origin and Evolution of Pacific Island biotas, New Guinea to Eastern Polynesia: Patterns and Processes. SPB Academic Publishing, Amsterdam.

Kitching, I. J., Forey, P. L., Humphries, C. J. and Williams, D. M. 1998. Cladistics: the Theory and Practice of Parsimony Analysis. Oxford University Press, in press.

Kluge, A. G. 1988. Parsimony in vicariance biogeography: a quantitative method and a Greater Antillean example. Systematic Zoology 37: 315-328.

Ladiges, P. Y., Humphries, C. J. and Martinelli, L. W. (Editors) 1991. Austral Biogeography. Australian Systematic Botany 4: 1-227.

Lees, D. C. and Smith, N. G. 1991. Foodplant associations of the Uraniinae and their systematic, evolutionary and ecological significance. Journal of the Lepidopterists' Society 45: 296-347/

Liebherr, J. K. (Editor) 1988. Zoogeography of Caribbean Insects. Cornell University Press, Ithaca.

Lynch, J. D. 1988. Refugia. In Analytical Biogeography: an integrated approach to the study of animal and plant distributions. pp. 311-342. Edited by A. A. Myers and P. S. Giller. Chapman and Hall, London.

MacArthur, R. H. and Wilson, E. O. 1967. The Theory of Island Biogeography. Princeton University Press, New Jersey.

Mani, M. S. 1974. Biogeographical evolution in India. In Ecology and Biogeography in India. Edited by M. S. Mani. Monographiae biologicae 23: 473-499. W. Junk, The Hague.

Marshall, L. G. 1988. Extinction. In Analytical Biogeography: an integrated approach to the study of animal and plant distributions. pp. 219-254. Edited by A. A. Myers and P. S. Giller. Chapman and Hall, London.

Martin, P. G. and Dowd, J. M. 1991. Application of evidence from molecular biology to the biogeography of the angiosperms. Australian Systematic Botany 4: 111-116.

Matthews, C. (Editor). 1989. Panbiogeography special issue. New Zealand Journal of Zoology 16: i-iv, 471-815.

McCall, R. A., Nee, S. and Harvey, P. H. 1996. Determining the influence of continental species-richness, island availability and vicariance in the formation of island-endemic bird species. Biodiversity Letters 3: 137-150.

Metcalfe, I. 1996. Pre-Cretaceous evolution of SE Asian terranes. In Tectonic evolution of Southeast Asia. pp. 97-122. Edited by R. Hall and D. J. Blundell. Geological Society of London Special Publication 106.

Miller, J. Y. 1989. The Taxonomy, Phylogeny and Zoogeography of the Neotropical Moth Subfamily Castniinae (Lepidoptera: Castnioidea: Castniidae) Ph.D. Thesis, University of Florida.

Minet, J. 1989. Nouvelles frontières, géographiques et taxonomiques, pour la famille des Callidulidae (Lepidoptera, Calliduloidea). Nouvelle Revue d'Entomologie (Nouvelle Série) 6: 351-368.

Munroe, E. G. 1996. Distributional patterns of Lepidoptera in the Pacific Islands. In The origin and evolution of Pacific island biotas, New Guinea to Eastern Polynesia: patterns and processes. pp. 275-295. Edited by A. Keast and S. E. Miller. SPB Academic Publishing, Amsterdam.

Myers, A. A. and Giller, P. S. (Editors). 1988. Analytical Biogeography: an integrated approach to the study of animal and plant distributions. Chapman and Hall, London.

Nässig, W. A., Lampe, R. E. J. and Kager, S. 1996. The Saturniidae of Sumatra (Lepidoptera). Heterocera Sumatrana 10: 3-110.

Nelson, G. and Ladiges, P. Y. 1996. Paralogy in cladistic biogeography and analysis of paralogy-free subtrees. American Museum Novitates, 3167: 1-58.

Nelson, G. and Platnick, N. 1981. Sytematics and Biogeography, Cladistics and Vicariance. Columbia University Press, New York.

Nelson, G. and Rosen, D. E. (Editors). 1981. Vicariance Biogeography: a critique. Columbia University Press, New York.

Parsons, M. J. 1996. A phylogenetic reappraisal of the birdwing genus *Ornithoptera* (Lepidoptera: Papilionidae: Troidini) and a new theory of its evolution in relation to Gondwanan vicariance biogeography. Journal of Natural History 30: 1707-1736.

Peterson, A. T. and Heaney, L. R. 1993. Genetic differentiation in Philippine bats of the genera *Cynopterus* and *Haplonycteris*. Biological Journal of the Linnean Society 49: 203-218.

Platnick, N. 1991. On areas of endemism. Australian Systematic Botany 4: vii-viii.

Raven, P. H. and Axelrod, D. I. 1974. Angiosperm biogeography and past continental movements. Annals of the Missouri Botanical Garden 61: 539-673.

Renvoize, S. A. 1979. The origins of Indian Ocean island floras. In Plants and Islands. pp. 107-129. Edited by D. Bramwell. Academic Press, London.

Rosen, B. R. 1984. Reef coral biogeography and climate through the Late Cainozoic: just islands in the sun or a critical pattern of islands? *In* Fossils and Climate. pp. 201-264. Edited by P. J. Brenchley. Geological Journal Special Issues 11.

Rosen, B. R. 1988a. Biogeographic patterns: a perceptual overview. In Analytical Biogeography: an integrated approach to the study of animal and plant distributions. pp. 23-55. Edited by A. A. Myers and P. S. Giller. Chapman and Hall, London.

Rosen, B. R. 1988b. From fossils to earth history: applied historical biogeography. In Analytical Biogeography: an integrated approach to the study of animal and plant distributions. pp. 437-481. Edited by A. A. Myers and P. S. Giller. Chapman and Hall, London.

Ruedas, L. A. and Kirsch, J. A. W. 1997. Systematics of *Maxomys* Sody, 1936 (Rodentia: Muridae: Murinae): DNA/DNA hybridization studies of some Borneo-Javan species and allied Sundaic and Australo-Papuan genera. Biological Journal of the Linnean Society 61: 385-408.

Ruedi, M. 1996. Phylogenetic evolution and biogeography of Southeast Asian shrews (genus *Crocidura*: Soricidae). Biological Journal of the Linnean Society 58: 197-219.

Scoble, M. J. and Edwards, E. D. 1989. *Paroepisparis* Bethune-Baker and the composition of the Oenochrominae (Lepidoptera: Geometridae). Entomologia Scandinavica 20: 371-399.

Sheppard. C. 1998. Coral reefs and corals of Chagos: their condition and their role in the Indian Ocean. *In* The Ecology of the Chagos Archipelago. Edited by M. C. D. Seaward and C. Sheppard. The Linnean Society, London, in press.

Smith, A. G., Briden, J. C. and Drewry, G. E. 1973. Phanerozoic world maps. *In* Organisms and Continents through Time: a Symposium. Edited by N. F. Hughes. Special Papers in Palaeontology 12: 1-42.

Sneath, P. H. A. 1967. Conifer distributions and continental drift. Nature, London 215: 467-470.

Sneath, P. H. A. and McKenzie, K. G. 1973. Statistical methods for the study of biogeography. Organisms and Continents through Time: a Symposium. Edited by N. F. Hughes. Special Papers in Palaeontology 12: 45-60.

Sommerer, M. 1995. The Oenochrominae (*auct.*) of Sumatra (Lep. Geometridae). Heterocera Sumatrana 9: 1-77.

Thorpe, R. S. and Malhotra, A. 1998. Molecular and morphological evolution within small islands. *In* Evolution on Islands. pp. 67-82. Edited by P. R. Grant. Oxford University Press.

Turner, H. 1995. Cladistic and biogeographic analyses of *Arytera* Blume and *Mischarytera* gen. nov. (Sapindaceae), with notes on methodology and a full taxonomic revision. Blumea, Supplement 9: 1-230.

Wagner, W. L. and Funk, V. A. (Editors) 1995. Hawaiian Biogeography. Evolution on a Hot Spot Archipelago. Smithsonian Institution Press, Washington.

Walton, C. Butlin, R. K. and Monk, K. A. 1997. A phylogeny for grasshoppers of the genus *Chitaura* (Orthoptera: Acridiidae) from Sulawesi, Indonesia, based on mitochondrial DNA sequence data. Biological Journal of the Linnean Society 62: 365-382.

Wang, Q., Thornton, I. W. B. and New, T. R. 1996. Biogeography of the phoracanthine beetles (Coleoptera: Cerambycidae). Journal of Biogeography 23: 75-94.

Weston, P. H. and Crisp, M. D. 1996. Trans-Pacific biogeographic patterns in the Proteaceae. *In* The origin and evolution of Pacific island biotas, New Guinea to Eastern Polynesia: patterns and processes. pp. 215-232. Edited by A. Keast and S. E. Miller. SPB Academic Publishing, Amsterdam.

Whitmore, T. C. 1975. Tropical Rain Forests of the Far East. Second Edition. Oxford University Press.

Whitmore, T. C. (Editor). 1981. Wallace's Line and Plate Tectonics. Oxford Monographs in Biogeography 1. Clarendon Press, Oxford.

Whitmore, T. C. (Editor). 1987. Biogeographical Evolution of the Malay Archipelago. Oxford Monographs in Biogeography 4. Clarendon Press, Oxford.

Wilson, E. O. 1961. The nature of the taxon cycle in the Melanesian ant fauna. American Naturalist 95: 169-193.

Williamson, M. 1988. Relationship of species number to area, distance and other variables. *In* Analytical Biogeography: an integrated approach to the study of animal and plant distributions. pp. 91-115. Edited by A. A. Myers and P. S. Giller. Chapman and Hall, London.

Woods, C. A. (Editor) 1989. Biogeography of the West Indies: past, present and future. Sandhill Crane Press, Gainesville, Florida.

Palaeozoic and Mesozoic geological evolution of the SE Asian region: multidisciplinary constraints and implications for biogeography

Ian Metcalfe
Division of Earth Sciences, School of Physical Sciences & Engineering, University of New England, Armidale NSW 2351, Australia

Key words: SE Asia, geological evolution, terranes, Gondwanaland, Palaeo-Tethys, Meso-Tethys, Ceno-Tethys, biogeography, palaeogeography

Abstract

East and SE Asia is a giant 'jigsaw puzzle' of continental blocks (terranes) which are bounded by faults, narrow mobile belts or sutures that represent the sites of former ocean basins. Comparative studies of the tectono-stratigraphy, palaeontology, and palaeomagnetism of the various terranes suggests that they were all derived directly or indirectly from Gondwanaland and that they formed part of a 'Greater Gondwanaland'. Rifting and separation of three continental slivers occurred on the northern margin of Gondwanaland, in the Devonian, Early-Middle Permian, and Late Triassic to Late Jurassic. The northwards drift of these terranes was accompanied by the opening and closing of three successive oceans, the Palaeo-Tethys, Meso-Tethys and Ceno-Tethys. Amalgamation and accretion of Asian terranes occurred progressively between the Late Devonian and the Cretaceous, beginning with the intra-Tethyan amalgamation of South China and Indochina to form Cathaysialand in the Late Devonian-Early Carboniferous, followed by the accretion of the Tarim terrane to Kazakhstan/Siberia in the Permian. Suturing of Sibumasu and Qiangtang to Cathaysialand and amalgamation of this super-terrane with North China occurred in the Permian-Triassic, and accretion to Laurasia was completed by Late Triassic-Early Jurassic times. The highly disrupted Kurosegawa terrane of Japan, possibly derived from Australian Gondwana, accreted to Japanese Eurasia in the Late Jurassic. The Lhasa, West Burma and Woyla terranes, which rifted from NW Australian Gondwana in the Late Triassic to Late Jurassic were accreted to proto-SE Asia in the Cretaceous. The SW Borneo and Semitau terranes were derived from the South China/Indochina margin by the opening of a marginal basin in the Cretaceous which was subsequently destroyed by southwards subduction during the rifting of the Reed Bank-Dangerous Grounds terrane from South China when the South China Sea opened. Following the final breakup of Gondwanaland, India travelled rapidly northwards to make its initial contact with Eurasia at the end of the Cretaceous. Reconstructions showing the postulated positions of the various terranes and the distribution of land and sea in the Palaeozoic and Mesozoic are presented.

Introduction

Geographic distributions of plants and animals in East and SE Asia show complex and evolving patterns that are the result of plate movements, shifting land/sea and continent/ocean configurations, shifting coastlines, and changing palaeoclimates and environments. Movements of continents and continental fragments and the development and destruction of oceanic basins during the evolution of the SE Asian region have resulted in the creation and destruction of biogeographic barriers at various times and the development, and disappearance of faunal and floral provinces. The dispersal and evolution of faunas and floras of SE Asia are intimately linked with the geological evolution of the region. Biogeographic data alone can help to constrain palaeogeographic reconstructions, but reconstructions based on other types of data can also elucidate observed biogeographic data which are difficult to understand in terms of present-day geography.

Mainland East and SE Asia is like a giant 'jigsaw puzzle' of continental fragments bounded by major geological discontinuities that represent the sites of former ocean basins (Fig.1). Some of these major discontinuities are now huge strike-slip faults, whereas others are actual suture zones that include remnants of oceanic crust (ophiolites), oceanic and continental-margin sedimentary rocks, accretionary complexes, melange, and sometimes volcanic arcs. Eastern SE Asia comprises a series of small continental fragments set in a 'matrix' of stretched continen-

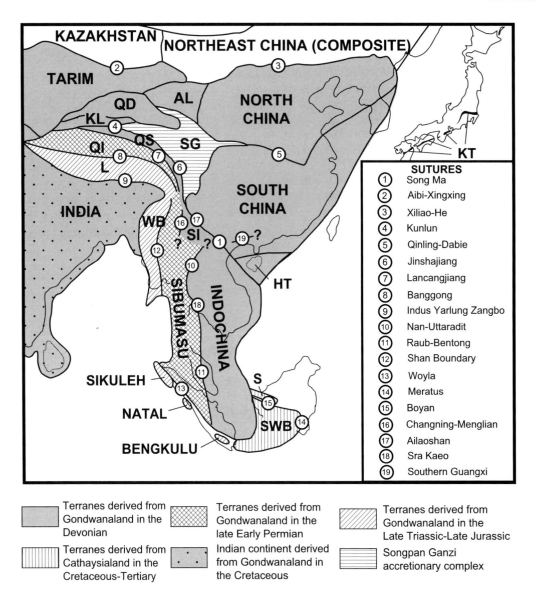

Fig.1. Distribution of principal continental terranes and sutures of East and SE Asia. WB = West Burma, SWB = South West Borneo, S = Semitau Terrane, HT = Hainan Island terranes, L = Lhasa Terrane, QI = Qiangtang Terrane, QS = Qamdo-Simao Terrane, SI= Simao Terrane, SG = Songpan Ganzi accretionary complex, KL = Kunlun Terrane, QD = Qaidam Terrane, AL = Ala Shan Terrane, KT = Kurosegawa Terrane.

tal crust or oceanic crust (marginal basins), accretionary complexes, ophiolites and volcanic arcs (Fig.2). The various continental fragments (terranes) have progressively assembled over the last 400 million years. In this paper, I shall discuss the origins and the Palaeozoic and Mesozoic evolution (540-65 million years ago) of these various continental terranes from a multidisciplinary viewpoint and emphasise the importance of, and implications for biogeographical data. The Cenozoic (65 million years ago to present) evolution of the region is discussed by Hall (1998 this volume).

In discussing the geological evolution of the region, a variety of multidisciplinary data (Table 1) is used here to constrain the origins of terranes; the timing of rifting and separation from their parent cratons; timing, directions and amount of drift; and timing of suturing (collision and welding) of terranes to each other. Some terranes sutured to each other (amalgamated) within a major ocean before they, as an amalgamated com-

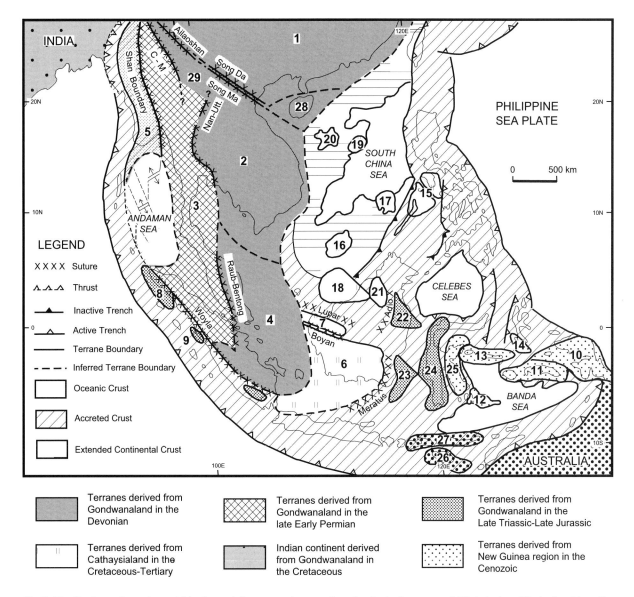

Fig.2. Distribution of continental blocks and fragments (terranes) and principal sutures of SE Asia (modified after Metcalfe, 1990). 1. South China 2. Indochina 3. Sibumasu 4. East Malaya 5. West Burma 6. SW Borneo 7. Semitau 8. Sikuleh 9. Natal 10. West Irian Jaya 11. Buru-Seram 12. Buton 13. Banggai-Sula 14. Obi-Bacan 15. North Palawan 16. Spratley Islands-Dangerous Ground 17. Reed Bank 18. Luconia 19. Macclesfield Bank 20. Paracel Islands 21. Kelabit-Longbowan 22. Mangkalihat 23. Paternoster 24. West Sulawesi 25. East Sulawesi 26. Sumba 27. Banda Allochthon 28. Hainan Island terranes. 29. Simao terrane.

posite terrane, sutured (accreted) to proto-Asia.

Origins of the East and SE Asian terranes

Using multidisciplinary data (Table 1), all the East and SE Asian continental terranes are interpreted to have had their origin on the margin of Gondwanaland, and probably on the India-N/NW Australian margin. Cambrian and Ordovician shallow-marine faunas of the North China, South China, and Sibumasu terranes have close affinities with those of eastern Gondwanaland, and especially Australian Gondwanaland (Burrett, 1973; Burrett and Stait, 1985; Metcalfe, 1988, Burrett *et al.*, 1990). This is observed in trilobites (Shergold *et al.*, 1988), brachiopods (Laurie and Burrett, 1992), corals and stromatoporoids (Webby et al., 1985; Lin and Webby, 1989), nautiloids (Stait and Burrett, 1982, 1984; Stait *et al.*, 1987), gastropods (Jell *et al.*, 1984), and conodonts (Burrett *et al.*, 1990; Nicoll and

Table 1. Multidisciplinary constraining data for the origins and the rift/drift/suturing of terranes.

Origin of Terranes	Age of Rifting and Separation	Drifting (Palaeo-positions of terranes)	Age of Suturing (Amalgamation/Accretion)
INDICATED BY	INDICATED BY	INDICATED BY	INDICATED BY
Palaeobiogeographic constraints (fossil affinities with proposed parent craton) Tectonostratigraphic constraints (similarity of gross stratigraphy with parent craton, presence of distinctive lithologies characteristic of parent craton, e.g. glacial lithologies) Palaeolatitude and orientation from palaeomagnetism consistent with proposed origin	Ocean floor ages and magnetic stripe data Divergence of Apparent Polar Wander Paths (APWPs) indicates separation Divergence of palaeolatitudes (indicates separation) Age of associated rift volcanism and intrusive rocks Regional unconformities (formed during pre-rift uplift and during block-faulting) Major block-faulting episodes and slumping Palaeobiogeography (development of separate biogeographic provinces after separation) Stratigraphy-rift sequences in graben/half graben	Palaeomagnetism (palaeolatitude, orientation) Palaeobiogeography (shifting from one biogeographic province to another due to drift) Palaeoclimatology (indicates palaeolatitudinal zone)	Ages of ophiolite and ophiolite obduction ages (pre-suturing) Melange ages (pre-suturing) Age of 'stitching' plutons (post-suturing) Age of collisional or post-collisional plutons (syn- to post-suturing) Age of volcanic arc (pre-suturing) Major changes in arc chemistry (syn-collisional) Convergence of Apparent Polar Wander Paths (APWPs) Loops or disruptions in APWPs (indicates rapid rotations during collisions) Convergence of palaeo-latitudes (may indicate suturing but no control on longitudinal separation) Age of blanketing strata (post-suturing) Palaeobiogeography (migration of continental animals/plants from one terrane to another indicates terranes have sutured) Stratigraphy/sedimentology (e.g. provenance of sedimentary detritus from one terrane onto another) Structural geology (age of deformation associated with collision)

Totterdell, 1990; Nicoll and Metcalfe, 1994). More recently, the Gondwanaland acritarch *Dicrodiacroium ancoriforme* Burmann has been reported from the Lower Ordovician of South China (Servais et al., 1996). Little is known of Cambrian-Ordovician faunas of Indochina, but Silurian brachiopods, along with those of South China, North China, Eastern Australia and the Tarim terrane belong to the Sino-Australian province characterised by the *Retziella* fauna (Rong et al., 1995). Lower Palaeozoic sequences and faunas of the Qaidam, Kunlun, and Ala Shan blocks are similar to those of the Tarim block and also to South and North China (Chen and Rong, 1992), and these blocks are regarded as disrupted fragments of a larger Tarim terrane by Ge et al. (1991). These biogeographic data suggest that North China, South China, Tarim (here taken to include the Qaidam, Kunlun and Ala Shan blocks), Sibumasu (with the contiguous Lhasa and Qiangtang blocks), and Indochina formed the outer margin of northern Gondwanaland in the Early Palaeozoic. The close faunal affinities, at both lower and higher taxonomic levels, suggest continental contiguity of these blocks with each other and with Gondwanaland at this time rather than mere close proximity.

Lower Palaeozoic palaeomagnetic data for the

Table 2. Suggested origins for the East and SE Asian continental terranes.

Terrane	Origin
North China	N Australia
South China	Himalaya-Iran Region
Indochina-East Malaya	E Gondwana
Sibumasu	NW Australia
West Burma	NW Australia
Lhasa	Himalayan Gondwana
Qiangtang	Himalayan Gondwana
Qamdo-Simao	E Gondwana? Extension of Indochina
Kunlun	NE Gondwana? Originally part of Tarim?
Qaidam	NE Gondwana? Originally part of Tarim?
Ala Shan	NE Gondwana? Originally part of Tarim?
Tarim	Australian Gondwana?
SW Borneo	Cathaysialand (S China/Indochina margin)
Hainan	NE Gondwana?
Kurosegawa	Australian Gondwana?

various East and SE Asian terranes are variable in both quantity and quality and often equivocal. This makes reconstructions based purely on palaeomagnetic data both difficult and suspect. However, sufficient data exist to be able in some cases to reasonably constrain palaeolatitudes (but not always the hemisphere) and in some cases the actual position of attachment to Gondwanaland. North China data (Zhao et al., 1996) provide a Cambrian to Late Devonian pole path segment, which, when rotated about a Euler pole to a position of fit with the Australian Cambrian to Late Devonian pole path, produces a good fit with North China positioned adjacent to North Australia (Klootwijk, 1996). This position is consistent with that proposed in reconstructions by Metcalfe (1993, 1996a, 1996b) and in this paper. Comparisons of the gross stratigraphies of North China and the Arafura basin (Fig.5 in Metcalfe, 1996b) show a remarkable similarity in the early Palaeozoic, also supporting the proposed position for North China. Positions for South China, Tarim, and Indochina are more equivocal, but latitudes of between 1 and 15 degrees are indicated for the Late Cambrian Early Ordovician for the South China Block (Zhao et al., 1996). Palaeolatitudes of between 6 and 20 degrees south are indicated for the Tarim block for the same time period, which is broadly consistent with a position on the Gondwanaland margin between the North and South China blocks. Comparisons of the Precambrian sequence on the northeast margin of the Tarim block with Australia led Li et al. (1996) to propose that this block had its origin outboard of the Kimberley region of Australia. Similar comparisons of the Precambrian of South China however led Li et al. (1995, 1996) to propose that South China was positioned between eastern Australia and Laurentia in the Late Proterozoic. This is rather different to the Early Palaeozoic position suggested here. Cambrian to Early Permian faunas of the Sibumasu terrane have strong Gondwanaland affinities, and in particular show close relationships with western Australian faunas (Metcalfe, 1988, 1996a,b; Burrett et al., 1990). In addition, Gondwanaland plants and spores are also reported from this terrane (Wang and Tan, 1994; Yang and Liu, 1996). Glacial-marine diamictites, with associated cold-water faunas and sediments, of Late Carboniferous to Early Permian age, are also found distributed along the entire length of Sibumasu and indicate attachment to the margin of Gondwanaland where substantial ice reached the sea. The most likely region for attachment of this terrane is NW Australia. Palaeomagnetic data for the Late Carboniferous suggest a palaeolatitude of 42 degrees south (Huang and Opdyke, 1991), which is consistent with such a placement. Comparison of the gross stratigraphy of Sibumasu with the Canning basin of NW Australia also reveals striking similarities in the Cambrian to Early Permian, and Sibumasu could easily have been positioned outboard of the Canning basin of western Australia during this time. Both the Qiangtang and Lhasa blocks of Tibet exhibit Gondwanaland faunas and floras up to the Early Permian, and also have glacial-marine diamictites, till, and associated cold-water faunas and sediments in the Late Carboniferous to Early Permian. Thus, all the East and SE Asian continental terranes appear to have had their origins on the margin of Gondwanaland (Table 2).

Carboniferous and younger faunas and floras of the North China, South China, Tarim, and Indochina terranes are typically Cathaysian in nature and they show no relationship with those of Gondwanaland (Metcalfe, 1988). These terranes were also situated at palaeolatitudes that indicate they were no longer attached to the margin of Gondwanaland from the Carboniferous onwards (Zhao et al., 1996) and that they separated from Gondwanaland in the Devonian

Table 3. Major Palaeozoic and Mesozoic tectonic events on the Australian NW Shelf (Colwell *et al.*, 1994).

Age	Event
Early Ordovician	Major intracontinental extension
Late Devonian-Early Carboniferous	Major upper crustal extension
Early Carboniferous-Early Permian	Extension: initiation of Westralian Superbasin
Late Permian	Uplift, extension, igneous activity
Middle Triassic	Structuring in parts of NW Shelf
Late Triassic-Early Jurassic	Regional structuring, compression, transpression, extension: Fitzroy Movement
Middle-Late Jurassic	Breakup in Argo Abyssal Plain

as previously suggested by Metcalfe (1994, 1996a, 1996b). Sibumasu, Qiangtang, and the Lhasa terrane continued to remain on the margin of Gondwanaland until the Permian, with the Lhasa terrane remaining attached to Gondwanaland possibly until the Late Triassic but with Sibumasu and Qiangtang separating in the late Early Permian.

The continental sliver which was located immediately outboard of NW Australia in the Triassic, and which must have rifted and separated in the Jurassic, is here considered to have comprised West Burma, the small Sikuleh, Natal and possibly Bengkulu terranes now located in SW Sumatra, and perhaps small continental fragments (West Sulawesi, Mangkalihat and the Banda allochthon) that now form parts of Borneo and eastern Indonesia. There is however little direct evidence to support this, apart from stratigraphic similarities between the Sikuleh block and the NW Australian shelf, and sparse palaeomagnetic data showing this block to be derived from the south in the Mesozoic (Haile, 1979; Görür and Sengör, 1992). The small Hainan Island terranes had their origin on the Early Palaeozoic margin of Gondwanaland but had probably separated by the Late Palaeozoic (Metcalfe, 1996b). The disrupted composite Kurosegawa terrane of Japan is believed to have been derived from Gondwanaland, and a position adjacent to eastern South China in the Silurian/Devonian has been suggested (Saito, 1992, Hisada *et al.*, 1994).

Carboniferous and Permian faunas of the South West Borneo and Semitau blocks do not appear to have any affinities with those of Sibumasu, but are very similar to those of South and North China (Vachard, 1990). Triassic and Jurassic floras and faunas have affinities with South China, Indochina and Japan and a South China/Indochina origin seems likely.

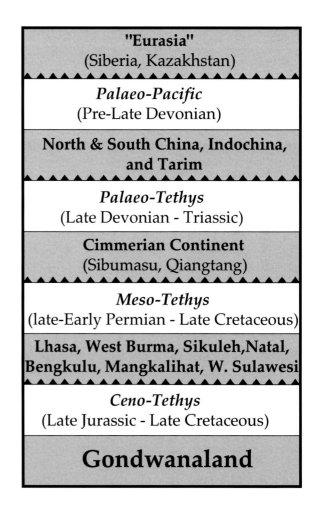

Fig.3. Schematic diagram showing the three continental slivers/collages of terranes, rifted from Gondwanaland and translated northwards by the opening and closing of three successive oceans, the Palaeo-Tethys, Meso-Tethys and Ceno-Tethys.

Table 4. East and SE Asian sutures and their interpreted ages and age constraints. For location of sutures see Fig.1 and Fig.2.

	Suture Name	Suture Age	Age constraints
1.	Song Ma	Late Devonian-Early Carboniferous	Large-scale folding, thrusting and nappe formation in Early to Middle Carboniferous. Middle Carboniferous shallow marine carbonates reported to blanket Song Ma suture in North Vietnam. Pre-middle Carboniferous faunas on each side of the Song Ma zone are different whilst the middle Carboniferous faunas are essentially similar. Carboniferous floras on the Indochina block in Northeast Thailand indicate continental connection between Indochina and South China in Carboniferous
2.	Aibi-Xingxing	Permian	Lower Carboniferous ophiolites. Major arc magmatism ceased in the Late Carboniferous. Late Permian post-orogenic subsidence and continental sedimentation in Junggar basin. Palaeomagnetic data indicate convergence of Tarim and Kazakhstan by the Permian. Upper Permian continental clastics blanket suture.
3.	Xiliao-He	Jurassic	Late Jurassic-Early Cretaceous deformation and thrust faulting. Widespread Jurassic-Cretaceous granites. Triassic-Middle Jurassic deep-marine cherts and clastics. Upper Jurassic-Lower Cretaceous continental deposits blanket suture.
4.	Kunlun	Permian-Triassic	Permo-Triassic ophiolites and subduction zone melange. Upper Permian calc-alkaline volcanics, strongly deformed Triassic flysch and Late Triassic granites
5.	Qinling-Dabie	Triassic-Jurassic	Late Triassic age of subduction-related granite. Late Triassic U-Pb dates of zircons from ultra-high pressure eclogites. Late Triassic-Early Jurassic convergence of APWPs and palaeolatitudes of South China and North China. Initial contact between South and North China is indicated by isotopic data in Shandong and sedimentological records along the suture. Widespread Triassic to Early Jurassic deformation in the North China block north of the suture.
6.	Jinshajiang	Late Permian-Late Triassic	Ophiolites are regarded as Upper Permian to Lower Triassic in age. Melange comprises Devonian, Carboniferous and Permian exotics in a Triassic matrix. Upper Permian to Jurassic sediments unconformably overlie Lower Permian ophiolites in the Hoh Xil Range.
7.	Lancangjiang	Early Triassic	Suture zone rocks include Devonian and Carboniferous turbiditic 'flysch'. Ocean-floor basalts of Permian age and Carbo-Permian melange. Carboniferous-Permian island arc rocks are developed along the west side of the suture. Upper Triassic collisional granites are associated with the suture. Suture zone rocks are blanketed by Middle Triassic continental clastics.
8.	Banggong	Late Jurassic-Earliest Cretaceous	Suture is blanketed in Tibet by Cretaceous and Paleogene rocks. Structural data indicate collision around Jurassic/Cretaceous boundary.
9.	Indus-Yarlung-Zangbo	Late Cretaceous-Eocene	Jurassic-Cretaceous ophiolites and ophiolitic melange with Jurassic-Lower Cretaceous radiolarian cherts. Eocene collision-related plutons. Palaeomagnetic data indicates initial collision around 60 Ma. Paleogene strata blanket the suture.
10.	Nan-Uttaradit-Sra Kaeo	Late Permian-Early Triassic	Pre-Permian ophiolitic mafic and ultramafic rocks with associated blueschists. Imbricate thrust slices dated as Middle Triassic by radiolarians (Sra Kaeo segment). Mafic and ultramafic blocks in the melange comprise ocean-island basalts, back-arc basin basalts and andesites, island-arc basalts and andesites and supra-subduction cumulates generated in Carboniferous to Permo-Triassic times. Limestone blocks in melange range from upper Lower Permian to middle Permian. Granitic lens has yielded a zircon U-Pb age of 486 ± 5 Ma. Permo-Triassic dacites and rhyolites associated with relatively unmetamorphosed Lower Triassic sandstone-shale turbidite sequence. Suture zone rocks are overlain unconformably by Jurassic redbeds and post-Triassic intraplate continental basalts.
11.	Raub-Bentong	Early Triassic	Melange includes Lower and Upper Permian limestone clasts. The Main Range 'collisional' 'S' Type granites of peninsular Malaysia range from Late Triassic (230 ± 9 Ma) to earliest Jurassic (207 ± 14 Ma) in age, with a peak of around 210 Ma. Within suture zone are Upper Devonian to Upper Permian deep-marine bedded cherts.
12.	Shan Boundary	Early Cretaceous	Cretaceous thrusts in the back-arc belt. Late Cretaceous age for the Western Belt tin-bearing granites.
13.	Woyla	Late Cretaceous	Cretaceous ophiolites and accretionary complex material.
14.	Meratus	Late Cretaceous	Subduction melange and ophiolite of middle Cretaceous age. Ophiolite obducted in Cenomanian. Suture overlain by Eocene strata.
15.	Boyan	Late Cretaceous	Upper Cretaceous melange.
16.	Changning-Menglian	Late Permian-Late Triassic	Oceanic ribbon-bedded chert-shale sequences have yielded graptolites, conodonts and radiolarians indicating ages ranging from Lower Devonian to Middle Triassic. Limestone blocks and lenses dominantly found within basalt sequence of suture and interpreted as seamount caps, have yielded fusulinids indicative of Lower Carboniferous to Upper Permian ages.
17.	Ailaoshan	Middle Triassic	Ophiolitic rocks are associated with deep-marine sedimentary rocks including ribbon-bedded cherts that have yielded some Lower Carboniferous and Lower Permian radiolarians. Upper Triassic sediments (Carnian conglomerates and sandstones, Norian limestones and Rhaetian sandstones) blanket suture.

Table 5. Palaeozoic and Mesozoic events and their ages in East and SE Asia.

Palaeozoic Evolution		Mesozoic Evolution	
PROCESS	AGE	PROCESS	AGE
1. Rifting of South China, North China, Indochina, Tarim and Qaidam from Gondwanaland.	Early Devonian	1. Suturing of South China with North China and final consolidation of Sundaland	Late Triassic to Early Jurassic
2. Initial spreading of the Palaeo-Tethys ocean.	Middle/Late Devonian	2. Rifting of Lhasa, West Burma and Woyla terranes	Late Triassic to Late Jurassic
3. Amalgamation of South China, Indochina and East Malaya to form Cathaysialand	Late Devonian to Early Carboniferous	3. Initial spreading of Ceno-Tethys ocean	Late Triassic (Norian) in west (North India) and Late Jurassic in east (NW Australia)
4. Rifting of Sibumasu and Qiangtang from Gondwanaland as part of the Cimmerian continent	Late Early Permian	4. Northward drift of Lhasa, West Burma and Woyla terranes	Jurassic to Cretaceous
5. Initial spreading of Meso-Tethys ocean	Middle Permian	5. Collision of the Lhasa Block with Eurasia	Cretaceous
6. Collision and suturing of Sibumasu to Indochina	Latest Permian to Triassic	6. Accretion of West Burma and Woyla terranes to Sibumasu	Late Early Cretaceous
7. Initial collision of South and North China and development of Tanlu Fault	Late Permian to Triassic	7. Suturing of Semitau to SW Borneo	Late Cretaceous

Rifting and separation of terranes from Gondwanaland

Multidisciplinary data suggest that the East and SE Asian terranes were successively rifted and separated from Gondwanaland as three continental slivers in the Devonian, late Early Permian and Late Triassic-Late Jurassic (Metcalfe, 1996a, 1996b, Fig.3). The separation of these slivers of continent was accompanied by the opening (and subsequent closing) of three ocean basins, the Palaeo-Tethys, Meso-Tethys and Ceno-Tethys, remnants of which are now to be found along the various suture zones of eastern Asia. The northwards drift of these three continental slivers is to some extent indicated by palaeomagnetic data for the various blocks (see Metcalfe, 1990, 1996a; Van der Voo, 1993; Zhao *et al.*, 1996 for details). In addition, recent studies of the NW Shelf of Australia (Colwell *et al.*, 1994) have identified major extensional events that can be broadly correlated with the rifting and separation of the Asian terranes (Table 3).

Devonian rifting and separation

South China, North China, Tarim and Indochina were attached to Gondwanaland in the Cambrian to Silurian, but by Carboniferous times were separated from the parent craton (see Metcalfe, 1996a, 1996b for details), suggesting a Devonian rifting and separation of these blocks. This timing is also supported by the presence of a conspicuous Devonian unconformity in South China, and a subsequent Devonian-Triassic passive margin sequence along the southern margin of South China (Nie, 1994). Devonian basin formation in South China has also been shown to be related to rifting (Zhao Xun *et al.*, 1996). The splitting of the Silurian Sino-Australian brachiopod province into two sub-provinces and the apparent loss of links between Asian terranes and Australia in the Early Devonian (Rong *et al.*, 1995) may be the result of the northwards movement and separation of the Chinese terranes from Gondwanaland.

Carboniferous-Permian rifting and Permian separation

There is now substantial evidence for rifting along the northern margin of Gondwanaland in the Carboniferous to Permian (Stöcklin, 1974; Powell, 1976; Falvey and Mutter, 1981; Bird, 1987; Boulin, 1988; Pogue *et al.*, 1992; Pillevuit, 1993; Wopfner, 1994; Metcalfe, 1996a, 1996b) accompanied by rift-related magmatism (Veevers and Tawari, 1995). This rifting episode led to the

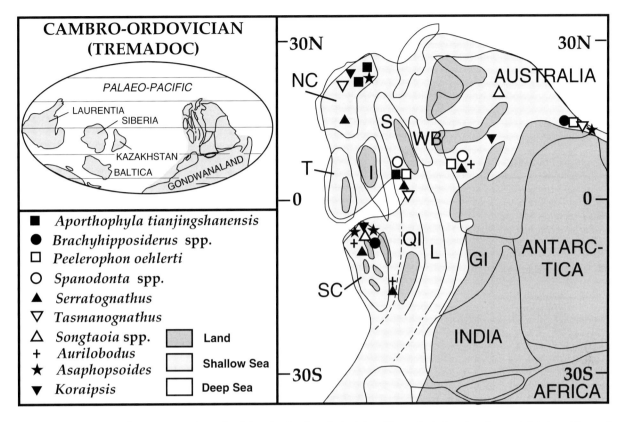

Fig.4. Reconstruction of eastern Gondwanaland for the Cambro-Ordovician (Tremadoc) showing the postulated positions of the East and SE Asian terranes, distribution of land and sea, and shallow-marine fossils that illustrate Asia-Australia connections at this time. NC = North China, SC = South China, T = Tarim, I = Indochina, QI = Qiangtang, L = Lhasa, S = Sibumasu, WB = West Burma, GI = Greater India. Present day outlines are for reference only. Distribution of land and sea for Chinese blocks principally from Wang (1985). Land and sea distribution for Pangaea/Gondwanaland compiled from Golongka *et al.* (1994), Smith *et al.* (1994); and for Australia from Struckmeyer and Totterdell (1990).

late Early Permian separation of the Sibumasu and Qiangtang terranes, as part of the Cimmerian continent, from the Indo-Australian margin of Gondwanaland.

Late Triassic to Late Jurassic rifting and separation

The separation of the Lhasa block from Gondwanaland has been proposed by different authors as occurring either in the Permian or Triassic. A Permian separation has been advocated, either as a part of the Cimmerian continent (Allègre *et al.*, 1984; Metcalfe 1988, 1990) or as a 'Mega-Lhasa' block which included Iran and Afghanistan (Baud *et al.*, 1993). Permian rifting on the North Indian margin and in Tibet (Baud, 1994) is here regarded as being related to the separation of the Cimmerian continental strip which included Iran, Afghanistan and the Qiangtang block of Tibet, but not the Lhasa block. Recent sedimentological and stratigraphical studies in the Tibetan Himalayas and Nepal (Liu, 1992; Liu and Einsele, 1994; von Rad *et al.*, 1994; Ogg and von Rad, 1994) have documented the Triassic rifting and Late Triassic (Norian) separation of the Lhasa Block from northern Gondwanaland. This Late Triassic episode of rifting is also recognised along the NW Shelf of Australia (Colwell *et al.*, 1994) where it continued into the Late Jurassic, resulting in the separation of West Burma and the Woyla terranes (Metcalfe, 1990, 1994, 1996a, 1996b; Görür and Sengör, 1992).

Amalgamation and accretion of terranes

The continental terranes of East and SE Asia have progressively sutured to one another during the Palaeozoic to Cenozoic. Most of the major terranes had coalesced by the end of the Cretaceous and proto SE Asia had formed. The age of welding of one terrane to another can be de-

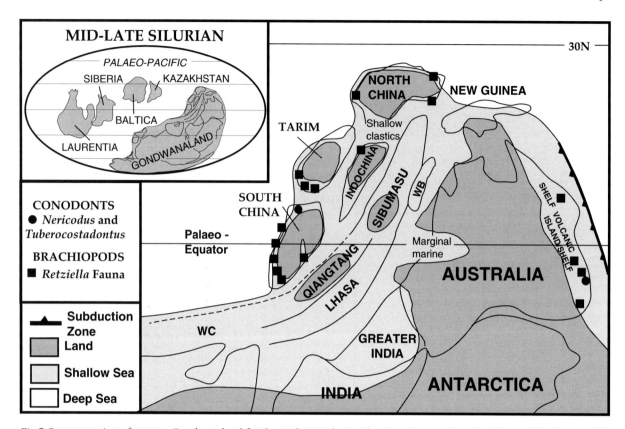

Fig. 5. Reconstruction of eastern Gondwanaland for the Mid-Late Silurian showing the postulated positions of the East and SE Asian terranes, distribution of land and sea, and shallow-marine fossils that appear to define an Australasian province at this time. WC = Western Cimmerian Continent, WB = West Burma. Present day outlines are for reference only. Distribution of land and sea for Chinese blocks principally from Wang (1985). Land and sea distribution for Pangaea/Gondwanaland compiled from Golongka *et al.* (1994), Smith *et al.* (1994); and for Australia from Struckmeyer and Totterdell (1990).

termined using the various criteria given in Table 1. When these criteria are applied to the various sutures and terranes of East and SE Asia the interpreted ages of suturing (amalgamation/accretion) are determined as given in Table 4.

Palaeozoic and Mesozoic regional evolution

The Palaeozoic and Mesozoic evolution of SE Asia involved three successive episodes of rifting and separation of continental terranes from the margin of Gondwanaland, their northwards drift and their amalgamation/accretion to form proto East and SE Asia. The various Palaeozoic and Mesozoic evolutionary events of the region are summarised in Table 5. In order to illustrate the various processes and the changing continent-ocean configurations during the Palaeozoic and Mesozoic, palaeogeographic reconstructions for the Palaeozoic and Mesozoic are presented and briefly discussed below.

Cambro-Ordovician (Tremadoc) reconstruction

This reconstruction (Fig.4) shows the proposed relative positions of the East and SE Asian continental terranes on the Indian-Australian margin of Gondwanaland forming a 'Greater Gondwanaland'. The proposed positions are based on palaeobiogeographic, tectonostratigraphic and palaeomagnetic data. The reconstruction also shows the distribution of land, shallow sea and deep sea, compiled from numerous sources. Late Cambrian and Ordovician shallow-marine faunas define an 'Australasian Province' and suggest continental connection between the Asian blocks and Australia at this time. Some of the genera and species providing close links between the Asian blocks and Australia are also plotted on this figure. The deep-marine gulf between South China and Tarim/Indochina was possibly the result of rifting at this time, also recorded by major intracontinental extension on the Australian NW Shelf (Colwell *et al.*, 1994).

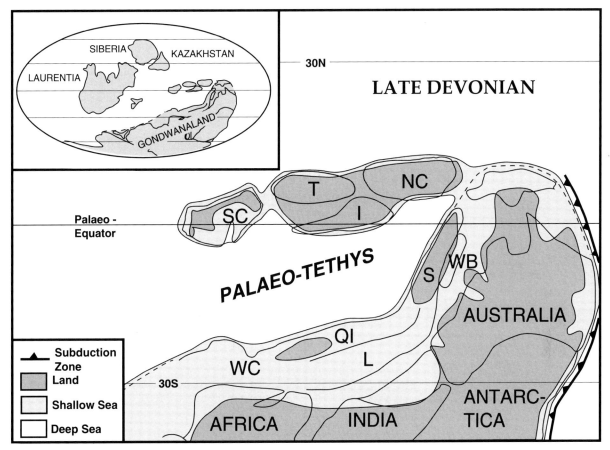

Fig.6. Reconstruction of eastern Gondwanaland for the Late Devonian showing the postulated positions of the East and SE Asian terranes, distribution of land and sea, and opening of the Palaeo-Tethys ocean at this time. Present day outlines are for reference only. Distribution of land and sea for Chinese blocks principally from Wang (1985). Land and sea distribution for Pangaea/Gondwanaland compiled from Golongka *et al.* (1994), Smith *et al.* (1994); and for Australia from Struckmeyer and Totterdell (1990). Symbols as for Figs.4 and 5.

Mid-Late Silurian reconstruction

The East and SE Asian terranes were still located in the same relative positions on the margin of Gondwanaland (Fig.5). However, northern Gondwanaland was relatively more emergent with the trans-Australian Larapintine seaway disappearing, and the previously completely or partly submerged North China, Tarim, and South China regions becoming emergent and largely land areas. Shallow-marine faunas continue to represent an Australasian province, exemplified by the *Retziella* brachiopod fauna (Rong *et al.*, 1995). The distribution of the distinctive conodonts *Nericodus* and *Tuberocostadontus* also indicates connections between South China and Australia at this time (Nicoll and Metcalfe, 1994).

Late Devonian

By Late Devonian times (Fig.6) the Palaeo-Tethys ocean had already opened between a continental sliver comprising North China, Indochina, Tarim and South China, and northern Gondwanaland. This is evident from the presence of Upper Devonian ribbon-bedded cherts in the Palaeo-Tethyan suture segments of Asia, and the fact that the separating continental terranes show no post-Devonian Gondwanaland affinities. The proposed clockwise rotation and separation of the main Chinese terranes and Indochina from Gondwanaland, and the opening of the Palaeo-Tethys, are consistent with a reported rapid anticlockwise rotation of mainland Gondwanaland at this time (Chen *et al.*, 1993). It is suggested here that an eastern conti-

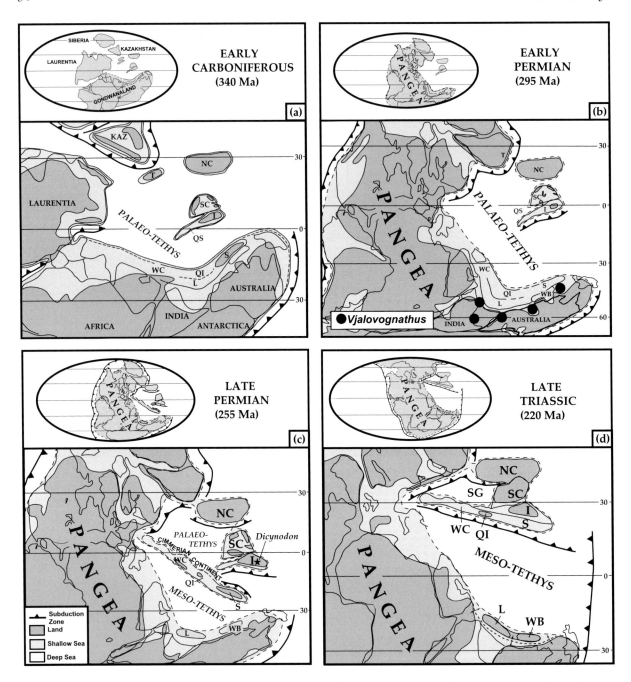

Fig. 7. Palaeogeographic reconstructions of the Tethyan region for (a) Early Carboniferous, (b) Early Permian, (c) Late Permian and (d) Late Triassic showing relative positions of the East and SE Asian terranes and distribution of land and sea. The distribution of the Lower Permian cold-water tolerant conodont genus *Vjalovognathus*, and the location of the Late Permian *Dicynodon* from Laos are also shown. Present day outlines are for reference only. Distribution of land and sea for Chinese blocks principally from Wang (1985). Land and sea distribution for Pangaea/Gondwanaland compiled from Golongka *et al.* (1994), Smith *et al.* (1994); and for Australia from Struckmeyer and Totterdell (1990). Symbols as for Figs. 4 and 5.

nental connection was maintained between the rifting sliver and Gondwanaland which may explain the migration of and similarities between Late Devonian sinolepids of South China and Australia. A Middle to Late Devonian connection between South China and Indochina is also possibly indicated by the recent discovery on the Indochina block of upper Middle Devonian yunnanolepid antiarchs (Tong-Dzuy Thanh *et al.*, 1996), formerly known only from the

Silurian-Lower Devonian of the South China block. The age disparity of these yunnanolepid faunas may also be interpreted as a possible Siluro-Devonian connection between South China and Indochina and subsequent isolation of Indochina and South China from each other by late Middle Devonian (Young and Janvier, 1997). The development of essentially endemic fish faunas on South China in the Early Devonian may have resulted from the formation of a deep-water marine barrier (but not necessarily continental separation) between South China and Tarim/Indochina during the rifting from Gondwanaland, the vestiges of this remaining as a shallow-marine seaway in the Late Devonian.

Early Carboniferous

Northeastern Gondwanaland was still in low southern latitudes (Fig.7a). Gondwanaland had rotated clockwise and initial collision between Gondwanaland and Laurentia was occurring. Siberia and Kazakhstan were nearing collision with each other and Tarim was about to accrete to them. South China and Indochina had amalgamated along the Song Ma suture zone but a narrow ocean still existed between the Qamdo-Simao extension of Indochina and South China. This narrow ocean closed in the Middle Triassic to form the Ailaoshan suture.

Early Permian

During the Late Carboniferous (Fig.7b) Gondwanaland continued to rotate clockwise sending eastern Gondwanaland into high southern palaeolatitudes and amalgamation with Laurentia, Siberia and Kazakhstan, and producing the supercontinent Pangaea. Gondwanaland was glaciated during the Late Carboniferous and Early Permian, and cold marine conditions with glacial-marine deposits were experienced on the northeastern margin of Gondwanaland. These cold conditions generally precluded the presence of warm water faunas on this part of the Gondwanaland margin, including the conodonts. Recently, however, some poor conodont faunas have been discovered in the Lower Permian rocks of Western Australia which include the cold-water tolerant genus *Vjalovognathus*, which defines an eastern peri-Gondwanaland cold-water conodont province (Metcalfe and Nicoll, 1995).

Late Permian

During the late Early Permian the Sibumasu and Qiangtang terranes, as part of the Cimmerian Continent, separated from Gondwanaland and the Meso-Tethys ocean basin opened behind it. The Palaeo-Tethys ocean continued to be subducted and destroyed beneath Laurasia, North China, and the amalgamated Indochina/South China. By Late Permian times (Fig.7c), South China was probably already in initial contact with North China (Yin and Nie, 1993), which itself may also have made initial contact with Laurasia. The Cimmerian continental sliver probably also had connections with Laurasia at its western end and also with the Qamdo-Simao part of Indochina. These various connections allowed the migration of the Pangaean dicynodont genus *Dicynodon* across to Indochina where its occurrence in the Late Permian has recently been confirmed (Battail *et al.*, 1995). The changing pattern of Permian brachiopod provincialism in the western Pacific, where blocks belonging to the Cimmerian continent contain distinct Cimmerian Province and Sibumasu Sub-Province faunas in the Sterlitamakian to Early Kungurian and then Cathaysian faunas in the Kazanian, is believed to be primarily the result of the separation and northwards drift of the Cimmerian continent in the Permian (Archbold and Shi, 1996).

Late Triassic

By Late Triassic times (Fig.7d) the main Palaeo-Tethyan branch between Sibumasu and Indochina had closed by their collision to form the Lancangjiang, Changning-Menglian, Uttaradit-Nan-Sra Kaeo, and Bentong-Raub suture zones of China, Thailand, and peninsular Malaysia. The Palaeo-Tethys oceanic lithosphere between the Cimmerian continent and Laurasia/North China continued to be subducted northwards during the Triassic, and South China collided with North China along the Qinling-Dabei suture, the resulting orogenic mountains providing huge amounts of sediment into the Songpan Ganzi accretionary complex constructed on the disappearing Palaeo-Tethyan oceanic crust (Nie *et al.*, 1995). The remnant narrow oceanic basin between Qamdo-Simao and South China was also closed during the Middle Triassic to form the Ailaoshan suture. The northwards drift of the Sibumasu terrane (and the Cimmerian continent) is also indicated by the available palaeomagnetic data (Van der Voo, 1993; Metcalfe,

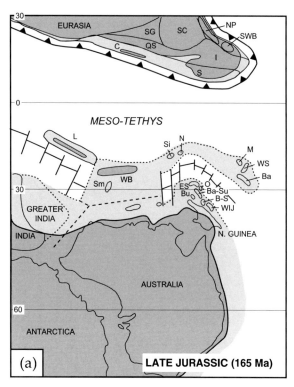

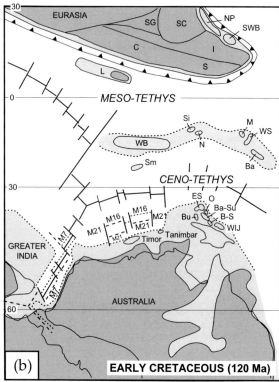

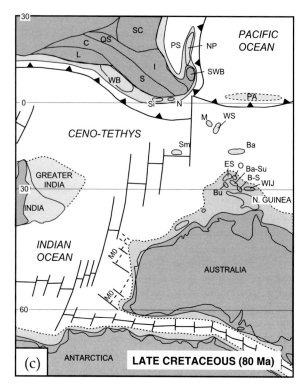

Fig.8. Palaeogeographic reconstructions for Eastern Tethys in (a) Late Jurassic, (b) Early Cretaceous, and (c) Late Cretaceous showing distribution of land and sea. SG = Songpan Ganzi accretionary complex, SWB = South West Borneo (includes Semitau), NP = North Palawan and other small continental fragments now forming part of the Philippines basement, Si = Sikuleh, N = Natal, M = Mangkalihat, WS = West Sulawesi, Ba = Banda allochthon, ES = East Sulawesi, O = Obi-Bacan, Ba-Su = Banggai-Sula, Bu = Buton, B-S = Buru-Seram, WIJ = West Irian Jaya, Sm = Sumba, PA = Incipient Philippine Arc, PS = Proto-South China Sea. M numbers represent Indian Ocean magnetic anomalies. Other terrane symbols as in Figs.4 and 5. Modified from Metcalfe (1990) and partly after Smith *et al.* (1981), Audley-Charles (1988) and Audley-Charles *et al.* (1988). Present day outlines are for reference only. Distribution of land and sea for Chinese blocks principally from Wang (1985). Land and sea distribution for Pangaea/Gondwanaland compiled from Golongka *et al.* (1994), Smith *et al.* (1994); and for Australia from Struckmeyer and Totterdell (1990).

1993, 1996; Zhao et al., 1996) and the changing provincial patterns of Permian brachiopod distributions (Archbold and Shi, 1996; Shi and Archbold, 1998 this volume).

Late Jurassic

Rifting of the northern margin of Gondwanaland had resumed by Late Triassic times and the Lhasa block separated from Gondwanaland by the Norian (Liu, 1992; Liu and Einsele, 1994; von Rad *et al.*, 1994; Ogg and von Rad, 1994). By Oxfordian times (Fig.8a) West Burma and other small continental terranes now located in south west Sumatra (Sikuleh, Natal and Bengkulu blocks), Borneo (Mangkalihat block) and

Sulawesi (West Sulawesi) had separated from northwest Australian Gondwanaland.

Early Cretaceous

During the Late Jurassic and Early Cretaceous (Fig.8b) the Lhasa block and the West Burma continental sliver drifted northwards towards proto-East and SE Asia. By late Early Cretaceous times, the Lhasa block collided with Eurasia forming the Banggong suture zone. Incipient ocean spreading also occurred between India and Australia during the Early Cretaceous as Gondwanaland began its final breakup.

Late Cretaceous

By Late Cretaceous times (Fig.8c) the West Burma, Sikuleh and Natal blocks had accreted to proto-Sundaland along the Shan Boundary and Woyla sutures. The SW Borneo block and the North Palawan block (plus other small continental fragments now forming part of the basement of the Philippines) were separating from Indochina/South China by back-arc spreading and opening of the Proto-South China Sea. The Philippine oceanic island arc was probably initiated at this time. India was well on its way northwards towards its collision with Eurasia, and Australia was moving slowly northwards and separating from Antarctica.

Acknowledgements

I would like to thank Professor Robert Hall and the Organising Committee of the Biogeography and Geological Evolution of SE Asia meeting for support to attend and present this paper at the meeting in London. I would also like to thank Professors Neville Haile and Robert Hall, and an anonymous referee for constructive review of the paper. I would also like to thank the Australian Research Council for continued funding under its Large Grant Scheme for research on East and SE Asia.

References

Allègre, C. J. and 34 others. 1984. Structure and evolution of the Himalaya-Tibet orogenic belt. Nature 307: 17-22.
Archbold, N. W. and Shi, G. R. 1996. Western Pacific Permian marine invertebrate palaeobiogeography. Australian Journal of Earth Sciences 43: 635-641.
Audley-Charles, M. G. 1988. Evolution of the southern margin of Tethys (North Australian region) from Early Permian to Late Cretaceous. In Gondwana and Tethys. pp. 79-100. Edited by M. G. Audley-Charles and A. Hallam. Geological Society Special Publication 37.
Audley-Charles, M. G., Ballantyne, P. D. and Hall, R. 1988. Mesozoic-Cenozoic rift-drift history of Asian fragments from Gondwanaland. Tectonophysics 155: 317-330.
Battail, B., Dejax, J., Richir, P., Taquet, P. and Veran, M. 1995. New data on the continental Upper Permian in the area of Luang-Prabang, Laos. Journal of Geology Series B No. 5/6: 11-15.
Baud, A. 1994. Late Permian sequence stratigraphy of the N Indian margin. International Symposium on Permian Stratigraphy, Environments and Resources, Guiyang, China, Abstracts: 1.
Baud, A., Marcoux, J., Guiraud, R., Ricou, L. E. and Gaetani, M. 1993. Late Murgabian (266 to 264 Ma). In Atlas Tethys Palaeoenvironmental Maps. Explanatory Notes. pp. 9-20. Edited by J. Dercourt, L. E. Ricou and B. Vrielynck. Gauthier-Villars, Paris,.
Bird, P. 1987. The geology of the Permo-Triassic rocks of Kekneno, west Timor. Ph.D. thesis, University of London.
Boulin, J. 1988. Hercynian and Eocimmerian events in Afghanistan and adjoining regions. Tectonophysics 148: 253-278.
Burrett, C. 1973. Ordovician biogeography and continental drift. Palaeogeography, Palaeoclimatology, Palaeoecology 13: 161-201.
Burrett, C. and Stait, B. 1985. South-East Asia as part of an Ordovician Gondwanaland — a palaeobiogeographic test of a tectonic hypothesis. Earth and Planetary Science Letters 75: 184-190.
Burrett, C., Long, J. and Stait, B. 1990. Early-Middle Palaeozoic biogeography of Asian terranes derived from Gondwana. In: Palaeozoic Palaeogeography and Biogeography. pp. 163-174. Edited by W. S. McKerrow and C. R. Scotese. Geological Society Memoir 12.
Chen, X. and Rong, J. 1992. Ordovician plate tectonics of China and its neighbouring regions. In Global Perspectives on Ordovician Geology. pp. 277-291. Edited by B. D. Webby and J. R. Laurie. A. A. Balkema, Rotterdam.
Chen, Z., Li, Z. X., Powell, C. McA. and Balme, B. E. 1993. Palaeomagnetism of the Brewer Conglomerate in central Austral, and fast movement of Gondwanaland during the Late Devonian. Geophysical Journal International 115: 564-574.
Colwell, J. B., Stagg, H. M. J., Symonds, P.A., Willcox, J. B. and O'Brien, G. W. (compilers for AGSO NW Shelf Study Group). 1994. Deep reflections on the North West Shelf: changing perceptions of basin formation. In The Sedimentary Basins of Western Australia. pp. 63-76. Edited by P. G. Purcell and R. R. Purcell. Petroleum Exploration Society of Australia.
Falvey, D. A. and Mutter, J. C. 1981. Regional plate tectonics and the evolution of Australia's passive continental margins. BMR Journal of Australian Geology and Geophysics 6: 1-29.
Ge Xiaohong, Duan Jiye, Li Cai, Yang Huixin and Tian Yushan. 1991. A new recognition of the Altun Fault Zone and geotectonic pattern of North-West China. In Proceedings First International Symposium on Gondwana Dispersion and Asian accretion — geological evolution of Eastern Tethys, Kunming, China. pp. 125-128. Edited by Ren Jishun and Xie Guanglian. China University of Geosciences, Beijing.
Golonka, J., Ross, M. I. and Scotese, C. R. 1994. Phanerozoic paleogeographic and paleoclimatic modeling maps. In

Pangaea: Global Environments and Resources. pp. 1-47. Edited by A. F. Embry, B. Beauchamp and D. J. Glass. Canadian Society of Petroleum Geologists Memoir 17.

Görür, N. and Sengör, A. M. C. 1992. Palaeogeography and tectonic evolution of the eastern Tethysides: Implications for the northwest Australian margin breakup history. pp. 83-106. Edited by U. von Rad, B. U. Haq *et al.*, Proceedings of the Ocean Drilling Program, Scientific Results 122.

Haile, N. S. 1979. Palaeomagnetic evidence for rotation and northward drift of Sumatra. Journal of the Geological Society of London 136: 541-545.

Hall, R. 1998. The plate tectonics of Cenozoic SE Asia and the distribution of land and sea. *In* Biogeography and Geological Evolution of SE Asia. Edited by R. Hall and J. D. Holloway (this volume).

Hisada, K., Arai, S. and Negoro, A. 1994. Devonian serpentinite protrusion confirmed by detrital chromian spinels in outer zone of SW Japan. *In* Proceedings of the International Symposium on Stratigraphic Correlation of Southeast Asia, pp. 76-80. Edited by P. Angsuwathana, T. Wongwanich, W. Tansathian, S. Wongsomsak and J. Tulyatid. Department of Mineral Resources, Bangkok, Thailand.

Huang, K. and Opdyke, N. D. 1991. Paleomagnetic results from the Upper Carboniferous of the Shan-Thai-Malay block of western Yunnan, China. Tectonophysics 192: 333-344.

Jell, P. A., Burrett, C. F., Stait, B. and Yochelson, E. L. 1984. The Early Ordovician bellerophontoid *Peelerophon oehlerti* (Bergeron) from Argentina, Australia and Thailand. Alcheringa 8: 169-176.

Klootwijk, C. T. 1996. Phanerozoic configurations of Greater Australia: Evolution of the North West Shelf. Australian Geological Survey Organisation Record 1996/52.

Laurie, J. R. and Burrett, C. 1992. Biogeographic significance of Ordovician brachiopods from Thailand and Malaysia. Journal of Paleontology 66: 16-23.

Li, Z. X., Zhang, L. and Powell, C. McA. 1995. South China in Rodinia: Part of the missing link between Australia-East Antarctica and Laurentia? Geology 23: 407-410.

Li, Z. X., Zhang, L. and Powell, C. McA. 1996. Positions of the East Asian cratons in the Neoproterozoic supercontinent Rodinia. Australian Journal of Earth Sciences 43: 593-604.

Lin Baoyu and Webby, B. D. 1989. Biogeographic relationships of Australian and Chinese Ordovician corals and stromatoporoids. Memoirs of the Association of Australasian Palaeontologists 8: 207-217.

Liu Guanghua. 1992. Permian to Eocene sediments and Indian passive margin evolution in the Tibetan Himalayas. Tubinger Geowissenschaftliche Arbeiten, Reihe A 13: 250 pp.

Liu Guanghua and Einsele, G. 1994. Sedimentary history of the Tethyan basin in the Tibetan Himalayas. Geologische Rundschau 83: 32-61.

Metcalfe, I. 1988. Origin and assembly of Southeast Asian continental terranes. *In* Gondwana and Tethys. pp. 101-118. Edited by M. G. Audley-Charles and A. Hallam. Geological Society of London Special Publication 37.

Metcalfe, I. 1990. Allochthonous terrane processes in Southeast Asia. Philosophical Transactions of the Royal Society of London A331: 625-640.

Metcalfe, I. 1993. Southeast Asian terranes: Gondwanaland origins and evolution. *In* Gondwana 8 — Assembly, Evolution, and Dispersal (Proceedings Eighth Gondwana Symposium, Hobart, 1991). pp. 181-200. Edited by R. H. Findlay, R. Unrug, M. R. Banks, and J. J. Veevers. A. A. Balkema, Rotterdam.

Metcalfe, I. 1994. Late Palaeozoic and Mesozoic Palaeogeography of Eastern Pangaea and Tethys. *In* Pangaea: Global Environments and Resources. pp. 97-111. Edited by A. F. Embry, B. Beauchamp and D. J. Glass. Canadian Society of Petroleum Geologists Memoir 17.

Metcalfe, I. 1996a. Pre-Cretaceous evolution of SE Asian terranes *In* Tectonic Evolution of Southeast Asia. pp. 97-122. Edited by R. Hall and D. J. Blundell. Geological Society Special Publication 106.

Metcalfe, I. 1996b. Gondwanaland dispersion, Asian accretion and evolution of Eastern Tethys. Australian Journal of Earth Sciences 43: 605-623.

Metcalfe, I. and Nicoll, R. S. 1995. Lower Permian conodonts from Western Australia, and their biogeographic and palaeoclimatological implications. Courier Forschungsinstitut Senckenberg 182: 559-560.

Nicoll, R. S. and Metcalfe, I. 1994. Late Cambrian to Early Silurian conodont endemism of the Sinian-Australian margin of Gondwanaland. *In* First Asian Conodont Symposium, Nanjing, China, Abstracts of Papers, p. 7. Edited by Wang Zhi-hao and Xu Fang-ming.

Nicoll, R. S. and Totterdell, J. M. 1990. Conodonts and the distribution in time and space of Ordovician sediments in Australia and adjacent areas. Tenth Australian Geological Convention, Geological Society of Australia, Abstracts 25: 46.

Nie, S. 1994. Devonian rifting of South China from Gondwana — a case study. 12th Australian Geological Convention, Perth, 1994, Geological Society of Australia, Abstracts 37: 319.

Nie, Y. S., Yin, A., Rowley, D. B. and Jin, Y. 1995. Exhumation of the Dabie Shan ultra-high pressure rocks and accumulation of the Songpan-Ganzi flysch sequence, central China. Geology 22: 999-1002.

Ogg, J. G. and von Rad, U. 1994. The Triassic of the Thakkhola (Nepal). II: Paleolatitudes and comparison with other Eastern Tethyan margins of Gondwana. Geologische Rundschau 83: 107-129.

Pillevuit, A. 1993. Les Blocs Exotiques du Sultanat d'Oman Evolution paléogéographique d'une marge passive flexurale. Mémoires de Géologie (Lausanne) 17: 249 pp.

Pogue, K. R., Dipietro, J. A., Said Rahim Khan, Hughes, S. S., Dilles, J. H. and Lawrence, R. D. 1992. Late Paleozoic rifting in Northern Pakistan. Tectonics 11: 871-883.

Powell, D. E. 1976. The geological evolution and hydrocarbon potential of the continental margin off north-west Australia. Journal of the Australian Petroleum Exploration Association 16: 13-23.

Rong Jia-Yu, Boucot, A. J., Su Yang-Zheng and Strusz, D. L. 1995. Biogeographical analysis of Late Silurian brachiopod faunas, chiefly from Asia and Australia. Lethaia 28: 39-60.

Saito, Y. 1992. Reading geologic history of Japanese Islands. Iwanami-shoten, Tokyo, 147 pp. (in Japanese with English abstract).

Servais, T., Brocke, R. and Fatka, O. 1996. Variability in the Ordovician acritarch *Dicrodiacrodium*. Palaeontology 39: 389-405.

Shergold, J., Burrett, C., Akerman, T. and Stait, B. 1988. Late Cambrian trilobites from Tarutao Island, Thailand. New Mexico Bureau of Mines and Mineral Resources Memoir 44: 303-320.

Smith, A. G., Hurley, A. M. and Briden, J. C. 1981. Phanerozoic palaeocontinental world maps. Cambridge: Cambridge University Press, 102 pp.

Smith, A. G, Smith, D. G. and Funnell, B. M. 1994. Atlas of Mesozoic and Cenozoic coastlines. Cambridge, Cambridge University Press, 99 pp.

Stait, B. and Burrett, C. F. 1982. *Wutinoceras* (Nautiloidea) from the Setul Limestone (Ordovician) of Malaysia. Alcheringa 6: 193-196.

Stait, B. and Burrett, C. F. 1984. Ordovician nautiloid faunas of Central and Southern Thailand. Geological Magazine 121: 115-124.

Stait, B., Wyatt, D. and Burrett, C. F. 1987. Ordovician nautiloid faunas of Langkawi Islands, Malaysia and Tarutao Island, Thailand. Neues Jahrbuch für Geologie und Palaontologie Abhandlungen 174: 373-391.

Stöcklin, J. 1974. Possible ancient continental margins in Iran. In The Geology of Continental Margins. pp. 873-887. Edited by C. A. Burk and C. L. Drake. Springer-Verlag, Berlin.

Struckmeyer, H. I. M. and Totterdell, J. M. (Coordinators) and BMR Palaeogeographic Group. 1990. Australia: Evolution of a continent. Bureau of Mineral Resources, Australia, 97 pp.

Tong-Dzuy Thanh, Janvier, P. and Ta Hoa Phuong. 1996. Fish suggest continental connections between the Indochina and South China blocks in Middle Devonian time. Geology 24: 571-574.

Vachard, D. 1990. A new biozonation of the limestones from Terbat area, Sarawak, Malaysia. CCOP Technical Bulletin 20: 183-208.

Van der Voo, R. 1993. Paleomagnetism of the Atlantic, Tethys and Iapetus oceans. Cambridge University Press 411 pp.

Veevers, J. J. and Tewari, R. C. 1995. Permian-Carboniferous and Permian-Triassic magmatism in the rift zone bordering the Tethyan margin of southern Pangaea. Geology 23: 467-470.

von Rad, U., Dürr, S. B., Ogg, J. G. and Wiedmann, J. 1994. The Triassic of the Thakkhola (Nepal). I: stratigraphy and paleoenvironment of a north-east Gondwana rifted margin. Geologische Rundschau 83: 76-106.

Wang, H. 1985. Atlas of the palaeogeography of China. 143 pp. maps, 85 pp. explanation (in Chinese), 27 pp. explanation (in English). Cartographic Publishing House, Beijing.

Wang, Z. and Tan, X. 1994. Palaeozoic structural evolution of Yunnan. Journal of Southeast Asian Earth Sciences 9: 345-348.

Webby, B. D., Wyatt, D. and Burrett, C. 1985. Ordovician stromatoporoids from the Langkawi Islands, Malaysia. Alcheringa 9: 159-166.

Wopfner, H. 1994. Late Palaeozoic climates between Gondwana and Western Yunnan. In IGCP 321 Gondwana Dispersion and Asian Accretion Fourth International Symposium and Field Excursion, Abstract Volume. pp.127-131. Edited by M. Cho and J. H. Kim.

Yang, W. and Liu, B. 1996. The finding of Lower Permian Gondwana-type spores and pollen in western Yunnan. In Devonian to Triassic Tethys in Western Yunnan, China. pp. 128-135. Edited by N. Fang and Q. Feng. University of Geosciences Press.

Young, G. C. and Janvier, P. 1997. Early-middle Palaeozoic vertebrate faunas in relation to Gondwana dispersion and Asian accretion. In Gondwana dispersion and Asian accretion. Edited by I. Metcalfe. A .A. Balkema (in press).

Yin, A. and Nie, S. Y. 1993. An indentation model for the North and South China collision and the development of the Tan-Lu and Honam Fault systems, eastern Asia. Tectonics 12: 801-813.

Zhao, X., Coe, R. S., Gilder, S. A. and Frost, G. M. 1996. Palaeomagnetic constraints on the palaeogeography of China: implications for Gondwanaland. Australian Journal of Earth Sciences 43: 643-672.

Zhao Xun, Allen, M. B., Whitham, A. G. and Price, S. P. 1996. Rift-related Devonian sedimentation and basin development in South China. Journal of Southeast Asian Earth Sciences 14: 37-52.

Biogeography and palaeogeography of the Sibumasu terrane in the Ordovician: a review

R. A. Fortey and L. R. M. Cocks
Department of Palaeontology, The Natural History Museum, Cromwell Road, London SW7 5BD, UK

Key words: biogeography, palaeogeography, Sibumasu, Shan-Thai, Thailand, Burma, China, Australia, Ordovician, trilobites, brachiopods

Abstract

Following an extensive review of previous work, and a consideration of new faunas from Thailand, the integrity of the Sibumasu palaeocontinent (Sumatra, Malaysia, West Thailand and Burma) in the early Palaeozoic is reinforced. Its lower Ordovician faunas have been claimed to show affinity with North China/Australia, but for the upper Ordovician, we describe undoubted similarity to South China, both in fossils and sedimentary facies. The possible resolution of these difficulties is discussed, and the provisional conclusions reached that (a) Sibumasu was most similar to South China; (b) South China and North China may not have been so far apart as has been suggested; and (c) Indochina carries a different faunal signal from any of these other terranes.

Introduction

The palaeogeography of Australasia is becoming progressively better understood for the 150 Ma period following the breakup of Pangaea; but there remain many unsolved problems in the Palaeozoic. The period with which we are concerned, the Ordovician (495-443 Ma), was a time of general continental dispersal (Scotese and McKerrow, 1990) long predating the assembly of Pangaea, and the disposition of the continental masses at this time is still under investigation. SE Asia has been divided into a number of terranes (*e.g.,* Mitchell, 1981; Burrett *et al.,* 1990; Metcalfe, 1992), each bounded by major faults. These terranes have histories which are variably decoupled from that of their neighbours — that is, their contiguity today is no guarantee of proximity in the past.

The Sibumasu (sometimes called Shan-Thai, as on Figs.1 and 2) terrane embraces much of the Malay peninsula, extending northwards into West Thailand and Burma and southwards into western Indonesia (Sumatra). It is bounded to the east by the Uttaradit-Nan to Raub-Bentong sutures, and to the west by what Metcalfe (1992) termed the Shan boundary (also possibly an ancient suture). Thus it comprises a tract of land some 4000 km long, clearly a continental entity of sufficient size to have acquired its own palaeogeographic signature. Several major neighbouring terranes have been referred to in relation to the history of Sibumasu. These are (Fig.1): the Indochina terrane immediately to the east; the South China terrane, today comprising many of the most populous regions of China south of the Tsinling (Chin Ling) suture; the North China terrane (including the northern part of China and associated regions of Korea); and the Australian continent, which was already constituted in something like its present form, although the Tasman fold-belt to the south (including New South Wales and much of Victoria) and montane Queensland were mobile belts in the Ordovician.

Here we review published evidence for the palaeogeographic position of the Sibumasu terrane in the Ordovician, referring to evidence derived from fossils and also in relation to the sedimentary sequences developed over this important area. These fossil faunas have been described in disparate publications over the last seventy years (many listed in Ingavat-Heimche, 1994). This review also provides a salutary reminder of the uncertainties that still apply to

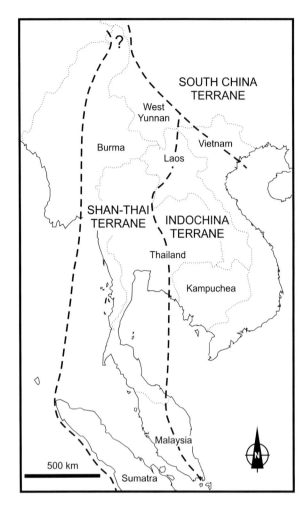

Fig.1. Modern map showing the boundaries of the various terranes in SE Asia.

understanding much of Lower Palaeozoic geography, and of the sometimes ambiguous signals given by biogeographic evidence.

Biogeographic methods and materials

The principal evidence for Ordovician biogeography used in this account comes from trilobites and brachiopods, with important additional contributions from fossil Mollusca, especially nautiloids. Trilobites have been used for many years to discriminate biogeographic areas (*e.g.*, Whittington and Hughes, 1972), and they are particularly useful as a highly speciose group with distinctive morphologies that tend towards endemicity which is related to, and dependent upon, past continental configurations and climatic zones. Brachiopods are sessile organisms with greater sensitivity to substrate, but whose distribution patterns frequently track those of the trilobites. Taken together with various other indicators, these fossils provide a powerful way of discriminating ancient biogeographic provinces — and hence an indication of ancient geography and plate distribution.

Both trilobites and brachiopods had different communities related, for example, to water depth, within any given palaeocontinent. These have received different descriptions according to biogeographic fashion: in this paper we shall simply refer to them as biofacies. The shelf biofacies are, in general most useful to recognise faunal similarities of palaeogeographic significance — thus, two shelf faunas identical as to species are likely to have been closely related geographically (Cocks and Fortey, 1982). Even among shelf faunas there were evidently taxa which (at least at generic level) were distributed widely within a single palaeoclimatic belt. Since Sibumasu occupied a tropical position in the Ordovician, such pandemic taxa are also distributed in other tropical contemporary sites — for example, in Laurentia, Australia and Siberia — and their palaeogeographical usefulness is limited.

Deeper water faunas, for example the Ordovician trilobite cyclopygid biofacies, may have even wider distribution in global terms (Fortey and Cocks, 1986; Cocks and Fortey, 1990) and thus be of little significance in discriminating which continent was close to another. However, they are important in a different way because they allow recognition of the *margins* of these ancient cratonic areas, to which they were confined. Also, their orientation should make geographic sense, because deep-water faunas should align with others of their kind, and generally face outwards towards ancient oceans. Thus the faunal approach we use to deduce the former geography of Sibumasu is twofold: (1) comparison of Ordovician shelf faunas with their contemporaries in other terranes; (2) orientation of biofacies into onshore-offshore gradients. In all comparisons identity of *species* is considered more important than generic-level similarity — although the latter has often proved useful for more global studies.

This approach is complemented by comparing sedimentary sequences in several areas in different terranes; if there are striking lithological and/or sequence similarities this supplies additional evidence for geographical continuity. However, there are many areas of Sibumasu for which the details of the sequences are imper-

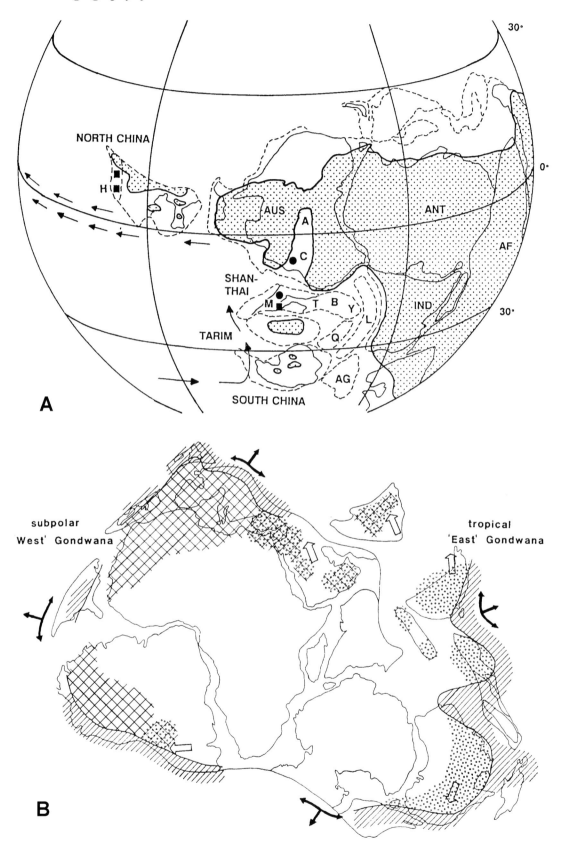

Fig. 2. Differing previous palaeogeographies of Gondwana in the Early Ordovician. A, from Laurie and Burrett, 1992; B, from Cocks and Fortey, 1988 (in the latter North China is not shown; it was imagined then as some distance away).

fectly known, and we naturally concentrate upon those areas that we have studied in the field and which are the best documented. The most fully known Ordovician succession is that in southern Thailand, Satun Province, of which many details have only recently been available. Several different palaeogeographies have been published (*e.g.,* Fig.2).

Ordovician sequence and palaeontology in southern Thailand

The palaeontology and stratigraphy of the Ordovician of southern Thailand and the adjacent areas of northern Malaysia is now comparatively well-known. Ordovician rocks crop out over a discontinuous north-south belt for several hundred kilometres, and sections have been studied particularly in the southern area, to either side of the Thai-Malaysia border. Outcrops extend on to two offshore islands: Tarutao Island on the Thai side of the border, and Langkawi Island on the Malaysian side. Earlier studies by Kobayashi and his colleagues (summarised in Kobayashi and Hamada, 1984) tended to be descriptions of 'spot faunas' — small collections of a few invertebrate fossils published as separate notes — and their mutual relationships are not obvious. However, Wongwanich *et al.* (1983, 1990) have published more complete accounts of the Palaeozoic stratigraphy in Satun Province, based on successions in Tarutao Island and the adjacent mainland, from which a better picture of the regional stratigraphy can be gained. This succession provides the principal factual base on which to hang our biogeographical assessment herein.

On Tarutao Island the Cambrian-Ordovician boundary lies within the upper part of the Tarutao Formation, a thick siliciclastic formation of sandstones and siltstones. An upper Upper Cambrian trilobite fauna described by Shergold *et al.* (1988; also Kobayashi, 1957) provides a lower limit for our discussion. They comment (p. 305) that the fauna "more closely resembles assemblages from Vietnam, western Yunnan, Sichuan and into Gangsu and eastern Qinghai Provinces of China, and central Australia" being inhabitants of an inshore detrital environment. The fauna is distinguished from those of the Sino-Korean platform by the absence of some typical genera found there, rather than by the presence of diagnostic endemic genera; furthermore, the species listed in Shergold *et al.* (1988) are not known outside Sibumasu.

Lower to lower Middle Ordovician faunas

An Ordovician (probably upper Tremadoc) trilobite fauna from the top 100 m of the Tarutao Formation (Stait *et al.,* 1984) is the earliest Ordovician yet discovered in Sibumasu. Five genera were recognised, and of two named species both were newly named. All the genera identified are widely distributed in Tremadoc strata, and comparative species were discussed from both palaeotropical China and Laurentia. The fauna is not, therefore, biogeographically critical.

The clastic rocks of the Tarutao Formation are succeeded by a thick carbonate succession grouped in the Lower to basal Middle Ordovician Thung Song Formation (the equivalent strata on Langkawi Island are known as the Lower Setul Limestone). These limestones are shallow-water, peritidal, even partially lagoonal in origin. The fossil faunas are sparse, in spite of excellent exposures around the coast of Tarutao Island. The limestones resemble in lithology many other lower-mid Ordovician carbonate platform sequences, *e.g.,* from North China/Korean plate, central Australia, much of Laurentia and the San Juan terrane of Argentina. From the base of the Thung Song Formation Stait and Burrett (1984a) recorded a polyplacophoran mollusc, *Chelodes whitehousei,* the same species as from the Ninmaroo Formation, central Queensland. This similarity is interesting, but we note that the genus *Chelodes* is actually pan-tropical in the Ordovician, since other species are known from as far away as Alabama which formed part of Laurentia (Runnegar *et al.,* 1979). The gastropod-bearing limestones of Pulau Langgon (Langkawi) yielded a distinctive gastropod operculum, *Teichiispira* (Yochelson and Jones, 1968), which has a similar pan-palaeotropical distribution. Here (Fig.3E-I) we figure some silicified brachiopods from the sparsely fossiliferous lower massive member of the Thung Song Formation; these, too are mainly genera (*Syntrophina* and *Archaeorthis?*) with global palaeotropical distribution in the early Ordovician (Ibexian) and one apparently endemic form of no biogeographical significance (gen. nov.?). While these examples indicate an interesting capacity for larvae of inshore, eurytopic genera to disperse around low palaeolatitudes, they do not contribute critically to arguments on palaeogeography. Fragments of the trilobite *Leiostegium,* similarly widespread, occur with them. Laurie and Burrett (1992) described two brachiopods (*Spanodonta floweri* and *Aporthophyla tiangjinshanensis?*)

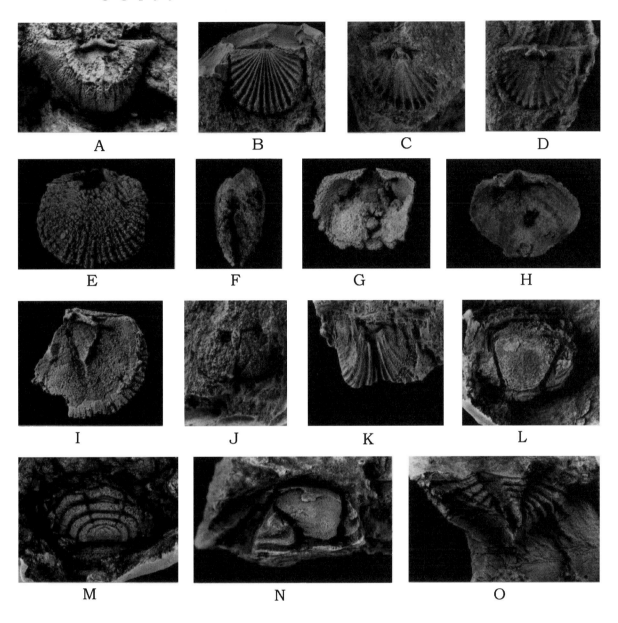

Fig.3. New faunas from Thailand and Burma. 3A-D, brachiopods from the Upper Naungkangyi Group, Linwe, Neyaungga, Southern Shan States, Burma. A, *Leptellina (Leptellina) minor* Cocks and Zhan, BC 52418, latex cast of a dorsal interior, x 5; B-D, *Saucrorthis irravadica* (Reed), B,BC 52411, latex cast of dorsal exterior, x 4, C, BB 37702, ventral interior mould, x 5, D, BB 37668, latex cast of dorsal interior, x 4. 3E-I, brachiopods from the Thung Song Formation (Upper Tremadoc), Langa Bay, north of Malaka inlet, Taratao island, Satun Province, south Thailand; 3E-G, *Archaeorthis?* sp. BC 51248, dorsal and side views of conjoined valves, and ventral interior of another valve, x 4; 3H, *Syntrophina* sp. BC 51247, dorsal interior, x 4; 3; 3I, BC 51249, orthide gen. nov?, dorsal interior, x 4; 3J-O, brachiopods and trilobites from the Satun Shale Formation (Llandeilo-Caradoc), track to Ban Pa Kae, 3.5km east of Route 4078, 10km north of Langu, Satun Province, Thailand; 3J, syntrophiid, BC 52142, internal mould of small ventral valve, x 6; 3K, eoorthid, BC 52141, internal mould of dorsal valve, x 4; 3L-M, *Ovalocephalus*, 3L, In 9300, pygidium, 3M, In 9301, a thoracic segment plus cranidium, both x 5; 3N, *Encrinurella* sp., In 9303, slightly distorted cephalon, x 5. 3O, *Hungioides* sp., In 9302, pygidium, x 2.

from the lower Ordovician which they regarded as "important confirmatory evidence for... ...Ordovician juxtaposition of the Shan Thai terrane and western Australia" (p. 16). *Spanodonta* is known from only two localities: on Langkawi Island and in the Canning basin, Australia. However, it would be impossible to make the same claim for *Aporthophyla*, because the genus has an almost pan-tropical Ordovician distribution (its type species comes from Nevada), particularly in marginal terranes (Cocks and Rong, 1989), and the Sibumasu species was not identical with any other and in fact compared with one (*A. tiangjinshanensis*) only previously described from North China. Jell *et al.* (1984) described a gastropod, *Peelerophon oehlerti*, from higher in the Thung Song Formation, from lower Arenig limestones (according to conodont evidence). This gastropod species has a long stratigraphic range (Tremadoc to Arenig) and a very wide geographical distribution: Argentina, Tasmania, western Australia (and a related species is in the Montagne Noire, France). Jell *et al.* (1984) made much of this distribution as reflecting Ordovician Gondwana, and noticed that, unusually, this gastropod cuts *across* palaeoclimatic zones, occurring from low to high palaeolatitudes but not, apparently, outside Gondwana: "it was eurythermal but unable to cross major oceanic barriers" (p. 169). Yu (1989) has subsequently recorded *P. oehlerti* from South China. The distribution is certainly different from the pan-tropical taxa noticed above. As will be noted from all our faunas, the Gondwanan link is well proven in general for Sibumasu, but does not help in identifying the closest Gondwana or peri-Gondwanan terranes to Sibumasu in Ordovician time.

Nautiloids from Tarutao and Langkawi Islands from the Thung Song Formation or its equivalents were described by Stait and Burrett (1984b) and Stait *et al.* (1987), who considered earlier literature. All are described as shallow water assemblages. The faunas range in age from the mid-Ibexian to Whiterockian (approximately Arenig-Llanvirn), and are quite diverse. Nautiloid-bearing limestones of this kind extend along the whole length of Sibumasu (Ingavat *et al.*, 1975, Kobayashi, 1961, 1984). The faunas have been compared at the generic level with those of the North China platform and the Georgina basin, western Queensland. In view of the lithological similarities between the palaeotropical limestones in these areas this is perhaps not surprising.

Mid-Ordovician faunas

In southern Thailand and adjacent Malaysia strata following the massive limestones are not well-exposed, faunas are small 'spot samples', and their stratigraphic relationships are still under investigation. Clastic rocks are predominant. Kobayashi and Hamada (1964) described a small fauna from the Satun Shale (?=Thung Wa Shale of Bunopas, 1983); the trilobite therein, *Basiliella satunensis*, belongs to a genus which is widely distributed in the mid-Ordovician and not palaeogeographically critical. Here we figure for the first time a small fauna from stratigraphically below the Pa Kae Formation, as described by Wongwanich *et al.* (1990, Figs.2, 3), collected from an isolated outcrop lying on the roadside 3 km to the east of the type section of the Pa Kae Formation. The fauna is from an impure, and now thoroughly decalcified shaly limestone. Rare brachiopods include a small ventral valve showing a spondylium and probably referable to a syntrophiid (Fig.3J) and two different fragmentary orthids including an eoorthid (Fig.3K) none of which are of much assistance in determining palaeogeography. The trilobites *Encrinurella*, *Ovalocephalus*, an agnostid (*Arthrorhachis* sp. not figured) and a dikelokephalinid (*Hungioides* sp.) are associated. The *Encrinurella* species (Fig.3N) is similar to the type species illustrated by Reed (1906); the genus occurs also in the upper Ordovician Tangtou Formation of the South China plate (Tripp *et al.*, 1989). *Ovalocephalus* (Fig.3L) is tentatively compared with *O. primitivus* Lu, which co-occurs with *Hungioides* in the Dawan Formation of Hubei Province The range of dikelokephalinids in general does not extend into the Upper Ordovician; *Hungioides* (Fig.3O) has an unusually wide peri-Gondwanan distribution, extending as far west as Portugal and Germany. A different species of *Ovalocephalus* is very common in the overlying Pa Kae Formation. This genus is considered absolutely characteristic of, and confined to, the South China plate before the Ashgill (Lu, 1975).

Finally, a cyclopygid faunule described from further south in Malaysia (Kobayashi and Hamada, 1971) probably also belongs within the mid-Ordovician time interval. Cyclopygid biofacies is a deep-water association almost independent of palaeocontinent distribution, but invariably associated with sites near former continental margins (Cocks and Fortey, 1990). The brachiopods in general are poorly known from Sibumasu at this time: the fauna described by Hamada (1964) from calcareous shales near Satun on the west coast of southern Thailand is recorded as three different species of *Cyrtonotella*, *Multicostella?* sp. and new species of *Rafinesquina* and *Oepikina?* and needs revision, for example one of the '*Cyrtonotella*' is a plectambonitoid and the '*Rafinesquina*' is a glyptomenid.

Upper Ordovician faunas

Upper Ordovician faunas of southern Thailand have recently been described for the first time by Fortey (1997) and Cocks and Fortey (1997). The Pa Kae Formation (Caradoc) is a striking series of bedded red limestones with abundant syneresis cracks. The upper part of the Lower Setul Limestone on Langkawi Island is regarded as probably correlative with the lower part of the Pa Kae Formation on the basis of shared trilobite species (Kobayashi and Hamada, 1978) among a rich trilobite fauna comprising nearly forty species (Fortey, 1997). These species are almost all identical to those from the Pagoda Formation, which is extensively exposed over several provinces within the South China platform (Ji, 1986); in fact Fortey (1997) described only two species from Thailand as new, and there are three genera which have so far only been reported from the South China plate and Thailand. The similarity is all the more significant because the distinctive and unusual lithologies of the Pa Kae and Pagoda Formations are strikingly similar — including the colour and the syneresis cracks. The latter had formerly been taken as 'mudcracks' in China, and said to be indicative of shallow-water deposition, but trilobites and other fossils support instead an open shelf setting, and the re-interpretation of the syneresis cracks is consistent with this (Ji, 1985). Cocks and Rong (1988) recorded a depauperate *Foliomena* brachiopod fauna from the same beds, which is also present in the correlative rocks in South China. Faunas of the same age are unknown from central Australia and rare and poorly known from the North China platform, while contemporary faunas from New South Wales are completely different even at generic level (Webby, 1973, 1974). Thus there is good evidence of faunal and lithological comparison with South China for the Pa Kae Formation. This is continued with the overlying Wang Tong Formation, which marks a return to clastic deposition, and completes Ordovician sedimentation. The lower part is entirely graptolitic, but higher in the section at the top of the Ordovician there is a thin development of the latest Ordovician *Hirnantia* fauna — including some seven brachiopod genera (making up 74% of the total abundance) and the typical trilobite *Mucronaspis mucronata* (Cocks and Fortey, 1997). The *Hirnantia* fauna *sensu lato* is practically pandemic. It is associated with a widespread climatic decline more or less coincident with the end-Ordovician glaciation. However, at the species level Cocks and Fortey (1997) showed that some of the brachiopods (*e.g., Aegiromena planissima*) were identical (a) with those from the Panghsa-pyé Beds of the Northern Shan States (Reed, 1915) and (b) with those from the Hirnantian of the South China platform. Hence, the closest faunal comparison for Sibumasu in the latest Ordovician is once again with South China. The similarity continues between the two areas into the early Silurian, with the two sequences becoming different only in late Llandovery times.

Northern Thailand and Burma

Much less is known about the Ordovician succession in Thailand north of the peninsula, and the deposits there are generally more tectonised than in the South. However, it is likely that the Lower Ordovician limestones were part of a single, if now discontinuous sedimentary belt, as shown by, for example, the similarity of the nautiloid faunas and lithologies (Kobayashi, 1961; Ingavat *et al.,* 1975; Stait and Burrett, 1984b). It has not been questioned that only one terrane is represented. Northwards into Burma, the Ordovician successions are known from the Southern Shan States, which seems to be a northward continuation of the Thai sequence (Garson *et al.,* 1976), including a thick development of early Ordovician limestones. However, in the Northern Shan States, the early Ordovician limestones are not present (Mitchell *et al.,* 1977), and the whole sequence is dominated by clastics. Given the otherwise striking continuity of the Sibumasu terrane, this difference is significant. Ordovician palaeontology in Burma still relies heavily on the descriptions of Reed (1906, 1915, 1936). These faunas have never been fully revised, and their palaeogeographical value is accordingly to be treated with some caution, although we have revised brachiopod faunas (Cocks and Zhan, 1998) from the Shan States collected by Mitchell *et al. (*1977) in their survey. Our comments below on the trilobites are preliminary, but important to illustrate the problems involved. There is some evidence that Reed misassociated some pygidia and cephalic parts. For example, *Calymene liluensis* Reed, 1915 has a pygidium assigned to it which actually probably belongs to an *Ovalocephalus* species (mentioned above as a common element of the Pa Kae Formation of Thailand), and is certainly not that of any calymenid; on the other hand, a cephalon illustrated as *Encrinurella insangensis*

(Reed, 1915, Plate 8, Fig.15) is a plausible candidate for that of *Ovalocephalus*. The type specimens of *E. insangensis* illustrated by Reed (1906) are quite different from those he assigned to the same species in 1915, and represent *Encrinurella* in the strict sense. The two main trilobite faunas of the Northern Shan States are those of the Hwe Mawng Beds and Upper Naungkangyi Beds respectively — both clastic deposits. The former includes '*Ogygites*' (now *Birmanites*) *birmanicus*; the genus is well known in the Middle Ordovician of central and southwest China, but it also occurs in Vietnam and Kazakhstan. The asaphid species described by Reed (1915) as *Ptychopyge thebawi* invites comparison with what Kobayashi and Hamada (1964) described as *Basiliella satunensis* from Thailand. An *Ampyx* and *Illaenus* species contribute nothing palaeogeographically because both are pandemic.

The Upper Naungkangyi fauna includes an obvious *Neseuretus* (*Neseuretinus*) species (named *Calymene birmanica* Reed by Reed in 1915), a taxon of exclusively inshore Gondwanan distribution, and confined to warm temperate to tropical palaeolatitudes (Turkey and eastwards); it, too, is a familiar component of Middle Ordovician faunas of the South China plate (Shensi and Szechuan Province: Lu, 1975). Dean (1967) noted that Reed's original (1906) type specimen of '*Calymene*' *birmanica* is different from those he later assigned to the same species (see also Lu et al., 1965, Pl. 128, Fig.13), and Fortey and Morris (1982) pointed out that the specimen is close to *Neseuretus* (*Neseuretus*) *tristani*, a widespread species of Llanvirn-Llandeilo age in Europe. *Ovalocephalus* occurs in the same fauna as Reed considered in 1915, as mentioned previously. We think it likely that the cranidium of Reed's *Calymene (Pharostoma) liluensis* is now referable to *Xuanenia* (see Tripp et al., 1989), a genus otherwise known only from the South China platform. Reed's *Cheirurus submitis* is very similar to *Hadromeros xiushanensis* (Sheng), a species widespread in the later Ordovician over the South China platform (see Ji, 1986). The species Reed described as *Holometopus wimani* would now be assigned to *Dulanaspis* Chugaeva, a genus also described from the Caradoc Gondwanan marginal terranes of Kazakhstan and Kirgizia. In general, the species comparisons of the trilobites suggest to us that the Upper Naungkangyi Beds may be of Llandeilo to Caradoc age, rather than somewhat older as Reed suggested. Virtually all the forms named by Reed require reconsideration in modern taxonomic terms; there are another five trilobite species from the Naungkangyi Beds which require examination before a modern interpretation can be made. Nonetheless it is clear that the species named can be compared with faunas from South China, and/or Gondwana margin faunas from central Asia. They are completely different from those described from New South Wales by Webby (1973, 1974 and Percival, 1991). This has been further reinforced by recent work on the Shan States brachiopods by Cocks and Zhan (1988), which shows that both Llanvirn/Llandeilo and also early to middle Caradoc faunas are present, with the main Upper Naungkangyi fauna from the Caradoc. That fauna includes two endemic strophomenoid genera, but the remaining 29 genera are more nearly similar to the contemporary Shihtzupu Formation fauna from South China (Xu et al., 1974) and to a lesser extent North China, and very different from contemporary brachiopod faunas in Australia and Laurentia. For example, the orthoid *Saucrorthis* (Fig.3B-D) is known only from South China and Burma. The Shan States fauna is again different from that of uncertain late Ordovician age from southern Thailand (Hamada, 1964), which is poorly characterised and needs revision, particularly since the latter is the only other known possible Caradoc fauna in addition to the Shan States from the Sibumasu palaeocontinent.

The uppermost Ordovician (Hirnantian) Panghsa-pyé beds were discussed by Cocks and Fortey (1997), and considered identical to correlative beds in southern Thailand. Reed's trilobite species *Dalmanites hastingsi* was regarded as a likely synonym of the widespread species *Mucronaspis mucronata*; and several of the brachiopods were equally cosmopolitan; but some species were less widespread globally, and only occur in Burma, Thailand and South China.

Ordovician palaeogeographical reconstructions of SE Asia

It is generally recognised that China, Australia and its peripheral terranes were at low palaeolatitudes, and allied to a large Ordovician Gondwana continent, which already included Africa, India, Antarctica and much of southern Europe (Cocks and Fortey, 1988). The south pole, and cooler-water faunas, lay in the west of Gondwana. What is at issue is the relative positions of North and South China, and the various other terranes which were part of, or marginal to,

Gondwana. Reconstructions of Sibumasu and adjacent areas in the Ordovician have been presented several times by Clive Burrett and his coworkers (Burrett and Stait, 1985, 1987; Burrett *et al.*, 1990; Laurie and Burrett, 1992). These authors repeatedly emphasise the similarity of the Ordovician faunas to those of North China and central Australia, and adduce evidence which places the Sibumasu terrane adjacent to, and to the west of Australia. The most recent version (Fig.2) of this Ordovician palaeogeography is that of Laurie and Burrett (1992, Fig.3). This reconstruction shows South China removed by some 40 degrees of latitude from North China, and Shan-Thai (Sibumasu) approximately half way between the two and adjacent to Australia.

Our review of the faunal evidence from Sibumasu as interpreted by previous workers emphasises the following points:

1. There is a number of pan-palaeotropical forms (brachiopods *Syntrophina* and *Aporthophyla*, molluscs *Teichiispira* and probably *Chelodes*). The detailed palaeobiogeographical use of these is limited, but they do require that the terranes on which they are found could not have been too far from the palaeoequator.

2. There is a small number of peri-Gondwanan forms (the gastropod *Peelerophon*; trilobite *Hungioides*); interesting though these are (and potentially important in correlating rocks), they cannot 'fine tune' the biogeography.

3. In the Lower Ordovician to basal mid-Ordovician limestones (and to a lesser extent inshore sandstones) of Sibumasu there is a fauna of up to nine nautiloid species which are supposedly strongly similar to those of North China (and platform Australia); this is augmented by one brachiopod (*Spanodonta*). The cephalopod faunas of North China ('Yellow River Province') are different from those of South China ('Yangtze River Province') according to Chen and Teichert (1988, Fig.2) "related, respectively, to low latitudes or tropical areas, and mid-latitudes or temperate zones" (ibid. p. 153). These differences may lie behind the supposed placement of North and South China plates at quite different palaeolatitudes (*e.g.*, Burrett *et al.*, 1990; Laurie and Burrett, 1992).

4. Mid- to upper Ordovician trilobite faunas from Sibumasu are strikingly similar to those of the *South* China plate. The fauna of the Pa Kae Formation of southern Thailand is exactly similar to that of the Pagoda Limestone; Fortey (1997) listed many species in common, and three genera *(Quyania, Elongatanileus, Hanjiangaspis)* are not yet recorded outside these two areas. Although revision of Reed's trilobite faunas from Burma is overdue, the brachiopods from there compare with South China rather than North (Cocks and Zhan, 1988). Burmese trilobite faunas include *Neseuretus (Neseuretinus)* with a Gondwanan distribution extending to temperate palaeolatitudes. Latest Ordovician sequence and faunas from the whole Sibumasu terrane are also like those of South China.

5. The oceanward edge of the Sibumasu terrane is probably marked by a cyclopygid biofacies fauna in Malaysia, that is, this edge should line up with marginal sites on other plates.

There is thus a difference between earlier and later Ordovician biogeographic signals, the former providing apparently strong evidence for similarity between North China and Australia (and suggesting a placement of Sibumasu next to Western Australia), the latter rather indicating a close relationship with the South China plate. The North and South China plates have been placed far apart on Ordovician continental reconstructions (*e.g.*, Burrett *et al.*, 1990). This leads to a paradox. How can Sibumasu apparently 'switch' between one and the other? Three explanations can be considered with regard to the biogeographic history of Sibumasu:

1. Either the North China, or the South China biogeographic 'signal' is misleading.

2. Sibumasu moved from being adjacent to North China to being adjacent to South China during the Ordovician.

3. The wide separations of North and South China plates given in previous reconstructions are misleading; Sibumasu 'bridges' the two.

To attempt to discriminate between these possibilities we consider the whole sequences (thickness, lithology and faunas) for those areas under contention.

Sequences of SE Asian Ordovician compared

Direct comparisons of sedimentary successions

In a brief review we cannot consider all the sequences developed in the possibly adjacent palaeoplates (Fig.4). We have compared the best-known areas of Sibumasu (S. Thailand and northern Shan States, Burma) discussed in some detail above, with representative sections from North China, South China, and central Australia. Chinese Ordovician stratigraphy has recently been extensively reviewed by Chen *et al.* (1995), wherein many sections from each province of China were described; from South China we have selected the sections at Yichang and Huayinshan as most typical of platform deposition. Chen et al. (1995, Fig.2) show a shallow to deep water gradient across the South China plate — cyclopygid biofacies lie at the edge of the basin to the South. For North China we have selected the Zibo section, Shandong Province (column 26 of Chen et al.); for central Australia the stratigraphy and section is taken from the southern Georgina basin (Shergold, 1985). In general, the succession in southern Thailand appears to be similar to that of the Yangtze platform; thick, calcareous lower Ordovician is followed by a comparatively 'condensed' younger Ordovician succession; the Pagoda Limestone lithology is identically developed in both areas; and the 'Hirnantia fauna' is a thin intercalation within graptolitic deposition which closes the Ordovician. However, Yangtze successions are, in general, much thinner particularly by comparison with the southern Shan States. But the sequence evidence would overall support the notion of Sibumasu as adjacent to, if not actually docked with, South China. Aligning deep-water marginal biofacies would imply that the Malaysian end of Sibumasu lay close to what is now Hainan Island.

Ambiguous palaeogeographic interpretations

The major problem remains of how to account for the previously claimed alignment of Sibumasu with North China/Australia rather than South China (Laurie and Burrett, 1992). It is compounded by the intercalation, on present geography, of the Indo-China terrane between Sibumasu and South China. This terrane has not been discussed previously in this review. The Ordovician succession in the Indo-China terrane is dominated by clastic rocks, and lacks thick, platform carbonates of Thung Song type. A recent review (Zhou et al., 1997) shows that the trilobite faunas are dominated by genera (even species) known from cool water palaeolatitudes of *western* Gondwana. Thus it seems probable that the Indo-China terrane had an Ordovician position well distant and to the west of where it is now, only interpolating itself in its present position between Sibumasu and China by subsequent transcurrent movement along the margin of Gondwana. The close juxtaposition today of trilobite species similar to those found near the Ordovician pole in Indo-China with lithologies and faunas of undoubted Ordovician tropical type in Sibumasu would seem to *demand* considerable transport of the former terrane. Since Sibumasu lies outboard of the Indo-China terrane it becomes less remarkable to postulate an extensive movement of this terrane also. Under this scenario, Sibumasu moved from adjacent to North China in the early Ordovician to close to South China in the later Ordovician. Even so, the change from one 'provincial' type to another seems to be abrupt and coincident with the ending of deposition of the previously stable carbonate platform.

However, there is recent evidence that the supposed contrast between faunas of North China and South China may not be as profound as claimed, for example, by Chen and Teichert (1988). Zhou and Fortey (1986) reviewed the Ordovician trilobites of the North China platform and found considerable similarity between the North China and the Yangtze platforms in the Tremadoc, even including species in common. Differences increased in the Arenig. However, Chen Ting-en (*in* Chen et al., 1995) listed similar nautiloid faunas in the Arenig of the Hunghuayuan Formation (South China) and the Liangchiashan Formation (North China). These include several genera listed (e.g., by Laurie and Burrett, 1992, Fig.2) from the lower Ordovician of Sibumasu, and presumably regarded by them as typical of the North China platform. Of the *Chisiloceras-Cochilioceras* Assemblage Zone Chen (in Chen et al., 1995 p. 16) states "it not only is widely distributed in the different plates of China, but also in North America and NW Europe" (*i.e.*, Baltica). This implies that some of the nautiloids may have been very widespread, as was the case of the pan-palaeotropical molluscs and brachiopods mentioned previously, which are not useful as more than indicators of tropical climates. In the Ashgill brachiopods, Zhan and Cocks (1998) have demonstrated close links between North and South China, with endemic genera known only from those two palaeoplates. In sum, it may not be necessary to invoke such wide separation of North and South China to explain such faunal differences as there are. The change in the faunas in Sibumasu in the mid-Ordovician may be more allied to a biofacies shift, perhaps associated with a widespread deepening event accompanied by the appearance of clastic rocks. Finally, it is worth adding that over North China, as in central Australia, the latest Ordovician is not present — if deposited at all it may have subsequently been eroded away. This, of course, is entirely negative evidence, and it would be ac-

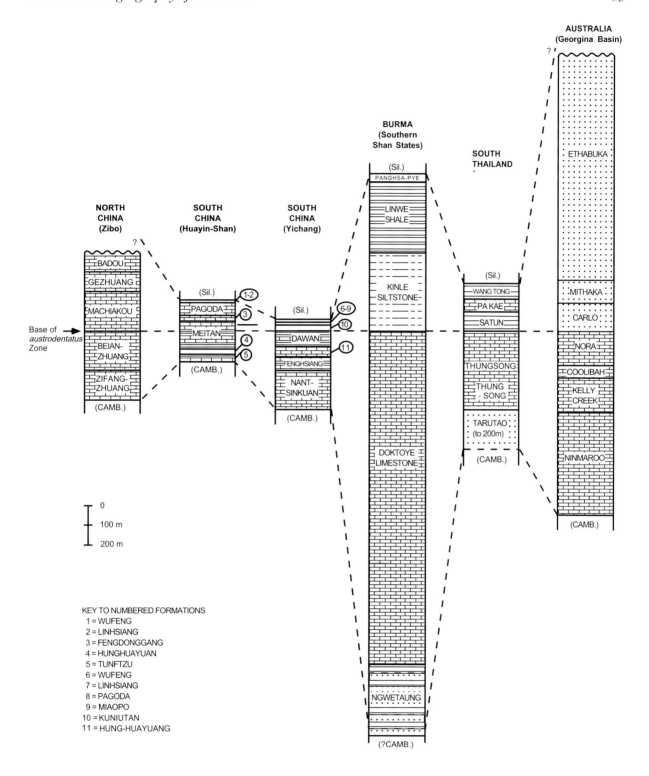

Fig.4. Comparative Ordovician successions from North China, South China, Burma, South Thailand and central Australia. Vertical dimension thickness. This demonstrates that the overall thickness and characters of the successions of South Thailand compares most closely with South China. Datum line shown is the base of the Middle Ordovician, taken here as the base of the *austrodentatus* graptolite biozone.

cordingly difficult to disprove the notion that Sibumasu preserved the entire sequence of that plate — which would itself have become more like South China as the Ordovician progressed.

Conclusions

A review of faunal evidence for the position of Sibumasu (Shan-Thai) terrane in the Ordovician reveals an apparent paradox. Lower to lower Middle Ordovician nautiloid faunas have been claimed to suggest the proximity of Sibumasu to the North China-Korean-Australian part of Gondwana on the palaeoequator. On the other hand, the nature of the whole sedimentary sequence in southern Thailand, and the taxonomic composition of the later Ordovician faunas, compare very closely with sections on the South China (Yangtze) plate. North China and South China have been placed widely apart in Ordovician continental reconstructions (*e.g.,* Laurie and Burrett, 1992). The question is how to explain this claimed 'switch' in faunal affinities? The possibility that Thailand/Sibumasu may have moved, away from the palaeoequator, from North China/Australia towards South China as the Ordovician proceeded, receives some support from evidence of considerable translation of the neighbouring Indo-China terrane. However, it seems that the differences between North and South China plates may have been overemphasised by previous workers; very similar lower Ordovician nautiloid faunas have recently been recorded from *both* areas.

At the present state of knowledge, therefore, a close biogeographical and physical proximity of South China and Sibumasu is favoured, based upon numerous specific identities among later Ordovician faunas, and several remarkable similarities in sedimentary features. However, this is far from certain; there remain many ambiguities in the evidence, which illustrate well the problems of integrating biogeography and palaeogeography for suspect terranes in the Lower Palaeozoic. Further planned work should take the story further.

References

Bunopas, S. 1983. Palaeozoic succession in Thailand. *In* Stratigraphic correlation of Thailand and Malaysia. pp. 37-96. Edited by P. Nutalaya. Geological Society of Thailand, Bangkok.

Burrett, C. F., and Stait, B. A. 1985. South East Asia as a part of an Ordovician Gondwanaland – a palaeobiogeographic test of a tectonic model. Earth and Planetary Science Letters 75: 184-190.

Burrett, C. F., and Stait, B. A. 1987. China and Southeast Asia as part of the Tethyan margin of Cambro-Ordovician Gondwanaland. *In* Shallow Tethys 2. pp. 65-77. Edited by K. G. McKenzie. Balkema, Rotterdam.

Burrett, C. F., Long, J. A. and Stait, B. A. 1990. Early-Middle Palaeozoic biogeography of Asian terranes derived from Gondwana. Memoirs of the Geological Society of London 12: 163-174.

Chen J.-Y. and Teichert, C. 1988. The Ordovician Suborder Cyrtocerinina (Order Ellesmeroceratida). Palaeontologia Cathayana 3: 145-229.

Chen Xu *et al.,* 1995. Correlation of the Ordovician rocks of China. International Union of Geological Sciences Publication 31: 1-104.

Cocks, L. R. M. and Fortey, R. A. 1982. Faunal evidence for oceanic separations in the Palaeozoic of Britain. Journal of the Geological Society of London 139: 465-478.

Cocks, L. R. M. and Fortey, R. A. 1988. Lower Palaeozoic facies and faunas around Gondwana. Geological Society Special Publication 37: 183-200.

Cocks, L. R. M. and Fortey, R. A. 1990. Biogeography of Ordovician and Silurian faunas. Geological Society Memoir 12: 97-104.

Cocks, L. R. M. and Fortey, R. A. 1997. A new *Hirnantia* fauna from Thailand and the biogeography of the latest Ordovician. Geobios 20: 111-120.

Cocks, L. R. M. and Rong Jia-Yu 1988. A review of the late Ordovician *Foliomena* brachiopod fauna with new data from China, Wales and Poland. Palaeontology 31: 53-67.

Cocks, L. R. M. and Rong Jia-Yu 1989. Classification and review of the brachiopod superfamily Plectambonitacea. Bulletin of the British Museum (Natural History) Geology 45: 77-163.

Cocks, L. R. M. and Zhan Ren-bin 1998. Caradoc brachiopods from the Shan States, Burma (Myanmar). Bulletin of the Natural History Museum (Geology) 54 (in press).

Dean, W. T. 1967. The distribution of Ordovician shelly faunas in the Tethyan region. *In* Aspects of Tethyan Biogeography. pp. 11-44. Edited by C. G. Adams and D. V. Ager. Special Publications of the Systematics Association 7.

Fortey, R. A. 1997. An Upper Ordovician trilobite fauna from Thailand. Palaeontology 40: 397-449.

Fortey, R. A. and Cocks, L. R. M. 1986. Marginal faunal belts and their structural implications, with examples from the Lower Palaeozoic. Journal of the Geological Society of London 143: 151-160.

Fortey, R. A. and Morris, S. F. 1982. The Ordovician Trilobite *Neseuretus* from Saudi Arabia, and the palaeogeography of the Neseuretus fauna related to Gondwanaland in the earlier Ordovician. Bulletin of the British Museum (Natural History) Geology 36: 63-75.

Garson, M. S., Amos, B. J. and Mitchell, A. H. G. 1976. The geology of the area around Neyaungga and Ye-ngan, Southern Shan States, Burma. Institute of Geological Sciences Overseas Memoir 2: 1-70.

Hamada, T. 1964. Some Middle Ordovician brachiopods from Satun, Southern Thailand. Contributions to Geology and Palaeontology of Southeastern Asia 1: 279-289.

Ingavat, R., Muenlek, S. and Udomrain, C. 1975. On the discoveries of some Permian fusulinids and Ordovician cephalopods of Banrai, West Thailand. Journal of the Geological Society of Thailand 1: 81-9.

Ingavat-Heimche, R. 1994. Paleozoic palaeontological evidence of Thailand. *In* Stratigraphic correlation of Southeastern Asia. pp. 43-54. Edited by P. Angusuwathana, T. Wongwanich, W. Tansathien, S. Wongsomsak and J. Tulyatid. IGCP Project 306 and Department of Mineral Resources, Thailand.

Jell, P. A., Burrett, C. F., Stait, B. and Yochelson, E. L. 1984. The Early Ordovician bellerophontid *Peelerophon oehlerti* (Bergeron) from Argentina, Australia and Thailand. Alcheringa 8: 169-176.

Ji, J.-L. 1985. On the depositional environment of the Pagoda Formation in central and southwestern China. Professional Papers in Palaeontology and Stratigraphy 12: 87-93 [In Chinese].

Ji, J.-L. 1986. Upper Ordovician (Middle Caradoc-Early Ashgill) trilobites from the Pagoda Formation in South China. Professional Papers in Stratigraphy and Palaeontology 15: 1-33 [In Chinese].

Kobayashi, T. 1957. Upper Cambrian fossils from peninsular Thailand. Journal of the Faculty of Science University of Tokyo 10: 367-82.

Kobayashi, T. 1961. On the occurrence of Ordovician nautiloids in North Thailand. Japanese Journal of Geology and Geography 32: 79-84.

Kobayashi, T. 1984. Older Palaeozoic gastropods and cephalopods of Thailand and Malaysia. Geology and Palaeontology of Southeastern Asia 25: 195-199.

Kobayashi, T. and Hamada, T. 1964. On the Middle Ordovician fossils from Satun, the Malaysian frontier of Thailand. Japanese Journal of Geology and Geography 35: 205-211.

Kobayashi, T. and Hamada, T. 1971. A cyclopygid-bearing Ordovician faunule discovered in Malaysia with a note on the Cyclopygidae. Geology and Palaeontology of Southeastern Asia 8: 1-18.

Kobayashi, T. and Hamada, T. 1978. Upper Ordovician trilobites from the Langkawi Islands, Malaysia. Geology and Palaeontology of Southeastern Asia 19: 1-28.

Kobayashi, T. and Hamada, T. 1984. Trilobites of Thailand and Malaysia. Geology and Palaeontology of Southeastern Asia 25: 273-284.

Laurie, J. R. and Burrett, C. 1992. Biogeographic significance of Ordovician brachiopods from Thailand and Malaysia. Journal of Paleontology 66: 16-23.

Lu, Y.-H. 1975. Ordovician trilobite faunas of central and southwest China. Palaeontologia Sinica (New Series, B) 11: 1-463.

Lu, Y.-H. Chang, W.-T, Chu, C.-L., Chien, Y.-Y. and Hsiang, L.-W. 1965. Trilobites of China. Science Press, Beijing. 766 pp. [in Chinese]

Metcalfe, I. 1992. Ordovician to Permian evolution of Southeast Asian terranes: NW Australian Gondwanan connections. *In* Global perspectives on Ordovician Geology. pp. 293-305. Edited by B. D. E. Webby and J. R. Laurie. Balkema, Rotterdam.

Mitchell, A. H. G. 1981. Phanerozoic plate boundaries in mainland S.E. Asia, the Himalayas and Tibet. Journal of the Geological Society of London 138: 109-122.

Mitchell, A. H. G., Marshall, T. R., Skinner, A. C., Baker, M. D., Amos, B. J. and Bateson, J. H. 1977. Geology and exploration geochemistry of the Yadanatheingi and Kyaukme-Longtawkno areas, Northern Shan States, Burma. Overseas Geology and Mineral Resources Report, Institute of Geological Sciences, London 51: 1-35.

Percival, I. G. 1991. Late Ordovician articulate brachiopods from central New South Wales. Association of Australasian Palaeontologists Memoir 11: 107-177.

Reed, F. R. C. 1906. The Lower Palaeozoic fossils of the Northern Shan States, Burma. Palaeontologia Indica (new series), 2: 1-154.

Reed, F. R. C. 1915. Supplementary memoir on new Ordovician and Silurian fossils from the Northern Shan States. Palaeontologia Indica (new series) 6 (1): 1-122.

Reed, F. R. C. 1936. The Lower Palaeozoic faunas of the Southern Shan States. Palaeontologia Indica (new series), 21 (3): 1-130.

Runnegar, B., Pojeta, J., Taylor, M. E. and Collins, D. H. 1979. New species of the Cambrian and Ordovician chitons *Matthevia* and *Chelodes* from Wisconsin and Queensland: evidence for the early history of polyplacophoran molluscs. Journal of Paleontology 53: 1374-1394.

Scotese, C. R. and McKerrow, W. S. 1990. Palaeozoic palaeogeography and biogeography. Geological Society Memoir 12: 1-435.

Shergold, J. H. 1985. Notes to accompany the Hay River-Mount Whelan Special 1:250 000 Geological Sheet, southern Georgina Basin. Bureau of Mineral Resources, Geology and Geophysics Report 251: 1-47.

Shergold, J. H. Burrett, C. F., Akerman, T. and Stait, B. 1988. Late Cambrian trilobites from Tarutao Island, Thailand. Memoir of the New Mexico Bureau of Mines and Mineral Resources 44: 303-320.

Stait, B. and Burrett, C. F. 1984a. Early Ordovician polyplacophoran *Chelodes whitehousei* from Tarutao Island, southern Thailand. Alcheringa 8: 112.

Stait, B. and Burrett, C. F. 1984b. Ordovician nautiloid faunas of Central and Southern Thailand. Geological Magazine 121: 115-124.

Stait, B., Burrett, C. F. and Wongwanich, T. 1984. Ordovician trilobites from the Tarutao Formation Southern Thailand. Neues Jahrbuch fur Geologie und Paläontologie Monatshefte 1984(1): 53-64.

Stait, B., Wyatt, D. and Burrett, C. F. 1987. Ordovician nautiloid faunas of Langkawi Islands, Malaysia and Tarutao Island, Thailand. Neues Jahrbuch fur Geologie und Paläontologie Abhandlungen 174: 373-391.

Tripp, R. P., Zhou, Z.-Y. and Pan, Z.-Q. 1989. Trilobites from the Upper Ordovician Tangtou Formation, Jiansu Province, China. Transactions of the Royal Society of Edinburgh, Earth Sciences 80: 25-68.

Webby, B. D. E. 1973. *Remopleurides* and other Upper Ordovician trilobites from New South Wales. Palaeontology 16: 445-75.

Webby, B. D. E. 1974. Upper Ordovician trilobites from central New South Wales, Palaeontology 17: 486-510.

Whittington, H. B. and Hughes, C. P. 1972. Ordovician geography deduced from trilobite distributions. Philosophical Transactions of the Royal Society, Series B 263: 265-278.

Wongwanich, T., Burrett, C. F., Tansathien, W. and Chaodumrong, P. 1990. Lower to Mid Palaeozoic stratigraphy of mainland Satun Province, Thailand. Journal of Southeast Asian Earth Sciences 4: 1-9.

Wongwanich, T., Burrett, C. F., Wyatt, D. J. and Stait, B. A. 1983. Correlations between the Ordovician of Tarutao Island, Satun Province (Thailand) and Langkawi Islands (Malaysia). *In* Stratigraphic Correlations between Thailand and Malaysia. pp. 77-95. Edited by P. Nutayala. Geological Society of Thailand, Bangkok.

Xu, H.-K., Rong, J.-Y. and Liu, D.-Y. 1974. Ordovician brachiopods. *In* Handbook of Stratigraphy and Palaeontology in Southwest China. pp.144-154. Academia Sinica, Beijing.

Yochelson, E. and Jones, C. R. 1968. *Teichiispira*, a new Ordovician gastropod genus. Professional Papers of the U.S. Geological Survey 613-B: 1-15.

Yu Wen 1989. The occurrence of *Peelerophon oehlerti* (Bergeron) from southeast China. Journal of Paleontology 63: 697.

Zhan Ren-bin and Cocks, L. R. M. 1998. Late Ordovician brachiopods from the South China Plate and their palaeogeographical significance. Palaeontology 41: (in press).

Zhou, Z.-Y. and Fortey, R. A. 1986. Ordovician trilobites from North and Northeast China. Palaeontographica Abteilung A 192: 157-210.

Zhou, Z.-Y., Dean, W.-T. and Luo, H.-L. 1997. Ordovician trilobites from the Hsiangyang Formation, Dali, western Yunnan, China. Palaeontology 40: (in press).

Permian marine biogeography of SE Asia

G. R. Shi and N. W. Archbold
School of Aquatic Science and Natural Resources Management, Deakin University, Rusden Campus, 662 Blackburn Road, Clayton, Victoria 3168, Australia

Key words: SE Asia, Permian, marine, invertebrate, vicariance biogeography

Abstract

Permian marine sequences and invertebrate faunas are widely distributed in all mainland terranes of SE Asia. A review of the spatial and temporal distributions of all major Permian marine invertebrate groups in this region, reinforced by the results of recent Permian stage-by-stage statistical analyses of western Pacific brachiopods, reveals that three biotic provinces are present in SE Asia during the Permian. The Cathaysian province occupied the Simao, Indo-China and East Malaya blocks throughout the Permian. The Sibumasu province of the Shan-Thai terrane (*s.s.*), Tengchong and Baoshan blocks developed in Late Sakmarian and continued to exist until, probably, the end of Midian when the same blocks joined the Cathaysian province. From Asselian to Early Sakmarian, the Shan-Thai terrane, Tengchong and Baoshan blocks belonged to the short-lived Indoralian province, which then also included Australia, India, the Himalayan and Lhasa terranes. The marked change of marine provinciality of the Shan-Thai terrane (*s.s.*), Tengchong and Baoshan blocks cannot be explained by the tectonic vicariance (rift-drift) model alone, nor can it be accounted for solely by migration of climatic zones. An interplay of both of these factors during the Permian is considered to be the most likely cause of this marked change of marine provinciality of these blocks.

Introduction

Tectonically, continental SE Asia is a collage of allochthonous terranes bounded by Siberia to the north and Kazakhstan to the west. The Palaeozoic-Mesozoic framework of the tectonic evolution of these and adjacent East Asian terranes has now been broadly established, with a dominant view that most of these continental terranes had their origins in northern Gondwana (Metcalfe, 1996, 1998 this volume, and references therein). In brief, the history of continental growth of East and SE Asia may be regarded as a process of step-wise accumulation of continental slivers drifted off from northern Gondwana. This process is considered step-wise because it involved several episodes of intensified rifting (*e.g.*, end Devonian and Permo-Carboniferous) and accretion/subduction (*e.g.*, Early Carboniferous and Permo-Triassic) intercalated by relatively longer intervals of drifting; thus, such concepts as Palaeo-Tethys and Neo-Tethys have been created in the literature. The former represented an ocean or a shallow seaway created in the Middle Palaeozoic after the separation of South China and Indo-China from Gondwana, while the latter corresponded to a younger ocean or seaway created during the Late Palaeozoic by the separation of a continental strip, or Cimmerian microcontinents (Sengör, 1979). The two seaways appear to have coexisted for much of the Permian and early Triassic before the Cimmerian microcontinents were finally amalgamated with southern Eurasia (Li *et al.*, 1995).

Thus, it is logical to reason from the above 'dispersion and accretion' model that for much of the Permian, the Cimmerian microcontinents were located between Gondwana in the south and Cathaysia to the north and, as such, must have played a unique and important role in the dispersal of shallow marine benthos (and plants) across the eastern and central Palaeo-Tethys. If we assume the validity of this tectonic hypothesis, the rift-drift-amalgamation history of the SE Asian sector of the Cimmerian microcontinents would provide an ideal model to demonstrate the vicariant evolution of the provincialism of these terranes through the Permian. To test this assumption and to provide constraints on the existing tectonic models, it is necessary to evaluate the spatial-temporal distribution of Permian marine faunas in the SE Asian region. In this paper, we present a review of the distribution of Permian marine invertebrate faunas of SE Asia in an attempt to unravel the relationship between provincial patterns of marine benthos and tectonic palaeogeography and climates during the Permian.

Tectonostratigraphic terranes in mainland SE Asia

In this paper, the broader SE Asian region is extended to include western Yunnan, Burma, Thailand, peninsular Malaysia, Vietnam, Laos, Cambodia, and Sumatra. In this region, at least eight tectonostratigraphic terranes or blocks have been recognised; they are West Burma, Shan-Thai (*s.s.*), East Malaya, Indo-China, Tengchong, Baoshan, Changning-Menglian, and Simao (Fig.1). The Permian stratigraphy and lithological successions of these terranes (except the Changing-Menglian Belt and Simao block) are shown in Fig.2, in comparison with coeval sequences of South China and Western Australia.

Continental SE Asian terranes have been generally classified into two broad categories in terms of their Late Palaeozoic tectonic history (Hutchison, 1993): terranes with Gondwanan affinities (West Burma, Shan-Thai, Tengchong, and Baoshan) and terranes with Cathaysian affinities (East Malaya, Indo-China, and Simao). The Permian of the former group is characterised by predominantly terrigenous, tilloid-bearing sediments in the lower part and carbonate formations in the upper part, in contrast to the blocks of Cathaysian affinities which are dominated by limestones and, in the case of the Pahang sequence of East Malaya, tuffaceous sandstones and siltstones (Fig.2). The tectonic suture between these two groups of terranes is identified by the Changning-Menglian Belt, which itself may be regarded as an accretionary prism comprising continental shelf, oceanic and magmatic deposits (Fang *et al.*, 1992; Liu *et al.*, 1993; Wu *et al.*, 1995). The Changning-Menglian Belt can be traced northwards to the western Yunnan 'Three River' (Shanjiang) syntaxes zone where it links with the Lancang River suture (Jin,

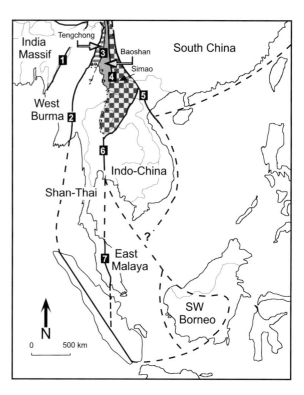

Fig.1. Map of SE Asia showing main continental terranes referred to in this paper (note that the Tengchong and Baoshan blocks are shown separated from the Shan-Thai terrane; see text for more discussion). Explanation of main sutures/faults: 1: Yarlung-Zangbu suture; 2: Shan Boundary suture; 3: Lujiang Fault; 4: Changning-Menglian suture; 5: Song Ma suture; 6: Uttaradit-Nan suture; 7: Raub-Bentong suture.

1994; Wu *et al.*, 1995). By contrast, the southern extension of the Changning-Menglian Belt across the boarder area between Burma, Thailand, western Yunnan and Laos is problematic; it could be linked to the Chiang Mai Volcanic belt in northwest Thailand, as preferred by Wu *et al.* (1995); or it may have been offset by major left-lateral strike-slip faults then connect with the Nan-Uttaradit suture and further south with the Raub-Bentong suture, as suggested by Metcalfe (1996). Due to this uncertainty, in this paper the term Simao terrane is restricted to the part in western Yunnan between the Lancang River suture and the Son Ma suture. Therefore, northwest Thailand (including the Chiang Mai Volcanic belt) is herein tentatively treated as part of the Shan-Thai terrane (*s.s.*).

Also controversial is the southern extension of the Raub-Bentong suture beyond peninsular Malaysia (Hutchison, 1993). The proposal of an almost north-south striking extension of the

Permian marine biogeography of SE Asia 59

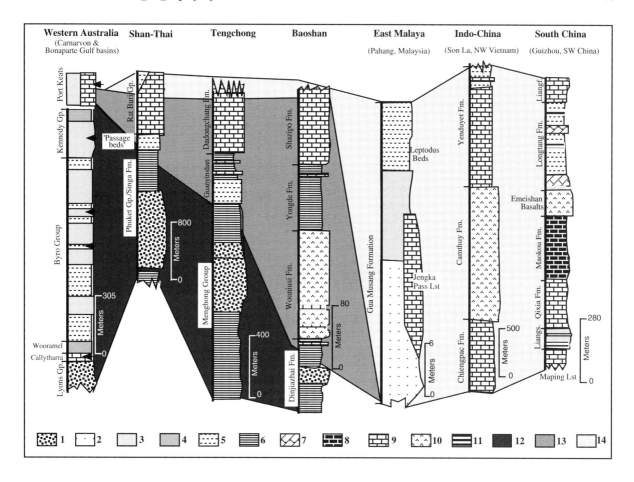

Fig.2. Permian marine stratigraphic columns and biogeographical development of main SE Asian terranes in comparison with the Gondwanan sequence of Western Australia and the Cathaysian sequence of South China (compiled from various sources). The solid black arrows in the Western Australian column indicate principal horizons of influxes of Sibumasu (and Cathaysian) faunal elements into Western Australia. Explanation of legend keys: 1: tillites, tilloids, diamictites, and/or pebbly mudstones; 2: coarse sandstone and greywacke; 3: sandstone; 4: fine sandstone and siltstone; 5: siltstone and shale; 6: predominantly shale and siltstone, containing pebbles locally; 7: marl and dolomitic limestone; 8: Limestone with chert concretions; 9: limestone; 10: basalts; 11: coal measures; 12: Gondwanan provinces (cold-water Gondwanan-type fauna); 13: Sibumasu province (transitional fauna); 14. Cathaysian province (palaeotropical, warm-water fauna).

Raub-Bentong suture through central Sumatra by Tjia (1989) is favoured here.

With respect to the tectonic affiliation of the Baoshan block, previously most workers have placed this block in the Shan-Thai (or Sibumasu) terrane which also includes the Tengchong block (Metcalfe, 1986, 1996; Fang, 1991; 1994; Shi and Archbold, 1995a; Brookfield, 1996). However, a comparison of the Permian stratigraphy and palaeontological assemblages, and recent field work by GRS in western Yunnan (July 1996), indicate distinct differences (Fig.3). Although glacigenic diamictites and associated pebbly mudstones are present in the Baoshan block, they are generally much thinner (up to 200 metres) in comparison with similar facies found in the Tengchong block reaching more than 2000 metres, and some 1600 metres in Langkawi Islands off northwest peninsular Malaysia (Stauffer and Lee, 1986). In addition, the Lower Permian of the Baoshan block has a distinctive suite of volcanic rocks (Woniusi Basalt) up to 1000 metres thick overlying the glacigenic diamictites and pebbly mudstones (Fig.3). This volcanic horizon is absent from the Permian sequences of Tengchong or Shan-Thai blocks. As will be discussed below, Upper Sakmarian fossil assemblages from the Shan-Thai terrane (*s.s.*) and the Baoshan block are also very different. The former is characterised by a diverse mixed fauna comprising Gondwanan, Cathaysian and endemic genera and species (the *Spinomartinia*

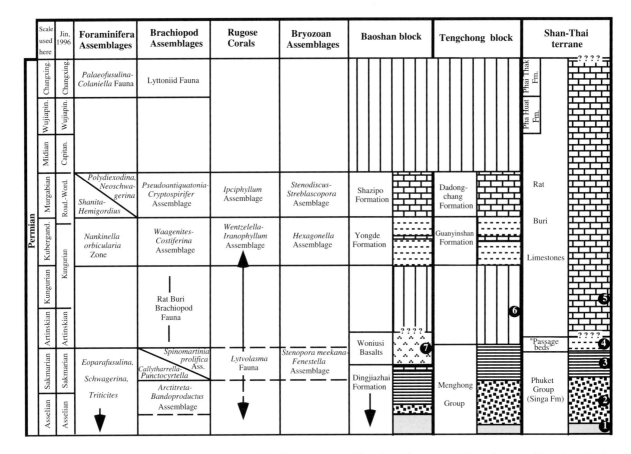

Fig.3. A schematic diagram comparing the stratigraphical successions of the Shan-Thai terrane, Tengchong and Baoshan blocks and also showing the main marine invertebrate fossil assemblages. Back circles with numbers indicate main lithologies as follows: 1: sandstones to coarse sandstones; 2: diamictites and pebbly sandstones; 3: pebbly mudstones: shale and siltstones; 4: siltstones and mudstones; 5: limestones; 6: hiatus; 7: basalts.

prolifica assemblage in Thailand and Malaysia), in contrast to the *Globiella* fauna from the upper Dingjiazhai Formation of the Baoshan block, which is dominated by Gondwanan taxa (Shi *et al.*, 1995). For these reasons and for the purpose of this paper, the Permian faunas and the biogeographical affinities of the Shan-Thai terrane, Tengchong and Baoshan blocks (hereafter referred to as STB blocks) are treated separately, and the term Shan-Thai terrane (*s.s.*) is accordingly restricted to include only eastern and peninsular Burma, northwest and peninsular Thailand, western peninsular Malaysia, and northeast Sumatra (Fig.1). It is noted, however, that the Shan-Thai terrane, in its broadest sense, could be regarded as a composite terrane comprising Shan-Thai (*s.s.*) as defined above, Tengchong and Baoshan blocks (cf., Brookfield, 1996). Based on stratigraphical and sedimentological analysis, Jin (1994) has also recognised the distinction between the Tengchong and Baoshan blocks and proposed that the two blocks must have been geographically separated from each other during the Permian.

Data, methods and previous studies

Permian marine rocks and invertebrate fossils are widely distributed in the SE Asian region and have been extensively studied. Important reviews on the stratigraphical and geographical distributions of these faunas may be found in Archbold *et al.* (1982), Archbold and Shi (1995, 1996), Basir Jasin (1991), Fang (1991, 1994), Fang and Fan (1994), Fontaine (1986, 1994), Fontaine *et al.* (1993), Fontaine and Suteethorn (1988), Ingavat (1984), Ingavat *et al.* (1980), Nakamura *et al.* (1985), Sakagami (1976), Shi and Archbold (1995a), Shi, Archbold and Zhan (1995), Smith (1988), Waterhouse (1973), and Yanagida (1976, 1984). These reviews provide important data for the interpretations presented in this paper.

In addition, the recently compiled databases of Permian brachiopod distributions in the western Pacific region (Shi and Archbold, 1993b-c, 1995d-e; Shi, Shen and Archbold, 1996) provide original lists and our revisions of Permian brachiopod species and genera described from this region. Based on these databases we have recently carried out a sequence of statistical analyses on the spatial and temporal distributions of Permian brachiopods in the western Pacific region (Shi and Archbold, 1993a, 1995b-c; 1996). This sequence of studies was conducted with a view that faunal similarities between tectonic blocks could be interpreted as a function of palaeogeographical distances between the blocks; therefore, a change of biotic similarities over time and space may be interpreted as an indication of changes in palaeogeography. Four Permian time slices have been analysed: Asselian-Early Sakmarian, Late Sakmarian-Early Artinskian, Late Artinskian-Early Kungurian, and Kazanian-Midian. Distributional data of brachiopod genera and species from a fifth time interval, the Changxingian, have also been compiled with a preliminary biogeographical assessment obtained (Shi, Shen and Archbold, 1996). In each of these quantitative studies, the shared and mutually exclusive brachiopod genera of each of the Asian-western Pacific tectonostratigraphic terranes was counted and the similarities based on the presence and absence of genera between the terranes calculated using one or two chosen similarity coefficients (see Shi (1993) for discussion on the criteria used in choosing the similarity coefficients). The mutual binary similarities were then ranked in a hierarchical fashion by cluster analysis or scaled in a two or three dimensional space by ordination methods to reveal the inter-relationships between the terranes (see Shi, 1993, 1995) for detailed discussion on the application of relevant multivariate statistical methods in palaeobiogeography). In these quantitative approaches, each tectonic unit is compared against another tectonic unit in the region in terms of the brachiopod genera shared and/or lacking between them. Such quantitatively derived biotic similarities were then interpreted as crude approximations of palaeogeographical distances.

A number of biotic provinces have been identified through the quantitative studies (see Archbold and Shi (1996) for a general review), three of which occurred in SE Asia: Indoralian, Cathaysian, and Sibumasu. The Indoralian province (Shi and Archbold, 1993a) was short-lived and occupied the STB blocks only in the Asselian-Early Sakmarian and reflects a proximity to the peak period of glaciation in Gondwana proper. The Cathaysian province (Fang, 1985) manifested itself prominently throughout the Permian in East Malaya, Simao, and Indo-China; it also incorporated the STB blocks during the Late Permian (Wujiapingian and Changxingian). The Sibumasu province was proposed by Fang (1991) to represent marine faunas from the Shan-Thai terrane (*s.s.*), Tengchong and Baoshan blocks. This name is followed here in preference to the term 'Cimmerian province' of Archbold (1983) as the latter will be raised to a regional rank – the Cimmerian Region – by Grunt and Shi (1997) and now includes 3 provinces: Iranian, Himalayan, and Sibumasu (see also Archbold, 1987 for discussions on the potential for subdivisions of the Cimmerian region). However, as discussed by Shi and Archbold (1995a) and further elaborated below, we restrict the Sibumasu province from the Late Sakmarian to the Midian, in contrast to Fang (1991, 1994), who considered it to span the entire Carboniferous and Permian.

A brief note on the Cathaysian province in SE Asia

In SE Asia, the Cathaysian province occurs in the Simao block, East Malaya and Indo-China. Permian stratigraphy and marine faunas remain little known but available data indicates a strong, persistent Cathaysian affinity throughout the Permian (Fontaine, 1986; Metcalfe, 1988), as confirmed by our stage-by-stage statistical analyses (Archbold and Shi, 1996). A few important and relatively better known Cathaysian marine invertebrate faunas are noted here. From north-central Thailand (Indo-China block), Yanagida (1967, 1976) described a brachiopod fauna which he compared closely with the Asselian-Sakmarian brachiopods of the south Urals, the Maping Limestone of South China, and the Taiyuan Formation of North China, with no characteristic Gondwanan links. Associated with the brachiopods are fusulinids dominated by *Triticites* and *Paraschwagerina* species. In west Cambodia, Permian limestones and pyroclastic beds with typical Cathaysian rugose corals (*Waagenophyllum*, *Wentzelella* and *Polythecalis*), brachiopods (*e.g.*, *Tyloplecta*, *Monticulifera*, *Leptodus*, *Oldhamina* and *Permophricodothyris*) and fusulinids (*e.g.*, *Nankinella*, *Chusenella*,

Verbeekina, Neoschwagerina and *Yabeina*) have been identified by Ishii *et al.* (1969), indicating an Artinskian to Midian age. Uppermost Permian marine sequences and faunas are now known from a number of localities in both the East Malaya and Indo-China blocks. Mohd Shafeea Leman (1993, 1994) has reported a diverse Changxingian lyttoniid fauna from the Leptodus Shale beds in the 'Central province' of the East Malaya block. This fauna can be closely correlated with the Changxingian brachiopods of South China. Similar lithofacies and biofacies have also been observed by the senior author from the Son La area in northwest Vietnam (Indo-China block) during a pre-conference field trip in November 1995 (see section in Fig.2). In a review of Permian corals of Thailand, Fontaine (1994) also noted great affinities of Permian corals between southeast, central and northeast Thailand (Indo-China block) with those of South China, Vietnam, Laos, Cambodia, east Malaysia and south Sumatra. In the Simao block of western Yunnan, a full Cathaysian faunal succession has been recorded (Geological Survey of Yunnan, henceforth GSY, 1990), with fusulinids ranging from the Asselian *Pseudoschwagerina* zone, through the middle Permian *Misellina, Cancellina, Neoschwagerina* and *Yabeina* zones, to the uppermost Permian *Palaeofusulina* zone.

Permian biogeographical evolution of the Shan-Thai terrane, Baoshan and Tengchong (STB) blocks

Unlike the Simao block, East Malaya and Indo-China, which collectively formed part of the Cathaysian province throughout the Permian, the biogeographical development of the STB blocks was highly dynamic during the Permian and is hence of great interest. Currently, there exist two different views on the identity and history of these blocks. Fang (1991, 1994) considers that the STB blocks had strong Gondwanan affinities during the Cambro-Ordovician time, then demonstrated a close relationship with the Rhenish-Bohemian province of the Old World Realm during the Siluro-Devonian, and finally formed an independent Sibumasu province in the Carboniferous and Permian. On the other hand, we (Shi and Archbold, 1995a) proposed that the STB blocks belonged to the Indoralian province during the Asselian to Early Sakmarian, which then also included peninsular India, the Himalayan terrane, the Lhasa terrane and Australia. The STB blocks were then occupied by an independent transitional province from Late Sakmarian to Midian, and finally joined the Cathaysian province by the Late Permian (Wujiapingian and Changxingian). This view is elaborated further below with additional new data.

Asselian to Early Sakmarian

The oldest Permian marine fossil assemblage is known from the upper Phuket Group in southern Thailand (Waterhouse, 1982) and contains a small, brachiopod-dominated fauna. Some 50% of its total genera are endemic to the Gondwanan realm, 37% are anti-tropically distributed (genera that are found in both Gondwanan and Boreal realms but completely absent from the intervening Palaeo-Equatorial realm, see discussion in Shi, Archbold and Zhan, 1995), and 13% (one genus, *Rhynchopora*) appear to be a wide-ranging element (Shi and Archbold, 1995a). The age of the brachiopod fauna is most likely to be Late Asselian to Early Sakmarian in view of its close comparisons with faunas from Australia and peninsular India. Genera that indicate a Gondwanan relationship include *Bandoproductus, Sulciplica, Lamniplica,* and *Elasmata*.

Late Sakmarian

Stratigraphically higher, above the Phuket 'cool-water' brachiopod fauna, is a very richly fossiliferous horizon (or horizons) widely found in the STB blocks. Brachiopods are particularly common and have been documented from the Dingjiazhai Formation of the Baoshan block (Fang, 1994; Shi, Fang and Archbold, 1996), the Ko Yao Noi Formation of southern Thailand (Waterhouse *et al.*, 1981), the upper Singa Formation and Kubang Pasu Formation of northwest Malaysia (Shi, personal observations), and the Nam Loong No.1 Mine beds of western peninsular Malaysia (Shi and Waterhouse, 1991). In terms of species composition, the Dingjiazhai brachiopod assemblage of the Baoshan block is significantly different from the coeval assemblages of the Shan-Thai terrane, in that the former is dominated by typical Gondwanan, especially Westralian, genera including *Arctitreta, Callytharrella, Globiella, Bandoproductus, Punctocyrtella, Trigonotreta,* and *Elivina* (named '*Stepanoviella*' (=*Globiella*) cool-water fauna by Fang, 1994). *Spinomartinia* appears to be absent from this assemblage, as do typical coeval

Cathaysian elements. In contrast, the brachiopods from the Shan-Thai terrane are dominated by *Spinomartinia prolifica* Waterhouse and *Spirelytha petaliformis* (Pavlova). Although Gondwanan elements are strongly represented in the Shan-Thai fauna, such as *Bandoproductus, Sulciplica,* and *Spirelytha,* the presence of warm-water palaeo-equatorial and endemic taxa is also significant. Among the palaeo-equatorial forms, *Urushtenia, Kutorginella, Karavankina, Brachythyrina, Spirigerella,* and *Purdonella* are of note. Another distinctive feature of the Shan-Thai assemblage is the presence of an endemic genus and species, *Spinomartinia prolifica,* which dominates the assemblage in terms of abundance of specimens.

In the Baoshan block, a distinctive cool-water *Lytvolasma* coral fauna has also been found in association with the Dingjiazhai *Globiella* brachiopod fauna; it is dominated by species of *Lytvolasma, Plerophyllum* and *Waanerophyllum* (Fang and Fan, 1994). The presence of *Lytvolasma* and allied genera is significant. These are thick-walled solitary corals adapted to cool- to cold-water environments and distributed in an essentially anti-tropical pattern (Wu, 1975).

A few marine fossils of possible Late Sakmarian age have also been reported from the upper Menghong Group or equivalents in the Tengchong block (Fang and Fan, 1994). They include *Phestia, Nuculopsis, Schizodus, Chonetinella,* '*Martinia*' (perhaps a *Spinomartinia*), and *Ramipora,* mostly suggestive of Gondwanan affinities. Of these genera, if '*Martinia*' is proved to be a true *Spinomartinia,* it will provide a strong link to the *Spinomartinia prolifica* assemblage found in the Shan-Thai terrane. The strong Gondwanan aspect of the Tengchong fauna is also demonstrated by the palynological material recently extracted from the upper Menghong Group, strongly allied to the *Pseudoreticulatispora pseudoreticulata* zone of Western Australia (Yang and Liu, 1996).

Bryozoans of Sakmarian-Artinskian age are also widely present in the STB blocks, locally in association with brachiopods and the *Lytvolasma* fauna mentioned above. Geological Survey of Yunnan (1990) has designated this bryozoan horizon the *Stenopora meekana-Fenestella* assemblage. This assemblage has been described from the Dingjiazhai Formation of the Baoshan block, the Menghong Group of the Tengchong block, and the upper Phuket Group and Singa Formation of the Shan-Thai terrane (Basir Jasin *et al.*, 1992; Fan, 1993; Lu, 1993, Sakagami, 1976). According to these authors, the bryozoan fauna is most closely related to coeval bryozoans from Australia, the Urals and, to a less extent, northeast Asia (northeast China, Siberia and the Russian Far East).

Artinskian-Kungurian

This time interval in SE Asia is characterized by a moderately diverse fauna commonly found in western and peninsular Thailand, from isolated limestone outcrops collectively known as the Rat Buri Limestone (or Group). Lateral equivalents in peninsular Malaysia are called the Chuping Limestone but its faunas remain little known. Northwards, the 'Rat Buri horizon' appears to be missing in western Yunnan (but see Fang, Z. J. (1994) for a different view).

The full age range of the Rat Buri Group remains poorly constrained but is likely to span from Sakmarian to Early Triassic in view of Fontaine *et al.* (1993). Several groups of fossils have been documented from various isolated limestone outcrops, but few can be used confidently for dating due either to their long stratigraphical range or poor taxonomic control. Shi and Archbold (1995a) preferred a Late Artinskian to Early Kungurian age for the Rat Buri brachiopod fauna (Waterhouse and Piyasin, 1970; Yanagida, 1970; Grant, 1976) in view of its close correlation with those of the Bitauni beds of Timor, the upper Amb Formation of Salt Range, and the Cundlego and Wandagee Formations of the Carnarvon basin of Western Australia. Recently, in a preliminary report Angiolini *et al.* (1996) reported Upper Permian brachiopods (Lower Murgabian) from south Oman believed to be closely related to those of the Rat Buri limestones. Archbold (1981) also demonstrated close links of the Rat Buri brachiopods with faunas of western Irian Jaya. At generic level, 35% of the Rat Buri brachiopod fauna suggests palaeotropical Cathaysian affinities and 11% indicates Gondwanan links. The observed Gondwanan links are of great interest and appear to represent an influx of the Shan-Thai taxa into the Westralian province (Archbold and Shi, 1995). Of particular note among the Gondwanan links is the presence of *Trigonotreta* discovered from an outcrop assigned to the Rat Buri Group in northern peninsular Thailand (Archbold in Baird *et al.*, 1993). *Trigonotreta* is a typical Gondwanan genus, restricted to the Austrazean and Westralian provinces and parts of the Cimmerian Region.

Probably corresponding to the Artinskian-

Kungurian time interval, a distinct fusulinid fauna also occurred in the Shan-Thai terrane and also possibly in western Yunnan. This fusulinid assemblage, typified by species of *Monodiexodina*, has been described from a number of localities in the Shan-Thai terrane (Basir Jasin, 1991) and also recently reported from western Yunnan, west of the Lancang River suture (Han *et al.*, 1991; Fang and Fan, 1994). As discussed by Shi, Archbold and Zhan (1995), the distribution of the *Monodiexodina* fauna is of considerable interest, being restricted to the northern and southern margins of the Tethys; therefore the fauna characterizes the Sino-Mongolian province and the Cimmerian Region, respectively (Shi, Archbold, and Zhan, 1995). Much like the *Lytvolasma* coral fauna, *Monodiexodina* and allies have been generally regarded to indicate cool-temperate climatic conditions (Han, 1980).

Kubergandian

The Kubergandian of the STB blocks is represented by a distinct brachiopod assemblage so far only known from the Guanyinshan Formation of the Tengchong block and the Yongde Formation of the Baoshan block in western Yunnan. Fang (1983, 1995) has described and named these brachiopods the *Waagenites-Costiferina* fauna. Personal inspection of this fauna by Shi in July 1996 at the Yunnan Institute of Geology and Mineral Resources, Kunming, China, revealed a mixed composition, consisting of wide-ranging, palaeo-tropical Cathaysian and characteristic Gondwanan genera. The latter group is represented by such taxa as *Chonetinella*, *Costiferina* (note that dictyoclostid species described as *Costiferina* by Fang (1983) probably belong to a new dictyoclostid genus allied to *Stereochia*, see discussion by Shi and Archbold, 1995a), and thick-shelled, strongly plicate *Neospirifer* with truncated cardinal extremities (close to the Western Australian *N. postplicatus* lineage). As noted by Fang (1983) and discussed by Shi and Archbold (1995a), the *Waagenites-Costiferina* fauna is very close to those from the Kalabagh Member of the Wargal Formation of Salt Range and, to a less extent, to the brachiopods from the Basleo beds of Timor, the Selong Formation of central Tibet, and to the *Liveringa magnifica* zone of the Canning basin of Western Australia. In this latter correlation, *Waagenites* plays a significant role because *W. stani* Archbold from the *Liveringa magnifica* zone is closely similar to *W. yunnanensis* Fang from western Yunnan (Archbold, 1988).

Shi and Archbold (1995a) assigned a Kazanian-Midian age to the *Waagenites-Costiferina* fauna based on brachiopod correlations. However, according to fusulinids, the age could be slightly older, being Upper Qixian (or Kubergandian/?Ufimian). Fang and Fan (1994) listed typical fusulinids of the Upper Qixian *Nankinella orbicularia* zone from the Guanyinshan Formation in the Tengchong block, which in turn may be correlated with the Kubergandian (?Ufimian).

The *Waagenites-Costiferina* fauna is associated with bryozoans ascribed to the *Hexagonella* assemblage (Fang and Fan, 1994). This bryozoan fauna is dominated by wide-ranging genera but also contains a significant proportion of diagnostic Gondwanan or peri-Gondwanan elements such as *Hexagonella, Coscinotrypa, Acanthotrypa, Ascopora, Streblascopora* and *Ogbinopora*. Also found with the brachiopods and bryozoans are solitary rugose corals including species of *Waanerophyllum* and *Lophophyllidium* along with characteristic Cathaysian genera *Wentzelella* and *Iranophyllum* (GSY, 1990; Fang and Fan, 1994).

Murgabian-Midian

This time interval is characterized by a predominantly carbonate unit of argillaceous limestone grading upwards into dolomitic and oolitic limestones. This unit is well represented in both the Tengchong and Baoshan blocks of western Yunnan (the Dadongchang and Shazipo Formations, respectively). At least part of the Rat Buri Group may be correlated to these formations (Fig.3). In western Yunnan, the lower part of the carbonate unit has yielded a characteristic Murgabian-Midian foraminiferal assemblage known as the *Shanita-Hemigordius (Hemigordiopsis)* fauna (Shen and He, 1983). The same fauna has also been recorded from some Rat Buri limestone localities in peninsular Thailand (Dawson *et al.*, 1993). This foraminiferal fauna has restricted geographical and stratigraphical distributions, being confined to the Murgabian-Midian along the Tethyan margin of northern Gondwanan (from Tunisia through Middle East, southwestern China eastwards to peninsular Thailand). This geographical distributional pattern broadly mirrors the southern belt (south Tethys) of *Monodiexodina* discussed previously.

Fusulinids typical of the Lower Maokouan Stage of South China have also been reported from the Shazipo Formation, consisting mainly of *Polydiexodina, Neoschwagerina, Chusenella, Nankinella,* and *Yanchienia* species. Shen and Jin (1994) correlated these fusulinids with the *Neoschwagerina craticulifera* zone of Lower Maokouan (Murgabian to Midian).

A small brachiopod fauna has been collected by one of the authors (GRS) from the limestone beds equivalent to the Shazipo Formation in northern Baoshan block. This well preserved brachiopod fauna, currently under study by GRS, contains a very interesting mix of Cathaysian and peri-Gondwanan elements. The Cathaysian constituents are represented by such typical Lower Maokouan South Chinese genera as *Spinomarginifera, Squamularia* and *Cryptospirifer*, while the Gondwanan aspect is identified by *Pseudoantiquatonia*, a genus so far only known from the Xiala Formation of the Lhasa terrane of central Tibet (Zhan and Wu, 1982). In addition to its occurrences in South China, *Cryptospirifer* is known only from north Iran and Turkey (Nakamura and Golshani, 1981).

Wujiapingian-Changxingian

Wujiapingian to Chanxingian marine sequences appear to be missing from western Yunnan (GSY, 1990) but are represented at a few localities in the Shan-Thai terrane. Carey *et al.* (1995) have recently recorded the *Neogondolella bitteri* conodont assemblage of Wujiapingian age from the limestone beds of the Pha Huat Formation in northwest Thailand. Overlying this unit is a siltstone-shale unit from which Waterhouse (1983) described a rich lyttoniid brachiopod fauna. Shi and Archbold (1995a) noted that up to 77% of the recorded brachiopod genera were characteristic Cathaysian taxa and the remaining constituents wide-ranging, with no diagnostic Gondwanan or peri-Gondwanan representatives. *Oldhamina squamosa* Huang is particularly abundant in the fauna. This species is a characteristic form of the Changxingian in South China (He and Shi, 1996) and is also present in great abundance in the lyttoniid fauna of the Leptodus Shale in East Malaya (Mohd Shafeea Leman, 1994). The close South Chinese correlation of the Thai lyttoniid fauna is reinforced by the occurrences of Changxingian *Palaeofusulina-Colaniella* foraminiferid fauna (Ingavat, 1984; Ueno and Sakagami, 1991), a *Paratirolites nakornsurii* ammonoid assemblage (Ishibashi and Chonglakamni, 1990), and a rich 'sphinctozoan' sponge fauna (Senowbari-Daryan and Ingavat-Helmcke, 1993) from the same formation in the area.

Change of provinciality of the Shan-Thai, Tengchong and Baoshan blocks: caused by tectonic vicariance or shifting of climatic zones?

It is clear from the preceding description of faunal successions that the STB blocks experienced significant changes in marine provinciality during the Permian. These changes may be summarised in three stages as follows:

Asselian-Early Sakmarian Gondwanan stage. The manifestation of this stage is the cool-water brachiopod fauna from the upper Phuket Group in southern peninsular Thailand of the Shan-Thai terrane. The fauna is associated with pebbly mudstones, which have been interpreted by many workers as of glaciomarine origin (Stauffer and Lee, 1986; Hutchison, 1993; Jin, 1994).

Late Sakmarian-Midian transitional stage (Sibumasu province stage). This stage corresponds to the entire duration of the Sibumasu province as previously defined. In spite of variations demonstrated by the various faunas included within this interval, they all exhibit a mixed or transitional nature in that taxa suggestive of Gondwanan and Cathaysian affinities coexist, in addition to wide-ranging and endemic taxa. Furthermore, as observed by Shi and Archbold (1995a), the ratio of Gondwanan to Cathaysian brachiopod genera in these mixed assemblages decreased with time as the reciprocal increased (Fig.4). Among these mixed faunas, there exist some notable differences between the Baoshan and Shan-Thai blocks, particularly with respect to their Late Sakmarian brachiopod assemblages. The *Globiella* fauna of the Baoshan block has striking similarities with the coeval faunas of the Callytharra Formation of Western Australia and the Bisnain assemblage of Timor, implying a strong Gondwanan affinity. This is in contrast to the *Spinomartinia prolifica* assemblage of the Shan-Thai block, which shows only a moderate similarity to coeval Gondwanan faunas as compared with a significant percentage of genera in common with the Cathaysian and the broader palaeo-Equatorial faunas. This relatively low Gondwanan affinity is surprising, given the modern geographical greater proximity of the Shan-Thai terrane to Australia and Timor. Of course, the present ori-

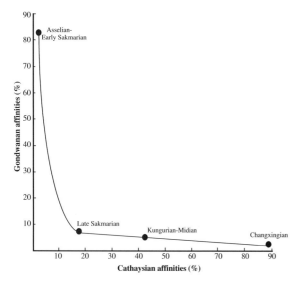

Fig.4. Marked change of marine provinciality of the Shan-Thai, Tengchong and Baoshan blocks through the Permian, as demonstrated by the variation of the ratio of Gondwanan-type Permian brachiopod genera over the Cathaysian-type genera (from Shi and Archbold, 1995a).

entation of the Shan-Thai block and its spatial relationship with the Baoshan terrane may not be the same as it was during the Permian; hence, it could be possible that the Baoshan block was situated closer to northwest Australia, either as an independent block or a part of the greater Shan-Thai terrane. In the latter scenario, it is speculated that the combined Shan-Thai-Tengchong-Baoshan block must have rotated 180° since the Sakmarian.

Wujiapingian-Changxingian Cathaysian stage. The Sibumasu province as an independent biogeographical unit probably had diminished by the end of Midian. In Wujiapingian to Changxingian, the faunal aspect of the Shan-Thai terrane (and presumably the Tengchong and Baoshan blocks as well) is identical to that of the Cathaysian province of South China and Indo-China, as evidenced by the presence of such typical warm-water lyttoniid brachiopods, palaeofusulinids, and sponges.

The question therefore arises as to why the marine provinciality of the STB blocks changed during the Permian? This change appears to be quite abrupt from Early Sakmarian (Gondwanan affinity) to Late Sakmarian (Sibumasu transitional province), although the transition from the Sibumasu province to the Cathaysian province seems more gradational (Fig.4). Shi, Archbold and Zhan (1995) discussed two alternative models in accounting for this marked change of provinciality. These models are elaborated further below.

The tectonic vicariance model (Fig.5). This interpretation was first suggested by Shi and Waterhouse (1990). They speculated that the Shan-Thai and other SE Asian Cimmerian terranes were probably located proximal to northern Gondwanan during the Asselian-Tastubian (Early Sakmarian), therefore sharing an uniform Indoralian fauna. Then, a rifting event sweeping away these terranes from northern Gondwana may have started at the Tastubian-Sterlitamakian (Late Sakmarian) boundary. It was thought that the initial rifting was followed by rapid northward drifting of the STB blocks from southern high latitudinal zones to lower latitudinal settings, resulting in a progressive decrease of Gondwanan faunal elements and concurrent increase of Cathaysian taxa, hence forming a characteristic transitional province (Sibumasu province) from the Late Sakmarian to Midian. By the Late Permian (Wujiapingian-Changxingian), the STB blocks may have drifted close enough to the Cathaysian massifs (South China and Indo-China) to form a single province (the Cathaysian province). Clearly, in this model the change of provinciality was treated as a consequence of tectonic vicariance, and the variation of the Gondwanan/Cathaysian faunal ratio as a function of the palaeo-distance between the STB blocks and Australia.

This vicariance model can only be tested by palaeomagnetic data. However, these data are still not sufficient to provide a palaeolatitudinal signature through the entire Permian for the blocks in question (see discussions by Shi, Archbold and Zhan, 1995), although an inferred palaeolatitudinal curve based on both observations and predictions (Metcalfe, 1996) does indicate a rapid, some 40 degrees in latitude, northward drift of these blocks through the Permian. Nevertheless, an Early Permian rifting event inferred by the biogeographical data seems possible at least for the Baoshan block, as evidenced by the eruption of Late Sakmarian-?Artinskian basalts (the Woniusi Basalt). This eruption may have signalled a wide-spread, more or less synchronous, rifting event along the northern margin of Gondwana, sweeping away a large strip of northern Gondwana. To date, this event has been documented from the Zanskar area, northwest India (Gaetani and Garzanti, 1991), northern Karakorum (Gaetani *et al.*, 1990), and northern Pakistan (Pogue *et al.*, 1992), and has also

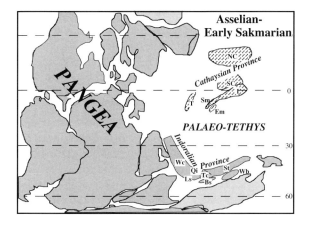

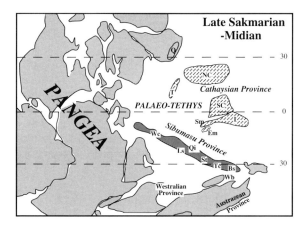

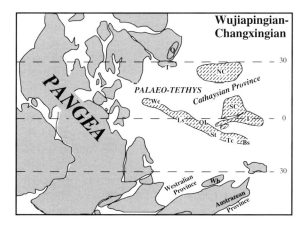

Fig.5. A tectonic vicariant model to interpret how the marked change of marine provinciality of the Shan-Thai terrane, Tengchong and Baoshan blocks was achieved during the Permian (base map from Metcalfe, 1996). Explanation of tectonic blocks as follows: Bs: Baoshan; Em: East Malaya; I: Indo-China; NC: North China; Q: Qaidam; Qi: South Qiangtang; SC: South China; Sm: Simao; St: Shan-Thai (s.s.); T: Tarim; Tc: Tengchong; Wb: Western Burma; Wc: western Cimmerian continent.

been linked to the inception of the Westralian superbasin in Western Australia (Veevers, 1988; Görür and Sengör, 1992).

However, unless the rifted STB blocks drifted at an 'enormous' speed at the time of the Tastubian-Sterlitamakian boundary (lasting about 3 Ma, estimated from Jones, 1996), the seaway (or Neo-Tethys) created by the drifting away of the STB blocks would not be wide enough to form an independent faunal province. Using a modern sea-floor spreading rate of 10 cm/yr, the width of the initial Neo-Tethys south of the STB blocks, from Tastubian to Sterlitamakian, would be no more than 300 km. Shi (1996) has recently argued that for a biotic province of a small island (like the STB blocks drifting away) to develop adjacent to a mainland (in this case, Gondwana), the minimum distance between the island and the mainland would have been 1500 km. Therefore, even although there is good evidence to suggest early Permian rifting along northern Gondwana creating a shallow seaway between the STB blocks and Gondwana, this factor alone is unlikely to have caused the marked change of provinciality of the STB blocks.

Change of provinciality induced by lateral shifting of climatic zones and global warming. Alternatively, as proposed by Shi and Archbold (1995a), the change of marine provinciality of the STB blocks may be explained by lateral shifting of climatic zones without necessarily shifting continental masses. The tectonic vicariance model discussed above may explain well the marked change of marine provinciality experienced by the STB blocks, but it cannot be invoked to account for a broadly similar biogeographical signature of the Westralian province in Western Australia. As noted by Archbold and Shi (1995) and Archbold *et al.* (1996), there is also a sharp boundary between Asselian-Tastubian and Sterlitamakian faunas in Western Australia in terms of diversity and biogeographical composition. Here, the Asselian and Tastubian faunas are closely related to contemporary faunas of eastern Australia, forming the single Indoralian province which also embraces the STB blocks (Shi and Archbold, 1993a). The Sterlitamakian faunas of Western Australian are diverse and contain many Sibumasu and some Cathaysian elements. Younger Permian Westralian faunas demonstrate a variable degree of endemism intercalated with several influxes of Sibumasu faunal elements (Fig.2; Archbold and Shi, 1995). By the Wujiapingian, typical warm-water Cathaysian brachiopod genus *Leptodus* had appeared in the Bonaparte Gulf basin of northwestern Australia (Thomas, 1957). These intermittent invasions of Sibumasu and Cathaysian taxa into the Westralian province indicate that only narrow seaways existed between Australian and the Sibumasu province and that there must have been episodes of climatic amelioration accompanied by southward expansion of the palaeotropical belt, facilitating the migration of Sibumasu and Cathaysian faunal elements into Western Australia.

Lateral shifts of climatic zones can be also invoked to explain why the STB blocks changed from a predominantly Gondwanan-type fauna through a transitional fauna (Sibumasu province) to a Cathaysian-type fauna through the Permian. In this interpretation, we assume that the STB blocks were located somewhere between Gondwana in the south and Cathaysia in the north during the Permian (in this sense, this model is comparable to the interpretation of Fang, 1991, 1994). During the Asselian-Tastubian, extensive glaciation prevailed on Gondwana, as evidenced by glacial-marine tilloids, diamictites and associated pebbly mudstones extensively found across the STB blocks and other Cimmerian continents. The cold Gondwanan climatic conditions may have extended towards the palaeo-equator to, perhaps, as low as 30°S, covering the STB blocks. This glaciation event was probably accompanied by a global cooling regime, under which boundaries between latitude-parallel climatic zones were sharp with no or only very weakly-developed transitional (mesothermal) climatic zones. As a consequence, the palaeotropical belt would have been restricted (narrow) and the STB blocks accordingly developed a Gondwanan-type fauna.

The Gondwanan glaciation probably had ended by the end of Tastubian on most of Gondwanan continents (including the STB blocks) and was followed by a rapid climatic amelioration during the Sterlitamakian (Dickins, 1985). The deglaciation, probably accompanied by global warming, would have resulted in the withdrawal of the palaeo-polar conditions from the STB blocks and the simultaneous expansion of and gradual replacement by the palaeo-tropical climatic belt. Probably also during this period of climatic amelioration, the boundaries between large-scale climatic zones were becoming less sharply defined and distinct mesothermal belts (temperate zones) emerged. It was probably due to this unique transitional climatic regime and the assumed intermediate palaeogeographical position of the STB blocks between Cathaysia and Gondwana that some eurothermal marine invertebrate taxa migrated from the palaeo-Tethys to the STB blocks and some of them, apparently accompanied by some endemic Sibumasu forms, may have invaded even further south into Western Australia. As a consequence, the mixed Gondwanan/Cathaysian fauna of the Sibumasu province developed on the STB blocks. As the climates ameliorated further and the STB blocks became more warm-temperate or subtropical, increasingly more Gondwanan taxa were extinguished and replaced by increasingly more Cathaysian taxa, resulting in a progressively 'Cathaysianized' fauna to emerge on the STB blocks. By the Wujiapingian, the warm tropical to subtropical conditions may have been firmly established in the STB blocks, eliminating all the remaining Gondwanan forms.

The climatic model just outlined does not envisage any significant lateral movements of the STB blocks but assumes that climatic zonation and its meridional migration plays a primary role in controlling the distribution and dispersal of marine invertebrates. This model seems to explain well the crudely parallel development of provinciality in both the STB blocks and Western Australia. A comparison of the lithologies and faunal sequences of the STB blocks and Western Australia (Fig.2) indicates a general trend of increase of palaeotemperature through the Permian although the general trend is complicated by second-order temperature drops in Western Australia as inferred from brachiopod diversity and composition (Archbold and Shi, 1995). The progressive warming is most conspicuous in the STB blocks where glaciomarine diamictites and pebbly mudstones grade upwards to mudstones, shale with interbdedded bioclastic limestones and finally to reddish/purplish beds (the Yongde Formation or equivalents in Baoshan block) and massive limestone (Shazipo Formation). A similar, although more varied, lithological sequence is also present in Western Australia. This correlation of both lithological and biogeographical sequences between Western Australia and the STB blocks would imply that a similar climatic or tectonic process was responsible for these sequential changes. A tectonic mechanism is plausible if we assume a continued northward drift (perhaps, a rotational drift) of the entire Gondwana including the STB blocks since Tastubian. This process would produce an increasingly warming temperature curve the same as that observed for the STB blocks and Western Australia just discussed. This scenario is supported by a sequence of four Permian reconstructions (Sakmarian, Artinskian, Kazanian, and Tatarian) recently produced by Ziegler et al. (1997), which showed that Gondwana moved northwards some 15 degrees in latitude since the Sakmarian.

Summarising the two models discussed above, there are arguments in favour of both scenarios. On the one hand, there exists good stratigraphical and structural evidence to suggest an early Permian rifting event between the STB blocks and Gondwana, but given the short time span (from Tastubian to Sterlitamakian), it is unlikely that this rifting event, unless superimposed and enhanced by climatic forcing, had created an effective, broad-enough biogeographical barrier between Gondwana and the STB blocks. The second, climatically driven, model assumes a relatively fixed position for the STB blocks but this is contradicted by the geological data. In view of the strength and weakness of each of these scenarios, we therefore prefer an integrated model combining both scenarios. This integrated interpretation has been briefly discussed by Shi, Archbold and Zhan (1995). In this integrated solution, the warming effect produced by the northward drift of the STB blocks would have been superimposed and hence intensified by the contemporary post-Tastubian global warming. Abrupt climatic amelioration accompanied by rapid expansion of the palaeo-tropical zone at the Tastubian-Sterlitamakian boundary may have been primarily responsible for the marked change of provinciality in both eastern Gondwana and the STB blocks. Subsequent, more gradational, changes in the marine provinciality of the STB blocks may be explained by the continuing northward drift of the STB blocks (possibly superimposed by a comparatively slower northward drifting of Gondwana), accompanied by more gradual warming, shrinking of the palaeo-polar to palaeo-temperate climatic zones and southwards expansion of the palaeotropical belt. As the STB blocks were approaching Cathaysia and their climatic conditions ameliorated further, their cool-water-adapted Gondwanan taxa were eliminated and replaced progressively by more Cathaysian taxa. By the Late Permian, the STB blocks may have drifted to the vicinity of the Cathaysian massifs and, as a result, warm palaeotropical conditions may have been firmly established, resulting in the final demise of the Sibumasu province and its incorporation into the Cathaysian province.

Conclusions

Three Permian marine biotic provinces may be recognised in SE Asia: Cathaysian, Indoralian and Sibumasu. The Cathaysian province includes the Indo-China, Simao and East Malaya blocks throughout the Permian and the Shan-Thai, Tengchong and Baoshan blocks (STB blocks) in the Late Permian (Wujiapingian and Changxingian). The Indoralian province occupied the STB blocks only in the earliest Permian (Asselian-Early Sakmarian). The Sibumasu province is restricted to the STB blocks and ranges from the Late Sakmarian to Midian. This province is transitional (mixed) in character, with its Gondwanan aspect fading progressively from Late Sakmarian to Midian as its Cathaysian affinity strengthened. *Monodiexodina, Shanita* and *Hemigordius (Hemigordiopsis)* are among the most characteristic taxa of this middle Permian transitional province.

The change of marine provinciality of the STB blocks from Early Sakmarian (Tastubian) to Late Sakmarian is notably abrupt. This may be explained by a rapid post-Tastubian climatic amelioration and southward expansion of the warm palaeotropical zone, superimposed by a rifting event that separated the STB blocks from northern Gondwana. In view of the stronger Gondwanan links of its Upper Sakmarian faunas, the Baoshan block (and, presumably, the Tengchong block as well) is speculated to have been located closer to northwestern Australia than the Shan-Thai terrane for at least the Late Sakmarian. After the Sakmarian, the Sibumasu province, incorporating the STB blocks, became strengthened as its mixed nature of both Gondwanan and Cathaysian faunal elements intensified. This process may be explained by a possible interplay of a continued northward drift of the STB blocks, and gradual global warming. This interplay probably initiated and intensified a continued climatic amelioration on the STB blocks after the Sakmarian as the STB blocks moved from a high southern latitudinal zone to a low southern latitudinal zone, climatically enhanced by the superimposed global warming. This process in turn is presumed to have facilitated the migration of Cathaysian faunas into the STB blocks and, at the same time, eliminated the existing Gondwanan elements.

Acknowledgements

Our research on Late Palaeozoic marine faunas

and biogeography is supported by the Australian Research Council (Grant A39601699). Miss Elizabeth Stagg assisted in drafting.

References

Angiolini, L. and 10 others. 1996. Late Permian fauna from the Khuff Formation, South Oman. Permophiles 29: 62-64.

Archbold, N. W. 1981. Permian brachiopods from western Irian Jaya. Geological Research and Development Centre, Paleontology Series 2: 1-25.

Archbold, N. W. 1983. Permian marine invertebrate provinces of the Gondwanan Realm. Alcheringa 7: 59-73.

Archbold, N. W. 1987. South western Pacific Permian and Triassic marine faunas: their distribution and implications for terrane identification. In: Terrane Accretion and Orogenic Belts. Edited by E. C. Leitch. Geodynamics Series 19: 119-127.

Archbold, N. W. 1988. Studies on Western Australian Permian brachiopods 8. The Late Permian brachiopod fauna of the Kirkby Range Member, Canning Basin. Proceedings of the Royal Society of Victoria 100: 21-32.

Archbold, N. W. and Shi, G. R. 1995. Permian brachiopod faunas of Western Australia: Gondwanan-Asian relationships and Permian climate. Journal of SE Asian Earth Sciences 11: 207-215.

Archbold, N. W. and Shi, G. R. 1996. Western Pacific Permian marine invertebrate palaeobiogeography. Australian Journal of Earth Sciences 43: 635-641.

Archbold, N. W., Pigram, C. J., Ratman, N. and Hakim, S. 1982. Indonesian Permian brachiopod fauna and Gondwana-Southeast Asia relationships. Nature 296: 556-558.

Archbold, N. W., Shah, S. C. and Dickins, J. M. 1996. Early Permian brachiopod faunas from peninsular India: their Gondwanan relationships. Historical Biology 11: 125- 135.

Baird, A., Dawson, O. and Vachard, D. 1993. New data on biostratigraphy of the Permian Ratburi Limestone from north peninsular Thailand. In Proceedings of the International Symposium on Biostratigraphy of Mainland SE Asia: Facies and Paleontology, pp. 243-258. Chiang Mai, Thailand.

Basir Jasin. 1991. Significance of Monodiexodina (Fusulinacea) in geology of peninsular Malaysia. Geological Society of Malaysia Bulletin 29: 171-181.

Basir Jasin, Wan Fuad Wan Hassan and Mohd Shafeea Leman. 1992. The occurrence of bryozoan bed in the Singa Formation, Bukit Durian Perangin, Langkawi. Warta Geologi 18 (2), 29-35.

Brookfield, M. E. 1996. Paleozoic and Triassic Geology of Sundaland. In The Phanerozoic Geology of the World 1. The Palaeozoic, B. pp. 181-264. Edited by M. Moullade and A. E. M. Nairn. Elsevier, Amsterdam.

Carey, S. P., Burrett, C. F., Chaodumrong, P., Wongwanich, T. and Chonglankmani, C. 1995. Triassic and Permian conodonts from the Lampang and Ngao Groups, northern Thailand. Courier Forschungsinstitut Senckenberg 182: 497-513.

Dawson, O., Racey, A. and Whittaker, J. E. 1993. The palaeoecological and palaeobiogeographical significance of Shanita (foraminifera) and associated foraminifera/algae from the Permian of peninsular Thailand. In International Symposium on Biostratigraphy of Mainland SE Asia: Facies and Paleontology. pp. 253-298. Chiang Mai, Thailand.

Dickins, J. M. 1985. Late Palaeozoic glaciation. BMR Journal of Geology and Geophysics 9: 163-169.

Fan, J. C. 1993. Bryozoans of Late Carboniferous-early Permian in Tengchong area of western Yunnan. Yunnan Geology 12 (4), 383-406.

Fang, R. S. 1983. Early Permian Brachiopoda from Xiaoxinzhai of Gengma, Yunnan and its geological significance. Contributions to Geology of Qinghai-Xizang Plateau 11: 93-120.

Fang, R. S. 1994. The discovery of cold-water brachiopod Stepanoviella fauna in Baoshan region and its geological significance. Yunnan Geology 13 (3): 264-277.

Fang, R. S. 1995. New study results of Brachiopoda of Early stage of Early Permian at Dadongchang, Tengchong. Yunnan Geology 14 (2): 134-152.

Fang, R. S. and Fan, J. C. 1994. Middle to Upper Carboniferous-Early Permian Gondwana Facies and Palaeontology in Western Yunnan. Yunnan Science and Technology Press, Kunming, 120 pp.

Fang, Z. J. 1985. Preliminary investigation into the Cathaysian faunal province. Acta Palaeontologica Sinica 24 (3): 344-348.

Fang, Z. J. 1991. Sibumasu biotic province and its position in Paleotethys. Acta Palaeontologica Sinica 30 (4): 511-532.

Fang, Z. J. 1994. Biogeographic constraints on the rift-drift-accretion history of the Sibumasu block. Journal of SE Asian Earth Sciences 9 (4): 375-385.

Fang, Z. J., Zhou, Z. C. and Lin, J. M. 1992. Some comments on the Changning-Menglian suture from a stratigraphical viewpoint. Journal of Stratigraphy 16 (4): 292-303.

Fontaine, H. 1986. The Permian of SE Asia. CCOP Technical Bulletin 18: 1-170.

Fontaine, H. 1994. Permian corals of Thailand. CCOP Technical Bulletin 24: 1-108.

Fontaine, H. and Suteethorn, V. 1988. Late Palaeozoic and Mesozoic fossils of west Thailand and their environments. CCOP Technical Bulletin 20: 1-212.

Fontaine, H., Chonglakmani, C., Ibrahim Bin Amanan and Piyasin, S. 1993. A well-defined Permian biogeographical unit: peninsular Thailand and northwest peninsular Malaysia. Journal of SE Asian Earth Sciences 9 (1/2): 129-151.

Gaetani, M. and Garzanti, E. 1991. Multicyclic history of the Northern India continental margin (northwest Himalaya). The American Association of Petroleum Geologists Bulletin 75 (9): 1427-1446.

Gaetani, M., Garzanti, E., Jadoul, F., Nicora, A., Tintori, A., Pasini, M. and Kanwar Sabir Ali Khan. 1990. The north Karakorum side of the central Asia geopuzzle. Geological Society of America Bulletin 102: 54-62.

Geological Survey of Yunnan (GSY). 1990. Regional Geology of Yunnan Province. P. R. China, Ministry of Geology and Mineral Resources Geological Memoirs Series 1 21: 728 pp.

Görür, N. and Sengör, A. M. C. 1992. Paleogeography and tectonic evolution of the eastern Tethysides: implications for the northwest Australian margin breakup history. Proceedings of the Ocean Drilling Program, Scientific Results 122: 83-106.

Grant, R. A. 1976. Permian brachiopods from northern Thailand. Paleontological Society Memoir 9: 1-269.

Grunt, T. A. and Shi, G. R. 1997. A hierarchical framework of Permian global marine biogeography. Proceedings of the 30th International Geological Congress 26 (in press).

Han, J. X. 1980. The morphology, evolution and distribution of the genus Monodiexodina and its allied genera. Bulletin of Shenyang Institute of Geology and Mineral Resources 1 (1): 90-112.

Han, N. R., Oyang, C. P., Li, W. H. and Li, Y. L. 1991. New

observations on the Permo-Carboniferous stratigraphy of Laochang in Lancang, Yunnan. Journal of Stratigraphy 15 (1): 56-58.

He, X. L. and Shi, G. R. 1996. The sequence of Permian brachiopod assemblages in South China. In Brachiopods. pp. 111-115. Edited by P. Copper and J. S. Jin. Balkelma, Rotterdam.

Hutchison, C. S. 1993. Gondwana and Cathaysian blocks, Palaeotethys sutures and Cenozoic tectonics in Southeast Asia. Geologische Rundschau 82: 388-405.

Ingavat, R. 1984. On the correlation of the Permian foraminiferal faunas of the western, central and eastern provinces of Thailand. Memoir Société Géologique de France, N.S. 147, 93-100.

Ingavat, R., Toriyama, R. and Pitakpaivan, K. 1980. Fusuline zonation and fauna characteristics of the Ratburi Limestone in Thailand and its equivalents in Malaysia. Geology and Palaeontology of SE Asia 21: 43-62.

Ishibashi, T., and Chonglakmani, C. 1990. Uppermost Permian ammonoids from northern Thailand. Journal of SE Asian Earth Sciences 4 (3): 163-170.

Ishii, K. I., Kato, M. and Nakamura, K. 1969. Permian limestones of west Cambodia. Lithofacies and biofacies. Special Paper of the Palaeontological Society of Japan 14: 41-55.

Jin, X. C. 1994. Sedimentary and paleogeographic significance of Permo-Carboniferous sequences in western Yunnan. Geologisches Institut der Universitaet zu Koeln Sonderveroeffentlichungen 99: 1-136.

Jin, Y. G. 1996. A global chronostratigraphic scheme for the Permian System. Two decades of the Permian Subcommission. Permophiles 28: 4-9.

Jones, P. J. 1996. AGSO Phanerozoic Timescale 1995. Explanatory Notes. Oxford University Press Melbourne 32 pp.

Li, C., Cheng, L. R., Hu, K., Yang, Z. R. and Hong, Y. R. 1995. Study of the Paleo-Tethys suture zone of the Lungmu Co-Shuanghu, Tibet. Geological Publishing House Beijing, 131 pp.

Liu, B. P., Feng, Q. L., Fang, N. Q., Jia, J. H. and He, F. X. 1993. Tectonic evolution of Palaeo-Tethys poly-island ocean in Changning-Menglian and Lancangjiang, southwestern Yunnan. Earth Science 18: 529-539.

Lu, L. H. 1993. Study of Upper Carboniferous-Lower Permian bryozoans from west Yunnan. Acta Palaeontologica Sinica 32 (1): 64-84.

Metcalfe, I. 1986. Late Palaeozoic palaeogeography of SE Asia: some stratigraphical, palaeontological and palaeomagnetic constraints. Geological Society of Malaysia Bulletin 19: 153-164.

Metcalfe, I. 1988. Origin and assembly of SE Asian continental terranes. In Gondwana and Tethys. Edited by M. G. Audley-Charles and A. Hallam. Geological Society Special Publication 37: 101-118.

Metcalfe, I. 1996. Gondwanaland dispersion, Asian accretion and evolution of eastern Tethys. Australian Journal of Earth Sciences 43: 605-623.

Metcalfe, I. 1998. Palaeozoic and Mesozoic geological evolution of the SE Asian region: multi-disciplinary constraints and implications for biogeography. In Biogeography and Geological Evolution of SE Asia. Edited by R. Hall and J. D. Holloway (this volume).

Mohd Shafeea Leman. 1993. Upper Permian brachiopods from northwest Pahang, Malaysia. In Proceedings of the International Symposium on Biostratigraphy of Mainland SE Asia: Facies and Paleontology (Bangkok), pp. 203-218.

Mohd Shafeea Leman. 1994. The significance of Upper Permian brachiopods from Merapoh area, northwest Pahang. Geological Society of Malaysia Bulletin 35: 113-121.

Nakamura, K. and Golshani, F. 1981. Notes on the Permian brachiopod genus Cryptospirifer. Journal of Faculty of Science, Hokkaido University, Series IV 20 (1): 67-77.

Nakamura, K., Shimizu, D, and Liao, Z. T. 1985. Permian palaeobiogeography based of brachiopods based on the faunal provinces. In The Tethys, Her Paleogeography and Paleobiogeography from the Paleozoic to Mesozoic. pp. 185-199. Edited by K. Nakazwa and J. M. Dickins. Tokai University, Tokyo.

Pogue, K. R., DiOietro, J. A., Khan, R. S., Hughes, S. S., Dilles, J. H. and Lawrence, R. D. 1992. Late Paleozoic rifting in northern Pakistan. Tectonics 11 (4): 871-883.

Sakagami, S. 1976. Paleobiogeography of the Permian bryozoa on the basis of Thai- Malaya district. Geology and Palaeontology of Southeast Asia 17: 155-172.

Sengör, A. M. C. 1979. Mid-Mesozoic closure of Permo-Triassic Tethys and its implications. Nature 279: 590-593.

Senowbari-Daryan, B. and Ingavat-Helmcke, R. 1993. Upper Permian sponges from Phrae province (northern Thailand). In International Symposium on Biostratigraphy of Mainland SE Asia: Facies and Paleontology (Bangkok). pp 439-451. Chiang Mai, Thailand.

Shen, J. Z. and He, Y. 1983. Permian Shanita-Hemigordius (Hemigordiopsis) (Foraminifera) Fauna in western Yunnan. Acta Palaeontologica Sinica 22: 55-59.

Shen, J. Z. and Jin, Y. G. 1994. Correlation of Permian deposits in China. Palaeoworld 4: 14-113.

Shi, G. R. 1993. Multivariate data analysis in palaeoecology and palaeobiogeography — a review. Palaeogeography, Palaeoclimatology, Palaeoecology 105: 199-234.

Shi, G. R. 1995. Spatial aspects of palaeobiogeographical data and multivariate analysis. Memoir of the Association of the Australasian Palaeontologists 18: 179-188.

Shi, G. R. 1996. A model of quantitatively estimating marine palaeobiogeographical provinciality. A case study of the Permian marine provincialism. Acta Geologica Sinica 70 (4): 351-360.

Shi, G. R. and Archbold, N. W. 1993a. Distribution of Asselian-Tastubian (Early Permian) circum-Pacific brachiopod faunas. Memoir of the Association of Australasian Palaeontologists 15: 343-351.

Shi, G. R. and Archbold, N. W. 1993b. A compendium of the Permian brachiopod faunas of the western Pacific region. 1. Asselian-Tastubian. Deakin University, School of Aquatic Science and Natural Resources Management Technical Paper, 1993/1.

Shi, G. R. and Archbold, N. W. 1993c. A compendium of the Permian brachiopod faunas of the western Pacific region. 2. Baigendzhinian-Early Kungurian. Deakin University, School of Aquatic Science and Natural Resources Management Technical Paper, 1993/2.

Shi, G. R. and Archbold, N. W. 1995a. Permian brachiopod faunal sequence of the Shan-Thai terrane: biostratigraphy, palaeobiogeographical affinities and plate tectonic/palaeoclimatic implications. Journal of SE Asian Earth Sciences 11 (3): 177-187.

Shi, G. R. and Archbold, N. W. 1995b. A quantitative analysis on the distribution of Baigendzhinian-Early Kungurian (Early Permian) brachiopod faunas in the western Pacific region. Journal of SE Asian Earth Sciences 11 (3): 189-205.

Shi, G. R. and Archbold, N. W. 1995c. Palaeobiogeography of Kazanian-Midian (Late Permian) Western Pacific brachiopod faunas. Journal of SE Asian Earth Sciences 12 (3): 129-141.

Shi, G. R. and Archbold, N. W. 1995d. A compendium of the Permian brachiopod faunas of the western Pacific region.

3. Kazanian-Midian (Late Permian). Deakin University, School of Aquatic Science and Natural Resources Management Technical Paper 1995/2.

Shi, G. R. and Archbold, N. W. 1995e. A compendium of the Permian brachiopod faunas of the western Pacific region. 4. Sterlitamakian-Aktastinian (Early Permian). Deakin University, School of Aquatic Science and Natural Resources Management Technical Paper 1995/3.

Shi, G. R. and Archbold, N. W. 1996. A quantitative palaeobiogeographical analysis on the distribution of Sterlitamakian-Aktastinian (Early Permian) Western Pacific brachiopod faunas. Historical Biology 11: 101-123.

Shi, G. R., Archbold, N. W. and Zhan, L. P. 1995. Distribution and characteristics of mixed (transitional) mid-Permian (Late Artinskian-Ufimian) marine fauna in Asia and their palaeogeographical implications. Palaeogeography, Palaeoclimatology, Palaeoecology 114: 241-271.

Shi, G. R., Fang, Z. J. and Archbold, N. W. 1996. An Early Permian brachiopod fauna of Gondwanan affinity from the Baoshan block, western Yunnan. Alcheringa 20: 81-101.

Shi, G. R., Shen, S. Z. and Archbold, N. W. 1996. A compendium of the Permian brachiopod faunas of the western Pacific region. 5. Changxingian. Deakin University, School of Aquatic Science and Natural Resources Management Technical Paper 1996/2.

Shi, G. R. and Waterhouse, J. B. 1990. Sakmarian (Early Permian) brachiopod biogeography and constraints on the timing of terrane rifting, drift, and amalgamation in SE Asia, with reference to the nature of the Permian 'Tethys'. In The Australasian Institute of Mining and Metallurgy Pacific Rim 90 Congress (Brisbane, Australia) 2: 271-276.

Shi, G. R. and Waterhouse, J. B. 1991. Early Permian brachiopods from Perak, west Malaysia. Journal of Southeast Asian Earth Sciences 6 (1): 25-39.

Smith, A. B. 1988. Late Palaeozoic biogeography of East Asia and palaeontological constraints on plate tectonic reconstructions. Philosophical Transactions of the Royal Society of London A326: 189-227.

Stauffer, P. H. and Lee, C. P. 1986. Late Paleozoic glacial marine facies in SE Asia and its implications. Geological Society of Malaysia Bulletin 20: 363-397.

Thomas, G. A. 1957. Oldhaminid brachiopods in the Permian of northern Australia. Journal of the Palaeontological Society of India 2: 174-182.

Tjia, H. D. 1989. Tectonic history of the Bentong-Bengkalis suture. Geologi Indonesia 12: 89-111.

Ueno, K. and Sakagami, S. 1991. Late Permian fusulinacean fauna of Doi Pha Phlung, north Thailand. Transactions and Proceedings of the Palaeontological Society of Japan, N.S. 164: 928-943.

Veevers. J. J. 1988. Morphotectonics of Australia's northwestern Margin — A review. In The North West Shelf Australia: Proceedings of Petroleum Exploration Society, Australia Symposium. pp.19-27. Edited by P. G. Purcell and R. R. Purcell. Petroleum Exploration Society of Australia, Perth.

Waterhouse, J. B. 1973. Permian brachiopod correlations for South-East Asia. Geological Society of Malaysia Bulletin 6: 187-210.

Waterhouse, J. B. 1982. An early Permian cool-water fauna from pebbly mudstones in south Thailand. Geological Magazine 119 (4): 337-354.

Waterhouse, J. B. 1983. A Late Permian lyttoniid fauna from northwest Thailand. Paper, Department of Geology and Mineralogy, University of Queensland 10 (3): 111-153.

Waterhouse, J. B. and Piyasin, S. 1970. Mid-Permian brachiopods from Khao Phrik, Thailand. Palaeontographica, Abt.A. 135 (3-6): 83-97.

Waterhouse, J.B., Pitakpaivan, K. and Mantajit, N. 1981. Early Permian brachiopods from Ko Yao Noi and near Krabi, southern Thailand. Department of Mineral Resources (Thailand), Geological Survey Memoir 4: 43-213.

Wu, H. R., Boulter, C. A., Ke, B. J., Stow, D. A. V. and Wang, Z. C. 1995. The Changning-Menglian suture zone; a segment of the major Cathaysian-Gondwana divide in SE Asia. Tectonophysics 242: 267-280.

Wu, W. S. 1975. The coral fossils from Mount Jolmo Lungma Region. In A Report of Scientific Expedition to the Mount Jolmo Lungma (Mt. Everest) Region (1966-1968). Palaeontology 1: 83-128.

Yanagida, J. 1967. Early Permian brachiopods from north-central Thailand. Geology and Palaeontology of SE Asia 3: 46-97.

Yanagida, J. 1970. Permian brachiopods from Khao Phrik, near Rat Buri, Thailand. Geology and Palaeontology of Southeast Asia 8: 69-96.

Yanagida, J. 1976. Palaeobiogeographical consideration on the Late Carboniferous and Early Permian brachiopods of central north Thailand. Geology and Palaeontology of SE Asia 17: 173-189.

Yanagida, J. 1984. Carboniferous and Permian brachiopods of Thailand and Malaysia with brief note on the Mesozoic. Geology and Palaeontology of SE Asia 25: 187-194.

Yang, W. P. and Liu, B. P. 1996. The finding of Lower Permian Gondwana-type spores and pollen in western Yunnan. In Devonian to Triassic Tethys in Western Yunnan, China. pp. 128-135. Edited by N. Q. Fang, and Q. L. Feng. China University of Geosciences Press Wuhan.

Zhan, L. P. and Wu, R. R. 1982. Early Permian brachiopods from Xainza district, Xizang (Tibet). Contributions to Geology of Qinghai-Xizang (Tibet) Plateau 7: 86-109.

Ziegler, A. M. and Hulver, M. L. and Rowley, D. B. 1997. Permian world topography and climate. In Glacial and Post-glacial Environmental Changes: Pleistocene, Permo-Carboniferous, Proterozoic. Edited by I. P. Martini. Oxford University Press, Oxford, in press.

Upper Palaeozoic floras of SE Asia

J. F. Rigby
School of Natural Resource Sciences, Queensland University of Technology, Box 2434, GPO Brisbane, 4001, Australia

Key words: SE Asia, floras, Upper Palaeozoic, Cathaysialand, Gondwanaland

Abstract

Small Carboniferous floras of limited significance occur in Thailand and West Malaysia. They are related to the pantropical Cathaysian and/or Laurasian floras. Extensive Permian floras are known from Thailand, Laos, West Malaysia, Sumatra and Irian Jaya and these floras were substantially related to Cathaysian floras. However, some floras from Irian Jaya and Thailand include one or more species of the Gondwanaland genus *Glossopteris*. When the distributions are plotted on a reconstruction for SE Asia prepared by Carey, requirements are met for proximity to centres for radiation of both Cathaysian and Gondwanan floras.

Introduction

Upper Palaeozoic floras have been described from SE Asia. The variety and distribution of these floras suggest that more occurrences may be found in the future as localities are rare, and spread over a very large area which has not been fully explored geologically. Permian floras have palaeogeographical significance for places outside the area because of the association of Cathaysian and Gondwanan species and the presence of endemic species.

Carboniferous floras

Despite the presence of extensive Carboniferous floras in neighbouring Yunnan and Guangxi, China, they are not well represented in SE Asia. This may reflect the smaller number of geological investigations undertaken in parts of SE Asia.

Edwards (1948) reported *Lepidodendron* and *Stigmaria* from Pahang, West Malaysia (Fig.1, near locality 6). Edwards (1926) reported *Pecopteris* sp. cf. *P. cyathea* (Schlotheim) Brongniart 1833 and *Cordaites* sp. from the Raub Series of Kelantan (NE state of West Malaysia).

Asama (1973) described two Lower Carboniferous (Visean) floral localities from Pahang, also within the Raub Series, which he named the Kuantan flora. His Locality A at Panching (NW of Kuantan, Fig.1, loc. 6) yielded only lycopods: *Lepidodendron acuminatum* Goeppert 1847; *Bergeria* sp.; *Lepidodendropsis vandergrachtii* Jongmans, Gothan and Darrah 1937; *Stigmaria* sp. His Locality B at Gambang (SW of Kuantan) yielded fern-like foliage and a seed: *Rhodea hsianghsiangensis* Sze 1953, ?*Adiantites* sp., ?*Neuropteris* sp., *Carpolithus* sp.

Asama *et al.* (1975) included some additional specimens from Trengganu (state adjoining Pahang to the north) and Pahang referring the identifications to Franks, but omitting the reference: *Lepidodendron* sp., *Rhodea* sp., *Sphenophyllum* sp. They compared the Kuantan flora to the floras of the Kaolishan Formation of Jiangsu and the Tseshui Formation of Hunan, both in the Visean of southern China, in Wu (1995). I consider the data to be somewhat too generalized for precise correlation, however the Early Carboniferous age is without doubt.

Laveine *et al.* (1993) described some plants from the Na Duang coal mine area of NE Thailand (Fig.1, loc. 2) which they considered as

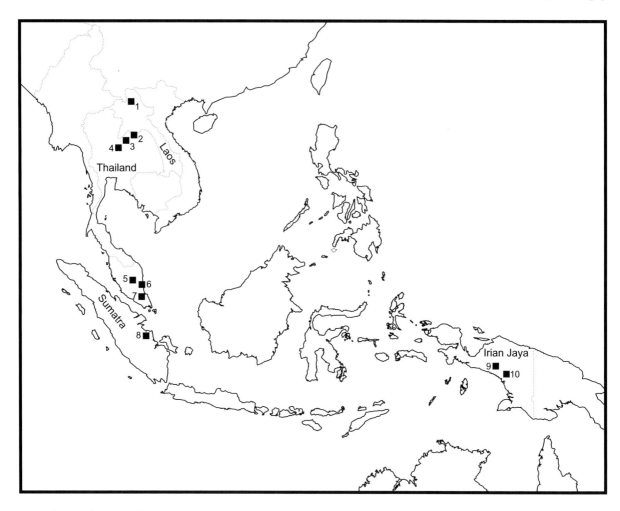

Fig.1. The distribution of localities throughout SE Asia. 1. Northern Laos: Nam-Ou (northerly), Phong-Saly (central), Bun-Tai (southern). The position of other localities in Laos is not known. 2. Na Duang. 3. Loei. 4. Phetchabun. 5. Jengka. 6. Kuantan. 7. Linggiu. 8. Jambi. 9. Rigby's 1997 localities, Irian Jaya. 10. Jongmans' 1940 localities, Irian Jaya. The quality of the locality data is variable, often based on data supplied by field workers.

representative of the Middle Carboniferous pantropical *Paripteris* flora. Their Fig.2 shows the flora's distribution. Their provisional species list included: *Stigmaria ficoides* Brongniart, *Lepidodendron* sp., arborescent lepidophytes with large ulodendroid scars, *?Bothrodendron* sp., *Archaeocalamites* sp., *Rhodeopteridium* sp. cf. *R. tenuis* (Gothan) Zimmermann, *?Rhodeopteridium* sp., *Adiantites* sp. cf. *A. spectabilis* Read, *Eusphenopteris* sp. cf. *E. hollandica* (Gothan and Jongmans) Novik, *?Cordaites* sp. This association confirms that northern Thailand lay within the Carboniferous tropics.

Permian floras

Floras from West Malaysia, Sumatra and Thailand

Extensive Permian floras occur in many parts of SE Asia. Asama *et al.* (1975, table 2) have assembled distribution data from occurrences in Malaya, Sumatra and Thailand. Table 1 is an edited version of some of their data which extend over 8 pages. Their table includes distribution within the Cathaysian Permian and comparison with extra-Chinese floras as well as the 66 species occurring only at localities near Jambi (Djambi) in Sumatra, which have been listed separately in Table 2, below. Permian floras from Irian Jaya (Western New Guinea) which are not closely related to the floras listed in Tables 1 and 2 are discussed separately under 'Permian floras of New Guinea'.

The Jambi flora

Many fossil plant localities were discovered in the Jambi district in central Sumatra, Indonesia during the Dutch administration. An earlier suite of almost 200 slabs collected by A. Tobler was described by Jongmans and Gothan (1925). A second, very much larger suite of 487 slabs was collected by O. Posthumus, and also described by Jongmans and Gothan (1935). Table 1 lists the species identified from Jambi also occurring elsewhere in SE Asia, whereas Table 2 lists species identified only from Jambi. A map showing the positions for the collecting localities spread over an area of 30 x 17 km was given by Jongmans and Gothan (1935, map 1). These authors have assembled the localities into six regions, generally along stream valleys within the area, listing the species distribution.

Most of these regions contributed floras of 20 or more species with most occurring in more than one region. Even if one re-identifies specimens into more broadly defined 'species', the total flora is still very large and varied. The most striking features are the considerable variety of ferns and fern-like foliage, the variety of *Taeniopteris* spp., and the number of *Cordaites* spp. These indicate a Cathaysian affinity. Asama et al. (1975) have also demonstrated a strong Laurasian affinity for the total flora. The first collection (Jongmans and Gothan, 1925) had affinity only with Laurasia, lacking any Cathaysian species. There was no suggestion of correlation with coeval floras of Gondwanaland.

Jongmans and Gothan dated the earlier collection (1925) as Carboniferous based on the flora's apparent Laurasian aspect. The much larger, second collection (1935) also appeared to be more Carboniferous than Permian, but a limestone underlying the two plant horizons yielded fusulinids dating the limestone as Middle Asselian to Sakmarian, most probably Late Asselian (Vachard, 1989).

Brouwer (1931) quoted Tobler as listing the occurrence of *Dadoxylon* above the plant-bearing horizons. Vozenin-Serra (1985) studied some woods from Jambi that may have included specimens from the earlier collections. She identified *Dadoxylon roviengense* Vozenin-Serra 1985 and *D. saxonicum* (Goeppert) Frentzen 1931, both of which lacked pronounced growth rings. She noted that this characteristic was typical of trees growing under a tropical or subtropical regime.

Upper Permian floras from Thailand

Kon'no (1963) described a small flora from Khlong Wang Ang, about 50 km SSW of Phetchabun (see Fig.1, locality 12, in Iwai et al., 1966, for location data), identified from "five small pieces of black shales". He commented that the specimens were "not very favourable in preservation". His identifications included: *Taeniopteris hallei* Kawasaki 1931, *T.* sp. cf. *T. serrulata* Kawasaki (*non* Halle) (both these species are present in the approximately Kungurian Upper Jido Series in Korea); *Sphenophyllum trapaefolium* Stockmans and Matthieu 1957 (Kon'no's photo (Plate 8, Fig.5) suggests the specimen was distorted: it occurs in the approximately Kazanian Kobosan Series of Korea); *Palaeovittaria parvifolia* Kon'no 1963; *Glossopteris* sp. cf. *G. angustifolia* Brongniart 1828 (the figured fragment is small, but has the distinctive glossopterid venation). The remaining species were endemic: *Bowmanites* sp., *Alethopteris thailandica* Kon'no 1963, and *Poacordaites phetchabunensis* Kon'no 1963.

Asama (1966) subsequently collected from the same locality identifying specimens as *Sphenophyllum phetchabunensis* Asama 1966, *Pecopteris* sp., *Compsopteris wongii* (Halle) Zalessky 1934 (= *Protoblechnum wongii*), *Taeniopteris thailandica* Asama 1966, *T. nystroemlii* Halle 1927, *T. konnoi* Asama 1966, *T. hallei* Kawasaki 1934, *T. iwaii* Asama 1966, *Taeniopteris* sp., *Cordaites principalis* (Germar) Geinitz 1855, *Poacordaites linearis* Grand'Eury 1877 (Asama included Kon'no's *P. phetchabunensis* as a synonym), *Psygmophyllum komalarjunii* Asama 1966, *Psygmophyllum?* sp., *Samaropsis* sp.

Asama (1966) drew attention to aspects of the flora, including the variety of the *Taeniopteris* species present, 3 large and 3 small leaved species. Taylor and Taylor (1993, p.599) have observed "In the Carboniferous (Pennsylvanian of Kansas), *Taeniopteris* foliage has been found with small seeds attached to the abaxial surface, suggesting affinities with the cycads. In the Mesozoic, however, the taxon has been associated with the bennettitalean *Williamsoniella*". It is possible that the two groups of *Taeniopteris*

Table 1. Distribution of Permian floras in Malaya, Sumatra and Thailand modified and edited from Table 2 of Asama *et al.* (1975). Additional species occurring only at Jambi (Djambi), Sumatra are listed on Table 2. I consider there are taxonomic differences between the Cathaysian and Gondwanan specimens currently identified as *Trizygia speciosa* *.

	Jambi flora Sumatra	Loei flora Thailand	Phetchabun flora Thailand	Jengka flora W Malaysia	Linggiu flora W Malaysia
LYCOPHYTA					
Lepidodendron sp. cf. *L. chosenense* Kawasaki					+
NOEGGERATHIOPHYTA					
Tingia subcarbonica Kon'no and Asama					+
SPHENOPHYTA					
Sphenophyllum trapaefolium Stockmans and Mathieu			+		
Sphenophyllum sp.		+			
Parasphenophyllum phetchabunense (Asama) Asama			+		
Trizygia oblongifolia (Germar and Kaulfuss) Asama	+	+			
T. sinocoreana (Yabe) Asama					+
T. speciosa Royle					+
Paratrizygia glossopteroides forma *minor* (Kawasaki) Asama					+
P. koboensis (Kobatake) Asama				+	+
Calamites suckowii Brongniart	+				cf.
Annularia shirakii Kawasaki				+	+
Lobatannularia fujiyamae Kon'no and Asama					+
L. johorensis Kon'no and Asama				+	+
L. johorensis subsp. *minor* Kon'no and Asama					+
L. suntharalingamii Kon'no and Asama					+
FERNS AND FERN-LIKE FOLIAGE					
Ptychocarpus malaianus Kon'no					+
Ptychocarpus sp.		+			
Rajahia (Pecopteris) bifurcata Kon'no					+
R. linggiuensis Kon'no					+
R. pseudohemitenoides Kon'no					+
R. rajahii Kon'no					+
R. sengensis Kon'no					+
Shirakiopteris loeiensis Asama		+			
Pecopteris arcuata Halle	+			+	+
P. hemitenoides Brongniart	+	+			
P. lativenosa Halle		+			
P. yinii Kon'no and Asama					+
Pecopteris sp.		+	+		
Cladophlebis ozakii Yabe and Oishi					+
Fascipteris sinensis (Stockmans and Mathieu) Gu and Zhi					+
Aphlebia sp.	+				+
Neuropteris sp.				+	+
Neuropteridium yokoyamae Kon'no and Asama				+	+
Alethopteris thailandica Kon'no			+		
Alethopteris sp.		+			
Protoblechnum wongii Halle		+	+	cf.	
Cathaysiopteris whitei (Halle) Koidzumi					cf.
Taeniopteris crassicaulis Jongmans and Gothan	+				+
T. hallei Kawasaki		+	+		+
T. iwaii Asama			+	+	
T. konnoi Asama			+		
T. latecostata Halle	+			+	
T. multinervis Weiss	cf.		+	cf.	
T. nystroemii Halle	+	+	+		+
T. shansiensis Halle				+	
T. taiyuanensis Halle	+			+	
T. thailandica Asama			+	cf.	
Taeniopteris sp. cf. *T. serrulata* Kawasaki			+		
Taeniopteris sp.			+	+	
Gigantonoclea lagrehi (Halle) Koidzumi		+			+
Bicoemplectopteris hallei Asama		+		+	+
Tricoemplectopteris taiyuanensis Asama					+
Bicoemplectopteridium longifolium (Kodaira) Asama		+			+
Gigantopteris nicotianaefolia Schenk			+		+

Table 1. Continued.

	Jambi flora Sumatra	Loei flora Thailand	Phetchabun flora Thailand	Jengka flora W Malaysia	Linggiu flora W Malaysia
Glossopteris sp. cf. *G. angustifolia* Brongniart			+		
Palaeovittaria parvifolia Kon'no			+		
Psygmophyllum komalarjunii Asama			+		
Psygmophyllum sp.			?		
Psaronius johorensis Ogura					+
CORDAITES					
Cordaites principalis (Germar) Geinitz	+		+		
C. schenkii Halle				+	
C. simplicinervius Jongmans and Gothan	+			cf.	
Poacordaites linearis Grand'Eury			+		
Poacordaites sp.	+	+			
Cordaianthus volkmannii (Ettingshausen) Zeiller				cf.	
Cordaicarpus cordai forma *elongata* Jongmans and Gothan	+				+
CYCADOPHYTA					
Sphenozamites sp.					+
GINKGOPHYTA					
Rhipidopsis baieroides Kawasaki and Kon'no					+
SEEDS					
Carpolithus sp.	+				+
Gigantospermum posthumii Jongmans and Gothan	+				+
Samaropsis sp.		+	+		
Trigonocarpus sp.	+			+	

Table 2. Permian plants whose only SE Asian occurrence is in the Jambi district of Sumatra. The authorship of species designated as J and G was by Jongmans and Gothan 1935; most or all of these were endemic. See Table 1 for species occurring both at Jambi and elsewhere in SE Asia.

LYCOPHYTA
Lepidodendron mesostigma J and G, *L. molle* J and G, *L. posthumii* J and G, *Stigmaria asiatica* J and G, *S. ficoides* Brongniart, *?Lycopodites* sp., *Maroesta rhomboidea* J and G.

SPHENOPHYTA
Sphenophyllum sp. cf. *S. emarginatum* Brongniart, *S. verticillatum* (Schlotheim) Brongniart, *Parasphenophyllum thonii* (Mahr) Asama, *Sphenophyllostachys* sp., *Calamites jubatus* Lindley and Hutton, *Annularia stellata* (Schlotheim) Wood, *Annularia* sp., *Palaeostachya incrassata* J and G.

FERNS AND FERN-LIKE FOLIAGE
Asterotheca sp., *Monocarpia posthumii* J and G, *Sphenopteris zwierzycki* J and G, *S.* sp. cf. *S. gothanii* Halle, *S.* sp. cf. *S. grabaui* Halle, *S.* sp. cf. *S. matheti* Zeiller, *S.* sp. cf. *S. tingii* Halle, *Sphenopteris* sp., *Pecopteris arborescens* Schlotheim, *P. candolleana* Brongniart, *P. densifolia* Goeppert, *P. djambiensis* J and G, *P. mengkarangensis* J and G, *P. oreopteridia* Schlotheim, *P. polymorpha* Brongniart, *P. unita* Brongniart, *P. unitaeformis* J and G, *P. verbeekii* J and G, *P.* sp. cf. *P. cistii* Brongniart, *P.* sp. cf. *P. daubreei* Zeiller, *Aphlebia ?acanthoides* Zeiller, *A. dimorpha* J and G, *A. minor* J and G, *A.* sp. cf. *A. crispa* Gutbier, *Neuropteris* sp. cf. *N. gleichenoides* (Stur) Sterzel, *Nemejcopteris feminaeformis* (Schlotheim) Barthel, *Cyclopteris* sp., *Alethopteris strictinervis* J and G, *Macralethopteris hallei* J and G, *Callipteridium mengkarangense* J and G, *C. sumatranum* J and G, *Dictyocallipteridium sundiacum* J and G, *Asterophyllites* sp., *Palaeogoniopteris mengkarangensis* (J and G) Koidzumi, *Taeniopteris camptoneura* J and G, *T. densissima* Halle, *T. incrassata* J and G, *T.* sp. cf. *T. norinii* Halle, *Gothanopteris bosschana* (J and G) Koidzumi.

CORDAITES
Cordaites lingulatus Grand'Eury, *Cordaianthus* sp., *Cordaicarpus cordai* Geinitz, *Cordaicarpus ovalis* J and G, *Cordaicladus* sp.

SEEDS
Artisia sp., *Carpolithus coffeoides* J and G, *C. granulosus* J and G, *C. multigranosus* J and G, *Rhynchogonium permocarbonicum* J and G, *Schuetzia* sp., *Tobleria bicuspis* J and G.

spp. at Phetchabun each belong to different groups of plants. Alternatively, the two groups of species may represent variants from two natural species as it is known from elsewhere that *Taeniopteris* species can be quite variable. *Psygmophyllum* is present commonly here, and also in many parts of Cathaysialand. *Compsopteris wongii* occurs only in Cathaysialand where it ranges from Kungurian to Kazanian.

Asama (1966, pp.194-196, Plate 1, Figs.4-6) named some leaves as *Sphenophyllum phetchabunense* Asama 1966 to which he suggested assignment of leaves previously included in *Zamiopteris glossopteroides* forma *minor* by Kawasaki (1939) who had used this name for some isolated leaves having a pronounced midrib. One of Asama's specimens has three leaves attached to a stem. I consider these leaves to be gymnospermous conforming to the now understood, tufted habit of many Late Palaeozoic gymnosperms. He also discussed the presence of *Glossopteris* in Thailand suggesting that the most probable explanation was that the genus arose by parallelism. An unfortunate confusion has arisen from calling floras including species from more than one phytological province 'mixed floras'. In this case, a plant having leaves of *Glossopteris* has grown alongside typical Cathaysian species, implying a mixture of the Cathaysian flora with the Gondwanan flora. This could only occur if there were a land connection between Cathaysialand and Gondwanaland across New Guinea in the Permian, also if the climate at the time favoured the growth of the tropical Cathaysian flora in the same place as the temperate Gondwanaland flora. This would occur near the subtropical zone.

Permian floras of West Malaysia

Two floras from the Permian of West Malaysia were described within a few months of each other in 1970, namely the Linggiu flora from Johore (Kon'no *et al.*, 1970) and the Jengka flora from Pahang (Kon'no and Asama, 1970). The species occurring are listed on Table 1. Considering the total number of species occurring in each flora (Jengka = 20, Linggiu = 40) it is significant how few species they have in common (5). These authors' discussion of the five species shows that even these may not have many features in common.

Neuropteris sp. — Specimens in each collection were compared with *Neuropteris* sp. cf. *N. gleichenoides* from Jambi (in Jongmans and Gothan, 1935), namely one pinnule from Jengka, and a number from Linggiu being smaller and having a difference in venation which they think may eventually prove to specifically significant.

Neuropteridium yokoyamae — Linggiu is the type locality; Jengka, one specimen identified with reservation.

Pecopteris arcuata — all specimens were identified positively.

Taeniopteris multinervis — the only specimen from Linggiu is "too incomplete" to add anything to a correct identification with this Laurasian species.

Bicoemplectopteris hallei — was abundant at Linggiu, but rare at Jengka. It is one of the index species for the Cathaysian flora.

Kon'no and Asama (1970) and Kon'no *et al.* (1970) conclude both floras equate to Stage P_2^1 of China, however the Linggiu flora was slightly older. The Jengka flora came from an interbedded marine/non-marine sequence dated as Late Permian, having an age supported by faunas. The Linggiu flora was found in arenites associated with shallow seas. Full data have been given by Rajah (in Kon'no *et al.*, 1970). The environmental difference is sufficient to account for the difference in floral composition.

Permian floras of Laos

Many of the original references are not available to me. To overcome this, data have been taken directly from Vozenin-Serra (1979). She identified and figured specimens in the columns headed Sap-Pong and Laos. Other references have not been seen by me, but citations have been extracted from Vozenin-Serra (1979) for completeness.

Permian floras of New Guinea

Permian floras occur in Irian Jaya (Western New Guinea) along the southern or Australasian plate side of the main suture zone. All occurrences for which I have data were found in the Lower Permian Aiduna Formation. The earliest published account of the fossil floras was by Jongmans (1940) who reported *Sphenophyllum verticillatum, Pecopteris* sp. cf. *P. arcuata, P.* sp. cf. *P. paucinervis, P.* sp. cf. *P. orientalis, P. unita, Taeniopteris* sp. cf. *T. multinervis, T.* sp. cf. *T. taiyuanensis* and *Vertebraria*. Other than

Vertebraria which was typical of Gondwanaland, most other species were typical of Carboniferous floras of both Cathaysialand and Laurasia. Lehrner *et al.* (1955) reported plant fossils from a number of localities in the Aiduna Formation, however I cannot comment as the paper is not referred to in other publications, and I do not have the citation, and therefore it is omitted from the references.

Hopping and Wagner (1962) figured and briefly described species from the Aiduna Formation from a number of localities identifying *Sphenophyllum* sp. cf. *S. speciosum*, *Pecopteris monyi*, *Cladophlebis* sp. cf. *C. australis*, '*Validopteris*' sp., *Glossopteris* sp. cf. *G. browniana*, *G.* sp. cf. *G. indica*, *G.* sp. aff. *G. retifera*, *Vertebraria* sp., *Taeniopteris* sp. cf. *T. hallei*. These species include both Gondwanaland and Cathaysialand representatives.

Rigby (1997) revised previous identifications and identified new collections from the Aiduna Formation. He noticed that Jongmans' figured *Taeniopteris* spp. showed evidence of occasional cross-veins in the secondary venation and therefore more correctly were *Glossopteris* spp., also proposing *G. iriani* and the Cathaysian-related *Gigantonoclea iriani* both occurring on the same rock slab. Together with revisions of previous workers' identifications, the total flora included *Trizygia speciosa* (*Sphenophyllum speciosum* in Hopping and Wagner, above); three *Pecopteris* spp., fern frond, *Ptychocarpus* sp, *Cladophlebis* sp, *Fascipteris aidunae* ('*Validopteris*' sp. of Hopping and Wagner); *Glossopteris iriani*, eight other *Glossopteris* spp.; *Vertebraria indica*; *Gigantonoclea iriani*; and *Koraua hartonoi* (probably a gymnosperm). In the case of *Trizygia speciosa* there are problems with both the specimens referred to the species and their generic attribution. Specimens from Cathaysialand and Gondwanaland differ, as pointed out by Li and Rigby (1995), which means they probably should belong in different species. The other problem is whether the species belongs in *Sphenophyllum* or in the Gondwanaland genus *Trizygia* as the type species. Further work is necessary.

The wood, *Planoxylon stopesii*, from the Aimau Formation of Vogelkop, Irian Jaya, was described as showing characters of both araucarian and abietinian wood (Prasad, 1981), common wood types from the Upper Palaeozoic of Gondwanaland (Marguerier, 1973).

The significance of these floras is their endemism, *i.e.*, constrained to species not found elsewhere. Jongmans' (1940) flora comprised largely fern or fern-like fronds and resembled floras from Jambi, however species differed. Hopping and Wagner (1962), and Rigby (1997) reported predominantly species of *Glossopteris* with some ferny foliage and having a superficial appearance to floras from the warmer parts of Gondwanaland such as India. The most significant aspect of these floras was the recognition by Rigby of seed plants linking the region to both Gondwanaland (*Glossopteris*) and Cathaysialand (*Gigantonoclea*) as there was no animal vector available to transport seeds across the Tethys Sea which had an opening of about 45° based on the more popular plate tectonics palaeogeographical reconstructions. However, using a reconstruction based on the earth expansion theory, the Tethys was a narrow seaway, and sometimes dry land (Dickins, 1996). This reconstruction would have allowed terrestrial contact between Irian Jaya and Cathaysialand along which seed plants could have migrated. Carey (1996) has proposed a reconstruction of the Permian land masses forming the SE Asia of today. It is used as a basis for the reconstruction herein (Fig.2). The data have also been plotted on a recent plate tectonic reconstruction (Fig.3) by Metcalfe (1996).

The only reference to Permian floras in Papua New Guinea is by Mackay and Little (1911) who reported the occurrence of *Glossopteris* in coal measures from central Papua New Guinea. At present the coal measures are considered to be Tertiary, and this record is considered a misidentification.

Conclusions

Carboniferous floras

Plant remains from West Malaysia are nondescript, however, they do suggest that the region formed part of the same continental block that included southern China. The only floras of this general age that includes specifically identifiable plants occurs in the Middle Carboniferous at the Na Duang coal mine of NE Thailand. The flora was within the distribution of the pan-tropical *Paripteris* flora.

Permian floras

Permian floras are all tropical Cathaysian although some from Thailand include the Gondwanan genus *Glossopteris*. The floras from Irian

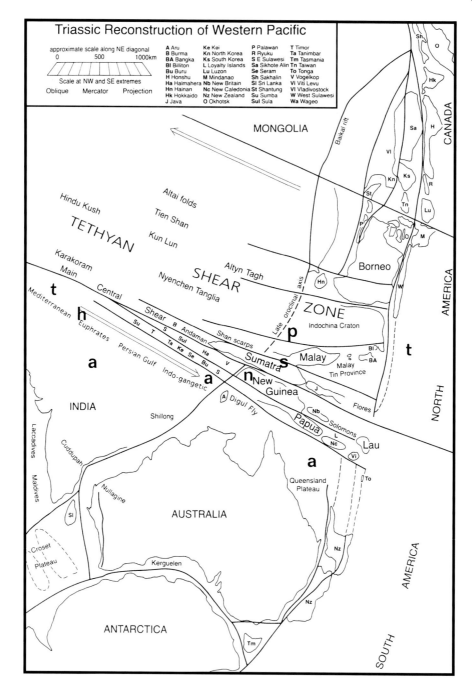

Fig.2. Permian plant localities indicated on a Triassic recontsruction by Carey (1996, Fig.44) based on the earth expansion theory. The Late Permian palaeogeography of SE Asia was almost identical with the Triassic restoration as no major earth movements took place over this period. The bold letters signify: south of a a a, typical Gondwanaland floras. Transitional floras at Hazro, Turkey by h, and in Irian Jaya by n. Pantropical floras north of the line joining t t, with the Thai and Laotian floras near p. The Jambi flora of Sumatra occurs at s.

Jaya (West New Guinea) are predominantly Gondwanan but with the addition of two Cathaysian genera each with one endemic species, viz. *Fascipteris aidunae* and *Gigantonoclea iriani*. The large and varied flora occurring around Jambi, Sumatra, had Cathaysian affinities although it had more than 30 species known only from Jambi. All floras included a number of fern species, often *Pecopteris*. The floras were tropical, but where the Gondwanan species also occurred, they may have been sub-tropical or warm temperate.

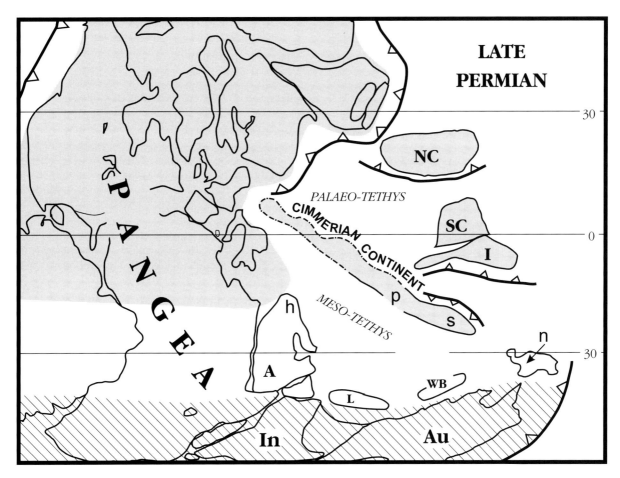

Fig.3. Distribution of Permian floras on a Late Permian restoration from Metcalfe (1996, Fig.13, lower left) based on the plate tectonics theory. Letters on the map are: A: Arabian peninsula, Au: Australia, C: Cimmerian continent, I: Indo Chinese region, In: India, L: Lhasa block, NC: Northern China, P: Pangea, SC: Southern China, WB: Western Burma. Pure pantropical floras of SE Asia occurred on block I, these floras were related to the floras on block SC and P. Floras combining species from both the pantropical region and Gondwanaland (In and Au) occurred on the southern margin of block I and on New Guinea. Transitional floras at Hazro, Turkey by h, and in Irian Jaya by n. The Thai and Laotian floras were situated near p. The Jambi flora of Sumatra occurs at s.

Acknowledgements

Professor S. W. Carey, Department of Geology, University of Tasmania, Australia, for the base map on which the floral distribution has been plotted. Professor Li X.-X., Nanjing Institute of Geology and Palaeontology, Academia Sinica, China, for discussion concerning aspects of the Permian Cathaysian floras. Professor R. Hall, Royal Holloway University of London, and my referees for improvements in the text.

References

Asama, K. 1966. Permian plants from Phetchabun, Thailand and problems of floral migration from Gondwanaland. Bulletin of the National Science Museum 9(2): 171-211. pls. 1-6.

Asama, K. 1973. Lower Carboniferous Kuantan flora, Pahang, West Malaysia. Geology and Palaeontology of Southeast Asia 9: 109-118, pls. XIV-XVI.

Asama, K., Hongnusonthi, A., Iwai, J., Kon'no, E., Rajah, S. and Veeraburus, M. 1975. Summary of the Carboniferous and Permian plants from Thailand. Geology and Palaeontology of Southeast Asia 15: 77-101.

Brouwer, H. A. 1931. De stratigraphie van Nederlandsch Oost-indië. 18. Paleozoic. Leidsche Geologische Mededeelingen 5: 552-566.

Carey, S. W. 1996. Earth, Universe, Cosmos. Geology Department, University of Tasmania, pp. xii + 231.

Dickins, J. M. 1996. The southern margin of Tethys. Ninth International Gondwana Symposium 2: 1125-1134.

Edwards, W. N. 1926. Carboniferous plants from the Malay States. Journal of the Malayan Branch of the Royal Asiatic Society 4: 171-172, pl. 1.

Edwards, W. N. 1948. Lepidodendroid remains from Malaya.

Appendix 9 in H. M. Muir-Wood, Malayan Lower Carboniferous fossils. British Museum (Natural History) London, pp. 78-81, pl. 10.

Hopping, C. A. and Wagner, R. H. 1962. Photographs of fossils. Enclosure 17 *In* W. A. Visser and J. J. Hermes. Geological results of the exploration for oil in Netherlands New Guinea. Koninklijk Nederlands Geologisch Mijnbouwkundig Genootschap. Geologische Serie, speciaal nummer 20.

Iwai, J. Asama, K., Veeraburas, M. and Hongnasonthi, A. 1966. Stratigraphy of the so-called Khorat Series and a note on the fossil plant-bearing Palaeozoic strata in Thailand. Japanese Journal of Geology and Geography 37 (1): 21-38.

Jongmans, W. and Gothan, W. 1925. Beiträge zur Kenntnis der Flora des Oberkarbons von Sumatra. Verhandelingen koninklijke nederlandse Geologisch Mijnbouwkundig Genootschap 8: 231-287, 5 pls.

Jongmans, W. and Gothan, W. 1935. Die Ergebnisse der paläobotanischen Djambi-Expedition 1925. 2. Die paläobotanischen Ergebnisse. Jaarboek van het Mijnwezen in Nederlandsch-indië 50: 71-201, pls. 1-58.

Jongmans, W. 1940. Beiträge zur Kenntnis der Karbonflora von niederländisch Neu-Guinea. Mededeelingen Geologische Stichting 1938-1939: 263-274, pls. 1-3.

Kawasaki, K. 1939. Addition to the flora of the Heian System. Bulletin of the Geological Survey of Tyôsen 6 (5): 1-39, pls. 1-9.

Kon'no, E. 1963. Some Permian plants from Thailand. Japanese Journal of Geology and Geography 34 (2-4): 139-159, pl. 8.

Kon'no, E., and Asama, K. 1970, Some Permian plants from the Jengka Pass, Pahang, West Malaysia. Geology and Palaeontology of Southeast Asia 8: 97-132, pls. 17-24.

Kon'no, E., Asama, K. and Rajah, S. S. 1970. The Late Permian Linggiu flora from the Gunong Blumut area, Johore, Malaysia. Bulletin of the National Science Museum 13(3): 491-580, pls. 1-17.

Laveine, J.-P., Ratanasthien, B. and Sithirach, S. 1993. The Carboniferous flora of Northeast Thailand, its paleogeographic importance. Comptes Rendus des Académie des Sciences de Paris 317 (2): 279-285.

Li X.-X. and Rigby, J. F. 1995. Further contributions to the study of the Qubu flora from southern Xizang (Tibet). Palaeobotanist 44: 38- 47.

Mackay, D. and Little, W. S. 1911. The Mackay-Little Expedition in southern New Guinea. Geographical Journal 38: 483-487, 1 map, 2 pls.

Marguerier, J. 1973. Paleoxylologie du Gondwana africain. Etude et affinités du genre *Australoxylon*. Palaeontographica Africana 16: 37-58.

Metcalfe, I. 1996. Pre-Cretaceous evolution of SE Asian terranes. *In* Tectonic evolution of Southeast Asia. pp. 97-122. Edited by R. Hall and D. J. Blundell. Geological Society of London Special Publication 106.

Prasad, M. N. V. 1981. New species of fossil wood *Planoxylon* from Late Paleozoic of Irian Jaya, Indonesia. Bulletin of the Geological Research and Development Centre 5: 37-40.

Rigby, J. F. 1997. The significance of a Permian flora from Irian Jaya (West New Guinea) containing elements related to coeval floras of Gondwanaland and Cathaysialand. Palaeobotanist 45: 295-302.

Taylor, T. N. and Taylor, E. L. 1993. The biology and evolution of fossil plants. Prentice-Hall, Englewood Cliffs. xxii + 982 pp.

Vachard, D. 1989. A rich algal microflora from the Lower Permian of Jambi Province. *In* The Pre-Tertiary fossils of Sumatra and their environments. pp. 59-69. pls. 6-9. Edited by H. Fontaine and S. Gafoer. Committee for coordination of joint prospecting for mineral resources in Asian offshore areas. Coop. Technical Section, Bangkok.

Vozenin-Serra, C. 1979. Étude de quelques empreintes de végétaux fossiles du paléozoïque supérieur du Laos. Comptes Rendus du 104 Congrès national des Sociétés savantes 1: 155-174.

Vozenin-Serra, C. 1985. Bois homoxylés du permien inférieur de Sumatra. Implications paléogógraphiques. Actes du 110 Congrès national des Sociétés. savantes Section des sciences 5: 55-63

Wu X.-Y. 1995. Carboniferous floras. *In* Fossil floras of China through the geological ages (English edition). pp. 78-126. Edited by Li X.-X. and others. Guangdong Science and Technology Press, Guangzhou.

The biogeographical significance of the Mesozoic vertebrates from Thailand

Eric Buffetaut[1] and Varavudh Suteethorn[2]
[1] UMR CNRS 5561, Université de Bourgogne, France. Present address: 16 cour de Liégat, 75013 Paris, France
[2] Geological Survey Division, Department of Mineral Resources, Rama VI Road, Bangkok 10400, Thailand

Key words: Thailand, Mesozoic, vertebrates, amphibians, reptiles, biogeography

Abstract

Mesozoic non-marine vertebrate faunas from Thailand (mainly from the Indochina block) range in age from Late Triassic to late Early Cretaceous. The oldest known assemblage, from the Huai Hin Lat Formation, is Norian. It includes fishes, amphibians and reptiles very similar to those from the Upper Triassic of Central Europe, and indicates that dispersal of continental vertebrates across Eurasia was easy in the Late Triassic. A slightly younger (Rhaetian) prosauropod dinosaur is biogeographically less significant, although it may be related to Chinese forms. Jurassic vertebrates are known from both the Indochina block (Phu Kradung Formation) and the Shan-Thai block. Resemblances between the assemblages from both areas suggest that they were already in contact in the Jurassic. Similarities with Jurassic vertebrates from China indicate links with more northerly parts of Asia. The Lower Cretaceous (pre-Aptian) Sao Khua Formation contains a vertebrate fauna which seems to be older than most Cretaceous assemblages known from other parts of Asia, and probably corresponds to a phase of relative isolation of Asia. It includes dinosaurs which may be close to the ancestry of groups which later played an important part in Asian and North American assemblages, such as tyrannosaurids and ornithomimids. Some of the Sao Khua sauropods may be related to the poorly known Upper Cretaceous sauropods of Mongolia and China. The Aptian-Albian Khok Kruat Formation has yielded a freshwater shark also known from the Lhasa block of Tibet, as well as the ceratopsian dinosaur *Psittacosaurus*, which was widespread in Asia during the Early Cretaceous. It also contains remains of iguanodontid dinosaurs which are probably immigrants from Europe and show that by the late Early Cretaceous the isolation of Asia had ended.

Introduction

In the last twenty years, Thailand has yielded a number of Mesozoic non-marine fossil vertebrates, ranging in age from Late Triassic to late Early Cretaceous (for a recent review of the dinosaurs, see Suteethorn *et al.*, 1995). This is by far the best record for that time interval in SE Asia, and it allows a reconstruction of vertebrate faunal history in that part of the world over a period of more than 100 million years. The purpose of this brief review is to discuss the palaeobiogeographical significance of the various Mesozoic vertebrate assemblages hitherto discovered in Thailand.

General geological setting

It is now generally accepted that Thailand consists of two continental blocks or microcontinents (Fig.1), the eastern part (with, notably, the Khorat Plateau where most Mesozoic vertebrate localities are) belonging to the Indochina block, while the western part (including the southern peninsula) is part of a terrane variously called 'Shan-Thai' or 'Sibumasu' (see Metcalfe, 1996, for a recent review). The Indochina block seems to have sutured to South China in the Carboniferous (Metcalfe, 1996). The timing of the collision of the Shan-Thai block with Indochina and South China is still debated (Late Permian to Triassic according to Metcalfe, 1996; as late as Late Jurassic according to Stokes *et al.*, 1996). The Mesozoic non-marine vertebrates found in Thailand (Table 1) can thus provide some evidence concerning the late stages of the accretion history of these terranes.

Upper Triassic: the vertebrate assemblage from the Huai Hin Lat Formation

The Huai Hin Lat Formation of the Khorat Plateau consists mainly of lacustrine bituminous limestones and shales. It is Upper Triassic, probably Norian, on the basis of plant macro-remains (Konno and Asama, 1973), palynomorphs (Haile, 1973; Racey *et al.*, 1996), conchostracans (Kobayashi, 1975) and vertebrates (review in Buffetaut *et al.*, 1993).

The vertebrate assemblage from the Huai Hin Lat Formation, which consists of both isolated elements and partly articulated skeletons, includes actinopterygian fishes (Martin, 1984), lungfishes (Martin and Ingavat, 1982), amphibians (*Cyclotosaurus*: Ingavat and Janvier, 1981; plagiosauroid indet: Suteethorn *et al.*, 1988), turtles (*Proganochelys*: Broin *et al.*, 1982, Broin, 1984), and phytosaurs, including a *Belodon*-like form and *Mystriosuchus* (Buffetaut and Ingavat, 1982; Buffetaut *et al.*, 1993).

From a biogeographical point of view, the freshwater vertebrate assemblage from the Huai Hin Lat Formation is remarkable in that it closely resembles the classic vertebrate fauna from the Norian of Germany (Stubensandstein). Genera of fishes, amphibians, turtles and phytosaurs are common to the Stubensandstein and the Huai Hin Lat Formation. This clearly suggests that in the Norian, dispersal of non-marine vertebrates was possible across Eurasia between Central Europe and the Indochina block of SE Asia. Interestingly, no assemblages closely similar to that from the Huai Hin Lat Formation are currently known from China (where, admittedly, Upper Triassic vertebrates are very poorly represented; Sun, 1989). Possible dispersal routes across Late Triassic Eurasia are still poorly known, but they must have existed.

Upper Triassic: a prosauropod dinosaur from the Nam Phong Formation

The Nam Phong Formation, which unconformably overlies the Huai Hin Lat Formation, consists of sandstones and mudstones of mainly fluvial origin. It has yielded very few fossils, and its age has long remained uncertain. Racey *et al.* (1994, 1996) reported palynomorphs indicating an age no younger than Rhaetian and no older than Ladinian. Since the underlying Huai Hin Lat Formation is well dated as Norian, a Rhaetian age is likely for the Nam Phong Formation.

The only recognisable vertebrate element

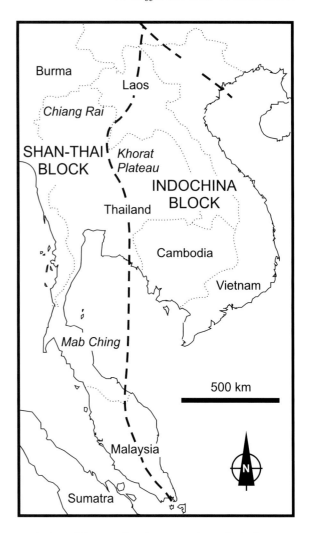

Fig.1. Map of SE Asia showing approximate limits between the Shan-Thai and Indochina blocks, and the location of fossiliferous areas or localities mentioned in text (in italics).

hitherto found in the Nam Phong Formation consists of the fused distal ends of the ischia of a large prosauropod dinosaur (Buffetaut *et al.*, 1995). A more accurate identification is not possible on the basis of the available material. As prosauropods had a nearly world-wide distribution, on both Gondwana and Laurasia, in the Late Triassic and Early Jurassic, the occurrence of an indeterminate prosauropod on the Indochina block in the Rhaetian is not especially surprising or significant. Abundant prosauropod remains, belonging to several taxa, have been found in the Lower Lufeng Beds of Yunnan in southern China, but, after much controversy over their age (for a recent review, see Dong, 1992), the Lower Lufeng Beds are now usually dated as Lower Jurassic (on the basis of paly-

Table 1. Main Mesozoic vertebrate-bearing formations of northeastern Thailand with their possible ages and significant fossils

Khok Kruat Formation	Aptian-Albian	*Thaiodus, Psittacosaurus,* Iguanodontidae
Phu Phan Formation	Barremian?	Theropod footprints
Sao Khua Formation	Valanginian-Hauterivian?	*Siamotyrannus, Phuwiangosaurus,* Ornithomimosauria
Phra Wihan Formation	Berriasian?	Theropod footprints
Phu Kradung Formation	Late Jurassic	Temnospondyls, euhelopodid sauropods
Nam Phong Formation	Rhaetian	Prosauropods
Huai Hin Lat Formation	Norian	Lungfish, *Cyclotosaurus, Proganochelys,* phytosaurs

nomorphs and molluscs), although their lower part is sometimes referred to the Rhaetian. In this connection, it is worth mentioning that an especially large prosauropod from the lower part of the Lower Lufeng Beds has been described as *Jingshanosaurus xinwaensis* by Zhang and Yang (1994), who consider it to be Upper Triassic. Whether the apparently even larger prosauropod from Thailand was more or less closely related to *Jingshanosaurus* is uncertain on the basis of the available material.

Jurassic: vertebrates from the Phu Kradung Formation of the Khorat Plateau and similar formations of the Shan-Thai block

The Phu Kradung Formation of the Khorat Plateau was once dated as Lower Jurassic. Because all the overlying formations of the Khorat Group are now considered as Lower Cretaceous (Racey *et al.*, 1994, 1996), rather than Jurassic and Cretaceous as previously supposed, constraints on the age of the Phu Kradung Formation have changed. Palynological evidence from the Phu Kradung Formation is inconclusive. Racey *et al.* (1996) conclude that the Phu Kradung Formation is "probably Late Jurassic or Early Cretaceous in age", but there is no factual palynological evidence for that conclusion. On the basis of its vertebrate fauna, a Late Jurassic age seems likely.

One of the first vertebrates to be reported from the Phu Kradung Formation was the crocodilian *Sunosuchus thailandicus* (Buffetaut and Ingavat, 1980, 1984), which belongs to a genus previously known from the Upper Jurassic of Gansu, northwestern China. Besides supporting a Late Jurassic age for the Phu Kradung Formation, the occurrence of *Sunosuchus* can be considered as evidence for faunal links with North China in the Jurassic.

More recently, other potentially important, albeit fragmentary, vertebrate remains have been collected from the Phu Kradung Formation on the Khorat Plateau. Among them are the first dinosaur remains from the Jurassic of Thailand, which consist of isolated teeth of sauropods and theropods. The theropod teeth do not exhibit any particular features which could point to their biogeographical affinities. The sauropod teeth are broad and spoon-shaped, and completely different from those referred to *Phuwiangosaurus,* the most frequent sauropod in the Lower Cretaceous Sao Khua Formation of the Khorat Plateau (see below). They are more reminiscent of those of the euhelopodid dinosaurs from the Jurassic of China, such as *Euhelopus* (Wiman, 1929) or *Mamenchisaurus* (Russell and Zheng, 1993). Although the evidence is admittedly still slight, these teeth suggest that the Jurassic sauropods of SE Asia belonged to the same group as the Chinese ones.

Other interesting vertebrate fossils from the Phu Kradung Formation are vertebral elements of temnospondyl amphibians (Buffetaut, Tong and Suteethorn., 1994). Temnospondyls were long believed to have become extinct at the end of the Triassic, until they were discovered in the Jurassic of Asia (China, Mongolia) and Australia, and in the Lower Cretaceous of Australia. Although the Thai temnospondyls cannot be identified with great accuracy on the basis of the available material, their occurrence is probably further evidence of faunal links with more northerly parts of Asia in the Jurassic, since there is no evidence of faunal links between the Indochina and Shan-Thai blocks and Australia in the Mesozoic. The SE Asian temnospondyls could also be considered as relics of a distribution predating the collision of the SE Asian blocks with mainland Asia.

Interestingly, temnospondyl vertebral elements have also been found in rocks of roughly the same age as the Phu Kradung Formation in western Thailand, *i.e.*, in areas which belong to the Shan-Thai block. One was found in northwestern Thailand, near Chiang Rai, in rocks very similar in lithology to the Phu Kradung Formation. Several others were found at Mab Ching, in the southern peninsula, in lacustrine beds which have been dated as Middle Jurassic on the basis of charophytes (Buffetaut, Raksaskulwong, Suteethorn and Tong, 1994). Although, again, the available material is too scanty to warrant detailed comparisons and correlations, the widespread occurrence of temnospondyls in Jurassic rocks in various parts of Thailand suggests that similar faunas were present on the Indochina block and on the Shan-Thai block in the Jurassic, which in turn suggests that these terranes were already in contact. Alternatively, these vertebrae could be considered as unrelated fragments of a former vast distribution of temnospondyls, encompassing both Australia and Asia. More complete material will be needed before a definite choice can be made between these competing hypotheses, but current geodynamic reconstructions seem to favour the former.

The Mab Ching locality has also yielded good turtle material (Buffetaut, Tong, Suteethorn and Raksaskulwong, 1994; Tong *et al.*, 1996). The Mab Ching turtles may be related to the Chengyuchelyidae, a family known from the Jurassic of China (Sichuan and Xinjiang), which would indicate biogeographical links between the Shan-Thai block and the Chinese blocks in the Jurassic.

Lower Cretaceous: vertebrates from the Sao Khua Formation

Vertebrate fossils from the Phra Wihan Formation, which overlies the Phu Kradung Formation on the Khorat Plateau and is dated as Lower Cretaceous (Berriasian to Barremian) on the basis of palynomorphs (Racey *et al.*, 1994, 1996), are restricted to dinosaur footprints, which have no clear biogeographical implications.

The Sao Khua Formation, which overlies the Phra Wihan Formation, consists of red clays, sandstones and conglomerates, indicating deposition in a floodplain with meandering rivers. Although it was long considered as Upper Jurassic, its stratigraphic position above the palynologically dated Phra Wihan Formation shows that it is at least Lower Cretaceous. Since the Aptian-Albian Khok Kruat Formation overlies the Phu Phan Formation, which itself overlies the Sao Khua Formation (see below), the Sao Khua Formation must be pre-Aptian and Lower Cretaceous.

The Sao Khua Formation has yielded the richest and most diverse Mesozoic vertebrate assemblage hitherto found in Thailand. It contains freshwater hybodont sharks, actinopterygian fishes, turtles, crocodilians, and dinosaurs (theropods and sauropods).

The only crocodilian taxon so far described from the Sao Khua Formation is *Goniopholis phuwiangensis*, based on a lower jaw fragment (Buffetaut and Ingavat, 1983). A relatively complete skeleton, possibly of the same form, is currently under study, and apparently does not belong to *Goniopholis*. An assessment of the real biogeographical significance of the Sao Khua crocodilians will be possible only after the new skeleton has been fully described.

Some of the theropod dinosaurs from the Sao Khua Formation have interesting biogeographical implications. *Siamotyrannus isanensis*, described on the basis of a pelvis and part of the vertebral column (Buffetaut *et al.*, 1996), is considered as the oldest and most primitive known tyrannosaurid. The Tyrannosauridae are otherwise known from the Upper Cretaceous of Asia and North America. No well-ascertained Lower Cretaceous tyrannosaurids have so far been recorded outside Thailand. The occurrence of *Siamotyrannus* in the Sao Khua Formation suggests that the family Tyrannosauridae probably originated in Asia, and later spread to North America (via Beringia), possibly at the end of the Early Cretaceous, during a phase of faunal interchange which has already been postulated by Russell (1993).

The Sao Khua Formation has also yielded remains of a still unnamed new ornithomimosaur. This small 'ostrich-dinosaur' exhibits an interesting foot structure, with a proximally 'pinched' third metatarsal which is more advanced than those of the primitive ornithomimosaurs *Harpymimus okladnikovi* (from the Aptian-Albian of Mongolia) and *Garudimimus brevipes* (from the Cenomanian-Turonian of Mongolia), in which the third metatarsal is less compressed proximally. The Thai ornithomimosaur is however slightly less advanced than *Archaeornithomimus asiaticus* (from the Cenomanian of Inner Mongolia), in that the proximal end of the third metatarsal is still visible, as a thin sliver of bone, between the second and fourth metatarsals, whereas in the Mongolian form it is hidden by the anterior contact between the second and fourth metatarsals. This new ornithomimosaur suggests both that the group started to diversify early in the Cretaceous, and that advanced ornithomimosaurs may have originated in Asia. Their biogeographical history may have been similar to that of tyrannosaurids, since their Late

Cretaceous distribution also included Asia and North America (reports of Upper Cretaceous ornithomimosaurs from other continents are debatable).

Sauropod dinosaurs are particularly abundant in the Sao Khua Formation. The best known of them is *Phuwiangosaurus sirindhornae*, described on the basis of an incomplete skeleton (Martin *et al.*, 1994). More complete specimens have now been discovered and provide some additional evidence about this form. *Phuwiangosaurus* is clearly different from the Euhelopodidae, a group of sauropods including such genera as *Euhelopus*, *Mamenchisaurus* and *Omeisaurus*, which was widespread in China during the Jurassic. Its narrow, lanceolate teeth (Suteethorn *et al.*, 1995) are unlike the broad, spoon-shaped teeth of the Euhelopodidae (and unlike the above-mentioned teeth from the Phu Kradung Formation). The post-cranial skeleton also shows many differences (Martin *et al.*, 1994), in particular in the shape of the cervical vertebrae, which are broad and dorsoventrally compressed, whereas those of the Euhelopodidae are narrow and laterally compressed. The neural spines of the posterior cervical vertebrae of *Phuwiangosaurus* are deeply bifurcated, whereas those of the euhelopodids only show a shallow bifurcation. There is therefore no evidence of close links between this Early Cretaceous form and the older Asian sauropods. *Phuwiangosaurus* shows a combination of characters which separate it from most previously described families of sauropods. In the shape of its teeth, however, it very closely resembles *Nemegtosaurus mongoliensis*, from the Upper Cretaceous of the Gobi desert (Nowinski, 1971), which is sometimes placed in a family of its own, the Nemegtosauridae (Upchurch, 1994). However, *Nemegtosaurus* is known only by its skull, while *Phuwiangosaurus* is represented mainly by postcranial elements, so that comparisons are difficult. Further evidence concerning the possible affinities of *Phuwiangosaurus* may be afforded by a recently reported sauropod skeleton from the Upper Cretaceous of Shanxi, China (Pang *et al.*, 1995), which is said to have *Nemegtosaurus*-like teeth, but detailed comparisons will have to await a more complete description of this Chinese form. Although admittedly the position of *Phuwiangosaurus* in sauropod evolution is still unclear, it may be close to the origin of at least some of the Upper Cretaceous sauropods of Asia, which would point to a biogeographical history similar to the above-mentioned tyrannosaurids and ornithomimosaurs, with the exception that Asian sauropods apparently did not disperse to North America in the Late Cretaceous.

Upper Lower Cretaceous: vertebrates from the Khok Kruat Formation

The only vertebrate fossils from the Phu Phan Formation, which overlies the Sao Khua Formation, are dinosaur footprints which do not provide any biogeographical information. The overlying Khok Kruat Formation has, however, yielded remains of various vertebrates which can be compared with forms from other parts of the world.

The Khok Kruat Formation consists of red clays, sandstones and conglomerates, and its depositional environment is supposed to have been similar to that of the Sao Khua Formation. The Khok Kruat Formation is relatively precisely dated. The freshwater hybodont shark *Thaiodus ruchae* is known only from the Khok Kruat Formation and from the Takena Formation of the Lhasa block of Tibet, which is dated as Aptian-Albian on the basis of foraminifera (Cappetta *et al.*, 1990), thus suggesting a similar age for the Khok Kruat Formation. This is consistent with the occurrence of Albian-Cenomanian palynomorphs in the overlying Maha Sarakham Formation (Sattayarak *et al.*, 1991).

The vertebrate fauna from the Khok Kruat Formation includes hybodont sharks, turtles, crocodilians, theropods, ceratopsians and ornithopods. As mentioned above, the hybodont shark *Thaiodus* indicates faunal resemblances with the Lhasa block of Tibet, which is supposed to have collided with mainland Asia in the Late Jurassic. It is likely that *Thaiodus*, being known both in Tibet and in SE Asia, had a larger distribution in Asia during the late Early Cretaceous.

The remains of turtles, crocodilians and theropods so far found in the Khok Kruat Formation are too fragmentary to warrant a biogeographical interpretation. Remains of ornithischians are more significant in this respect, all the more so that, curiously enough, no ornithischian remains have so far been found among the abundant dinosaur material from the Sao Khua Formation.

The first ornithischian dinosaur to have been reported from the Khok Kruat Formation is the small ceratopsian *Psittacosaurus* (Buffetaut *et al.*, 1989), represented by jaws for which the species *P. sattayaraki* was erected (Buffetaut and Suteethorn, 1992). The genus *Psittacosaurus* was previously known from a number of localities in northern Asia (Mongolia, Siberia, northwestern and northeastern China, and possibly Japan). The discovery of *Psittacosaurus* some 3000 km south of the Chinese localities has caused some surprise among palaeontologists (Dodson, 1996), because the so-called *Psittacosaurus* fauna was generally held to be characteristic of a northern Asiatic faunal province (Dong, 1993), but it is not surprising from a stratigraphic point of view, since *Psittacosaurus* is a common dinosaur in the Lower Cretaceous of Asia (Dong, 1992; Jerzyckiewicz and Russell, 1991), and the apparently considerable geographical distance between the Thai and northern Asian localities mainly shows how little is known about the Lower Cretaceous dinosaurs of southern China (Dong, 1993). It is likely that further research on the Khok Kruat Formation will increase the number of Lower Cretaceous vertebrate taxa common to SE Asia and northern Asia.

This has been confirmed to some extent by the recent discovery in the Khok Kruat Formation of iguanodontid remains. The material is still scanty, consisting of a caudal vertebra and a few shed and worn teeth. The latter, however, are quite typical of iguanodontids. The Iguanodontidae had a wide geographical distribution in the Early Cretaceous, both in Laurasia and Gondwana. However, their history in Asia seems to be relatively complex. In the Gobi basin, according to Jerzykiewicz and Russell (1991), iguanodontids do not appear until their Khukhtekian 'Age', which they consider as upper Aptian to lower Albian, at a time when faunal interchange between 'Central Asia' (which probably meant most of central and eastern Asia) and both North America and Europe became possible again after a fairly long period of isolation of Asia (Russell, 1993). The Thai record is in general agreement with what has been observed in the Gobi basin: the Sao Khua faunal assemblage, which is pre-Aptian, contains no ornithopods, whereas the Aptian-Albian Khok Kruat assemblage contains iguanodontids. However, the Japanese record suggests a somewhat different chronological pattern. There, iguanodontids are present in the Tetori Group (Manabe and Hasegawa, 1991), which is Neocomian. Iguanodontid teeth resembling those from Thailand have been described (Hasegawa *et al.*, 1995) from the upper part of the Itoshiro Subgroup, which is overlain by the Kitadani Formation, considered as upper Barremian to lower Aptian on the basis of the non-marine bivalve *Nippononaia ryosekiana* (Isaji, 1993). Although there are uncertainties concerning the exact age of some of the Lower Cretaceous non-marine formations of Japan, the occurrence of marine intercalations makes correlations easier than in the Gobi basin or northeastern Thailand. In addition, iguanodontid teeth are known in Japan from a horizon lying below a volcanic tuff which has been dated (by fission-track studies) as 135 ± 7 Ma BP, which would indicate a Valanginian/Hauterivian age (S. E. Evans, pers. comm., 1997). If the dating of the Japanese iguanodontids is correct, it would appear that this group of dinosaurs reached eastern Asia well before the Aptian, in which case their absence in the 'Shinkhudukian' localities of the Gobi basin, which Jerzykiewicz and Russell (1991) consider as Aptian, and in the Sao Khua Formation of Thailand, is surprising. According to Manabe and Hasegawa (1991) the Iguanodontidae migrated to East Asia before the Berriasian. The Thai and Mongolian records suggest a later immigration, because iguanodontids were apparently not present in those regions at the very beginning of the Cretaceous. Obviously, uncertainties remain about the date of arrival of iguanodontids in various parts of Asia.

In connection with the discussion of the fauna from the Khok Kruat Formation, it should be mentioned that the dinosaur localities of southern Laos first reported by Hoffet (1936, 1942, 1944) are probably in an equivalent of the Khok Kruat Formation on the east side of the Mekong, and are not Upper Cretaceous, contrary to Hoffet's original interpretation (Buffetaut, 1991). The ornithopods ascribed by Hoffet (1944) to the Upper Cretaceous genus *Mandschurosaurus* are in fact more primitive (Buffetaut, 1991), and probably iguanodontids (Taquet *et al..*, 1995) and the sauropod, contrary to Hoffet's opinion, is not a titanosaurid (Buffetaut, 1991; Taquet *et al.*, 1995). This poorly known Lao assemblage seems to suport the faunal history outlined above, in which iguanodontids did not reach SE Asia until relatively late in the Early Cretaceous.

Conclusion

The Thai record of non-marine Mesozoic vertebrates is, at least stratigraphically, one of the best in Asia, covering as it does a long time interval extending from the Late Triassic to the end of the Early Cretaceous. The main biogeographical conclusions to be drawn from that record can be summarised as follows:

In the Late Triassic, links between the vertebrate fauna of the Indochina Block and the western part of Eurasia are clearly apparent. Because of the dearth of Upper Triassic vertebrate faunas from China, dispersal routes across Eurasia are not easily reconstructed, but they must have existed.

Resemblances between the Jurassic assemblages from the Indochina Block and the Shan-Thai Block strongly suggest that these terranes were already in contact at that time. The SE Asian assemblage shows some resemblances to Chinese Jurassic faunas.

The Lower Cretaceous fauna from the Sao Khua Formation represents a period which is otherwise very poorly known in Asia. It seems to correspond to the late stage of a period of isolation of Asia, and includes forms which may be ancestral to (or close to the ancestry of) dinosaur groups which became prominent in later Cretaceous assemblages both in Asia and North America (in particular, tyrannosaurids and ornithomimids). The sauropod *Phuwiangosaurus* may be related to some of the poorly known Late Cretaceous sauropods of Asia (such as *Nemegtosaurus*).

The upper Lower Cretaceous (Aptian-Albian) assemblage from the Khok Kruat Formation is characterised by the presence of *Psittacosaurus*, a dinosaur which is widespread in Asia at that time, and iguanodontids. The latter seem to have been absent in older Cretaceous formations such as the Sao Khua Formation, and they probably indicate dispersal from Europe after the isolation of Asia ceased some time during the Early Cretaceous.

Acknowledgements

This work was supported by the Department of Mineral Resources, Bangkok, and by the French Ministry of Foreign Affairs (Mission paléontologique française en Thaïlande). We thank all the Thai and French participants who took part in our field work in Thailand, as well as all the colleagues in various institutions who made specimens in their care available for comparison. Thanks to Anne-Marie Lézine and Sylvain Duffaud for help in the preparation of the map.

References

Broin, F. de 1984. *Proganochelys ruchae* n.sp., Chélonien du Trias supérieur de Thaïlande. Studia Palaeocheloniologica 1: 87-97.

Broin, F. de, Ingavat, R., Janvier, P. and Sattayarak, N. 1982. Triassic turtle remains from northeastern Thailand. Journal of Vertebrate Paleontology 2: 41-46.

Buffetaut, E. 1991. On the age of the Cretaceous dinosaur-bearing beds of southern Laos. Newsletters on Stratigraphy 24: 59-73.

Buffetaut, E. and Ingavat, R. 1980. A new crocodilian from the Jurassic of Thailand, *Sunosuchus thailandicus* n.sp. (Mesosuchia, Goniopholididae), and the palaeogeographical history of Southeast Asia in the Mesozoic. Geobios 13: 879-889.

Buffetaut, E. and Ingavat, R. 1982. Phytosaur remains (Reptilia, Thecodontia) from the Upper Triassic of northeastern Thailand. Geobios 15: 7-17.

Buffetaut, E. and Ingavat, R. 1983. *Goniopholis phuwiangensis* nov.sp., a new mesosuchian crocodile from the Jurassic of northeastern Thailand. Geobios 16: 79-91.

Buffetaut, E. and Ingavat, R. 1984. The lower jaw of *Sunosuchus thailandicus*, a mesosuchian crocodilian from the Jurassic of Thailand. Palaeontology 27: 199-206.

Buffetaut, E. and Suteethorn, V. 1992. A new species of the ornithischian dinosaur *Psittacosaurus* from the Early Cretaceous of Thailand. Palaeontology 35: 801-812.

Buffetaut, E., Raksaskulwong, L., Suteethorn, V. and Tong, H. 1994. First post-Triassic temnospondyl amphibians from the Shan-Thai block: intercentra from the Jurassic of peninsular Thailand. Geological Magazine 131: 837-839.

Buffetaut, E., Sattayarak, N. and Suteethorn, V. 1989. A psittacosaurid dinosaur from the Cretaceous of Thailand and its implications for the palaeobiogeographical history of Asia. Terra Nova 1: 370-373.

Buffetaut, E., Suteethorn, V., Martin, V., Chaimanee, Y. and Tong-Buffetaut, H. 1993. Biostratigraphy of the Mesozoic Khorat Group of northeastern Thailand: the contribution of vertebrate palaeontology. In Proceedings of the International Symposium on Biostratigraphy of Mainland Southeast Asia: Facies and Paleontology, pp. 51-62. Edited by T. Thanasuthipitak. Department of Geological Sciences, University of Chiang Mai.

Buffetaut, E., Suteethorn, V., Martin, V., Tong, H., Chaimanee, Y. and Triamwichanon, S. 1995. New dinosaur discoveries in Thailand. In Proceedings of the International Conference on Geology, Geotechnology and Mineral Resources of Indochina, pp. 157-161. Edited by L. Wannakao. Department of Geotechnology, Khon Kaen University.

Buffetaut, E., Suteethorn, V. and Tong, H. 1996. The earliest known tyrannosaur from the Lower Cretaceous of Thailand. Nature 381: 689-691

Buffetaut, E., Tong, H. and Suteethorn, V. 1994. First post-Triassic labyrinthodont amphibian in South-East Asia: a temnospondyl intercentrum from the Jurassic of Thailand. Neues Jahrbuch für Geologie und Paläontologie Monatshefte 7: 385-390.

Buffetaut, E., Tong, H., Suteethorn, V. and Raksaskulwong, L. 1994. Jurassic vertebrates from the southern peninsula of Thailand and their implications. A preliminary report. In Proceedings of the International Symposium on Stratigraphic Correlation of Southeast Asia, pp. 253-256. Edited by P. Angsuwathana, T. Wongwanich, W. Tansathien, S. Wongsomak, S. and J. Tulyatid. Department of Mineral Resources, Bangkok.

Cappetta, H., Buffetaut, E. and Suteethorn, V. 1990. A new hybodont shark from the Lower Cretaceous of Thailand. Neues Jahrbuch für Geologie und Paläontologie Monatshefte 11: 659-66.

Dodson, P. 1996. The horned dinosaurs. Princeton University Press, Princeton.

Dong, Z. 1992. Dinosaurian faunas of China. China Ocean Press, Beijing, and Springer Verlag, Berlin.

Dong, Z. 1993. Early Cretaceous dinosaur faunas in China: an introduction. Canadian Journal of Earth Sciences 30: 2096-2100.

Haile, N. S. 1973. Note on Triassic fossil pollen from the Nam Pha Formation, Chulabhorn (Nam Pha) dam, Thailand. Geological Society of Thailand Newsletter 6: 15-16.

Hasegawa, Y., Manabe, M., Isaji, S., Ohkura, M., Shibata, I. and Yamaguchi, I. 1995. Terminally resorbed iguanodontid teeth from the Neocomian Tetori Group, Ishikawa and Gifu Prefecture, Japan. Bulletin of the National Science Museum C 21: 35-49.

Hoffet, J. H. 1936. Découverte du Crétacé en Indochine. Comptes Rendus de l'Académie des Sciences de Paris 202: 1867-1868.

Hoffet, J. H. 1942. Description de quelques ossements de titanosauriens du Sénonien du Bas-Laos. Comptes Rendus des Séances du Conseil des Recherches Scientifiques de l'Indochine, 51-57.

Hoffet, J. H. 1944. Description des ossements les plus caractéristiques appartenant à des Avipelviens du Sénonien du Bas-Laos. Bulletin du Conseil des Recherches Scientifiques de l'Indochine, 179-186.

Ingavat, R. and Janvier, P. 1981. *Cyclotosaurus* cf. *posthumus* Fraas (Capitosauridae, Stereospondyli) from the Huai Hin Lat Formation (Upper Triassic), northeastern Thailand, with a note on capitosaurid biogeography. Geobios 14: 711-725.

Isaji, S. 1993. *Nippononaia ryosekiana* (Bivalvia, Mollusca) from the Tetori Group in central Japan. Bulletin of the National Science Museum C19: 65-71.

Jerzykiewicz, T. and Russell, D. A. 1991. Late Mesozoic stratigraphy and vertebrates of the Gobi Basin. Cretaceous Research 12: 345-377.

Kobayashi, T. 1975. Upper Triassic estheriids in Thailand and the conchostracan development in Asia in the Mesozoic era. Geology and Palaeontology of Southeast Asia 16: 57-90.

Konno, E. and Asama, K. 1973. Mesozoic plants from Khorat, Thailand. Geology and Palaeontology of Southeast Asia 12: 149-171.

Manabe, M. and Hasegawa, Y. 1991. The Cretaceous dinosaur fauna of Japan. In Fifth Symposium on Mesozoic Terrestrial Ecosystems and Biota. Extended Abstracts, pp. 41-42. Edited by Z. Kielan-Jaworowska, N. Heintz, N. and H. A. Nakrem Contributions from the Paleontological Museum, University of Oslo, 364.

Martin, M. 1984.The actinopterygian scales and teeth (Pisces) from the continental Upper Triassic of Thailand, their paleogeographical significance. Mémoires de la Société géologique de France 147: 101-105.

Martin, M. and Ingavat, R. 1982. First record of an Upper Triassic ceratodontid (Dipnoi, Ceratodontiformes) in Thailand and its paleogeographical significance. Geobios 15: 111-114.

Martin, V., Buffetaut, E. and Suteethorn, V. 1994. A new genus of sauropod dinosaur from the Sao Khua Formation (Late Jurassic or Early Cretaceous) of northeastern Thailand . Comptes Rendus de l'Académie des Sciences de Paris II, 319: 1085-1092.

Metcalfe, I. 1996. Pre-Cretaceous evolution of SE Asian terranes. In Tectonic evolution of SE Asia. pp. 97-122. Edited by R. Hall and D. J. Blundell. Geological Society of London Special Publication 106.

Nowinski, A. 1971. *Nemegtosaurus mongoliensis* n.gen., n.sp. (Sauropoda) from the uppermost Cretaceous of Mongolia. Palaeontologia Polonica, 25: 58-81.

Pang, Q., Cheng, Z., Yang, J., Xie, M., Zhu, C. and Luo, J. 1995. The principal characters and discussion on its ages of dinosaur fauna in Tianzhen, Shanxi, China. Journal of Hebei College of Geology 18, supplement: 1-6.

Racey, A., Goodall, J. G. S., Love, M. A., Polachan, S. and Jones, P. D. 1994. New age data for the Mesozoic Khorat Group of Northeast Thailand. In Proceedings of the International Symposium on Stratigraphic Correlation of Southeast Asia, pp. 245-252. Edited by P. Angsuwathana, T. Wongwanich, W. Tansathien, S. Wongsomak, S. and J. Tulyatid. Department of Mineral Resources, Bangkok.

Racey, A., Love, M. A., Canham, A. C., Goodall, J. G. S., Polachan, S. and Jones, P. D. 1996. Stratigraphy and reservoir potential of the Mesozoic Khorat Group, NE Thailand. Part 1: Stratigraphy and sedimentary evolution. Journal of Petroleum Geology 19: 5-40.

Russell, D. A. 1993. The role of Central Asia in dinosaurian biogeography. Canadian Journal of Earth Sciences 30: 2002-2012.

Russell, D. A. and Zheng, Z. 1993. A large mamenchisaurid from the Junggar Basin, Xinjiang, People's Republic of China. Canadian Journal of Earth Sciences 30: 2082-2095.

Sattayarak, N., Polachan, S. and Charusirisawad, R. 1991. Cretaceous rocksalt in the northeastern part of Thailand. GEOSEA VII Abstracts, Geological Society of Thailand, Bangkok: 36.

Stokes, R. B., Lovatt, P. F. and Soumphonphakdy, K. 1996. Timing of the Shan-Thai-Indochina collision: new evidence from the Pak Lay foldbelt of the Lao PDR. In Tectonic evolution of SE Asia. pp. 225-232. Edited by R. Hall and D. J. Blundell. Geological Society of London Special Publication 106.

Sun, A. 1989. Before dinosaurs. China Ocean Press, Beijing.

Suteethorn, V., Buffetaut, E., Martin, V., Chaimanee, Y., Tong, H. and Triamwichanon, S. 1995. Thai dinosaurs: an updated review. In Sixth Symposium on Mesozoic Terrestrial Ecosystems and Biota, pp. 133-136.

Suteethorn, V., Janvier, P. and Morales, M. 1988. Evidence for a plagiosauroid amphibian in the Upper Triassic Huai Hin Lat Formation of Thailand. Journal of Southeast Asian Earth Sciences 2: 185-187.

Suteethorn, V., Martin, V., Buffetaut, E., Triamwichanon, S. and Chaimanee, Y. 1995. A new dinosaur locality in the Lower Cretaceous of northeastern Thailand. Comptes Rendus de l'Académie des Sciences de Paris, IIa 321: 1041-1047.

Taquet, P., Battail, B., Dejax, J., Richir, P., Sayarath, P. and Veran, M. 1995. First discovery of dinosaur footprints and new discoveries of dinosaur bones in the Lower Cretaceous of the Savannakhet Province, Laos. In Proceedings of the International Symposium on the Geology of SE Asia and adjacent areas, Hanoi: 167.

Tong, H., Buffetaut, E. and Suteethorn, V. 1996. Jurassic turtles from southern Thailand. Journal of Vertebrate Paleontology 16: supplement to 3, 48A.

Upchurch, P. 1994. Sauropod phylogeny and palaeoecology. Gaia 10: 249-260.

Wiman, C. 1929. Die Kreide-Dinosaurier aus Shantung. Palaeontologia Sinica C6, 1: 1-67.

Zhang, Y. and Yang, Z. 1994. A new complete osteology of Prosauropoda in Lufeng Basin, Yunnan, China. *Jingshanosaurus*. Yunnan Publishing House of Science and Technology, Kunming (in Chinese with English abstract).

The distribution of *Apsilochorema* Ulmer, 1907: biogeographic evidence for the Mesozoic accretion of a Gondwana microcontinent to Laurasia

Wolfram Mey
Museum für Naturkunde, Institut für Systematische Zoologie, Humboldt-Universität Berlin, Invalidenstrasse 43, D - 10115 Berlin, Germany

Key words: historical biogeography, Asia, Australia, Gondwana, zoogeography, aquatic insects, Trichoptera, *Apsilochorema*

Abstract

Apsilochorema is the only genus of the southern hemisphere family Hydrobiosidae that occurs on the Asian continent. The origin of the genus is obscure and enigmatic. A phylogenetic analysis of the species reveals the existence of 2 different subgenera: *Archichorema* and *Apsilochorema sensu stricto*. They show different distributional patterns on the Asian continent, and in the SW Pacific region including Australia. A hypothesis is proposed to explain the distribution based on the plate tectonic history of SE Asia. Ancestors of *Apsilochorema* reached Asia by rafting on terranes of Gondwana origin. This crossing of the Tethys ocean happened at least twice. The species migrated into Asia on the Lhasa and West Burma terrane in the Cretaceous (*Archichorema*) and on the Indian plate in the Tertiary (*Apsilochorema sensu stricto*).

Introduction

The distribution of the caddisfly family Hydrobiosidae has several times attracted the attention of Trichoptera specialists. One reason for this interest is the conspicuous distribution pattern (Fig.1) of the family, which comprises 47 genera, with a disjunction between South America and Australia. Only the genus *Apsilochorema* Ulmer occurs on the Asian continent. Secondly, the Hydrobiosidae are regarded as the sister group of the Rhyacophilidae (Frania and Wiggins, 1997), the most ancestral family of extant Trichoptera, which has a pronounced northern hemisphere distribution in contrast to the Hydrobiosidae (Fig.2). Within Hydrobiosidae *Apsilochorema* has a special position. It is regarded as representing the oldest evolutionary lineage of the family (Neboiss, 1962; Ross, 1951, 1956, 1967). The age of the family has been estimated as approximately 100 million years (Ross, 1951) meaning an early Cretaceous appearance. This assumption is corroborated by the discovery of a fossil forewing in Upper Cretaceous amber of North Siberia. It was assigned to Hydrobiosidae and described as *Palaeohydrobiosis simberambra* (Botosaneanu and Wichard, 1983). However, the family must be even older. The fossil record of the order indicates that the first true caddisflies appeared in the Permian (*e.g.*, Ivanov, 1988), and in the Triassic the families Philopotamidae and Hydropsychidae were already in existence (Sukatcheva, 1991). According to proposed phylogenies of the Trichoptera (Frania and Wiggins, 1997; Ross, 1956, 1967; Schmid, 1989; Weaver, 1984; Wiggins and Wichard, 1989) these families are younger clades than the Hydrobiosidae. Thus, the Hydrobiosidae should be at least Triassic in age. Therefore the family is old enough to have been in existence during the breakup of Pangaea in the Mesozoic. Accordingly, the complicated history of Gondwana and its terranes in the SE Asian and Australian region in the Mesozoic and Tertiary (Audley-Charles, 1987; Metcalfe, 1996) should have affected the distribution of Hydrobiosidae.

Biology of *Apsilochorema*

The species are inhabitants of small rivers and brooks. In the north *A. sutshanum* Martynov lives in mountain streams as well as lowland streams. In SE and East Asia the genus is found

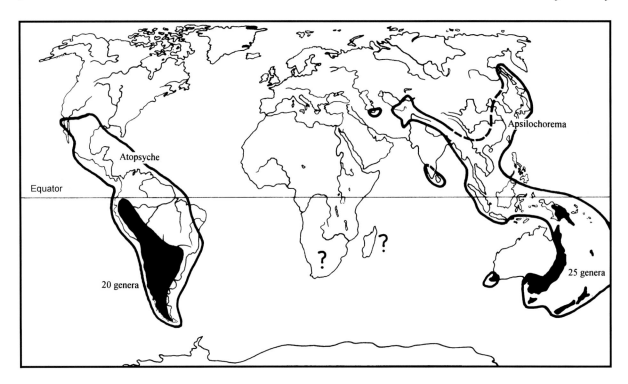

Fig. 1. Distribution area of Hydrobiosidae. The main or core areas in South America and Australia are filled with black, and their respective numbers of genera are indicated.

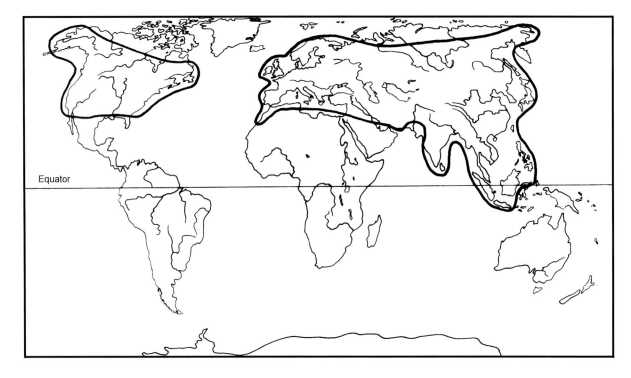

Fig. 2. Distribution area of Rhyacophilidae, the sister family of Hydrobiosidae.

in mountains at heights from 1500m to 4000m (Nepal). The larvae are free-living predators, with 5 instars (Kotcharina, 1986). Pupation takes place in a case, made of small stones and attached to larger stones in the current. The adults are active in the daytime. There are no pronounced flight periods.

The problem

The south hemisphere disjunction of the family Hydrobiosidae appears very pronounced, if the ranges of *Atopsyche* Banks and *Apsilochorema* Ulmer are omitted from the map of the distribution area of the family (Fig.1). The majority of genera is confined to certain parts of South America, Australia, New Guinea and the Southwest Pacific. In contrast, the areas of *Atopsyche* and *Apsilochorema* are large and seem to be simply northern extensions of the main range. If this were the case, we might expect the more primitive or ancestral forms to occur in the south and more derived ones in the north. Surprisingly, the situation in *Apsilochorema* is the reverse. The ancestral species live in the north. This exactly is the problem, because we have to find an interpretation for this mysterious pattern.

Up to now, two hypotheses have been proposed to explain the occurrence of *Apsilochorema* in Asia. Both asserted the ancestry of the Asian species. Ross (1951, 1956) assumed a SE Asian origin for the ancestor of Hydrobiosidae. According to this author the species spread into the southern continents via Sundaland or across an Alaska connection. After having reached Australia and South America, the ancestor started to differentiate. Some populations remained in Asia, became isolated and gave rise to the *Apsilochorema* line, which subsequently extended its range towards Australia. Neboiss (1962) agreed with Ross (1956) concerning the dispersal of *Apsilochorema* from the north into Australia. In contrast, Schmid (1989) proposed a Gondwana origin and Gondwana differentiation of the Hydrobiosidae. In coping with the problem of *Apsilochorema* he briefly discussed the possibility that the northward-moving Indian plate could have carried ancestral *Apsilochorema* species, which dispersed in Asia after the collision with Laurasia. However, he rejected this idea, because it implied a Jurassic or Cretaceous age of *Apsilochorema*. He was not prepared to accept such a great age. However, he admitted the differentiation of the genus had taken place in South Asia.

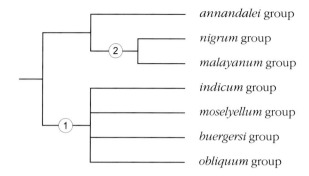

Fig.3. Cladogram of the phylogenetic relationships within *Apsilochorema* Ulmer. The numbers refer to suggested synapomorphies.

The arguments

These studies on *Apsilochorema* were started with an examination of the morphology of the species. The objective was to elucidate the phylogenetic relationships within the genus and to construct a cladogram. The characters were polarised based on comparison to all other genera of Hydrobiosidae as defined in Schmid (1989). The phylogenetic reconstruction is based on cladistic principles. The second issue was the documentation of the range of the genus and the production of a more detailed distribution map.

The phylogenetic relationships depicted in the cladogram of Fig.3 show a clear separation of the genus into 2 groups. For practical reasons they were given subgeneric rank and were named *Archichorema* and *Apsilochorema sensu stricto* (Mey, 1998). The morphological basis for this distinction is the presence or absence of a chorema on the underside of the male forewing (1 in Fig.3). The rather complex structure is illustrated in Figs. 4 and 5. It is regarded as a synapomorphy of *Apsilochorema sensu stricto*. Some additional 8 characters provide evidence for the further branching of the subgenera into 7 species groups. The character descriptions are listed in Mey (1998).

The mapping of the distribution of all *Apsilochorema* species revealed a wide range, extending from Tasmania in the south to the island of Sakhalin in the north, and from the Iranian Elburz Mountains in the west to the oceanic islands of Fiji in the east (Fig.6). However, the distribution is by no means homogeneous. It is a mixture of smaller and larger areas with continuous distribution and a number of widely scattered localities of a single species. Regions with

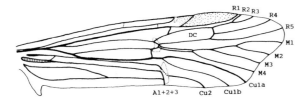

Archichorema annandalei

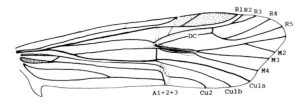

Apsilochorema indicum

Fig.4. Male forewings of *A. annandalei* Martynov (Himalaya) and *A. indicum* Ulmer (Tianshan). The chorema of *A. indicum* is situated below the discoidal cell (DC).

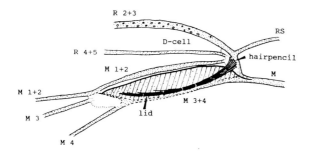

Fig.5. Structure of the chorema of *Apsilochorema indicum* Ulmer at the underside of male forewing.

the highest number of species are the southern slope of the Himalayas (8 species; Schmid, 1970), New Guinea (8 species; Schmid, 1989) and Borneo (5 species; Huisman, 1992).

After separating the distributional records according to the subgenera and plotting a map for each subgenus, the following picture emerges (Fig.7): *Archichorema* has a relatively compact range with its ancestral species in the Himalaya and Assam region. Surprisingly, the range of *Apsilochorema sensu stricto* shows a disjunction between India/Middle Asia and Sundaland/ Australia; there is a large gap in the distributions in SE Asia.

These are the new arguments. Besides the enigma of the ancestral species in South Asia we now have the additional problem of an explanation for the disjunction of the subgenus *Apsilochorema*. The progress of plate tectonic theory in the last decades has provided a firm geophysical framework against which the evolution and biogeography of *Apsilochorema* can be viewed (Hall and Blundell, 1996). Continental SE Asia is an assemblage of small and large terranes, which were rifted off the northern margin of Gondwanaland and were accreted to Laurasia and/or to Protochina. Similarly, a number of islands in the SW Pacific are continental terranes too, derived from Gondwana or Australia (Audley-Charles, 1987; Burrett *et al.*, 1991; Metcalfe, 1996). There are several models for how the SE Asian and SW Pacific region became assembled. They differ mainly from one another on the timing of rift, drift and collision histories of the terranes. However, all agree that the amalgamating process forming SE Asia was due to a series of arrivals of northward moving microcontinents of Gondwana origin. The earliest arrived blocks are of Palaeozoic age. Further accretion occurred during the Mesozoic and the Tertiary.

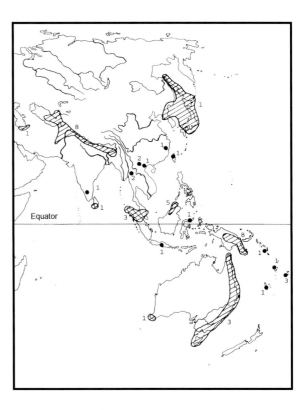

Fig.6. Distribution area of *Apsilochorema sensu lato*, Ulmer 1907. Hatched areas indicate continuous distributions. Numbers indicate the numbers of species in each area.

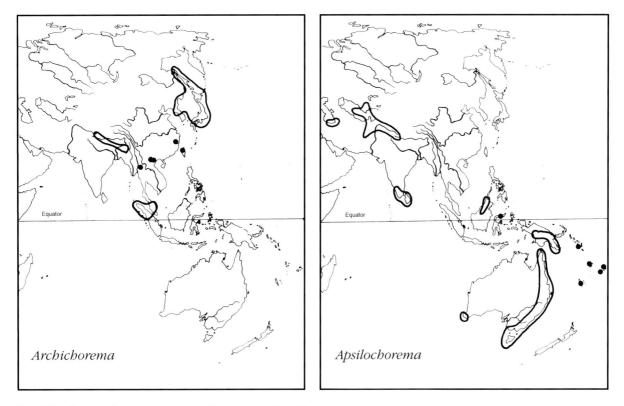

Fig. 7. Distribution of the subgenera *Archichorema* and *Apsilochorema sensu stricto.*

The model

Before the separation of India from Australia/Antarctica was accomplished in the late Jurassic (Veevers, 1991) a microcontinent or several continental fragments detached from the northern edge of eastern Gondwanaland and moved northwards. The rifting started in the late Triassic with the Lhasa block, followed by the West Burma and Woyla terranes (Metcalfe, 1996). They might have carried populations of the ancestor of *Apsilochorema*, which was possibly distributed in northeastern Gondwanaland in that time. Drifting with one of these microcontinents the species reached the Pan-Laurasian continent in the Cretaceous. This would be the most simple explanation for the occurrence of *Apsilochorema* in Asia. However, it does not provide an explanation for the disjunction of *Apsilochorema sensu stricto*. It is questionable if the differentiation of the species had occurred before or during the journey with the Gondwana terrane, or if it developed after the establishment of the species in SE Asia. Both possibilities appear to be plausible, but they offer an explanatory potential only for the splitting and spreading of *Archichorema*. The basal separation into the 2 subgenera must have happened in another way.

Returning to the rifting of terranes from the margin of northern Gondwana, the process can be regarded as removing only a part of the distributional area of the *Apsilochorema* ancestor (Fig. 8). The remaining part stayed on Gondwana. The subsequent isolation enabled the beginning of an independent evolution and divergence into *Archichorema* and *Apsilochorema sensu stricto*. In the Gondwana species the development of the chorema on the male forewings probably started. It could have been either an anagenetic or cladistic process. The species with this evolutionary novelty was distributed over the northern margin of Gondwana. In the late Jurassic/early Cretaceous a western part of the range was broken off with the Indian continent and started to move northwards. With the accretion of the Indian continent to Asia the second immigration of *Apsilochorema* species into Asia took place, this time by a species with a wing chorema, *i.e.*, a member of *Apsilochorema sensu stricto*. It remained isolated on the Indian subcontinent and the adjacent northwestern area. The species in eastern Gondwana were confined to the Australian continent and gave rise to the development of 2 species groups.

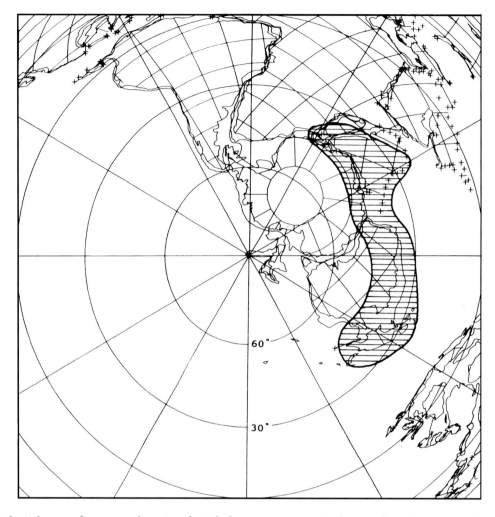

Fig. 8. Hypothetical range of an ancestral species of *Apsilochorema* in Eastern Gondwana. 140 Ma (Late Jurassic) map adapted from Smith *et al.* (1981).

This scenario gives an explanation for the stepwise formation of the distribution pattern of *Apsilochorema sensu lato*. In the light of this hypothesis the different dispersal routes of *Apsilochorema* species become recognisable. They are depicted in Fig. 9. According to this notion the species of *Apsilochorema sensu stricto* known from northern Borneo originated from Australian stock. Such a deep penetration of Australian elements into the Oriental region is rather unusual for the Trichoptera. However, the species in North Sulawesi is *A. gisbum* Mosely, which has a wide range in the Australian region including Tasmania. The high dispersal capacity of this species points possibly to an intrinsic character of the species of the subgenus. It enabled the colonisation of various islands in the SW Pacific (Fig. 6), as well as Borneo on the Sunda shelf. But in contrast to this dispersal-based interpretation it is possible to suggest an alternative explanation. The Gondwana terranes which have been recognised in Sundaland offered the same potential and possibility for transport of *Apsilochorema* species in the past. Unfortunately, faunistic research in the region is at quite a low level and does not permit the distinction between a mobilistic or vicariant explanation in the case of the Borneo species. The question is left open for future research.

Conclusions

The proposed hypothesis allows some predictions. The most interesting one is the possible occurrence of *Apsilochorema* in the southern part of Africa and especially on Madagascar. That island was very close to the Indian conti-

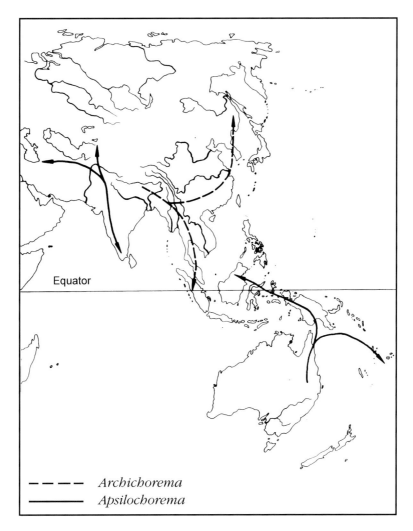

Fig.9. Presumed dispersal routes of *Apsilochorema sensu lato* in the SE Asian and SW Pacific/Australian region

nent and to western Gondwana at the time when the two plates separated. If an *Apsilochorema* species really survived, than it should belong to *Apsilochorema sensu stricto* and not to *Archichorema*.

The predatory larvae of *Apsilochorema* feed mainly on other aquatic insects (Kotcharina, 1986). Thus, the persistence of *Apsilochorema* requires a diverse and rich community of aquatic insects offering enough prey species in adequate quantities. The co-evolution of the aquatic biota should have resulted in a similar persistence of other species or groups, not only in *Apsilochorema*. In consequence, further congruent patterns of distribution should exist. The most promising insect groups, which could provide further examples, should be found within the ancient orders Ephemeroptera (Mayflies) and Plecoptera (Stoneflies).

In conclusion, if the hypothesis proves to be valid, we have an example of a very old genus with two subgenera of at least Cretaceous age. Further, the small morphological differences between nearly all *Apsilochorema* species points to a remarkable property of this evolutionary line: cohesion and stasis as successful mechanisms for survival through the Mesozoic and Cenozoic until today.

Acknowledgements

Special gratitude is extended to H. Duffels and R. Hall for their encouragement and help in preparing the manuscript. Thanks also to A. Neboiss, H. Malicky and L. P. Hsu for sending specimens of *Apsilochorema*. Most of the Asian material studied comes from field work supported by the Deutsche Forschungsgemeinschaft (Me 1085/1, Me 1085/3, Me 1085/5).

References

Audley-Charles, M. G. 1987. Dispersal of Gondwanaland: relevance to evolution of the angiosperms. *In* Biogeographical evolution of the Malay Archipelago. pp. 5-25. Edited by T. C. Whitmore.

Botosaneanu, L. and Wichard, W. 1983. Upper Cretaceous Siberian and Canadian amber caddisflies. Bijdr. Dierk. 53: 187-217.

Burrett, C., Duhig, N., Berry, R. and Varne, R. 1991. Asian and South-western Pacific continental terranes derived from Gondwana, and their biogeographic significance. Australian Systematic Botany 4: 13-24.

Frania, H. E. and Wiggins, G. B. 1997. Analysis of morphological and behavioural evidence for the phylogeny and higher classification of Trichoptera (Insecta). Royal Ontario Museum Life Sciences Contributions 160: 1-67.

Hall, R. and Blundell, D. J. 1996. Tectonic evolution of Southeast Asia. Geological Society Special Publication 106: 566 pp.

Huisman, J. 1992. New species of *Apsilochorema* (Trichoptera: Hydrobiosidae) from Sabah, East Malaysia. Zool. Med. Leiden 66: 127-137.

Ivanov, V. D. 1988. The structure of the Paleozoic caddisflies of the family Microptysmatidae (Insecta). Paleontological Journal 22: 63-69.

Kotcharina, S. 1986. The biology of three predacious caddisflies of the Maritime territory. Latvijas Entomologs 29: 134-143.

Metcalfe, I. 1996. Pre-Cretaceous evolution of SE Asian terranes. *In* Tectonic evolution of Southeast Asia. pp. 97-122. Edited by R. Hall and D. J. Blundell. Geological Society Special Publication 106.

Mey, W. 1998: Notes to the taxonomy of *Apsilochorema* Ulmer. (Trichoptera; Hydrobiosidae). Dtsch. entomol. Z., N.F. 45 (in press).

Neboiss, A. 1962. The Australian Hydrobiosinae (Trichoptera: Rhyacophilidae). Pacific Insects 4: 521-582.

Neboiss, A. 1984. Notes on New Guinea Hydrobiosidae (Trichoptera). Aquatic Insects 6: 177-184.

Ross, H. H. 1951. The origin and dispersal of a group of primitive caddisflies. Evolution 5: 102-115.

Ross, H. H. 1956. Evolution and classification of the mountain caddisflies. University of Illinois Press, Urbana, 211 pp.

Ross, H. H. 1967. The evolution and past dispersal of the Trichoptera. Annual Reviews of Entomology 12: 169-201.

Schmid, F. 1970. Sur quelques *Apsilochorema* orientaux (Trichoptera, Hydrobiosidae). Tijdschrift voor Entomologie 113: 261-271.

Schmid, F. 1989. Les Hydrobiosides (Trichoptera; Annulipalpia). Bull. Inst. royal Sci. nat. Belg., Entomol. 59, supplement: 154 pp.

Smith, A. G., Hurley, A. M. and Briden, J. C. 1981. Phanerozoic palaeocontinental world maps. Cambridge University Press, 102 pp.

Sukatcheva, I. D. 1991. Historical development of the order Trichoptera. *In* Proceedings of the 6th International Symposium on Trichoptera, Lodz-Zakopane 1989. Edited by C. Tomaszewski.

Wiggins, G. B. and Wichard, W. 1989. Phylogeny of pupation in Trichoptera, with proposals on the origin and higher classification of the order. J. N. Am. Benthol. Soc. 8: 260-276.

Veevers, J. J. 1991. Phanerozoic Australia in the changing configuration of Proto-Pangea through Gondwanaland and Pangea to the present dispersed continents. Australian Systematic Botany 4: 1-11.

Weaver, J. S. 1984. The evolution and classification of Trichoptera, Part I: The groundplan of Trichoptera. *In* Proceedings of the 4th International Symposium on Trichoptera, Clemson 1983. Edited by J. C. Morse.

The plate tectonics of Cenozoic SE Asia and the distribution of land and sea

Robert Hall
SE Asia Research Group, Department of Geology, Royal Holloway University of London, Egham, Surrey TW20 0EX, UK Email: robert.hall@gl.rhbnc.ac.uk

Key words: SE Asia, SW Pacific, plate tectonics, Cenozoic

Abstract

A plate tectonic model for the development of SE Asia and the SW Pacific during the Cenozoic is based on palaeomagnetic data, spreading histories of marginal basins deduced from ocean floor magnetic anomalies, and interpretation of geological data from the region. There are three important periods in regional development: at about 45 Ma, 25 Ma and 5 Ma. At these times plate boundaries and motions changed, probably as a result of major collision events.

In the Eocene the collision of India with Asia caused an influx of Gondwana plants and animals into Asia. Mountain building resulting from the collision led to major changes in habitats, climate, and drainage systems, and promoted dispersal from Gondwana via India into SE Asia as well as creating barriers between SE Asia and the rest of Asia. Continued indentation of Asia by India further modified Sundaland and created internal barriers affecting biogeographic patterns. From a biogeographic and tectonic viewpoint, the major Cenozoic tectonic event in SE Asia occurred about 25 million years ago, resulting in major changes in the configuration and character of plate boundaries, and caused effects which propagated westwards through the region. This event led to the progressive arrival of Australian microcontinental fragments in Sulawesi, providing possible pathways for migration of faunas and floras between Asia and Australia, but also creating new barriers to dispersal.

Tectonic reconstruction maps of lithospheric fragments cannot be translated simply into maps of land and sea which are of greater value to biogeographers. Determining the palaeogeography of the region is not yet possible, but an attempt is made to outline the main likely features of the geography of the region since the late Oligocene.

Evidence from all fields of biogeography is required to test different tectonic models and identify the origin of present biogeographic patterns but there must be a focus on plants and animals which have difficulty in dispersing, and for which non-geological controls are unimportant. The present distribution of plants and animals in SE Asia may owe much more to the last one million years than the preceding 30 million years.

Introduction

For the geologist, SE Asia is one of the most intriguing areas of the Earth. The mountains of the Alpine-Himalayan belt turn southwards into Indochina and terminate in a region of continental archipelagos, island arcs and small ocean basins. To the south, west and east the region is surrounded by island arcs where lithosphere of the Indian and Pacific oceans is being subducted at high rates, accompanied by intense seismicity and spectacular volcanic activity. Within this region we can observe collision between island arcs, between island arcs and continents, and between continental fragments. At the same time ocean basins are opening within this convergent region. SE Asia includes areas with the highest global rates of plate convergence and separation.

It is clear from the geology of the region that the snapshot we see today is no less complicated than in the past. The region has developed by the interaction of major lithospheric plates, principally those of the Pacific, India-Australia and Eurasia (Fig.1), but at the present day a description only in terms of these three plates is a very great oversimplification. Many minor plates need to be considered, and in some parts of the region the boundaries between these smaller plates are very uncertain. It is also clear that some of the deformation cannot be described in simple plate tectonic terms. Lithosphere has deformed internally, and material has been added by arc volcanic processes, which means that at least one of the axioms of

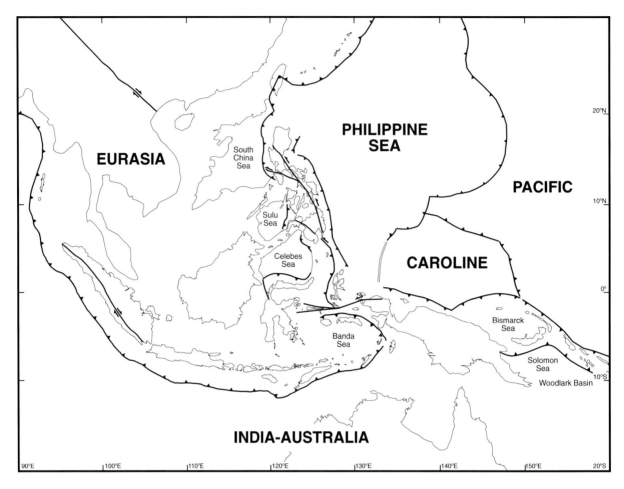

Fig.1. The larger tectonic plates of SE Asia and the SW Pacific. The many small ocean basins and the major strike-slip fault systems at the margins of SE Asia and Australia are manifestations of the complexity of plate tectonics in the region which requires a description in terms of many more plates than those shown.

plate tectonics, of rigid fragments moving on a sphere, cannot be assumed.

The complexity of the present-day tectonics of the region and the observable rates of plate motions (*e.g.*, Hamilton, 1979; McCaffrey, 1996) indicate that major oceans, or multiple small oceans, have closed during the Cenozoic. Several major island arcs have certainly formed during this time and some may have completely disappeared. At some plate boundaries strike-slip faulting has dismembered previously coherent regions, and along these boundaries there can be both major crustal subsidence and uplift due to deformation. During the past 50 million years the configuration of the region has therefore changed significantly in plate tectonic terms. Accompanying these large scale movements have been equally significant vertical movements, recorded in the sedimentary basins of the region, into which large volumes of sediments have been shed, removed from rising mountains. Thus the distribution of land and sea has changed during the Cenozoic, and many parts of the region have seen dramatic vertical movements of several kilometres, with mountains where once there were oceans, and deep marine regions where mountains had existed.

The abrupt division between Asian and Australian floras and faunas in Indonesia, first recognised by Wallace in the nineteenth century, has its origin in the rapid plate movements and reorganisation of land-masses in SE Asia. Wallace realised that the region had changed dramatically in the past without knowing the cause, and since his work there has been a general awareness that the present distribution of land and sea is not the same as that of the past, and that the changes are in some way implicated in biogeographic patterns. We are now confident that the geological changes are the result of plate movements and are not the consequence of an expanding Earth. However, although the very large-scale motions of major plates have been reasonably well known for the last 20 years or so, the detail necessary to reconstruct SE Asia has been lacking. Furthermore, because of the complexity of the geology of the region and because so much of it has been remote and difficult of access, geologists have not been able to clearly describe its long term development and it is only in the last few years that models explaining the development of the region have been produced.

Making tectonic reconstructions of SE Asia becomes more difficult as the age of the reconstruction becomes greater, and examination of the present tectonics of the region shows why this is so. Projecting motions that are known today into the past is very problematical; our observations of the present tectonics indicate that plates, plate boundaries and motions can be geologically ephemeral features. In some parts of the region, for example the Philippines and east Indonesia, it is not even certain that plate tectonics provides a suitable model for a detailed understanding of the development of the region. Despite these difficulties, a plate tectonic model does have value and by working back from the present-day we can, albeit with diffi-culty, make reconstructions; the known motions of major plates do impose limits on possibilities; and the resulting interpretations do offer a means of identifying important tectonic events and highlighting key problems. This paper explains the background to a plate tectonic model of SE Asia and the SW Pacific and discusses its implications for biogeography.

Mesozoic to Cenozoic background

In very general terms, the region owes its origin to the pre-Cenozoic break-up of the Gondwana super-continent (Fig.2), the subsequent movement of Gondwana fragments northwards, and their eventual collision with Eurasia. Metcalfe (1998 this volume) provides an account of present knowledge of the Palaeozoic and Mesozoic development of SE Asia. It is clear that many fragments separated from Gondwana and amalgamated in SE Asia over a considerable period of time. The process of rifting led to formation of new oceans, and the northward motion of Gondwana fragments required subduction of older oceanic crust at the edges of the growing Eurasian continent. By the Mesozoic, a region composed of fragments derived from Gondwana formed a Sundaland core surrounded by subduction zones.

Subduction meant that the Sundaland margins were complex. Island arcs at the margins may have been underlain by continental and oceanic crust, and there were probably many small ocean basins behind the arcs and above the subduction zones. The widespread ophiolites are fragments of oceanic lithosphere now found on land, and much of this lithosphere was formed in subduction-related settings, such as backarc basins and forearcs. Ophiolites are commonly emplaced at some stage during the convergence of two plates and convergence is ultimately completed by collision between arc and continent, or continent and continent. Throughout the Mesozoic there appear to have been collisions of fragments with the Sundaland margins, and by the beginning of the Cenozoic SE Asia was a composite mosaic of continental crust, island arc material and oceanic crust.

Two major fragments separated from Gondwana in the Cretaceous and moved northwards as parts of different plates: India and Australia. India completed its passage in the early Cenozoic and collided with the Asian continent about 50 million years ago (Fig.3). However, collision did not cause India to become fixed to Asia as

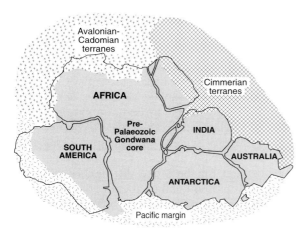

Fig.2. Reconstruction of Gondwana (after Unrug, 1997) in the early Palaeozoic showing outlines of the major continental fragments which separated during the Palaeozoic and Mesozoic. Many of the Cimmerian terranes had accreted to Laurasia to form the core of SE Asia by the end of the Mesozoic.

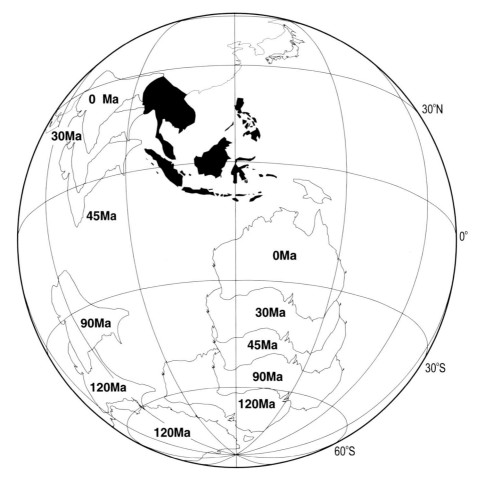

Fig.3. India and Australia separated from Gondwana in the Cretaceous. The map shows the late Cretaceous and Cenozoic movement of these two major continental fragments north with respect to Asia and SE Asia, both of which are shown in their present day positions for reference.

predicted by early plate tectonic models. Instead, India continued to move northwards, albeit at a slower rate than during the Cretaceous. There is considerable disagreement amongst geologists about how the continued northward movement was accommodated during the Cenozoic, and its consequences. According to Tapponnier and colleagues (*e.g.* Tapponnier *et al.*, 1982, 1986, 1990; Peltzer and Tapponnier, 1988; Briais *et al.*, 1993) the impact of a rigid Indian 'indentor' on an Asian margin weakened by subduction-related heating and magmatism caused eastward 'extrusion' and rotation of continental fragments, and opening of some of the small oceanic marginal basins of SE Asia. The progressive extrusion of continental fragments to the east and consequent rotation of crustal blocks has been simulated in laboratory experiments using plasticine (Fig.4), and the strike-slip faults which cut across Asia are considered to be zones of major displacements which link to marginal basins, such as the South China Sea, offshore. If correct, this hypothesis implies major changes in SE Asia linked to India's continued northward movement. In contrast, other workers (*e.g.* England and Houseman, 1986; Dewey *et al.* 1989; Houseman and England, 1993) dismiss the extrusion hypothesis, arguing that the displacement on the strike-slip faults has been small and that the continued convergence of India and Asia has been accommodated by crustal thickening with very little eastward movement of crust.

Australia separated from Gondwana, leaving Antarctica as its final remnant, at about the same time as India, but moved less quickly northwards. Instead of a direct collision with another continent Australia is now making a glancing

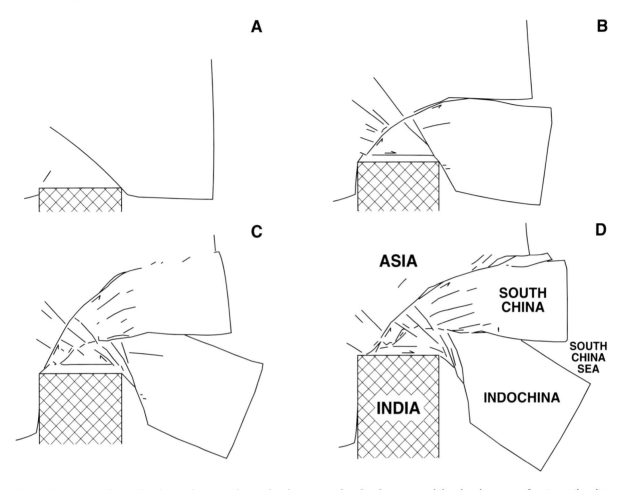

Fig. 4. The impact of a rigid Indian indentor with Asia has been postulated to have caused the development of major strike-slip faults as India progressively penetrated Asia (A to D). Analogue models produced during experiments (redrawn from Peltzer and Tapponnier, 1988) show striking similarities to the major features of SE Asia.

collision with a composite SE Asia, which includes some of the earlier Gondwana fragments to arrive, and also includes the island arcs formed due to the subduction of oceanic crust north of Australia. In east Indonesia the northward movement of Australia during the Cenozoic has been marked by arc-continent collision and major strike-slip motion within the north Australian margin. Further east, arc-continent collisions have been the result of elimination of marginal basins formed above subduction zones as Australia has moved north, and this system of arcs and marginal basins can be traced east along the margin of the Pacific plate in Melanesia.

During the late Mesozoic and Cenozoic there was subduction of the Pacific ocean to the east of Asia, although the eastern margin of Asia and SE Asia was probably a region of small plates and marginal basins as it is at the present day. At present the Philippine Sea plate and Philippine island arcs separate the east Asian margin from the Pacific plate south of Japan. Unlike the Indian ocean and Pacific oceans, the Philippine Sea plate lacks well defined sea-floor magnetic anomalies which normally provide the basis for reconstructing past plate motions. The Philippine Sea plate is also difficult to link to the movements of the other major plates because it is surrounded by subduction zones. For these reasons the history of movement of plate movements in the west Pacific north of New Guinea has been very uncertain, and consequently the eastern edge of SE Asia has been difficult to reconstruct. Palaeomagnetic data from east Indonesia (Hall *et al.*, 1995) have provided the basis for reconstructing the Philippine Sea plate and its motion since the early Cenozoic, and conse-

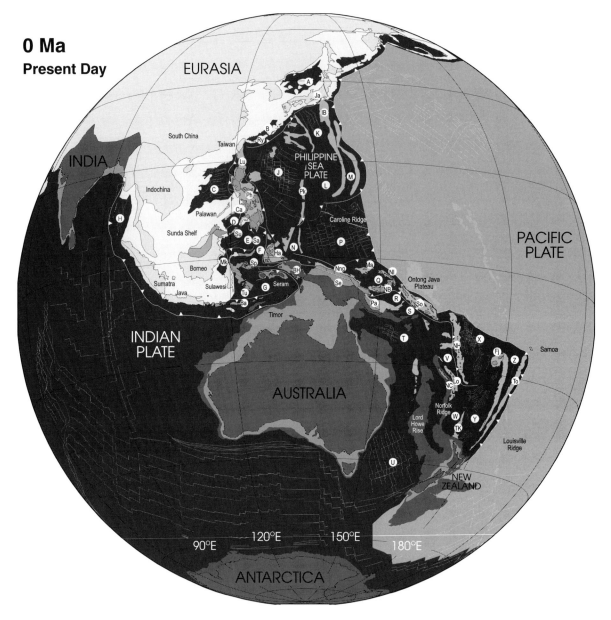

Fig.5. Present-day tectonic features of SE Asia and the SW Pacific. Light straight lines are selected marine magnetic anomalies. and active spreading centres. White lines are subduction zones and strike-slip faults. The present extent of the Pacific plate is shown in mid grey. Labelled filled areas are mainly arc, ophiolitic, and accreted material formed at plate margins during the Cenozoic, and submarine arc regions, hot spot volcanic products, and oceanic plateaus. Pale grey areas represent submarine parts of the Eurasian continental margins. Dark grey areas represent submarine parts of the Australian continental margins. See pages 126-131 for colour plates of Figs.5 to 10. Letters represent marginal basins and tectonic features as follows:

Marginal Basins

A	Japan Sea	N	Ayu Trough
B	Okinawa Trough	P	Caroline Sea
C	South China Sea	Q	Bismarck Sea
D	Sulu Sea	R	Solomon Sea
E	Celebes Sea	S	Woodlark Basin
F	Molucca Sea	T	Coral Sea
G	Banda Sea	U	Tasman Sea
H	Andaman Sea	V	Loyalty Basin
J	West Philippine Basin	W	Norfolk Basin
K	Shikoku Basin	X	North Fiji Basin
L	Parece Vela Basin	Y	South Fiji Basin
M	Mariana Trough	Z	Lau Basin

Tectonic features

Ba	Banda Arc	Mk	Makassar Strait	Ry	Ryukyu Arc
BH	Bird's Head	Mn	Manus Island	Sa	Sangihe Arc
Ca	Cagayan Arc	NB	New Britain Arc	Se	Sepik Arc
Fj	Fiji	NC	New Caledonia	So	Solomons Arc
Ha	Halmahera Arc	NH	New Hebrides Arc	Sp	Sula Platform
IB	Izu-Bonin Arc	NI	New Ireland	Su	Sulu Arc
Ja	Japan Arc	Nng	North New Guinea Terranes	TK	Three Kings Rise
Lo	Loyalty Islands	Pa	Papuan Ophiolite	To	Tonga Arc
Lu	Luzon Arc	Pk	Palau-Kyushu Ridge	Tu	Tukang Besi Platform

quently, for making reconstructions of the regions adjacent to the Philippine Sea plate.

Combining the Philippine Sea plate history with the known movements of the major plates, India, Australia, Pacific and Eurasia, provides some limits within which reconstructions of SE Asia, and parts of the west Pacific, can be attempted. Within this region there are recent interpretations of the South China Sea opening (Briais *et al.*, 1993) and other marginal basins which limit options still further and provide constraints of variable quality on modelling the tectonic history of the region. The importance of the marginal basins is that only they are likely to contain a clear record, based on ocean floor magnetic anomalies, of the motion history of some of the minor plates. However, many of the marginal basins of SE Asia completely lack magnetic anomalies, many have not been drilled during the ocean drilling campaigns, and their ages and character are still poorly known. Therefore much of the evidence which must be used in a regional tectonic model of SE Asia is based on *interpretation* of geological data from the small ocean basins, their margins, and from the geologically more complicated land areas around them. The reader should be aware that, as in other areas of science, geologists differ in their interpretations of these data, and much of the information does not lend itself to unambiguous reconstruction. Nonetheless, a complete tectonic history can *only* be deduced from the geology on land combined with data from the oceans. The account here is therefore my view of the development of Cenozoic SE Asia using plate tectonic reconstructions based on such deductions.

The model

The reconstructions were made using the ATLAS computer program (Cambridge Paleomap Services, 1993) and plate motion model for the major plates. The motion of Africa is defined relative to magnetic north and motions of the major plates are all relative to Africa. Eurasia has been close to its present-day position throughout the Cenozoic. The reconstructions in this paper add to this model for the major plates by including a large number of smaller fragments in SE Asia and the SW Pacific. More than 100 fragments are currently used, and most retain their current size in order that they remain recognisable. During the 50 Ma period fragments represented may have changed size and shape or may not have existed, both for arc and continental terranes. Thus, the plate model can only be an approximation. Some of the elements of the model are deliberately represented in a stylistic manner to convey the processes inferred rather than display exactly what has happened, for example, the motion of the terranes of north New Guinea.

Previous reconstructions which cover all or parts of the region discussed here include those of Katili (1975), Crook and Belbin (1978), Hamilton (1979), Briais *et al.* (1993), Burrett *et al.* (1991), Daly *et al.* (1991), Lee and Lawver (1994), Rangin *et al.* (1990), and Yan and Kroenke (1993) who also produced an animated reconstruction of the SW Pacific. The reader is referred to the original papers for accounts of the earlier models. Some differences between the model here and other models result from the choice of reference frames; some use the hotspot reference frame, and others use a fixed Eurasia, whereas these reconstructions use a palaeomagnetic reference frame. These choices result in different palaeolatitudes and can cause other differences. There have also been improvements in our knowledge of global plate motions since the earlier regional reconstructions. However, in many cases the principal differences between the different models result from different interpretations of geological data.

This paper gives an account of a plate tectonic model for the Cenozoic development of the region based on my interpretations of a large range of geological data. It summarises the regional tectonic development of SE Asia using a plate model which has been animated using 1 Ma time-slices. Below is a brief account of the model and its major features, which is followed by a discussion of its principal implications for biogeographers relating to the distribution of land and sea during the last 30 million years.

Reconstructions

The model discussed here includes that developed earlier for SE Asia (Hall, 1996) which has been extended to include the SW Pacific (Hall, 1997). Reconstructions of SE Asia and the SW Pacific (Fig.5) shown on a global projection are presented at 10 Ma intervals for the period 50-10 Ma (Figs.6-10). The reader is referred to Hall (1995, 1996) for a more complete account of the assumptions and data used in reconstructions of SE Asia and for maps showing only SE Asia but with more detail.

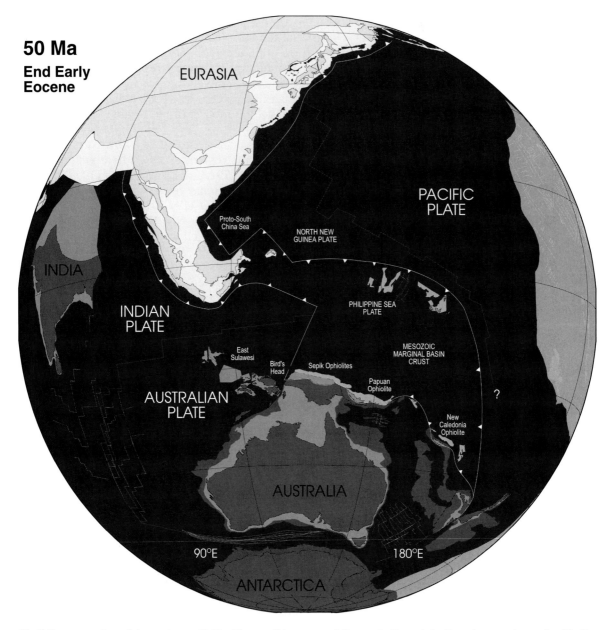

Fig.6. Reconstruction of the region at 50 Ma. The possible extent of Greater India and the Eurasian margin north of India are shown schematically. Shortly before 50 Ma collision between the north Australian continental margin and an island arc had emplaced ophiolites on the north New Guinea margin, and in New Caledonia, eliminating ocean crust formed at the former Australian-Indian ocean spreading centre. Double black arrows indicate extension in Sundaland.

Configuration at 50 Ma

At 50 Ma (Fig.6) India and Australia were separate plates although their motions were not greatly different. Transform faults linked the slow-spreading Australia-Antarctic and the fast spreading India-Australia spreading centres. Some of the ophiolites of Sulawesi probably formed at the India-Australia mid-ocean ridge.

India collided with Asia in the early Tertiary but there remains considerable controversy about the exact age of collision, and its consequences (Packham, 1996; Rowley, 1996). The position of the Eurasian margin and the extent of Greater India are major problems. The reconstruction shown in Fig.6 shows a conservative estimate and, since India-Asia collision began at about 50 Ma, this implies that the Asian margin extended south to at least 30°N. Many of the tectonic events in SE Asia are commonly attributed to the effects of Indian indentation into Asia and the subsequent extrusion of continental fragments eastwards along major strike-slip faults. Despite the great attraction of this hypothesis and the spectacular evidence of displacements on the Red River fault (Tapponnier et al., 1990) the predictions of major rotations, southeastward extrusion of fragments, and the timing of events (Tapponnier et al., 1982), remain poorly supported by geological evidence in SE Asia.

The east Eurasian continental margin was oriented broadly NE-SW. From Japan northwards Asia was bounded by an active margin. Taiwan, Palawan and the now extended crust of the South China Sea margins formed a passive margin, established during Cretaceous times. Sundaland was separated from Eurasia by a wide proto-South China Sea probably floored by Mesozoic ocean crust. The southern edge of this ocean was a passive continental margin north of a continental promontory extending from Borneo to Zamboanga. The Malay peninsula was closer to Indochina and the Malay-Sumatra margin was closer to NNW-SSE. Because rotation of Borneo is part of this model the reconstruction differs from those of Rangin et al. (1990) and Daly et al. (1991) who infer a margin oriented closer to E-W. I see no evidence to support the almost E-W orientation of the Sundaland margin in the region of Sumatra as shown on these and many other reconstructions (e.g. Briais et al., 1993; Hutchison, 1996). Furthermore, such models have major difficulties in explaining the amount, timing and mechanism of rotation required to move Sumatra from an E-W to NW-SE orientation. West Sumatra includes arc and ophiolitic material accreted in the Cretaceous. East Borneo and West Sulawesi appear to be underlain by accreted arc and ophiolitic material as well as continental crust which may be early-rifted Gondwana fragments. This material had been accreted during the Cretaceous and may have resulted in a highly thickened crust in this part of Sundaland, possibly sustained by subduction.

Australia was essentially surrounded by passive margins on all sides. To the west the passive margin was formed in the Late Jurassic, and Fig.6 postulates a failed rift, possibly floored by oceanic crust on the site of the present-day Banda Sea, partially separating the Bird's Head microcontinent from Australia. Mesozoic oceanic lithosphere was present north of the Bird's Head, south of the active Indian-Australian spreading centre. Further east in the Pacific, Indian and Australian oceanic lithosphere had been subducting northwards beneath the Sepik-Papuan arc in the early Tertiary. During the Paleocene and early Eocene the New Guinea Mesozoic passive margin collided with this intra-oceanic arc causing emplacement of the Sepik and Papuan ophiolites (Davies, 1971). Subsequently, most of the New Guinea margin was a passive margin during the Paleogene but the oceanic crust to the north is inferred to have formed during the Mesozoic in an intra-oceanic marginal basin behind the Sepik-Papuan arc. The position and character of the east Australia-Pacific margin is also uncertain. Tasman and Coral Sea opening had probably been driven by subduction but the site of subduction must have been considerably east of the Australian continent, beyond the Loyalty Rise and New Caledonia Rise. Spreading had ceased in both basins by about 60 Ma (Paleocene). By the Paleocene it appears that subduction east of New Caledonia was to the east not to the west (Aitchison et al., 1995). The history of this region remains poorly known since it is almost entirely submarine, and magnetic anomalies in this area are poorly defined.

Java and West Sulawesi were situated above a trench where Indian plate lithosphere was subducting towards the north. The character of this boundary is shown as a simple arc but may have included marginal basins and both strike-slip and convergent segments depending on its local orientation. Extending plate boundaries into the Pacific is very difficult. A very large area of the West Pacific has been eliminated by subduction since 50 Ma which will continue to cause major problems for reconstructions. However, there is clear evidence that this area resembled the present-day West Pacific in containing marginal basins, intra-oceanic arcs and subduction zones. The Java subduction system linked east into Pacific intra-oceanic subduction zones required by the intra-oceanic arc rocks within the Philippine Sea plate; parts of the east Philippines, the West Philippine basin and Halmahera include arc rocks dating back at least to the Cretaceous. North of the Philippine Sea plate there was a south-dipping subduction zone at the southern edge of a Northern New Guinea plate.

50-40 Ma

Whatever the timing of India-Asia collision, a consequence was the slowing of the rate of plate convergence after anomaly C21 and a major change in spreading systems between anomaly C20 and C19 at about 42 Ma. India and Australia became one plate during this period (Figs.6 and 7) and the ridge between them became inactive. Northward subduction of Indian-Australian lithosphere continued beneath the Sunda-Java-Sulawesi arcs although the direction of convergence may have changed. Rift basins formed throughout Sundaland, but the timing of their initial extension is uncertain because they contain continental clastics which are poorly dated, and their cause is therefore also uncertain. They may represent the consequences of oblique convergence or extension due to relaxation in the over-riding plate in response to India-Asia collision, enhanced by slowing of subduction, further influenced by older structural fabrics.

The Java-Sulawesi subduction system continued into the West Pacific beneath the east Philippines and Halmahera arcs. Further east, the direction of subduction was southward towards Australia and this led to the formation of a Melanesian arc system. During the Eocene the extended eastern Australasian passive margin had collided with the intra-oceanic arc already emplaced in New Guinea resulting in emplacement of the New Caledonia ophiolite (Aitchison et al., 1995; Meffre, 1995) followed by subduction polarity reversal. Subduction began beneath Papua New Guinea with major arc growth producing the older parts of the New Britain, Solomons and Tonga-Kermadec systems, leading to development of major marginal basins in the SW Pacific whose remnants probably survive only in the Solomon Sea. This model postulates the initial formation of these arcs at the Papuan-east Australian margin as previously suggested by Crook and Belbin (1978) following subduction flip, rather than by initiation of intraoceanic subduction within the Pacific plate outboard of Australia as suggested by Yan and Kroenke (1993). The evidence for either proposal is limited but this model has the simplicity of a single continuous Melanesian arc.

During this interval there were major changes in the Pacific. The Pacific plate is widely said to have changed its motion direction at 43 Ma, based on the age of the bend in the Hawaiian-Emperor seamount chain, although this view has recently been challenged by Norton (1995) who attributes the bend to a moving hotspot which became fixed only at 43 Ma. Subduction of the Pacific-Northern New Guinea ridge (Fig.7) led to massive outpouring of intra-oceanic volcanic rocks (Stern and Bloomer, 1992) which formed the Izu-Bonin-Mariana arc system, and the Philippine Sea plate was a recognisable entity by the end of this period. There was significant rotation of the Philippine Sea plate between 50 and 40 Ma and the motion history of this plate (Hall et al., 1995) provides an important constraint on development of the eastern part of SE Asia. The West Philippine basin, Celebes Sea, and Makassar Strait opened as single oceanic basin within the Philippine Sea plate although the reconstructions probably underestimate the width of the Makassar Strait and Celebes Sea, which may have been partially subducted in the Miocene beneath west Sulawesi.

The opening of the West Philippine-Celebes Sea basin required the initiation of southward subduction of the proto-South China Sea beneath Luzon and the Sulu arc. It is this subduction which caused renewed extension along the South China margin, driven by slab-pull forces due to subduction between eastern Borneo and Luzon, and later led to sea-floor spreading in the South China Sea, rather than indentor-driven tectonics.

40-30 Ma

In this interval (Figs.7 and 8) the spreading of the marginal basins of the West and SW Pacific continued. Indian ocean subduction continued at the Sunda-Java trenches, and also beneath the arc extending from Sulawesi through the east Philippines to Halmahera. Sea floor spreading continued in the West Philippine-Celebes Sea basin until about 34 Ma. This spreading centre may been linked to backarc spreading of the Caroline Sea which formed from about 40 Ma due to subduction of the Pacific plate. The Caroline Ridge is interpreted in part as a remnant arc resulting from Caroline Sea backarc spreading, and the South Caroline arc ultimately became the north New Guinea arc terranes. By 30 Ma the Caroline Sea was widening above a subduction zone at which the newly-formed Solomon Sea was being destroyed as the Melanesian arc system migrated north. The backarc basins in the SW Pacific were probably very complex, as indicated by the anomalies in the South Fiji basin, and will never be completely reconstructed because most of these basins have been subducted.

The Philippines-Halmahera arc was stationary, so spreading in the West Philippine-Celebes

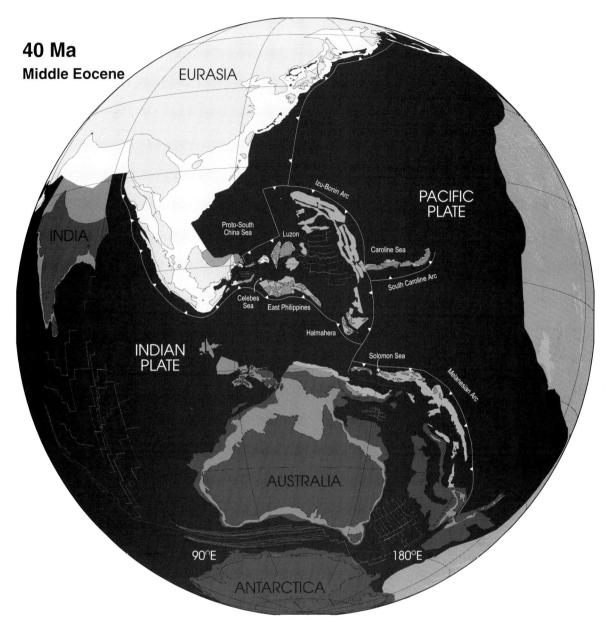

Fig. 7. Reconstruction of the region at 40 Ma. India and Australia were now parts of the same plate. An oceanic spreading centre linked the north Makassar Strait, the Celebes Sea and the West Philippine basin. Spreading began at about this time in the Caroline Sea, separating the Caroline Ridge remnant arc from the South Caroline arc. Spreading also began after subduction flip in marginal basins around eastern Australasia producing the Solomon Sea and the island arcs of Melanesia.

Sea basin maintained subduction between NE Borneo and north of Luzon. The pull forces of the subducting slab therefore account for stretching of the Eurasian margin north of Palawan, and later development of oceanic crust in the South China Sea which began by 32 Ma. In contrast, the indentor model does not account for stretching at the leading edge of the extruded blocks, such as Indochina, or the normal faulting east of Vietnam often shown as kinematically linked to the Red River fault system. There was approximately 500-600 km left-lateral movement on the Red River fault (Briais *et al.*, 1993) during the extrusion of Indochina (32-15 Ma).

The dextral Three Pagodas and Wang Chao faults are simplified as a single fault at the north end of the Malay peninsula. There are a host of

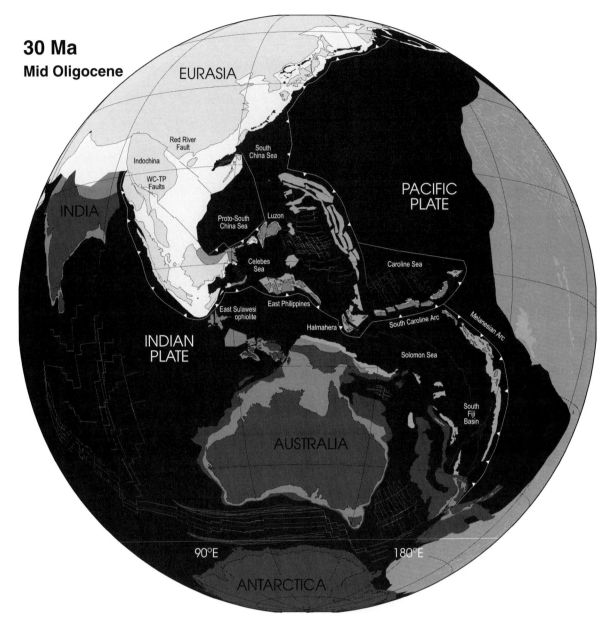

Fig.8. Reconstruction of the region at 30 Ma. Indentation of Eurasia by India led to extrusion of the Indochina block by movement on the Red River fault and Wang Chao-Three Pagodas (WC-TP) faults. Slab pull due to southward subduction of the proto-South China Sea caused extension of the South China and Indochina continental margin and the present South China Sea began to open. A wide area of marginal basins separated the Melanesian arc from passive margins of eastern Australasia, shown schematically between the Solomon Sea and the South Fiji basin.

faults through this region, and a plate tectonic model can only oversimplify the tectonics of the continental regions by considering large and simple block movements and broadly predicting regional stress fields. The implication of this simplified model is that basins such as the Malay and Gulf of Thailand basins have a significant component of strike-slip movement on faults controlling their development. However, they may have been initiated in a different tectonic setting, in which a pre-existing structural fabric influenced their development (Hutchison, 1996).

30-20 Ma

This period of time (Figs.8 and 9) saw the most important Cenozoic plate boundary reorganisation within SE Asia. At about 25 Ma, the New Guinea passive margin collided with the leading edge of the east Philippines-Halmahera-New Guinea arc system. The Australian margin, in the Bird's Head region, was also close to collision with the Eurasian margin in West Sulawesi and during this interval ophiolite was emplaced in Sulawesi.

By 30 Ma the Sulawesi margin may have been complex and included ocean crust of different types (mid-ocean ridge, backarc basin). Thus the Sulawesi ophiolite probably includes material formed within the Indian Ocean (Mubroto et al., 1994) as well as ocean basins marginal to Eurasia (Monnier et al., 1995). The arrival of the Australian margin at the subduction zone caused northward subduction to cease. The ocean crust trapped between Sulawesi and Halmahera first became part of the Philippine Sea plate and later the Molucca Sea plate. The Philippine Sea plate began to rotate clockwise and the trapped ocean crust began to subduct beneath Sulawesi in the Sangihe arc.

Soon afterwards the Ontong Java plateau collided with the Melanesian arc. These two major collisions caused a significant change in the character of plate boundaries in the region between about 25 and 20 Ma (Early Miocene). They also linked the island arcs of Melanesia to the New Guinea terranes at the southern margin of the Caroline plate, and to the Halmahera-Philippines arcs. This linkage seems to have coupled the Pacific to the marginal basins of the West Pacific, and the Caroline and Philippine Sea plates were subsequently driven by the Pacific. Both began to rotate, almost as a single plate, and the Izu-Bonin-Mariana trench system rolled back into the Pacific. Rifting of the Palau-Kyushu ridge began, leading first to opening of the Parece Vela basin and later to spreading in the Shikoku basin. The change in plate boundaries led to subduction beneath the Asian margin.

Subduction beneath the Halmahera-Philippines arc ceased and the New Guinea sector of the Australian margin became a strike-slip zone, the Sorong fault system, which subsequently moved terranes of the South Caroline arc along the New Guinea margin.

Advance of the Melanesian arc system led to widening of the South Fiji basin and Solomon Sea basin (now mainly subducted). At the Three Kings Rise subduction seems to have been initiated soon after ocean crust was formed to the east, allowing the rise to advance east and spreading to propagate behind the rise into the Norfolk basin from a triple junction to the north.

20-10 Ma

The clockwise rotation of the Philippine Sea plate necessitated changes in plate boundaries throughout SE Asia which resulted in the tectonic pattern recognisable today (Figs.9 and 10). These changes include the re-orientation of spreading in the South China Sea, and the development of new subduction zones at the eastern edge of Eurasia and in the SW Pacific. Continued northward motion of Australia caused the counter-clockwise rotation of Borneo. Northern Borneo is much more complex than shown. There was volcanic activity and build-out of delta and turbidite systems into the proto-South China Sea basin. Major problems include the source of sediment in the basins surrounding central Borneo and the location and timing of volcanic activity in Borneo. The remaining oceanic crust of the western proto-South China Sea, and thinned continental crust of the passive margin to the north, was thrust beneath Borneo thickening the crust, resulting in rapid erosion of sediments into the Neogene circum-Borneo deltas, and ultimately leading to crustal melting.

The rotation of Borneo was accompanied by counter-clockwise motion of west Sulawesi, and smaller counter-clockwise rotations of adjacent Sundaland blocks. In contrast, the north Malay peninsula rotated clockwise, but remained linked to both Indochina and the south Malay peninsula. This allowed widening of basins in the Gulf of Thailand, but the simple rigid plate model overestimates the extension in this region. This extension was probably more widely distributed throughout Sundaland and Indochina on many different faults. The Burma plate became partly coupled to the northward-moving Indian plate and began to move north on the Sagaing fault leading to stretching of the Sunda continental margin north of Sumatra, and ultimately to ocean crust formation in the Andaman Sea.

North Sumatra rotated counter-clockwise with south Malaya, and as the rotation proceeded the orientation of the Sumatran margin changed with respect to the Indian plate motion vector. The consequent increase in the convergent component of motion, taken up by subduction, may have increased magmatic activity in the arc and weakened the upper plate, leading to formation of the dextral Sumatran strike-slip fault system taking up the arc-parallel component of India-Eurasia plate motion.

East of Borneo, the increased rate of subduction caused arc splitting in the Sulu arc and the

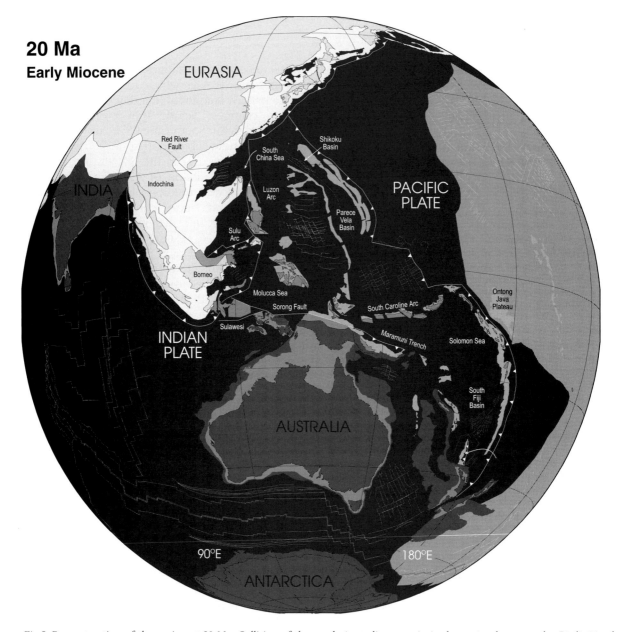

Fig.9. Reconstruction of the region at 20 Ma. Collision of the north Australian margin in the region between the Bird's Head microcontinent and eastern New Guinea occurred at about 25 Ma. The Ontong Java plateau arrived at the Melanesian trench at about 20 Ma. These two events caused major reorganisation of plate boundaries. Subduction of the Solomon Sea began at the eastern New Guinea margin. Spreading began in the Parece Vela and Shikoku marginal basins. The north Australian margin became a major left-lateral strike-slip system as the Philippine Sea-Caroline plate began to rotate clockwise. Movement on splays of the Sorong fault system led to the collision of Australian continental fragments in Sulawesi. This in turn led to counter-clockwise rotation of Borneo and related Sundaland fragments, eliminating the proto-South China Sea. The Sumatra fault system was initiated.

Sulu Sea opened as a backarc basin (Hinz *et al.*, 1991; Silver and Rangin, 1991) south of the Cagayan ridge. The Cagayan ridge then moved northwards, eliminating the eastern proto-South China Sea, to collide with the Palawan margin. New subduction had also begun at the west edge of the Philippine Sea plate below the north Sulawesi-Sangihe arc which extended north to south Luzon. This was a complex zone of opposed subduction zones linked by strike-slip faults. The Philippine islands and Halmahera were carried with the Philippine Sea plate to-

Cenozoic plate tectonics of SE Asia 113

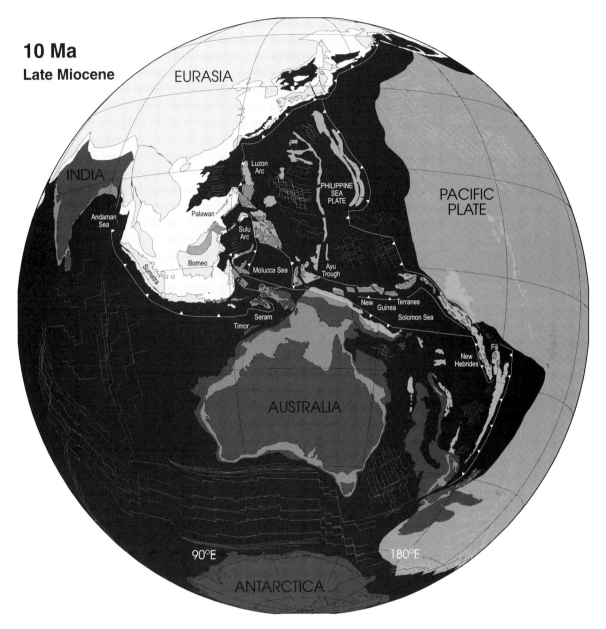

Fig. 10. Reconstruction of the region at 10 Ma. The Solomon Sea was being eliminated by subduction beneath eastern new Guinea and beneath the New Hebrides arc. However, continued subduction led to development of new marginal basins within the period 10-0 Ma, including the Bismarck Sea, Woodlark basin, North Fiji basins, and Lau basin. The New Guinea terranes, formed in the South Caroline arc, docked in New Guinea but continued to move in a wide left-lateral strike-slip zone. Further west, motion on strands of the Sorong fault system caused the arrival of the Tukang Besi and Sula fragments in Sulawesi. Collision events at the Eurasian continental margin in the Philippines, and subsequently between the Luzon arc and Taiwan, were accompanied by intra-plate deformation, important strike-slip faulting and complex development of opposed subduction zones. Rotation of Borneo was complete but motion of the Sumatran forearc slivers was associated with new spreading in the Andaman Sea.

wards this subduction zone. North of Luzon, sinistral strike-slip movement linked the subducting west margin of the Philippine Sea plate to subduction at the Ryukyu trench. Collision of Luzon and the Cagayan ridge with the Eurasian continental margin in Mindoro and north Palawan resulted in a jump of subduction to the south side of the Sulu Sea. Southward subduction beneath the Sulu arc continued until 10 Ma. The remainder of the Philippines continued to move with the Philippine Sea plate, possibly with intra-plate strike-slip motion and subduction resulting in local volcanic activity. At the east edge of the Philippine Sea plate spreading terminated in the Shikoku basin.

As a result of the change in plate boundaries, fragments of continental crust were emplaced in Sulawesi on splays at the western end of the Sorong fault system. The earliest fragment to collide is inferred to have been completely underthrust beneath West Sulawesi and contributed to later crustal melting (Polvé et al., 1997). Later, the Tukang Besi platform separated from the Bird's Head and was carried west on the Philippine Sea plate to collide with Sulawesi. Locking of splays of the Sorong fault caused subduction to initiate at the eastern margin of the Molucca Sea, thus producing the Neogene Halmahera arc. In this way the Molucca Sea became a separate plate as the double subduction system developed.

After the collision of the Ontong Java plateau with the Melanesian arc the Solomons became attached to the Pacific plate. Westward subduction began on the SW side of Solomon Sea, beneath eastern New Guinea, eliminating most of Solomon Sea and resulting in the formation of Maramuni arc system. As the Solomon Sea was eliminated, the South Caroline arc began to converge on the north New Guinea margin and the arc terranes were translated west in the major left-lateral shear zone, probably accompanied by some rotation. In the southern part of the Solomons Sea subduction was in the opposite direction (eastward) and created the New Hebrides arc system. Spreading ceased in the South Fiji basin.

10-0 Ma

At the beginning of this period SE Asia was largely recognisable in its present form (Fig.10). Rotation of Borneo was complete. This, with collision in the central Philippines and Mindoro, and continued northward movement of Australia, resulted in reorganisation of plate boundaries and intra-plate deformation in the Philippines. The Luzon arc came into collision with the Eurasian margin in Taiwan. This may be the cause of the most recent regional change in plate motions at about 5 Ma. The Philippine Sea plate rotation pole moved north from a position east of the plate; clockwise rotation continued but the change in motion caused re-orientation of existing, and development of new, plate boundaries. Subduction continued at the Manila, Sangihe and Halmahera trenches, and new subduction began at the Negros and Philippine trenches. These subduction zones were linked by strike-slip systems active within the Philippines, and this intra-plate deformation created many very small fragments which are difficult to describe using rigid plate tectonics.

The Molucca Sea continued to close by subduction on both sides. At present the Sangihe forearc has overridden the northern end of the Halmahera arc, and is beginning to over-thrust west Halmahera. In the Sorong fault zone, accretion of Tukang Besi to Sulawesi locked a strand of the fault and initiated a new splay south of the Sula platform. The Sula platform then collided with the east arm of Sulawesi, causing rotation of the east and north arms to their present position, leading to southward subduction of the Celebes Sea at the north Sulawesi trench.

The Eurasia-Philippine Sea plate-Australia triple junction was and remains a zone of microplates but within this contractional setting new extension began in the Banda Sea. The Bird's Head moved north relative to Australia along a strike-slip fault at the Aru basin edge. Mesozoic ocean crust north of Timor was eliminated at the eastern end of the Java trench by continued northern motion of Australia which brought the Australian margin into this trench as the volcanic inner Banda arc propagated east. Seram began to move east requiring subduction and strike-slip motion at the edges of this microplate. Since 5 Ma the southern Banda Sea has extended to its present dimensions, and continental fragments are now found in the Banda Sea ridges within young volcanic crust. The Banda Sea is here interpreted to be very young as suggested by Hamilton (1979) and others.

In west Sundaland, partitioning of convergence in Sumatra into orthogonal subduction and strike-slip motion effectively established one or more Sumatran forearc sliver plates. Extension on the strike-slip system linked to the spreading centre in the Andaman Sea (Curray et al., 1979). Within Eurasia reversal of motion on the Red River system may have been one consequence of the regional change in plate motions.

Opening of the Ayu trough separated the Caroline plate and Philippine Sea plate, although the rate of separation at this spreading centre was very low. North of the Bird's Head, and further east in New Guinea, transpressional movements were marked by deformation of arc and ophiolite slivers separated by sedimentary basins. Progressive westward motion of the South Caroline arc within the left-lateral transpressional zone led to docking of the north New

Cenozoic plate tectonics of SE Asia

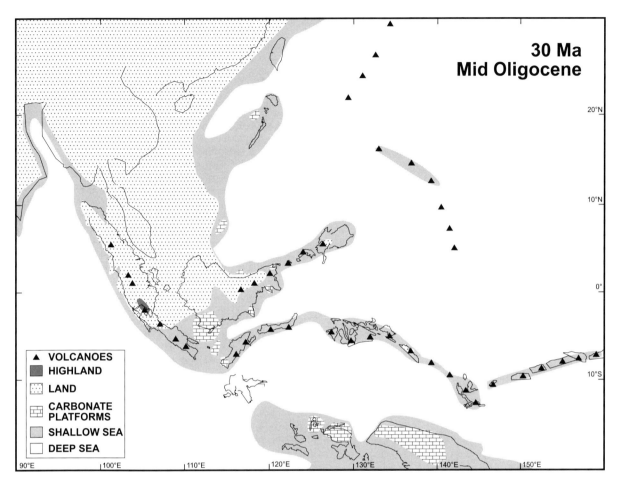

Fig.11. Postulated distribution of land and sea in SE Asia at 30 Ma. No attempt has been made to represent topography with Asia and Indochina. Much of the area north and east of the Indian collision zone must have been highlands.

Guinea terranes. This caused the cessation of southward subduction of the Solomon Sea plate but resulted in its northward subduction beneath New Britain. The New Britain subduction led to rapid spreading in Woodlark basin as a consequence of slab-pull forces and rapid ripping open of continental crust beneath the Papuan peninsula. Elimination of most of the remaining Solomons marginal basin by eastward subduction led to formation of the New Hebrides arc and ocean crust formation in the North Fiji basins.

Determining the extents of land and sea

For the biogeographer, the tectonic development of the region is only a starting point for understanding. In order to understand the distribution of most organisms it is also necessary to know where there was land and sea, where the sea was shallow and deep, and how wide were the seas. For the land, there needs to be some knowledge of topography, particularly where there were mountainous regions. The distribution and character of land and sea will have provided physical pathways and barriers to dispersal, and may well have influenced plant and animal distribution by effects on other controlling factors such as local and global climate, oceanic circulation patterns, and sea-level.

However, moving from tectonic reconstruction maps to detailed palaeogeographical maps involves further complexities. In many ways the geological record is a marine record. Most of Earth history is recorded in rocks deposited at the surface, and the areas where most sediments are deposited are close to or below sea-level, and mainly at continental margins. Dating of rocks is largely based on fossils, and marine or-

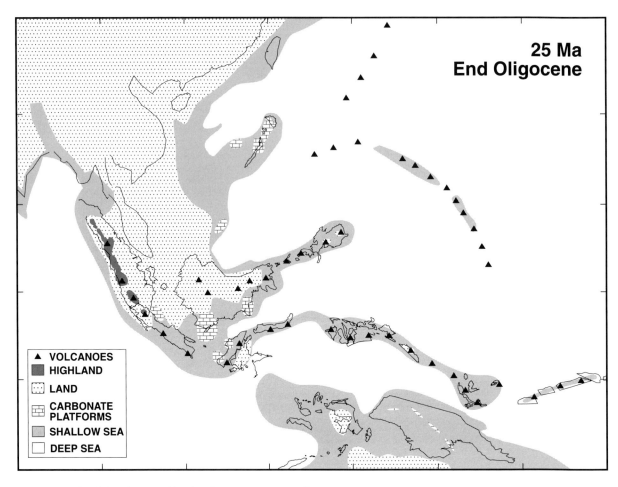

Fig.12. Postulated distribution of land and sea in SE Asia at 25 Ma.

ganisms generally provide the fossils of greatest biostratigraphic value which usually provide some insight into the environment of deposition. Geologists are therefore usually able to reconstruct the history of marine areas. In the deep oceans sedimentary rocks may lack fossils but the history of sediments deposited on ocean crust is known because ocean crust subsides with age due to lithospheric cooling and age-depth relationships are well established. Thus, many postulated land-bridges in oceanic regions can be dismissed with some confidence.

In contrast, mapping environments and physiography of former land areas is a great deal more difficult. Uplift, erosion and periods of emergence are mainly recorded by negative evidence, such as unconformities and stratigraphic incompleteness. Even when there is a rock record it will often be difficult to date because sediments deposited on land typically represent restricted types of environments, and usually contain few fossils which have limited biostratigraphic value. Unlike marine fossils, fossil assemblages from land rarely yield information about the history of their enclosing sediments relative to sea-level.

However, there are ways to solve some of these problems, and mapping palaeogeography onto the reconstructions is not, in principle, impossible although much of the information required is not yet available. It is possible to identify the positions of former coastlines, interpret the location of former river systems, and indirectly infer areas of mountains. In SE Asia some of the information can be compiled from the literature; an attempt to do this for the region of Wallace's Line is discussed by Moss and Wilson (1998, this volume). Some data, for example location of former coastlines, could be determined from records of oil companies acquired during

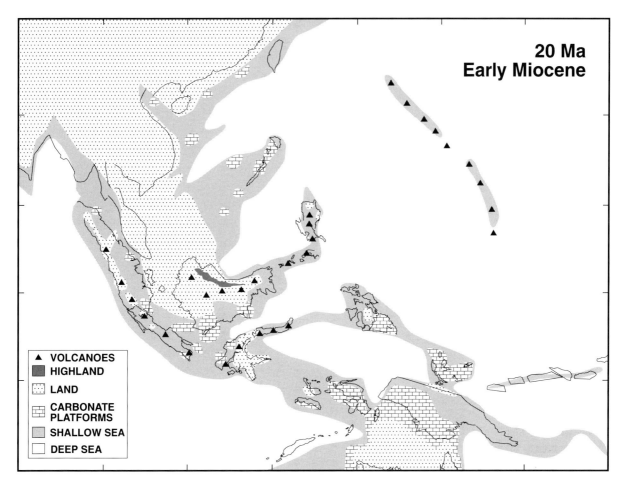

Fig.13. Postulated distribution of land and sea in SE Asia at 20 Ma.

extensive seismic surveys of SE Asia for hydrocarbons. New research could provide further detail and biogeographers themselves could also contribute by, for example, mapping distributions of fossil plants and interpreting their environments.

Land and sea for 30-0 Ma

Figs.11 to 16 are an attempt to compile the general features of land and sea onto maps of the tectonic reconstructions showing 5 million year intervals between 30 and 5 Ma for the region of SE Asia. The maps may be useful in indicating the likely geographical connections and barriers and the periods when these were in existence. There are few studies that compile this type of information and all cover limited parts of area for limited times. Thus these maps are based on those few sources, some proprietary information from oil companies, and a wide range of literature and maps. The sources are too numerous to cite and the quality of coverage is very variable. The task is a very large one, given the size of the area, and the results should therefore be regarded as a first order approximation only. I have not attempted to draw palaeogeographical maps for periods before 30 Ma. The period 30-0 Ma is of most interest to biogeographers; before then the separation between Asia and Australia was greater and the tectonic reconstructions are also more uncertain.

The limited ranges of environments and distributions shown are best estimates. Broadly speaking, each area shown should be regarded as a probability. For example, for an area shown as deep marine, the probability of that area being shallow marine is low, and of it being land is very low. Some of the assignments are educated

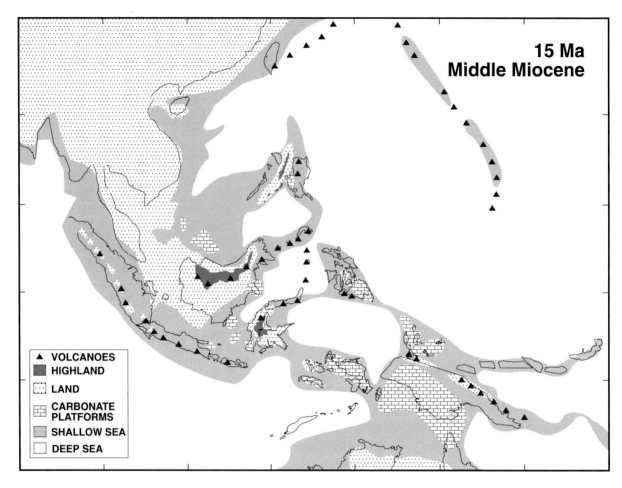

Fig.14. Postulated distribution of land and sea in SE Asia at 15 Ma.

guesses. For example, areas of long-lived island arcs develop thickened crust, implying relative shallow water areas and local emergence. When volcanoes are active, magma production, thermal expansion and crustal buoyancy can lead to emergence but individual volcanoes can be very short-lived on a geological time scale (typically less than one million years) even though an arc may have been a long-lived feature. It is usually not possible to identify precisely which areas were emergent, simply that there are likely to have been such areas.

The mid Oligocene (Fig.11) was the time of a major fall in global sea-level (Haq *et al.*, 1987). Very large areas of Sundaland and Sunda shelf were exposed and there were probably more emergent areas than at any subsequent time until the end of the Cenozoic. North of Sundaland, Asia was a persistent highland area, and large amounts of sediment moved south from central Asia down major river systems. Much of southern Sundaland was the site of deposition of alluvial, fluvial and deltaic sediments. There were major embayments in the eastern Asian margin formed by the South China Sea, the proto-South China Sea and the Celebes Sea-Makassar Strait. Separating these were elongate bathymetric features which were probably mainly shallow water with intermittent emergent areas, notably where arc volcanoes were active. The southernmost promontory was the Sulawesi-Philippines-Halmahera arc which could have provided a pathway into the Pacific, via volcanic island stepping stones, for organisms that could cross seawater. The other promontories terminated in the deep ocean area of the Pacific.

At about 25 Ma (Fig.12) the north Australian margin came into contact with Sulawesi and the Halmahera arc, and this could have created a discontinuous land connection via the island

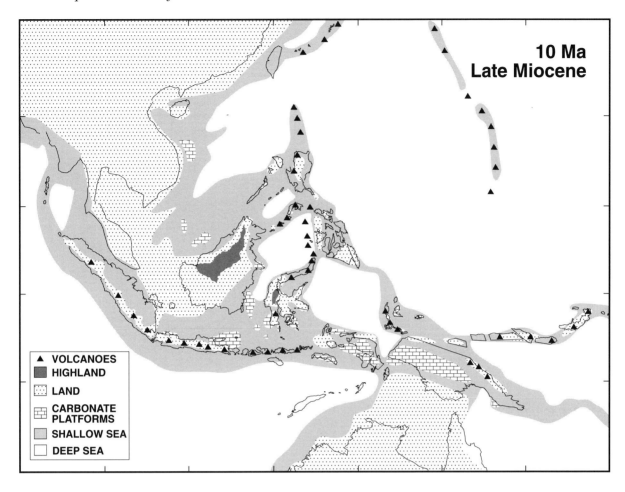

Fig.15. Postulated distribution of land and sea in SE Asia at 10 Ma.

arcs of Halmahera and the Philippines into Sulawesi. The arc-continent collision closed the deep water passage between the Pacific and Indian oceans (Kennett et al. 1985) by about 20 Ma (Fig.13) and there must have been major changes of oceanic currents (Fig.17 and 18) with implications for the distribution of many marine organisms, particularly those of shallow marine environments. North-central Borneo was uplifted and shed huge volumes of sediments into the deltas which formed in north and east Borneo.

From about this time there was probably always some land in the area of Sulawesi, and the extensive but poorly dated Celebes molasse (Kündig, 1956) represents the products of subaerial erosion, although there were no permanent land links to Sundaland nor to Australia. However, there were intermittently emergent areas between Australia and Sulawesi, and a broad zone of shallow water within which there could have been numerous islands. Furthermore, strike-slip fault movements led to the arrival of numerous fragments of continental crust in Sulawesi, sliced from the Bird's Head microcontinent. The northern Makassar Strait remained a deep water area, and presumably formed a barrier to migration for many plant and animals (Moss and Wilson, 1998 this volume).

From 15 Ma to 5 Ma (Figs.14, 15, 16) was a period in which emergent Sundaland reduced in area, while the deep marginal basins in the east were eliminated (proto-South China Sea) or reduced in size (Sulu, Celebes and Molucca Sea). Local collision and volcanic arc activity led to intermittent emergence in many of the arc regions but these probably always resembled the present Philippine and North Molucca arcs, with land separated by sea, which could locally have been quite deep. More of Borneo became emer-

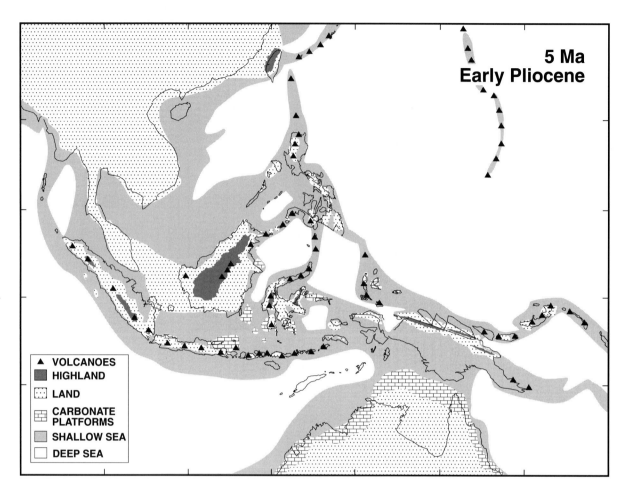

Fig.16. Postulated distribution of land and sea in SE Asia at 5 Ma.

gent and the central mountains on the Sarawak-Kalimantan border extending into Sabah became wider and higher with time. It is important to be aware that within this convergent setting deep basins also formed (*e.g.,* Sulu Sea, Banda Sea) which must have represented new barriers to dispersal which formed at the same time as new land pathways were established.

Conclusions

There are three important periods in regional development. At about 45 Ma plate boundaries changed, probably as a result of India-Asia collision. From a biogeographical viewpoint the arrival of India would have led to a movement of Gondwana plants and animals into Asia. Mountain building resulting from the collision led to major changes in habitats, and climate, accompanied by changes in land area and drainage systems. Huge volumes of sediment began to move south from central Asia into the sedimentary basins of the Sunda shelf. Ultimately all this would have driven dispersal from Gondwana via India into SE Asia (*e.g.* Harley and Morley, 1995), and later speciation centred in Sundaland which for many organisms became separated from Asia by climate and topography, and which remained separated from Australia by marine barriers. Continued indentation of Asia by India modified the Eurasian continent but much more knowledge is required of the timing of fault movements and the amounts of displacements before Sundaland can be adequately understood. The deformation within Asia and Sundaland is likely to have led to the formation of geographical barriers, principally mountains, some of which were associated with strike-slip faults and geologically short-lived.

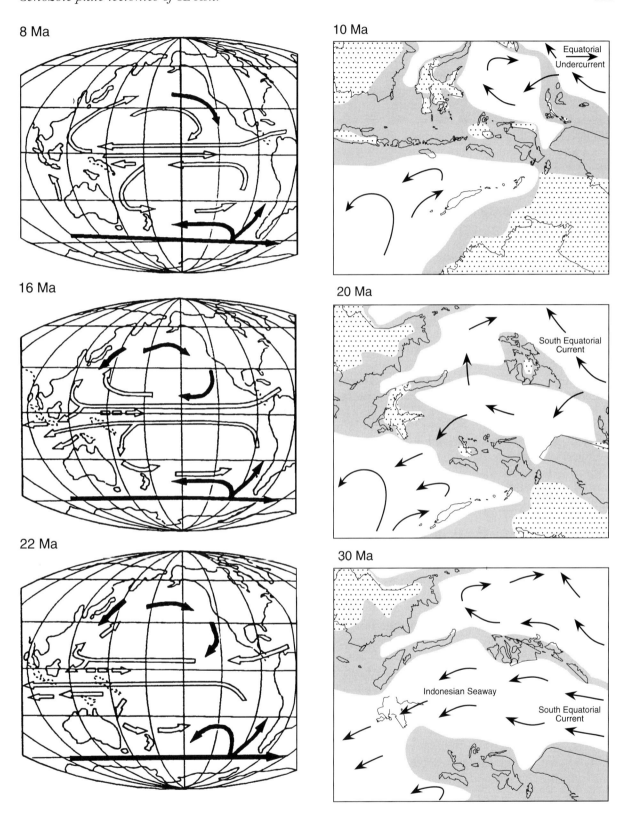

Fig. 17. Circulation patterns of surface and near-surface waters in the Pacific ocean inferred by Kennett *et al.* (1985) at three stages during the Neogene as the Indonesian sea-way closed. Black arrows indicate cold currents and unfilled arrows indicate warm currents

Fig. 18. Possible circulation patterns of surface and near-surface waters in eastern Indonesia shown on the tectonic reconstructions of this paper. The currents postulated are based on Kennett *et al.* (1985) and present-day circulation patterns (Fine *et al.*, 1994).

The second major period is around 25 Ma when plate boundaries and motions changed again, partly due to collision between the north Australian margin and arcs to the north. This, together with collision of the Melanesian arcs and the Ontong Java plateau, changed the tectonics of the oceanic-arc region east of Asia (Philippines, Celebes Sea, Sulu Sea, Philippine Sea, Caroline Sea, north New Guinea, New Britain, Solomons, Tonga). The 25 Ma event was probably the most important tectonic event from the biogeographical point of view as it led to new, albeit discontinuous, links between Australia via Sulawesi into SE Asia across areas which were mainly shallow marine and locally included land. It also resulted in a very long discontinuous island arc link between Asia and Melanesia. However, as the pathways between Australia and Sundaland came into existence, new barriers also formed. The central Borneo mountains began to rise in the early Miocene and became a regional drainage divide sending sediment north into the Sarawak basins and Baram delta, and southeast into the Tarakan and Mahakam deltas. North of Borneo, as the proto-South China Sea closed, the Oligo-Miocene South China Sea widened and the Sulu Sea opened. As the distance between Australia and Sulawesi closed, the deep Banda Sea opened. Thus, movement of plants and animals between Australia and Sundaland would have remained difficult. Perhaps it was this zone of barriers, close to a region of deep and former deep ocean barriers separating Borneo and Australia, which is the origin of Wallace's line. The narrow Makassar Strait, which at its south end terminates in a long-lived discontinuous carbonate platform, could not alone have been a major barrier to dispersal.

Plate motions and boundaries changed again at about 5 Ma, possibly as a consequence of arc-continent collision in Taiwan, and in the last 5 Ma there has been renewed tectonic activity and a significant increase in land and highlands all round the margins of SE Asia. A number of new dispersal pathways developed across the region, for example those linking Taiwan and New Guinea through the Philippines and North Moluccas, and connecting New Guinea to Thailand via the Banda and Sunda arcs. It is also probable that there was an increase in the range of habitats along these routes, due to elevation of mountains, and likely associated variations in rainfall.

Disentangling the contribution of geology to biogeographic patterns is not simple. Geology and tectonics could be a controlling factor in some cases. Cicada distributions in New Guinea suggest a geological control (Boer and Duffels, 1996), and slicing of crustal fragments from the Bird's Head could have caused influxes of faunas and floras into Sulawesi from Australia at intervals in the last 20 Ma. However, geology and tectonics also influence other variables which are more subtle controls on biogeographic patterns. Sea-level, elevation of land areas, soil, wind and water movements, and climate are all examples of factors upon which there is some geological influence. Climatic controls are too difficult to model at present, but at some time in the future it will be possible to use the tectonic models as the basis for simulation of ancient climates in SE Asia. It is notable that at present there are more highland areas, and a greater area of land than at any time during the last 30 million years. This is consistent with rather restricted areas of modern carbonate platforms which are limited in part by clastic sediment influx. The present distribution and size of shallow water carbonate areas may in part reflect a period of relatively low sea-level, but also record the recent rise of mountains due to tectonic forces as the region is compressed between Asia and Australia.

Some of the biogeographic patterns in SE Asia at present are difficult to relate simply to geology, for example, the distance between Borneo and Sulawesi (Wallace's line and equivalents) should have been as easy to cross as the barriers between Australia and Sulawesi. This raises the question of the longevity of biogeographic patterns, about which we currently lack adequate information. During the last million years there have been periods of low sea-level associated with glacial intervals when far greater areas of land were emergent than at present, and the present areas are significantly greater than those during the Neogene. Much of the Sunda shelf would have been emergent although in eastern Indonesia there are many narrow deep water areas (such as the Makassar Strait) which would have remained physical barriers. However, elsewhere large sea-level falls would have separated some formerly connected ocean basins as shallow water areas became emergent, changing oceanic circulation patterns and modifying weather and climate (*e.g.* Huang *et al.*, 1997). Fluctuations in temperatures and rainfall are likely to have been more extreme at intervals in the last million years than in the preceding 30 million years. Therefore, the last period of geological history, perhaps one million years or even much less, may have had a far greater influence on biogeographic patterns than the much longer period before.

To go further, detailed maps of land and sea, and palaeo-topography must be compiled from published maps and papers, and unpublished coastline, shelf edge, age and lithofacies information, much in oil company files. In particular the display of uplift and subsidence, and timing of magmatic events, on tectonic reconstructions would help in identifying underlying processes and give more confidence in mapping land and sea into areas where there is little direct evidence. Biogeographers must contribute, for example, distributions of fossil plants can provide information on palaeo-temperatures and environments. There is a need to focus on biogeographic patterns which are most likely to reveal the links to geology using plants and animals which have difficulty in dispersing, and for which non-geological controls are unimportant. It is for the biogeographers to identify such critical floral and faunal indicators. We still know little about rates of speciation and dispersal, and for most animals and plants the fossil record is poor or non-existent. DNA studies offer one way of determining a time-scale for biological development which could contribute to an explanation of biogeographic patterns. Another way forward is mathematical simulation of the biological variables, testing biogeographic patterns against predictions. It is certain that no single factor will account for the distribution of plants and animals in SE Asia; tectonic movements may be a control but their importance is still far from clear.

Acknowledgements

Financial support has been provided by NERC, the Royal Society, the London University Central Research Fund, and the London University SE Asia Research Group currently supported by Arco, Canadian Petroleum, Exxon, Lasmo, Mobil, Union Texas and Unocal. Work in Indonesia has been facilitated by GRDC, Bandung and Directors including H. M. S. Hartono, M. Untung, R. Sukamto and I. Bahar. I am grateful to Moyra Wilson, Steve Moss and Alistair Fraser for information. Clive Burrett, Bob Musgrave, Gordon Packham and Rupert Sutherland made helpful comments on the reconstructions and the manuscript. I thank Kevin Hill for considerable help and discussion in reconstructing the New Guinea and the SW Pacific.

References

Aitchison, J. C., Clarke, D. L., Meffre, S. and Cluzel, D. 1995. Eocene arc-continent collision in New Caledonia and implications for regional southwest Pacific tectonic evolution. Geology 23: 161-164.

Briais, A., Patriat, P. and Tapponnier, P. 1993. Updated interpretation of magnetic anomalies and seafloor spreading stages in the South China Sea: implications for the Tertiary tectonics of Southeast Asia. Journal of Geophysical Research 98: 6299-6328.

Boer, A. J. de and Duffels, J. P., 1996. Historical biogeography of the cicadas of Wallacea, New Guinea and the West Pacific: a geotectonic explanation. Palaeogeography, Palaeoclimatology, Palaeoecology 124: 153-177.

Burrett, C., Duhig, N., Berry, R. and Varne, R. 1991. Asian and South-western Pacific continental terranes derived from Gondwana, and their biogeographic significance. *In* Austral Biogeography. Edited by P. Y. Ladiges, C. J. Humphries and L. W. Martinelli, Australian Systematic Botany 4: 13-24.

Cambridge Paleomap Services. 1993. ATLAS version 3.3. Cambridge Paleomap Services, P.O. Box 246, Cambridge, U.K.

Crook, K. A. W. and Belbin, L. 1978. The Southwest Pacific area during the last 90 million years. Journal of the Geological Society of Australia 25: 23-40.

Curray, J. R., Moore, D. G., Lawver, L. A., Emmel, F. J., Raitt, R. W., Henry, M. and Kieckheffer, R. 1979. *In* Tectonics of the Andaman Sea and Burma. American Association of Petroleum Geologists Memoir 29: 189-198.

Daly, M. C., Cooper, M. A., Wilson, I., Smith, D. G. and Hooper, B. G. D. 1991. Cenozoic plate tectonics and basin evolution in Indonesia. Marine and Petroleum Geology 8: 2-21.

Davies, H. L. 1971. Peridotite-gabbro-basalt complex in eastern Papua: an overthrust plate of oceanic mantle and crust. BMR Journal of Australian Geology and Geophysics 128: 48 pp.

Dewey, J. F., Cande, S. and Pitman W. C. III 1989. Tectonic evolution of the India/Eurasia collision Zone. Eclogae Geologicae Helvetiae 82: 717-734.

England, P. and Houseman, G. 1986. Finite strain calculations of continental deformation 2. Comparison with the India-Asia collision zone. Journal of Geophysical Research 91: 3664-3676.

Fine, R. A., Lukas, R., Bingham, F. M., Warner, M. J. and Gammon, R. H. 1994. The western equatorial Pacific: a water mass crossroads. Journal of Geophysical Research 99 (C12): 25,063-25080.

Hall, R. 1995. Plate tectonic reconstructions of the Indonesian region. Proceedings of the Indonesian Petroleum Association 24th Annual Convention: 71-84.

Hall, R. 1996. Reconstructing Cenozoic SE Asia. *In* Tectonic Evolution of SE Asia. Edited by R. Hall and D. J. Blundell. Geological Society of London Special Publication 106: 153-184.

Hall, R. 1997. Cenozoic tectonics of SE Asia and Australasia. *In* Petroleum Systems of SE Asia and Australasia. pp. 47-62. Edited by J. V. C. Howes and R. A. Noble. Indonesian Petroleum Association, Jakarta.

Hall, R., Fuller, M., Ali, J. R. and Anderson, C. D. 1995. The Philippine Sea Plate: magnetism and reconstructions. American Geophysical Union Monograph 88: 371-404.

Hamilton, W. 1979. Tectonics of the Indonesian region. USGS Professional Paper 1078: 345 pp.

Haq, B. U., Hardenbol, J. and Vail, P. R. 1987. Chronology of

fluctuating sea levels since the Triassic. Science 235: 1156-1167.
Harley, M. M. and Morley, R. J. 1995. Ultrastructural studies of some fossil and extant palm pollen, and the reconstruction of the biogeographical history of subtribes Iguanurinae and Calaminae. Review of Palaeobotany and Palynology 85: 153-182.
Hinz, K., Block, M., Kudrass, H. R. and Meyer, H. 1991. Structural elements of the Sulu Sea. Geologische Jahrbuch A127: 883-506.
Houseman, G. and England, P. 1993. Crustal thickening versus lateral expulsion in the Indian-Asian continental collision. Journal of Geophysical Research 98: 12,233-12,249.
Huang, C.-Y., Liew, P.-M., Zhao, M., Chang, T.-C., Kuo, C.-M., Chen, M.-T., Wang, C.-H. and Zheng, L.-F., 1997. Deep sea and lake records of the Southeast Asian paleomonsoons for the last 25 thousand years. Earth and Planetary Science Letters 146: 59-72.
Hutchison, C. S. 1996. South-East Asian Oil, Gas, Coal and Mineral Deposits. Clarendon Press, Oxford. 265 pp.
Katili, J. A. 1975. Volcanism and plate tectonics in the Indonesian island arcs. Tectonophysics 26: 165-188.
Kennett, J. P., Keller, G., and Srinivasan, M. S. 1985. Miocene planktonic foraminiferal biogeography and paleogeographic development of the Indo-Pacific region. *In* The Miocene ocean: paleogeography and biogeography. Geological Society of America Memoir 163: 197-236.
Kündig, E. 1956. Geology and ophiolite problems of east Celebes. Verhandelingen van het Koninklijk Nederlandsch Geologisch-Mijnbouwkundig Genootschap, Geologische Serie 16: 210-235
Lee, T-Y. and Lawver, L. A. 1994. Cenozoic plate tectonic reconstruction of the South China Sea region. Tectonophysics 235: 149-180.
McCaffrey, R. 1996. Slip partitioning at convergent plate boundaries of SE Asia. *In* Tectonic Evolution of SE Asia. Edited by R. Hall and D. J. Blundell. Geological Society of London Special Publication 106: 3-18.
Meffre, S. 1995. The development of island arc-related ophiolites and sedimentary sequences in New Caledonia. Ph.D. thesis, University of Sydney, Australia, 239 pp.
Metcalfe, I. 1998. Palaeozoic and Mesozoic geological evolution of the SE Asian region: multi-disciplinary constraints and implications for biogeography. *In* Biogeography and Geological Evolution of SE Asia. Edited by R. Hall and J. D. Holloway (this volume).
Monnier, C., Girardeau, J., Maury, R. C. and Cotten, J. 1995. Back arc basin origin for the East Sulawesi ophiolite (eastern Indonesia). Geology 23: 851-854.
Moss, S. J. and Wilson, M. J. 1998. Tertiary evolution of Sulawesi-Borneo: implications for biogeography. *In* Biogeography and Geological Evolution of SE Asia. Edited by R. Hall and J. D. Holloway (this volume).
Mubroto, B., Briden, J. C., McClelland, E. and Hall, R. 1994. Palaeomagnetism of the Balantak ophiolite, Sulawesi. Earth and Planetary Science Letters 125: 193-209.
Norton, I. O. 1995. Tertiary relative plate motions in the North Pacific: the 43 Ma non-event. Tectonics 14: 1080-1094.
Packham, G. 1996. Cenozoic SE Asia: Reconstructing its aggregation and reorganisation. *In* Tectonic Evolution of SE Asia. Edited by R. Hall and D. J. Blundell. Geological Society of London Special Publication 106: 123-152.
Peltzer, G. and Tapponnier, P. 1988. Formation of strike-slip faults, rifts, and basins during the India-Asia collision: an experimental approach. Journal of Geophysical Research 93: 15085-15117.
Polvé, M. Maury, R. C., Bellon, H., Rangin, C., Priadi, B., Yuwono, S., Joron, J. L. and Soeria Atmadja, R. 1997. Magmatic evolution of Sulawesi (Indonesia): constraints on the Cenozoic geodynamic history of the Sundaland active margin. Tectonophysics 272: 69-92.
Rangin, C., Jolivet, L., Pubellier, M. 1990. A simple model for the tectonic evolution of southeast Asia and Indonesia region for the past 43 m. y. Bulletin de la Société géologique de France 8 VI: 889-905.
Rowley, D. B. 1996. Age of initiation of collision between India and Asia: a review of stratigraphic data. Earth and Planetary Science Letters 145: 1-13.
Silver, E. A. and Rangin, C. 1991. Leg 124 tectonic synthesis. *In* Proceedings of the Ocean Drilling Program, Scientific Results 124: 3-9. Edited by E. A. Silver, C. Rangin, M. T. von Breymann *et al.*
Stern, R. J. and Bloomer, S. H. 1992. Subduction zone infancy: examples from the Eocene Izu-Bonin-Mariana and Jurassic California arcs. Bulletin of the Geological Society of America 104: 1621-1636.
Tapponnier, P., Lacassin, R., Leloup, P., Scharer, U., Dalai, Z., Haiwei, W., Xiaohan, L., Shaocheng, J., Lianshang, Z. and Jiayou, Z. 1990. The Ailao Shan/Red River metamorphic belt: Tertiary left lateral shear between Indochina and South China. Nature 343: 431-437.
Tapponnier, P., Peltzer, G., LeDain, A., Armijo, R. and Cobbold, P. 1982. Propagating extrusion tectonics in Asia: new insights from simple experiments with plasticine. Geology 10: 611-616.
Tapponnier, P., Peltzer, G. and Armijo, R. 1986. On the mechanics of the collision between India and Asia. *In* Collision Tectonics. Edited by M. P. Coward and A. C. Ries. Special Publication of the Geological Society of London 19: 115-157.
Unrug, R. 1997. Rodinia to Gondwana: the geodynamic map of Gondwana supercontinent assembly. GSA Today 7 (1): 1-6.
Yan, C. Y. and Kroenke, L. W. 1993. A plate tectonic reconstruction of the SW Pacific 0-100 Ma. *In* Proceedings of the Ocean Drilling Program, Scientific Results 130: 697-709. Edited by T. Berger, L. W. Kroenke, L. Mayer. *et al.*

Colour plates for:

The plate tectonics of Cenozoic SE Asia and the distribution of land and sea

Robert Hall
SE Asia Research Group, Department of Geology, Royal Holloway University of London

Captions

Fig. 5. Present-day tectonic features of SE Asia and the SW Pacific. Yellow lines are selected marine magnetic anomalies. Cyan lines outline bathymetric features. Red lines are active spreading centres. White lines are subduction zones and strike-slip faults. The present extent of the Pacific plate is shown in pale blue. Areas filled with green are mainly arc, ophiolitic, and accreted material formed at plate margins during the Cenozoic. Areas filled in cyan are submarine arc regions, hot spot volcanic products, and oceanic plateaus. Pale yellow areas represent submarine parts of the Eurasian continental margins. Pale and deep pink areas represent submarine parts of the Australian continental margins. Letters represent marginal basins and tectonic features as follows:

Marginal Basins				**Tectonic features**					
A	Japan Sea	N	Ayu Trough	Ba	Banda Arc	Mk	Makassar Strait	Ry	Ryukyu Arc
B	Okinawa Trough	P	Caroline Sea	BH	Bird's Head	Mn	Manus Island	Sa	Sangihe Arc
C	South China Sea	Q	Bismarck Sea	Ca	Cagayan Arc	NB	New Britain Arc	Se	Sepik Arc
D	Sulu Sea	R	Solomon Sea	Fj	Fiji	NC	New Caledonia	So	Solomons Arc
E	Celebes Sea	S	Woodlark Basin	Ha	Halmahera Arc	NH	New Hebrides Arc	Sp	Sula Platform
F	Molucca Sea	T	Coral Sea	IB	Izu-Bonin Arc	NI	New Ireland	Su	Sulu Arc
G	Banda Sea	U	Tasman Sea	Ja	Japan Arc	Nng	North New Guinea Terranes	TK	Three Kings Rise
H	Andaman Sea	V	Loyalty Basin	Lo	Loyalty Islands				
J	West Philippine Basin	W	Norfolk Basin	Lu	Luzon Arc	Pa	Papuan Ophiolite	To	Tonga Arc
K	Shikoku Basin	X	North Fiji Basin			Pk	Palau-Kyushu Ridge	Tu	Tukang Besi Platform
L	Parece Vela Basin	Y	South Fiji Basin						
M	Mariana Trough	Z	Lau Basin						

Fig. 6. Reconstruction of the region at 50 Ma. The possible extent of Greater India and the Eurasian margin north of India are shown schematically. Shortly before 50 Ma collision between the north Australian continental margin and an island arc had emplaced ophiolites on the north New Guinea margin, and in New Caledonia, eliminating ocean crust formed at the former Australian-Indian ocean spreading centre. Double black arrows indicate extension in Sundaland.

Fig. 7. Reconstruction of the region at 40 Ma. India and Australia were now parts of the same plate. An oceanic spreading centre linked the north Makassar Strait, the Celebes Sea and the West Philippine basin. Spreading began at about this time in the Caroline Sea, separating the Caroline Ridge remnant arc from the South Caroline arc. Spreading also began after subduction flip in marginal basins around eastern Australasia producing the Solomon Sea and the island arcs of Melanesia.

Fig. 8. Reconstruction of the region at 30 Ma. Indentation of Eurasia by India led to extrusion of the Indochina block by movement on the Red River Fault and Wang Chao-Three Pagodas (WC-TP) Faults. Slab pull due to southward subduction of the proto-South China Sea caused extension of the South China and Indochina continental margin and the present South China Sea began to open. A wide area of marginal basins separated the Melanesian arc from passive margins of eastern Australasia, shown schematically between the Solomon Sea and the South Fiji basin.

Fig. 9. Reconstruction of the region at 20 Ma. Collision of the north Australian margin in the region between the Bird's Head microcontinent and eastern New Guinea occurred at about 25 Ma. The Ontong Java plateau arrived at the Melanesian trench at about 20 Ma. These two events caused major reorganisation of plate boundaries. Subduction of the Solomon Sea began at the eastern New Guinea margin. Spreading began in the Parece Vela and Shikoku marginal basins. The north Australian margin became a major left-lateral strike-slip system as the Philippine Sea-Caroline plate began to rotate clockwise. Movement on splays of the Sorong Fault system led to the collision of Australian continental fragments in Sulawesi. This in turn led to counter-clockwise rotation of Borneo and related Sundaland fragments, eliminating the proto-South China Sea. The Sumatra Fault system was initiated.

Fig. 10. Reconstruction of the region at 10 Ma. The Solomon Sea was being eliminated by subduction beneath eastern new Guinea and beneath the New Hebrides arc. However, continued subduction led to development of new marginal basins within the period 10-0 Ma, including the Bismarck Sea, Woodlark basin, North Fiji basins, and Lau basin. The New Guinea terranes, formed in the South Caroline arc, docked in New Guinea but continued to move in a wide left-lateral strike-slip zone. Further west, motion on strands of the Sorong Fault system caused the arrival of the Tukang Besi and Sula fragments in Sulawesi. Collision events at the Eurasian continental margin in the Philippines, and subsequently between the Luzon arc and Taiwan, were accompanied by intra-plate deformation, important strike-slip faulting and complex development of opposed subduction zones. Rotation of Borneo was complete but motion of the Sumatran forearc slivers was associated with new spreading in the Andaman Sea.

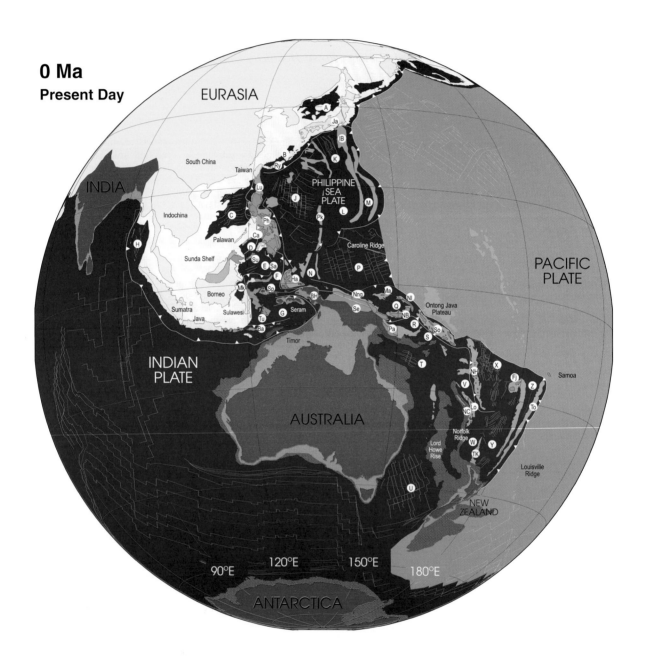

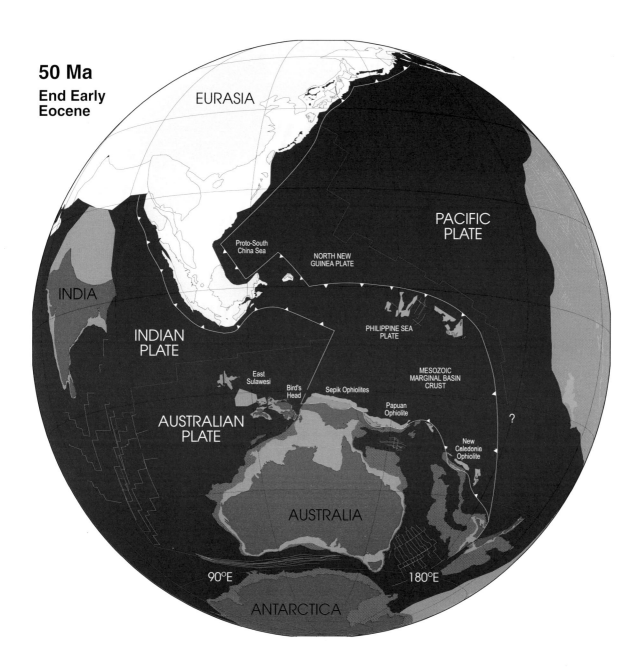

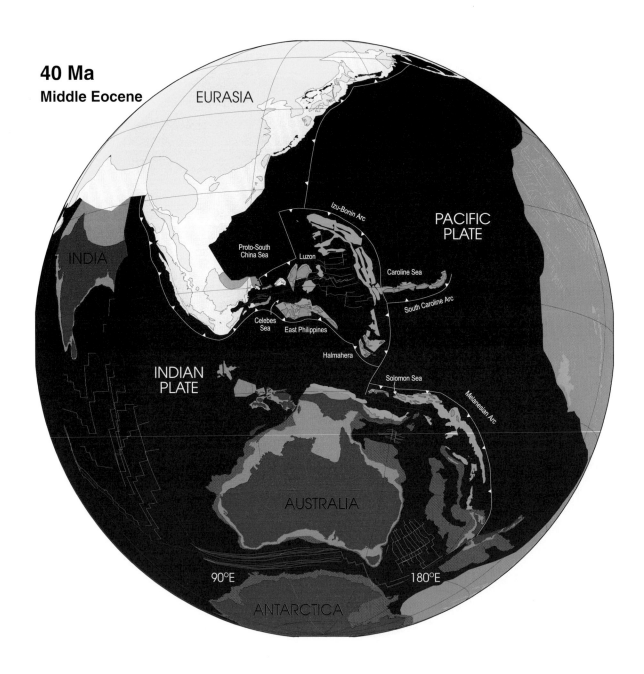

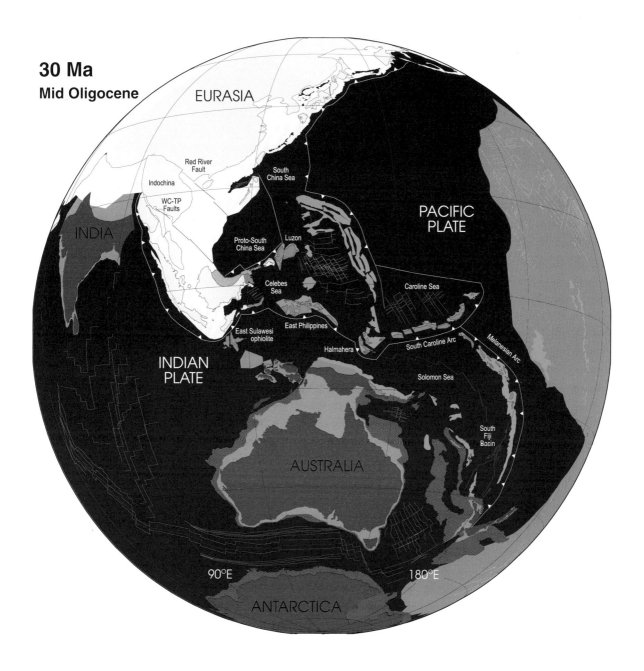

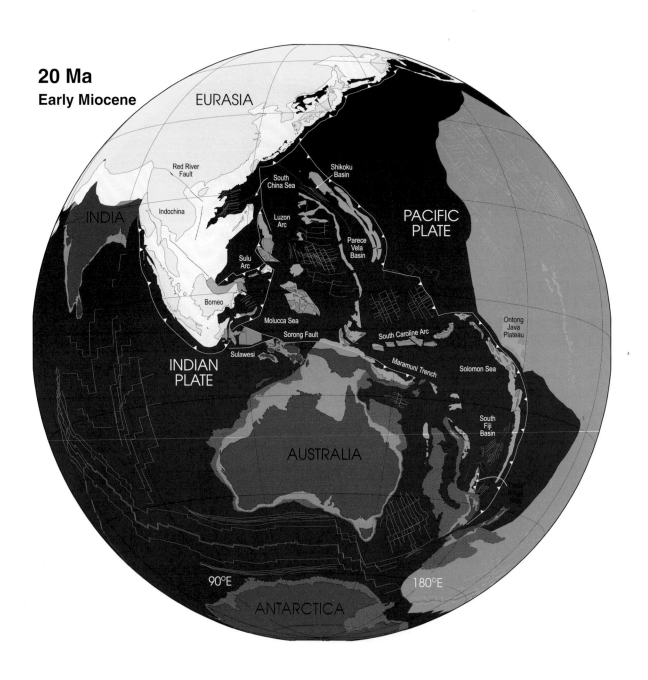

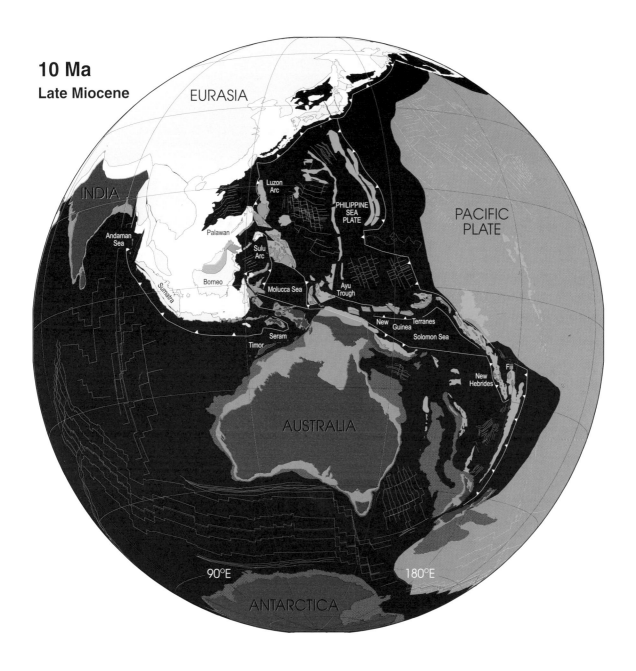

Biogeographic implications of the Tertiary palaeogeographic evolution of Sulawesi and Borneo

Steve J. Moss[1] and Moyra E. J. Wilson
SE Asia Research Group, Department of Geology, Royal Holloway University of London, Egham, Surrey, TW0 0EX, UK. Email moyra@gl.rhbnc.ac.uk [1]*Now at School of Applied Geology, Curtin University of Technology, Perth, 6845, W.A. Australia. Email steve@lithos.curtin.edu.au*

Key words: Wallacea, Wallace's Line, plants, animals, land, sea, maps

Abstract

Sulawesi and Borneo are located in a critical position for biogeography, bordering Wallace's faunal divide and in the middle of the Indonesian archipelago, an extremely active tectonic area throughout the Tertiary. Wallace's original line now marks the western boundary of Wallacea: a biogeographic zone with a high degree of species endemism between areas with Asiatic and Australian flora and fauna. Tectonic controls have strongly influenced the distribution of depositional environments and hence the past and present distribution of terrestrial and shallow marine biota.

Palaeogeographic maps presented, using plate tectonic reconstructions as a base, illustrate the evolution of the area and highlight important features for palaeobiogeography. The Tertiary geological history of eastern Kalimantan and Sulawesi is inextricably linked to the progressive accretion of continental and oceanic material from the east, onto the eastern margin of Sundaland (Eurasian margin), and to the resultant development of volcanic arcs. This westward drift of material throughout the Tertiary, particularly that of microcontinental blocks, may have provided a potential pathway which allowed rafting or island hopping of Australian biota towards Asia and vice versa. A land bridge existed between Borneo and mainland SE Asia for much of the Tertiary, whereas the formation of the Makassar Straits in the early Tertiary isolated small land areas in Sulawesi from those in Borneo. Both of these factors resulted in strong biogeographic differences between Borneo and Sulawesi and contributed to a high degree of endemism on Sulawesi. Chains of volcanic island arcs, related to subduction along the eastern edge of Sundaland during the Tertiary, may have presented island hopping routes to and from the Philippines, Borneo and Sulawesi, and possibly Java.

Introduction

The islands of Borneo and Sulawesi are of prime importance to the biogeography and palaeobiogeography of SE Asia. Wallace's (1863) faunal divide, originally thought to delineate regions of Asiatic and Australian flora and fauna, runs between the islands of Bali and Lombok and north through the Makassar Straits, which separate the islands of Borneo and Sulawesi (Fig.1; George, 1981). This faunal divide is now taken as the western boundary of Wallacea (Dickerson, 1928), a 'transitional' area between Asiatic and Australian biotas (Whitten *et al.*, 1987). Wallacea includes Sulawesi, the Moluccas and the Lesser Sunda islands as well as an extensive area of shallow sea, and its eastern margin is taken as Lydekker's line; the western boundary of the strictly Australian fauna (Fig.1). The transitional nature of biota within Wallacea is sometimes seen, for example, by the eastward increase in Australian representatives in the reptile fauna between the western and eastern boundary lines (Ziegler, 1983). However, for most organisms Wallacea does not represent a region of homogeneous biota, or a region of gradual change in species composition. In reality, Wallacea is best described as a biogeographic region, situated between areas with Asiatic and Australian floras and faunas, where organisms show a high degree of endemism (George, 1981; Whitten *et al.*, 1987).

The sharp contrast in fauna between areas bordering Wallace's original line is not reflected to the same extent by the flora (George, 1981; Balgooy, 1987). There are approximately 2,300 genera of flowering plants in total in the archipelago and for most Wallace's line is unimportant, although 297 genera, including some of the palms, do reach their eastern limit there

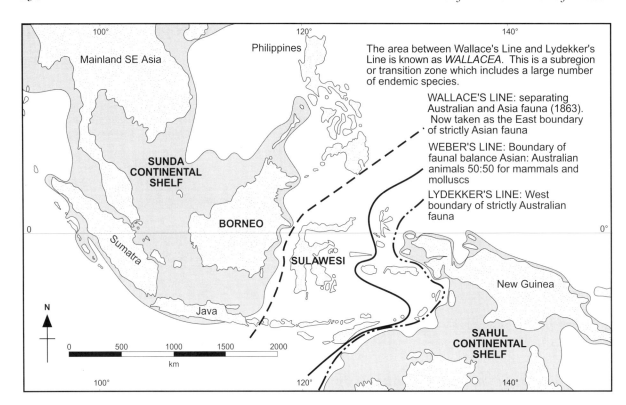

Fig.1. Map of SE Asia showing the location of Borneo and Sulawesi in the centre of the region. The original faunal divide of Wallace (1863) is shown. Areas of continental shelves are shown in grey.

(Dransfield, 1981; George, 1981). This may be because many plants are better at overseas dispersal (Briggs, 1987), because their range is often strongly related to suitable habitats and climate (Steenis, 1979; Balgooy, 1987; Takhtajan, 1987), or because certain floral elements were present on both sides of Wallace's line before the formation of the Makassar Straits. Multiple migration pathways, via northern and austral routes, have been suggested for some groups of plants after breakup of Gondwana (Dransfield, 1981, 1987; Whitmore, 1981b; Audley-Charles, 1987; Truswell *et al.*, 1987; Morley, 1998 this volume). For certain groups of animals the Makassar Straits appear to have been a barrier to dispersal, whereas organisms such as some oriental frogs, reptiles, birds and mammals occur on both sides of Wallace's line and in some cases their range extends as far as Australia. However, overall there is still a clear faunal change across the Makassar Straits (Cranbrook, 1981; Briggs, 1987; Musser, 1987).

The SE Asian region has been an extremely active tectonic area throughout the Cenozoic, and geological and geophysical evidence indicates that considerable lateral and vertical crustal, or plate, movements have occurred in the region. During the last twenty years several plate tectonic reconstructions have been postulated for the area (Carey, 1975; Peltzer and Tapponnier, 1988; Rangin *et al.*, 1990; Daly *et al.*, 1991; Lee and Lawver, 1994, 1995; Hall, 1996). There is now a growing consensus over some of the main points of the plate tectonic evolution of the area, although details of the reconstructions and the driving mechanisms are still under dispute. The reconstructions show the movements of lithospheric plates, and it is important to realise that these do not directly correspond to regions of land and sea.

A number of workers have related biogeography to the plate tectonic evolution of SE Asia. Audley-Charles *et al.* (1981) showed SE Asia in its world-wide context and identified major plate tectonic events, evaluating their importance for biogeographers. These included the separation of India from Antarctica-Australia, the separation of Australia-New Guinea from Antarctica, the juxtaposition of India with Asia, and most recently the convergence and subsequent juxtaposition of Australia with SE Asia. The regional evolution of SE Asia and its biogeographic implications, with particular reference to Sulawesi, were reviewed by Audley-Charles (1981). The formation of the Makassar Straits, and possible temporary land links across this seaway, and the juxtaposition of different tectonic fragments in Sulawesi, particularly those in the east with Australian affinity, were identified as major influences on the biogeography across Wallace's line. The evolution and dispersal of the angiosperms were discussed in the light of plate tectonic events by Audley-Charles (1987). Burrett *et al.* (1991) related rifting of fragments away from Australia and their subsequent convergence and collision with mainland SE Asia to biogeography. All authors note the importance of superimposing palaeoenvironmental data onto plate tectonic reconstructions to provide meaningful information to biogeographers, and the actual difficulties of undertaking such a task. Audley-Charles (1987) and Hall (1998 this volume) attempted to identify areas of past land, shallow and deep sea regions from the plate tectonic evolution of SE Asia.

This paper concentrates on the Tertiary evolution of Borneo and Sulawesi and synthesises palaeoenvironmental information obtained from the geological record with plate tectonic reconstructions. Areas of land, marginal marine, shallow marine and deep marine areas are shown on plate tectonic reconstructions of Borneo and Sulawesi for the Tertiary and provide a means to examine possible migrations or dispersals in the areas bordering Wallace's line.

Present-day distribution of organisms in Borneo and Sulawesi

Borneo and Sulawesi include a variety of habitats, such as high mountainous areas, lower rolling topography and flat coastal plains. A complex interplay of factors, including physical differences in environment (altitude, soil type and slope all influenced by local geology), local and regional variations in climate as well as regional geological evolution may all influence the nature and diversity of organisms found on Borneo and Sulawesi. Today both Borneo and Sulawesi have tropical climates, although some areas of Sulawesi may experience long dry periods and a more monsoonal climate than areas in Borneo. The present-day environments of Borneo and Sulawesi are described in detail in MacKinnon *et al.* (1996) and Whitten *et al.* (1987).

The biota of Borneo at both generic and species level is similar to mainland SE Asia and other islands on the Sunda Shelf, particularly Sumatra, from which Borneo is separated by about 220 km (Fig.1). There are, however, differences and in particular Borneo has a higher number of endemic plants and animals than adjacent areas to the west. In relation to its size the fauna of Borneo is less diverse than its western neighbour Sumatra, but more of its animals are exclusive with thirty nine land animals and thirty birds being endemic to the island (Whitten and Whitten, 1992; MacKinnon *et al.*, 1996). The mammal population of Borneo, represented by 222 species of land animals, is almost identical to that of mainland Asia at family level with primates, bears, cats, squirrels and rhinos all represented. Borneo has the highest number of species of primates and tree shrews within the SE Asian region (MacKinnon *et al.*, 1996). Borneo has some 450 resident species of birds, the third highest in the region after New Guinea and Sumatra, which is a reflection of the size of the island, the diversity of available habitats and proximity to mainland Asia. Borneo is one of the richest islands on the Sunda Shelf for freshwater fish (394 species of which 149 are endemic), amphibians (at least 100) and reptiles (at least 166 species of snake, MacKinnon *et al.*, 1996). Most of the freshwater fish species of Borneo are seen in Sumatra and elsewhere in SE Asia, although again Borneo has a higher number of endemic species. Of all the vertebrate groups the primary freshwater fish most clearly demarcate Wallace's original line and there is a total absence of primary division freshwater fish in Sulawesi, although a few occur in the Philippines (Kottelat *et al.*, 1993).

Borneo has the most diverse flora of the Sunda Islands, with some 10,000-15,000 species of flowering plants of which one third are endemic (Whitten and Whitten, 1992; MacKinnon *et al.*, 1996). Invertebrate species are extremely abundant on Borneo, although accurate figures are not available for the less well known groups. Much of the flora and fauna endemic to Borneo is confined to the high mountainous areas, particularly in Sabah and Sarawak, indicating a degree of niche partitioning related to local environmental conditions and to the lateral separation of similar habitats.

On Sulawesi there is an extremely high degree of endemism amongst the fauna, particularly of the mammals. The biota of Sulawesi, particularly at a generic and higher level, shows affinities with those of both Australia and Asia, although far fewer families are represented compared with Borneo or New Guinea (Whitten *et* al., 1987; Michaux, 1994; Holloway, 1997). Out of 127 species of indigenous mammals, 79 (62%) are endemic to Sulawesi, and this rises to 98% if the bats are excluded (Whitten et al., 1987). Much of the characteristic Sundaic fauna, including moles, flying lemurs, tree shrews, lorises, gibbons, pangolins, porcupines, dogs, otters, weasels, cats, elephants, tapirs, rhinoceroses and mouse-deer do not occur east of Wallace's line (Musser, 1987; Whitten et al., 1987), although Plio-Pleistocene elephant fossils have been found in Sulawesi (Cranbrook, 1981; Whitten et al., 1987; Aziz, 1994). Of the Sulawesian mammals many of the placentals have Sundaic affinities, some of the endemics, such as the anoa, have no Sundaic relatives, whereas the marsupials are clearly an Australian element (Musser, 1987; Michaux, 1994). A high degree of endemism also occurs within the indigenous amphibian (19 out of 25 species) and reptile (13 out of 40 lizard species and 15 out of 64 species of snake are endemic with one monotypic genus) populations and it is likely that more species of these groups have yet to be discovered (Whitten et al., 1987). About a quarter of the birds on Sulawesi (total 328) are endemic and although a dominant Sundaic influence is seen, some Australian elements also occur (Mayr, 1944; Whitten et al., 1987; Holmes and Phillipps, 1996; Michaux, 1996). Unlike Borneo, Sulawesi has no records of strictly freshwater fish (Cranbrook, 1981; Whitten et al., 1987; Kottelat et al., 1993). However, of the fish that live primarily in freshwater and show a little saltwater tolerance, and those with considerable marine tolerance, such as the diadromous fish, 68 species occur in Sulawesi and of these 52 are endemic (78%, compared with 38% in Borneo). Like Borneo there are 8 endemic genera (Kottelat et al., 1993).

The distribution of flora in Sulawesi appears to be strongly related to climatic and local physical conditions, including altitude and soil types (Whitten et al., 1987). Analysis of over 4000 species in 540 genera suggests that the Sulawesi flora is most closely related to other relatively dry islands in the region (Balgooy, 1987). Dransfield (1981), suggested that the paucity of palms on Sulawesi may be related to a drier Pleistocene climate. Balgooy et al. (1996), recognised 933 indigenous plant species on Sulawesi, and of these 112 were endemic to the island. Montane floras are similar to those in Borneo, whereas the flora of the lowlands and areas underlain by ultrabasic rocks have a strong affinity with that of New Guinea (Balgooy, 1987;

Whitten et al., 1987). Of the non-endemic plants in Borneo, about 50% do not occur in Sulawesi (Whitten et al., 1987).

The invertebrate fauna of Sulawesi in general has affinities with areas to the west, but in comparison is depauperate and displays a higher degree of endemism (Gressitt, 1961; Whitten et al., 1987; Vane-Wright, 1991). An Australian, or in some cases a Philippine, affinity has also been recognised within some of the invertebrates, including the Lepidoptera (de Jong, 1990; Holloway, 1987, 1990, 1997) and the cicadas. The cicadas are poor dispersers, show a high degree of endemism (73 out of 77 species in Sulawesi are endemic), and a number of species are restricted to one of Sulawesi's arms or to its central part, perhaps reflecting the Tertiary geological evolution of Sulawesi (Duffels, 1990).

Data sources

The plate tectonic reconstructions of Hall (1996) are used as a template in this paper upon which to draw palaeogeographic maps. Although these reconstructions differ from many of the earlier plate tectonic reconstructions for SE Asia (Rangin et al., 1990; Daly et al., 1991; Lee and Lawver, 1994; 1995) in terms of their detail, particularly in eastern SE Asia, the Borneo-Sulawesi parts of the reconstructions do not differ significantly for the purposes of plotting palaeogeographic data.

The series of palaeogeographic maps presented here illustrates the evolution of Borneo and Sulawesi during the Cenozoic. The data used in constructing these maps were derived from fieldwork by the London University SE Asia Research Group, and an extensive literature review of the islands of Borneo and Sulawesi. These maps have been based on available geological evidence, including facies data, stratigraphic information, biostratigraphy, igneous, metamorphic, structural and palaeomagnetic data. Significant gaps exist within the data set and particularly for some of the remoter areas there is limited information. Previous attempts at palaeogeographic reconstruction have either been limited to small areas, such as hydrocarbon exploration blocks (Wain and Berod, 1989) or very generalised palaeogeographies for very long time periods (Umbgrove, 1938; Beddoes, 1980; Rose and Hartono, 1978; Weerd and Armin, 1992). As far as we know this is the first attempt to synthesise palaeographic data with plate tectonic information for the whole area of

Borneo and Sulawesi.

The most important environmental elements distinguished are land areas, including regions of mountainous topography, low lying regions of fluvial deposition, major river systems and marginal marine/deltaic systems. Areas of shallow water and deeper water, fine grained and coarser redeposited clastic and carbonate deposits were also mapped. Volcanic centres, with regions of inferred subaerial and submarine volcanism, are also shown on the reconstructions. Thus the key environments for terrestrial and marine faunas and floras are emphasised. The outlines of the present coastlines are shown on the reconstructions for reference.

Geology and tectonics of Borneo and Sulawesi

Borneo and Sulawesi are situated in a tectonically complex region between three major plates (Eurasia, Indo-Australia and Pacific/Philippine Sea). The present day setting is mirrored by the complexity of the pre-Tertiary and Tertiary geology of these two islands. Large areas of eastern Kalimantan and western Sulawesi had been accreted onto southwestern Borneo, part of the eastern margin of Sundaland, by the Cenozoic (Hall, 1996; Metcalfe, 1998 this volume). Subduction of the Indian Ocean, Philippine Sea and Molucca Sea plates has been responsible for the progressive collision and accretion of fragments of continental and oceanic crust along the eastern margin of Sundaland throughout the Cenozoic. Within this overall compressional regime a number of sedimentary basins and deep marginal basins formed along the eastern and southern margins of Sundaland as a result of Tertiary extension and subsidence.

Borneo

Borneo is bounded by three marginal basins (South China, Celebes and Sulu Seas), microcontinental fragments of south China origin to the north, and mainland SE Asia (Indochina and peninsular Malaysia) to the west (Figs.1 and 2). Borneo has been interpreted as the product of Mesozoic accretion of oceanic crustal material (ophiolite), marginal basin fill, island arc material and microcontinental fragments onto the Palaeozoic continental core of the Schwaner Mountains in the SW of the island (Fig.2; Hutchison, 1989; Metcalfe, 1998 this volume). At the beginning of the Tertiary, Borneo formed a promontory of the Sundaland craton: the stable eastern margin of the Eurasian plate (Hall, 1996; Metcalfe, 1998 this volume). East of Borneo, separating it from Sulawesi, are the deep Makassar basins (Fig.2), formed during the Paleogene (Situmorang, 1982). Two NW-SE trending fault zones, the Adang and the Sangkulirang, bound the North Makassar basin to the south and north (Fig.2). Major tracts of eastern, central and northern Borneo are covered by Tertiary sediments (Fig.2) which were deposited in fluvial, marginal-marine or marine environments. These depocentres were often laterally interconnected through intricate and narrow links (Fig.2; Pieters *et al.*, 1987; Pieters and Supriatna, 1990). Tertiary sedimentation in these regions occurred contemporaneously with, and subsequent to, a period of widespread Paleogene extension and subsidence, which may have begun in the middle Eocene or earlier.

Sulawesi

Sulawesi is formed of distinct north-south trending tectonic provinces (Fig.3; Sukamto, 1975) which are thought to have been sequentially accreted onto Sundaland during the Cretaceous and Tertiary. In part due to a lack of information, and to different interpretations of the available data, the evolution and juxtaposition of fragments within Sulawesi remain highly contentious and an attempt has been made to describe different hypotheses and to evaluate how these might affect biogeographic studies.

The north and south arms of Sulawesi are composed of thick Tertiary sedimentary and volcanic sequences overlying pre-Tertiary tectonically intercalated metamorphic, ultrabasic and marine sedimentary rocks (Fig.3; Sukamto, 1975; Leeuwen, 1981). Central Sulawesi and parts of the SE arm of Sulawesi are composed of sheared metamorphic rocks and in the east there is a highly tectonised melange complex (Sukamto, 1975; Hamilton, 1979). Similarities between the pre-Tertiary rocks and dating of metamorphic rocks suggests that they were accreted onto the eastern margin of Sundaland before the Tertiary (Sukamto, 1975; Hasan, 1991; Parkinson, 1991; Wakita *et al.*, 1994). The Tertiary stratigraphy of western Sulawesi is similar to that of east Kalimantan and the East Java Sea because the whole area began to subside in the early middle Eocene and a large basin formed (Weerd and Armin, 1992).

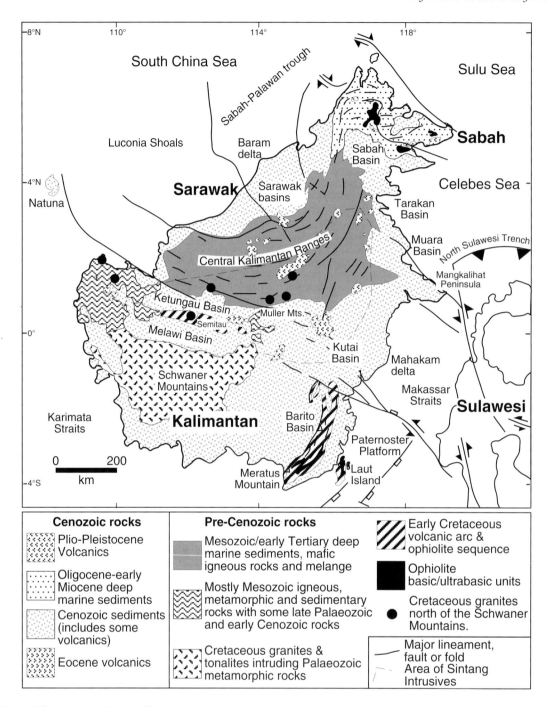

Fig.2. Simplified geological map of Borneo.

The eastern side of south Sulawesi and much of the eastern part of the east and southeast arms of Sulawesi are composed of tectonically intercalated marine sedimentary rocks and mafic and ultramafic igneous rocks (Sukamto, 1975; Silver *et al.*, 1978; Simandjuntak, 1990; Parkinson, 1991; Bergman *et al.*, 1996). These rocks are ophiolites inferred to represent oceanic lithosphere and overlying marine sediments accreted to Sulawesi. Dates varying between Cretaceous to Miocene have been obtained for the mafic and ultramafic rocks in eastern south Sulawesi (Yuwono *et al.*, 1987; Bergman *et al.*, 1996), although it is not clear which represent emplacement or deformation ages (Bergman *et al.*, 1996; Polvé *et al.*, 1997).

Tertiary palaeogeography of Sulawesi and Borneo

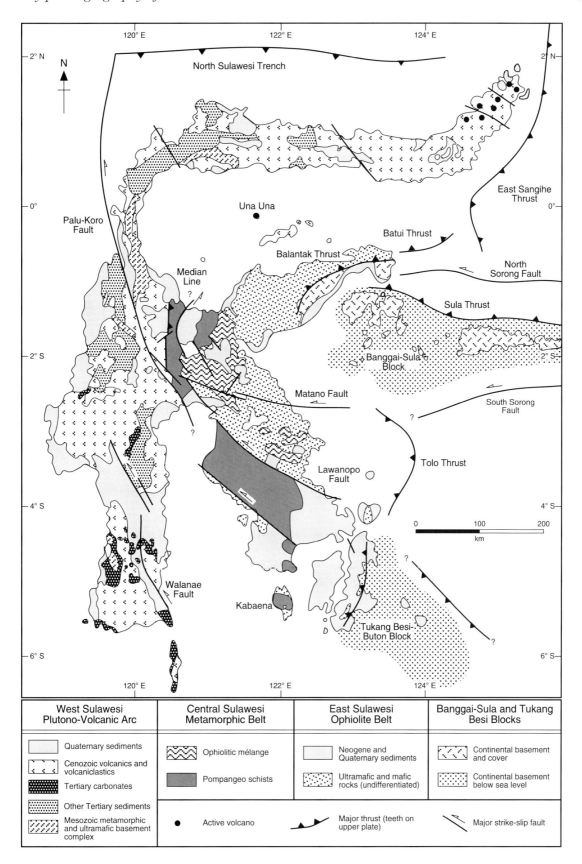

Fig.3. Simplified geological map of Sulawesi.

Central and part of southeast Sulawesi have been attached to western Sulawesi in the reconstructions. The East Sulawesi ophiolite was emplaced after the mid Oligocene. Debate on the location and origin of this oceanic crust (Parkinson, 1991; Mubroto *et al.*, 1994; Monnier *et al.*, 1995) need not concern biogeographers since before accretion and uplift it was situated in a deep marine setting. Palaeogene igneous lithologies in the north arm of Sulawesi may also represent part of an ophiolite sequence (Monnier *et al.*, 1995), although both shallow and deep marine origins for these rocks have been inferred (Carlile *et al.*, 1990).

On the islands of Buton-Tukang Besi and Banggai-Sula metamorphic and igneous lithologies of continental origin are exposed or are thought to underlie sediments of Palaeozoic and Mesozoic ages respectively. The Palaeozoic lithologies have Australian-New Guinea affinities, whereas deposition of shallow and deep marine sediments during rifting and drifting of the fragments is inferred for the Mesozoic (Audley-Charles, 1974; Hamilton, 1979; Pigram and Panggabean, 1984; Garrard *et al.*, 1988) Buton is thought to have collided with eastern Sulawesi during the early (Davidson, 1991) or middle Miocene (Smith and Silver, 1991), whereas latest Miocene to early Pliocene collision with the east arm of Sulawesi is inferred for Banggai-Sula (Garrard *et al.*, 1988; Davies, 1990). Fortuin *et al.* (1990) and Davidson (1991) suggested that Tukang Besi was a separate microcontinental block which was accreted to Buton in the Plio-Pleistocene, although not all authors recognise this as a separate microcontinental fragment (Smith and Silver, 1991).

Palaeogeographic maps

Eocene

During the Eocene (Figs.4 and 5) a land connection between southern Borneo and mainland SE Asia is inferred (Pupilli, 1973) which may have been present since the Jurassic (Lloyd, 1978). Little geological information is available for the Sunda Shelf, since few wells penetrate this area. A thin cover (<300 m) of Quaternary sediments is reported to overlie pre-Tertiary rocks in this area (Ben-Avraham and Emery, 1973). Although land is inferred for this area throughout much of the Tertiary, it is possible that marine sediments deposited during possible transgressions of this region may have been removed by later erosion.

In the NW tip of Borneo, Maastrichtian to early Eocene fluvial/marginal marine sands were deposited within an intra-montane basin (Tate, 1991). These sediments may have been part of a larger fluvial system, fed by river systems from Indochina, supplying material to turbidites in Sarawak, Sabah and parts of Kalimantan (Fig.4) from late Cretaceous to lower Eocene times (Moss, 1998). In the early to middle Eocene deep marine turbidites in Sarawak were uplifted and deformed by the 'Sarawak orogeny' (Hutchison, 1996). Late middle Eocene shallow and deep marine sediments unconformably overlie older rocks. By the early to middle Eocene much of Borneo appears to have been emergent. Eocene (50-45 Ma) rhyolitic lavas and ash occur in three localities in Kalimantan (Heryanto *et al.*, 1993; Pieters *et al.*, 1993a; Suwarna *et al.*, 1993; Moss *et al.*, 1997).

On the reconstructions, west, central and parts of the SE arm of Sulawesi are regarded as a region of microcontinental material forming a contiguous land area during the early Paleogene. Much of mainland SE Asia, southern Borneo and western Sulawesi appears to have been emergent during the Paleocene and the early Eocene, with a distinct lack of dated sediments recorded from these periods. Geochemistry and dating of calc-alkaline rocks and interbedded sediments in eastern South Sulawesi suggests there was a volcanic arc in this area during the Paleogene (Sukamto, 1975; Leeuwen, 1981; Sukamto and Supriatna, 1982). Paleogene basic volcanics and volcaniclastic lithologies are also present in western central and northern Sulawesi, although both shallow and deep marine origins have been suggested (Trail *et al.*, 1974; Carlile *et al.*, 1990; Polvé *et al.*, 1997).

There was widespread basin formation in middle Eocene times around the margins of Sundaland. Much of eastern Borneo, western Sulawesi, the Makassar Straits and the east Java Sea was an area of Tertiary sedimentation, in which the depositional environments varied between fluvial, deltaic, shallow marine clastic and carbonate shelves and areas of deeper water sedimentation. Evidence for Eocene extension, block-faulting and subsidence is seen on seismic lines crossing the Makassar Straits (Burollet and Salle, 1981; Situmorang, 1982; Guntoro, 1995; Bergman *et al.*, 1996), and this was the time when the land connection between Borneo and Sulawesi was severed. Sea floor spreading began in the marginal oceanic basin of the Celebes Sea in the mid-Eocene (Weissel, 1980; Rangin and Silver, 1990) and may have influenced basin

Tertiary palaeogeography of Sulawesi and Borneo

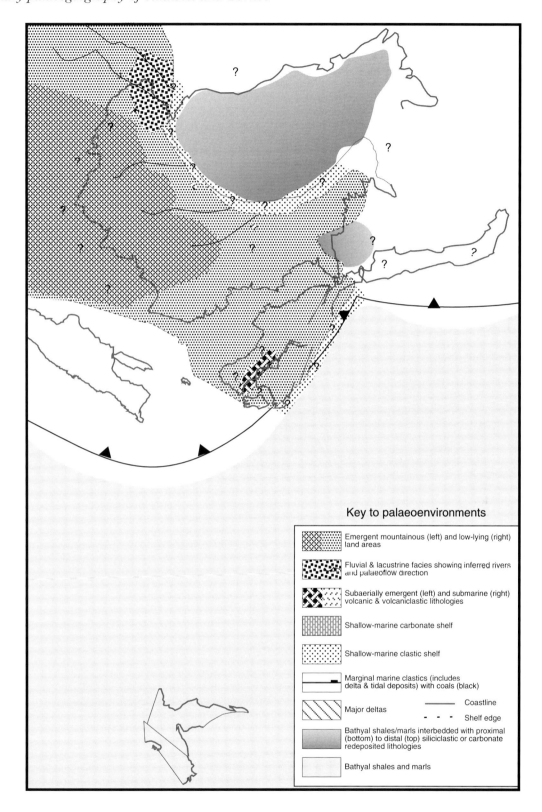

Fig.4. Palaeogeographic map for 50 Ma, early Eocene. A key to the environments is shown. Note that rivers are shown schematically for all time slices. See pages 157-164 for colour plates of Figs.4 to 9.

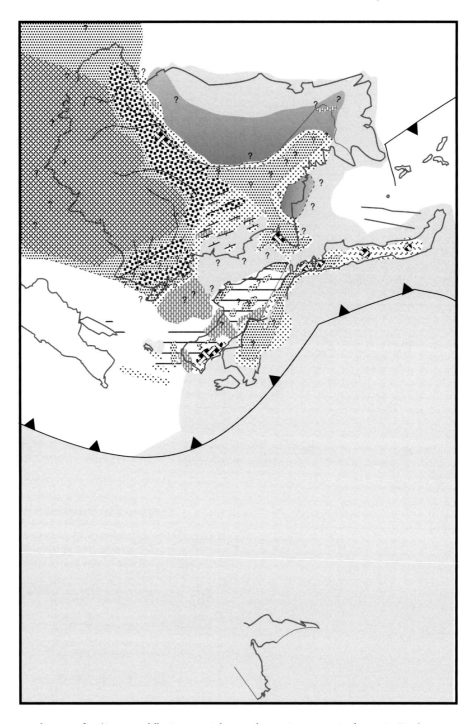

Fig.5. Palaeogeographic map for 42 Ma, middle Eocene. A key to the environments is shown in Fig.4.

initiation in Borneo and Sulawesi (Hall, 1996; Moss *et al.*, 1997). Rhyolitic volcanism, ash-falls and lava flows were partially contemporaneous with initiation of formation of the Tarakan (Netherwood and Wight, 1992) and Kutai basins (Leeuwen *et al.*, 1990; Moss *et al.*, 1997). Similar volcanism also occurred along the southern margin of the Mangkalihat peninsula (Sunaryo *et al.*, 1988) and in the Muller Mountains (Pieters *et al.*, 1993b).

In west Borneo fluvial and lacustrine sediments were deposited throughout much of the Eocene in elongate depocentres of the Melawi-Ketungau-Mandai area (Fig.2). It is in-

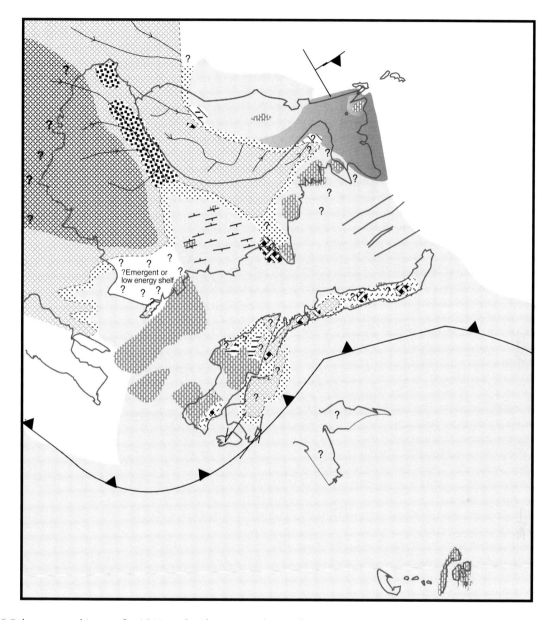

Fig.6. Palaeogeographic map for 34 Ma early Oligocene. A key to the environments is shown in Fig.4.

ferred that sediment deposited in these areas was supplied by rivers flowing from the south (Schwaner Mountains) or west (Indochina) since deep marine areas were present to the north and east. Fluvial and marginal marine sediments were also deposited in the Barito-Asem-Asem-Pasir area, although by late Eocene times shallow and deep marine clastic and carbonate sedimentation prevailed. Marginal marine lithologies with coals and intercalations of volcanic rocks occur with the Tarakan-Muara depocentres, although shallow and deeper water carbonate deposition was common by the late Eocene. By middle to late Eocene times (Moss and Finch, 1998), large parts of the Kutai depocentre and neighbouring western central Sulawesi were areas of deep, open oceanic sedimentation and this is inferred to have been the case for much of the north and south Makassar basins. Around the margins of this deep basin in the Kutai area, the northern part of the Tarakan basin and western, central and northern Sulawesi upper Eocene deltaic sands, coals and some shallow marine clastics were deposited (Kusuma and Darin, 1989; Coffield *et al.*, 1993; Moss *et al.*, 1997; Weerd and Armin, 1992). In western south Sulawesi marginal marine clastics and coals are conformably overlain by a thick shallow marine carbonate succession (Wilson and Bosence, 1997). By late Eocene times, shallow marine carbonate sedimentation had been established over much of south Sulawesi (Wilson, 1995) and southwestern central Sulawesi (Coffield *et al.*, 1993) although these areas were separated by a deep marine basinal area (Wilson and Bosence, 1996).

Oligocene

In the Oligocene (Fig.6) a land connection between Borneo and Indochina is inferred (Pupilli, 1973; Lloyd, 1978). Turbidites were deposited in Sabah in a deep marine basin. These may have been fed from major river systems of Indochina or Kalimantan. They pass laterally into shallow marine and deltaic sediments in Brunei (Tate, 1994) and at the margins of the basin in Sabah and Sarawak. There were several areas of shallow marine carbonates in this region throughout the Oligocene. During the Oligocene the Sarawak basin was progressively infilled. The west Sarawak coastline had a north-south orientation for the early Oligocene to early Miocene (Doust, 1981; Agostinelli *et al.*, 1990).

Flat-lying reflectors seen on seismic sections across much of the north and south Makassar basins suggest deep marine sedimentation occurred in a uniformly subsiding basin during the Oligocene (Situmorang, 1982; Guntoro, 1995). However, in other parts of the basin, particularly along the eastern margin of the Paternoster platform, seismic and borehole data suggests active faulting may have continued through the Oligocene and possibly into the Miocene (Situmorang, 1982; Guntoro, 1995; Wilson and Bosence, 1996). Oligocene deep marine sedimentation also occurred in much of the Kutai basin (Moss *et al.*, 1997) and in some areas of western central Sulawesi. An input of volcaniclastic material is also recorded in western central Sulawesi. In the Tarakan-Muara area, the Mangkalihat peninsula, Barito basin, offshore southern Barito and in South Sulawesi extensive shallow water carbonate platforms developed or continued to accumulate sediment during the Oligocene, while deeper water marls were deposited in adjacent areas (Fig.6; Bransden and Matthews, 1992; Bishop, 1980; Armin *et al.*, 1987; Netherwood and Wight, 1992; Saller *et al.*, 1992, 1993; Supriatna *et al.*, 1993; Weerd and Armin, 1992; Weerd *et al.*, 1987; Wilson, 1995).

The mainly fluviatile west Kalimantan basin (Melawi-Ketungau-Mandai area) had begun to diminish in size or had already been infilled by the early Oligocene (Fig.6). Seismic data and fission track dating of derived apatite grains suggest that the Semitau ridge began to rise in the early Oligocene (Moss *et al.*, 1997) and this uplift would have favoured erosion rather than deposition in this area. Deltaic and pro-delta environments appear in the western part of the Kutai basin toward the end of the Oligocene (Weerd and Armin, 1992; Tanean *et al.*, 1996; Moss *et al.*, 1997, 1998). This is thought to be related to uplift and erosion of the central ranges of Borneo at the end of the Oligocene which supplied sediment towards the Makassar Straits.

Tectonically disrupted ophiolitic rocks (Simandjuntak, 1986) comprising much of the east and SE arms of Sulawesi represent oceanic crust and overlying deep marine sediments. Metamorphic ages of 28-32 Ma obtained from at the base of the East Sulawesi ophiolite (Parkinson, 1991) suggest the ophiolites were detached in an oceanic setting at this time and emplaced later. The progressive emplacement of the ophiolitic sequence would have resulted in the development of more extensive land areas in Sulawesi.

The microcontinental blocks of Banggai-Sula, Buton and Tukang Besi, although drifting westwards towards Sulawesi, had yet to be accreted onto Sulawesi. These blocks rifted from the Australian-New Guinea continent during the late Mesozoic (Audley-Charles, 1974; Hamilton, 1979; Pigram and Panggabean, 1984; Garrard *et al.*, 1988; Davidson, 1991). During the Cenozoic, some were areas of shallow marine sedimentation (Garrard *et al.*, 1988; Smith and Silver, 1991; Davidson, 1991) and may have become emergent.

Miocene

During the Miocene (Figs.7 and 8) a switch in sedimentation style from extensive carbonate shelves to deltaic deposition and progradation occurred on the eastern side of Borneo, particularly in the Tarakan-Muara and Barito basins (Achmad and Samuel, 1984; Netherwood and Wight, 1992; Weerd and Armin, 1992; Siemers *et al.*, 1992; Carter and Morley, 1996; Stuart *et al.*, 1996). The predominance of deltaic sedimentation around the northern and eastern parts of Borneo, particularly the extremely deep Kutai

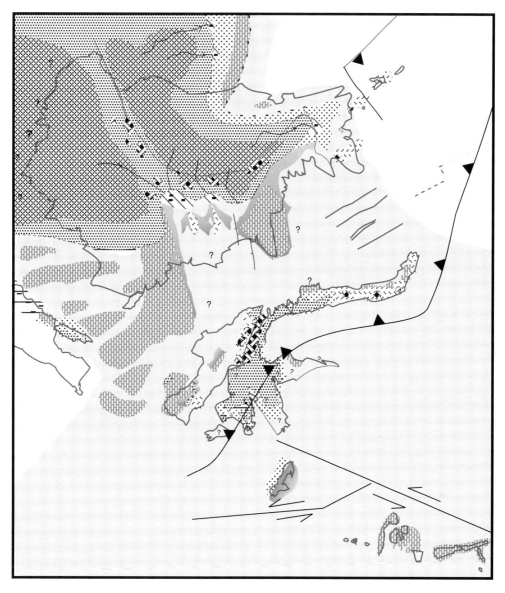

Fig. 7. Palaeogeographic map for 21 Ma, early Miocene. A key to the environments is shown in Fig.4.

basin, suggests that most of the major river systems were draining into these areas. Abundant detritus was supplied from the uplift and denudation of the centre of the island and coeval volcanism (Tanean *et al.*, 1996; Moss *et al.*, 1997, 1998). By the end of the Miocene the drainage system within Borneo was similar to the present-day. The Mahakam delta had prograded to near its present-day position by the late Miocene (Addison *et al.*, 1983; Land and Jones, 1987) and siliciclastic marginal marine and deltaic deposition predominated in this area. The Makassar Straits remained a deep water basin separating Sulawesi from Kalimantan, although as the land area increased in eastern Borneo due to the progradation of deltas, the distance across this seaway was progressively reduced.

A large area of carbonate deposition developed during the middle to late Miocene in the Luconia Shoals area, which was distant from areas of coastal siliciclastic deposition in Sarawak (Doust, 1981; Agostinelli *et al.*, 1990; Rice-Oxley, 1991). By the end of the Miocene the palaeo-Baram delta had begun to prograde to the NW, and the coastline had adopted an orientation similar to its present-day NE-SW orientation (Doust, 1981; Agostinelli *et al.*, 1991). In Sabah, shallow marine to marginal marine

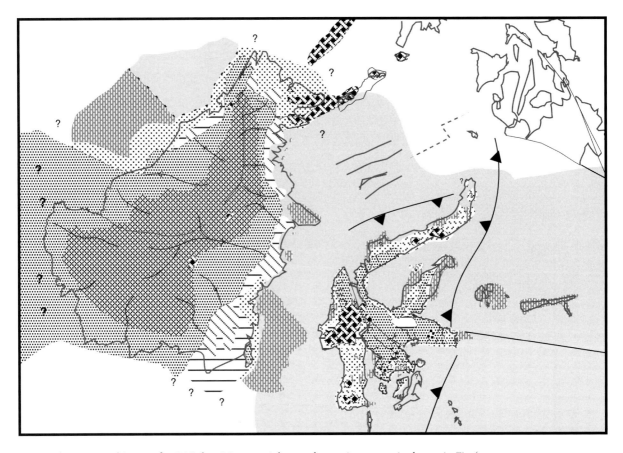

Fig.8. Palaeogeographic map for 8 Ma late Miocene. A key to the environments is shown in Fig.4.

sediments were deposited (Tjia et al., 1990; Clennell, 1992). Volcanic arc activity occurred in the Sulu and Cagayan arcs during the Miocene (Rangin et al., 1990) and this probably resulted in emergent chains of volcanic islands. During the late Miocene to the Pleistocene, basalts and trachy-andesites were extruded in central Borneo (Moss et al., 1997, 1998).

Shallow marine carbonate deposition continued on high blocks in southern and western central Sulawesi until the middle Miocene, surrounded by deep marine sedimentation (Wilson, 1995). The East Sulawesi ophiolite had been accreted onto western Sulawesi, and it is inferred that land areas were emergent in central Sulawesi during at least part of the Miocene as a result of this collision. In the north arm of Sulawesi island arc basalts were erupted during the Oligocene and Miocene. They are interbedded with shallow marine carbonate deposits, and volcanic islands are inferred to have been emergent. Between the middle Miocene through to the earliest Pliocene a volcanic arc developed along the length of western Sulawesi (Yuwono et al., 1987; Bergman et al., 1996).

The microcontinental blocks of Banggai-Sula and Buton-Tukang Besi were accreted onto eastern Sulawesi during the Miocene or earliest Pliocene. The inferred timing of collision of the various fragments onto eastern Sulawesi varies depending on the author. On Buton, obduction of ophiolitic material and the reworking of pre-Miocene strata into clastic deposits has been related to early to middle Miocene collision with SE Sulawesi (Fortuin et al., 1989; Smith and Silver, 1991; Davidson, 1991). Prior to collision, deep marine sediments were deposited on Buton, although uplift and thrusting associated with collision would have created emergent land areas in the middle Miocene. Pliocene block faulting on Buton, which created contemporaneous areas of deep and shallow marine sedimentation, may have been related to collision of Tukang Besi (Fortuin et al., 1989; Davidson, 1991). Eocene to middle Miocene limestones present on Banggai-Sula indicate shallow water depositional environments, possibly locally emergent, prior to collision in the lat-

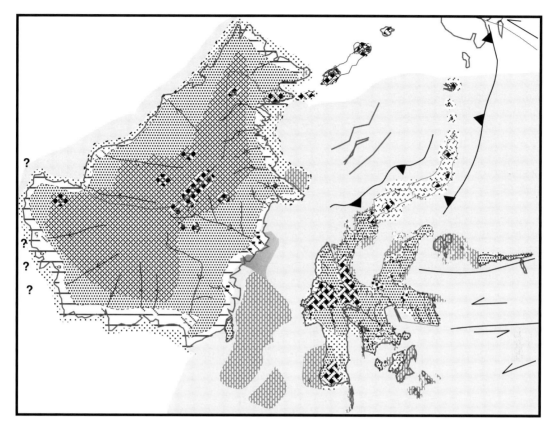

Fig.9. Palaeogeographic map for 4 Ma, early Pliocene. A key to the environments is shown in Fig.4.

est Miocene to early Pliocene (Garrard *et al.*, 1988; Davies, 1990). More extensive land areas would have formed after collision.

Pliocene-Recent

By the Pliocene (Fig.9) the coastlines of both Borneo and Sulawesi were similar to the present. Major shallow water carbonate areas persisted on the east and west sides of the Makassar Straits. The coastline of Borneo was dominated by deltaic and marginal marine depositional environments. The Mahakam river carries a large sediment volume which is deposited mainly in the delta and deltaic progradation occurred throughout the Pliocene and continues at the present day. At some point in the latest Miocene or Pliocene Borneo lost its land connection to mainland Indochina (Lloyd, 1978). Possible events responsible include reorganisation of plates, the end of a late Miocene glaciation, and/or global sea level changes in the late Miocene. The present-day Karimata Strait did not form until 7000 years ago, shown by the presence of a drowned Pleistocene river drainage system on the Sunda Shelf (Umbgrove, 1938). The Pleistocene river drainage system recognised by Umbgrove (1938) across the Sunda Shelf is related to a sea level lowstand during a glacial period.

The final juxtaposition of the fragments that comprise Sulawesi occurred between the Pliocene and the present. Although most authors infer that internal rotation and juxtaposition was achieved via a system of linked strike-slip faults and thrusts, the linkage and displacements along faults is still contentious. The Pliocene reconstruction shown is based on matching palaeoenvironments and assuming reasonable strike-slip displacements along major faults and some internal deformation of blocks within Sulawesi. Collisions and subduction east of Sulawesi caused transpression during the Neogene and Quaternary which resulted in uplift of extensive areas in Sulawesi and caused rapid uplift of a number of high mountain areas, particularly in central Sulawesi (Bergman *et al.*, 1996). The transpressive regime also resulted in extension and subsidence in other areas. Bone

Bay separates the south and SE arms of Sulawesi is suggested to have developed as an extensional feature in the late Oligocene (Davies, 1992). Seismic data suggests that the form of Bone Bay was then further modified during the Miocene/Pliocene by transpressive and transtensional movements (Davies, 1992; Guntoro, 1995). The nature of Quaternary sediments and historic legends suggest that until quite recently a seaway separated the south arm from the rest of Sulawesi (Bemmelen, 1949; Sartono, 1982).

Igneous activity continued in western Sulawesi until the Pliocene and Pleistocene and is similar in nature to the late Miocene volcanics in the same area (Yuwono et al., 1987; Bergman et al., 1996). Along the north arm of Sulawesi, Miocene to Pliocene volcanism is related to south-dipping subduction of the Celebes Sea oceanic crust under the north arm of Sulawesi (Carlile et al., 1990). A string of active Quaternary to Recent volcanoes dominate the Sangihe Islands and the eastern part of the Minahasa region, and are related to west-dipping subduction under this area.

Implications for palaeobiogeography and biogeography

Important elements of the palaeogeographic reconstructions for biogeographers are the extent of land areas and island chains, the timing and nature of material accreted onto the margin of the Sundaland craton and the distribution of different environments through time and space. The reconstructions show that the spatial and temporal distribution of land areas and habitats differed considerably during the Tertiary from those seen today in the Sulawesi-Borneo area.

Land migration of terrestrial organisms between Borneo, mainland SE Asia and some of the Sunda Islands, such as Sumatra, would have been possible throughout much of the Tertiary, since at least a transitory land connection is inferred to have existed between these areas until the Plio/Pleistocene. The marked similarity between the flora and fauna of Borneo and mainland SE Asia can be accounted for by the existence of this land bridge during the Tertiary. As noted earlier there is little information on the geology of the Sunda Shelf and evidence for marine incursions, perhaps associated with Cenozoic global sea level changes (Haq et al., 1987; 1988), may have been overlooked or removed by later erosion.

The Paleogene formation of the Makassar Straits, at a period when other basins were forming around the margins of the Sundaland craton, was one of the most important geological events to affect the biogeography of Borneo and Sulawesi. Western Sulawesi was accreted to eastern Borneo in the late Cretaceous, forming a continuous land area probably during part of the Paleocene and the early Eocene. During this period, prior to the formation of the Makassar Straits, the interchange of terrestrial biota between Borneo and western Sulawesi could have occurred unhindered across a continuous land bridge. However, the actual movement of organisms would have depended on habitat availability and dispersal rates of organisms across a land bridge which may have existed for as little as ten million years. Palynological evidence from the middle to upper Eocene Malawa Formation in South Sulawesi suggests that diverse angiosperms with Laurasian affinity were established in South Sulawesi prior to the rifting of the Makassar Straits (Morley, 1998 this volume). Distribution patterns and fossil evidence for some groups of plants (Truswell et al., 1987), such as some of the palms, the families of Magnoliaceae/Winteraceae and the genera *Fagus/Nothofagus*, have been used to infer various dispersal pathways via northern and austral routes after the breakup of Gondwana, with an ancient Gondwana origin being implicit (Dransfield, 1981; Whitmore, 1981b).

Subsidence resulting in the formation of the Makassar Straits and surrounding Tertiary basinal areas would have progressively increased the spatial separation of emergent land areas in Borneo and western Sulawesi from the Eocene to the Oligocene. Extensive shallow water carbonate platforms developed in South Sulawesi, SE Kalimantan and the Mangkalihat peninsula during the Tertiary. Although it is possible that small areas of these platforms were emergent, subsidence rates in the region were quite high and that any such islands would have been limited in size and only emergent for short periods of time.

Palaeogeographic information indicates that there were only small land areas in western Sulawesi between the middle Eocene and middle Miocene, and that for the most part western Sulawesi was isolated from major clastic input. However, a volcanic arc, part of which may have been emergent, is inferred to have extended down the east side of western Sulawesi into southern Java. The formation of the Makassar Straits and resultant isolation of localised, small land areas in western Sulawesi goes some way towards explaining the affinities at generic and higher level of Sulawesi's terrestrial biota with those of Borneo and mainland SE Asia, and the high degree of endemism amongst Sulawesi's fauna and flora at species level (Musser, 1987; Holloway, 1987). Musser (1987) noted that the non-volant mammals of Sulawesi have ancient origins and that because the fauna is an unbalanced, depauperate one (40% of species are bats) typical of oceanic islands, origin is believed to have been across sea.

Rivers supplying clastic material derived from the erosion of the central part of Borneo beginning in the Oligocene resulted in the progradation of large delta systems into northern and eastern Borneo. Sedimentation in the Kutai basin and Makassar Straits progressively reduced the distance between Borneo and Sulawesi to the present 200 km across the Makassar Straits. A compressive regime, which developed in the area in the middle to late Miocene and continues to the present day, resulted in uplift or reduced subsidence rates in the Makassar Straits region compared with the Paleogene. Eustatic sea level changes, with magnitudes of tens of metres, related to fluctuations in the size of the polar ice sheets have been recorded during the Plio-Pleistocene. This, together with the compressive regime, suggests that part of the shallow water areas separating Borneo and Sulawesi may have become emergent during the later part of the Neogene with only a narrow seaway remaining where deep water channels now occur. If such a situation developed, and assuming favourable conditions occurred, migration of terrestrial organisms would have been facilitated, particularly during the Plio-Pleistocene, whereas the migration of marine organisms may have been hindered. On the basis of palynological data Morley (1998 this volume) suggested that during the Neogene only a few plant taxa were able to disperse across the Makassar Straits, and it is likely that those that did were well adapted to dispersal.

A number of authors (Duffels, 1990; Musser, 1987; Balgooy, 1987) have noted that the flora and fauna of the south arm of Sulawesi is different from the rest of Sulawesi. For much of the Tertiary South Sulawesi was below sea level and it was probably only in the Pliocene when island areas in South Sulawesi became connected with those in central Sulawesi. Giant tortoise, elephant, stegodont and pig fossils have been found in Pleistocene deposits in South Sulawesi, but not in the rest of Sulawesi. It is not clear whether these animals were restricted to South Sulawesi (Musser, 1987) and only migrated into the area during the later part of the Neogene during periods of low sea level and emergence of more extensive land areas bordering a narrow seaway in the Makassar Straits. Today South Sulawesi has a dry climate compared with other areas in Sulawesi, and there are mostly limestones or volcanic rocks in this area. The climate and soil type, as well as palaeobiogeographic separation of this area, would all have influenced the flora and therefore to some degree the fauna of this area (Balgooy, 1987).

Volcanic arcs were important as potential land bridges and island hopping routes for organisms during the Tertiary. They are inferred to have existed along the north arm of Sulawesi and the eastern side of west Sulawesi, perhaps extending down through Java from the Eocene until the late Oligocene. During the Neogene a volcanic arc occurred along the north arm of Sulawesi and possible island chain connections may have existed to the Philippines. During the Miocene, the Sulu and Cagayan volcanic island chain may have formed an island hopping route for organisms between the northeast tip of Borneo and parts of the Philippines.

The nature and timing of material accreted onto western Sulawesi is of considerable importance for potential migrations of terrestrial organisms with Australian affinities. The East Sulawesi ophiolite, which now comprises much of the east and southeast arms of Sulawesi, formed in a deep marine area and this area would not have been subaerially emergent until its accretion onto western Sulawesi. Therefore the East Sulawesi ophiolite could not have formed a potential 'raft' for the dispersal of terrestrial organisms.

In comparison, parts of the microcontinental fragments which make up the Banggai-Sula and Buton-Tukang Besi islands, which were accreted onto eastern Sulawesi in the Miocene or Pliocene would have been islands or shallow water areas during their drift towards Sulawesi during the Tertiary. These microcontinental blocks were rifted from the margins of the Australian-New Guinea continent in the late Mesozoic, and land bridges would have been severed at this time. The stratigraphy of these microcontinental blocks indicates that the blocks subsided and became areas of deep marine sedimentation from the late Jurassic to Cretaceous. Therefore, although parts of these areas became emergent in the Tertiary, Australian affinity fauna and flora could not have been 'rafted' on these microcontinental blocks. However, during the Tertiary, the Australian continent drifted steadily northwards and some microcontinental blocks between the Australian and Sundaland cratons became emergent as islands. These blocks, some of which were accreted onto Sulawesi in the Miocene or Pliocene, may then have rafted or acted as an island hopping routes for Australian organisms, such as some of the marsupials, to colonise Sulawesi. Palms have a large fruit, suggesting long-distance dispersal might be limited and Dransfield (1981, 1987) noted that the disjunction of Papuasian and Sundaland palms occurred between Sulawesi and the Moluccas, consistent with relatively recent (Miocene) juxtaposition. Another possible island hopping route for Australian fauna to reach Sulawesi may have been via the Philippines and along the Sangihe volcanic arc. Palaeomagnetic evidence suggests that India rifted away from the Gondwanaland supercontinent at about 140 Ma, whereas Australia-New Guinea separated from Antarctica between 110-90 Ma. It may be possible that organisms, particularly some of the flora, which had evolved prior to these two separation events, were present in both India and Australia, and that after the interaction of these areas with the Sundaland craton that migrations could have occurred from both east and west into the Malay archipelago.

The wind and current direction as well as the distribution of land masses and their local climates during the Tertiary would have affected the distribution and migration of both terrestrial and marine biota. Although, in general, Borneo has a wetter climate than Sulawesi, both have tropical monsoonal climates and are affected by winds from the east and west at opposing times of the year, depending on when low and high pressure zones develop over the adjacent continental land masses of Australia and mainland SE Asia. These strong monsoonal winds would help to distribute wind blown biota east-west between some of the closer islands in the archipelago today. Although it is difficult to infer past wind directions and strengths, it would seem likely that during the Tertiary, since the Australian land mass was further to the south, the effects of one of the monsoons would have been less than today (Whitmore, 1981a). There is a growing amount of evidence that Indonesia, and particularly Borneo (from the pollen record), was more strongly seasonal during the late Tertiary and middle Pleistocene than today (Whitmore, 1981a; Morley and Flenley, 1987).

This would have favoured the migration of savannah plants and animals across the area between mainland SE Asia, Borneo and other islands on the Sunda Shelf (Morley and Flenley, 1987). Cooler conditions during glacial periods and lowered sea levels exposing more land areas (Whitmore, 1981) in the late Tertiary and Quaternary would have encouraged animals to move southwards over land bridges to Borneo. Also during glacial maxima, montane vegetation zones would have been lower (Hope, 1996) providing 'stepping stones' for the migration of montane and temperate taxa (Whitmore, 1981a; Morley and Flenley, 1987).

Today the current directions and strengths around Borneo and Sulawesi vary depending on the prevailing wind directions. Surface currents in the East Java Sea flow predominantly east or west at opposing times of year, whilst those in the Makassar Straits are predominantly south directed and around the rest of the Sulawesi coastline more variable. The deep water channels in the Makassar Straits form a pathway for waters from the Celebes Sea, through the East Java Sea to the Indian Ocean, principally through the Lombok Straits. During the Tertiary, the land areas of Borneo and Sulawesi would have been smaller than today and currents would have been less constrained, but still controlled by prevailing wind directions. The dispersal of marine organisms, especially those with a planktonic larval stage, such as corals or molluscs, would have been unhindered throughout the Tertiary and partly controlled by prevailing wind and current directions and the availability and spacing of suitable habits. Whitmore (1981a) noted that sea level fluctuations would alter current patterns in the region and thereby affect climates, with periods of more seasonal climates occurring during glacial periods when sea levels were lowered.

Conclusions

Borneo and Sulawesi, bordering Wallace's original line of faunal divide, are of prime importance to biogeographers, since these islands occur at the 'mixing zone' or area of overlap between Asian and Australian biota. Borneo and Sulawesi are located in the midst of a convergent zone of three major plates, and the tectonic evolution of the area, together with the climate and local factors, strongly influenced the past and present day distribution of fauna and flora.

The main palaeogeographic changes in the

Borneo-Sulawesi region and their implications for biogeography are:

1. A continuous land connection between Borneo and mainland SE Asia may have existed throughout much of the Tertiary and would have allowed fairly unhindered migration of terrestrial biota.
2. Western Sulawesi had been accreted to eastern Borneo by the late Cretaceous, and by the early Eocene there were continuous land areas in the Schwaner Mountains, NW Kalimantan, the Mangkalihat peninsula and parts of western Sulawesi. The dispersal of certain flora and fauna between Borneo and western Sulawesi could have occurred at this time.
3. Extension in the Makassar Straits region and the formation of surrounding Tertiary basinal areas in the Paleogene resulted in the progressive separation of locally emergent land or volcanic areas in western Sulawesi and Borneo, and this isolation may have contributed to the high degree of species endemism on Sulawesi.
4. The East Sulawesi ophiolite was accreted onto Sulawesi during or after the late Oligocene and resulted in the formation of more extensive land areas in Sulawesi. Since the ophiolite formed in a deep marine setting it would not have acted as a potential raft for biota.
5. Microcontinental fragments accreted onto eastern Sulawesi in the Miocene to Pleistocene may have been emergent as they drifted towards Sulawesi and allowed island hopping or rafting for biota of Australian affinity.
6. Island hopping routes for the dispersal of organisms between Borneo-Sulawesi and the Philippines may have existed along volcanic arcs, such as the long-lived North Sulawesi arc, the Sulu and Sangihe arcs, and the Cagayan arc.
7. The uplift and subsequent erosion of Borneo since the late Oligocene and the emergence of more extensive land areas in Sulawesi led to the progressive reduction in width of the Makassar Straits. This may have facilitated the interchange of biota during the later part of the Tertiary, particularly during periods of relative sea level lows when large areas of shallow marine shelves may have become emergent.
8. Although Borneo and western Sulawesi remained in near tropical latitudes throughout the Tertiary, the wide variety of micro-environments shown by the palaeogeographic maps would have led to niche partitioning. This partially explains the higher number of endemic species in Borneo compared with Sumatra, which otherwise does show a marked similarity in fauna and flora with Borneo.

Acknowledgements

Discussions with Tony Barber, John Chambers, Ian Cloke, Robert Hall, Neville Haile, Jon Noad and Chris Parkinson are gratefully acknowledged. Robert Hall is particularly thanked for allowing us to use the reconstructions. The helpful comments of reviewers Tony Whitten and Charles Hutchison were much appreciated. SJM's and MEJW's post-doctoral work with the University of London SE Asia Research Group are funded by a consortium of oil companies, Arco, Lasmo, Mobil Oil, Exxon, Canadian Petroleum, Union Texas and Unocal who are all gratefully acknowledged. Additional funding for fieldwork was provided by British Petroleum and Maersk Oil and these companies are also thanked. Drs. Irwan Bahar, Rab Sukamto and Kustomo Hasan of GRDC, Bandung are gratefully acknowledged for their help and encouragement with our research projects. Dharma Satria Nas, Alexander Limbong and the people of Sulawesi and Kalimantan are thanked for their invaluable help and friendship in the field.

References

Achmad, Z. and Samuel, L. 1984. Stratigraphy and depositional cycles in the N.E. Kalimantan Basin. Indonesian Petroleum Association Proceedings 13th Annual Convention 109-120.

Addison, R., Harrison, R. K., Land, D. H. and Young, B. R. 1983. Volcanogenic tonsteins from Tertiary coal measures, East Kalimantan, Indonesia. International Journal of Coal Geology 3: 1-30.

Agostinelli, E., Mohamad Raisuddin bin Ahmad Tajuddin, Antonielli, E. and Mohamad bin Mohd. Aris. 1990. Miocene-Pliocene palaeogeographic evolution of a tract of Sarawak offshore between Bintulu and Miri. Bulletin of the Geological Society of Malaysia 27: 117-135.

Armin, R. A., Cutler, W. G., Mahadi and Weerd, A. van de. 1987. Carbonate platform, slope and basinal deposits of Upper Oligocene, Kalimantan, Indonesia (abstract) American Association of Petroleum Geologists Bulletin 71: 526.

Audley-Charles, M. G. 1974. Sulawesi. In Mesozoic-Cenozoic Orogenic Belts. pp. 365-378. Edited by A. M. Spencer. Geological Society of London Special Publication 4.

Audley-Charles, M. G. 1981. Geological history of the region of Wallace's line. In Wallace's line and plate tectonics. pp. 24-35. Edited by T. C. Whitmore. Clarendon Press, Oxford Monographs on Biogeography 1.

Audley-Charles, M. G. 1987. Dispersal of Gondwanaland: relevance to evolution of the Angiosperms. In Biogeographical Evolution of the Malay archipelago. pp. 3-25. Edited by T. C. Whitmore. Clarendon Press, Oxford Monographs on Biogeography 4.

Audley-Charles, M. G., Hurley, A. M. and Smith, A. G. 1981. Continental movements in the Mesozoic and Cenozoic. In Wallace's line and plate tectonics. pp. 9-23. Edited by T. C. Whitmore. Clarendon Press, Oxford Monographs on Biogeography 1.

Aziz, F. 1994. Vertebrate faunal evolution of Sulawesi Island, Indonesia, during the late Neogene. *In* Pacific Neogene events in time and space. pp. 79-85. Edited by R. Tsuchi. IGCP 246.

Balgooy, M. M. J. 1987. A plant biogeographical analysis of Sulawesi. *In* Biogeographical Evolution of the Malay archipelago. pp. 94-102. Edited by T. C. Whitmore. Clarendon Press, Oxford Monographs on Biogeography 4.

Balgooy, M. M. J., Hovenkamp, P. H. and Welzen, P. C. van. 1996. Phytogeography of the Pacific — floristic and historical distribution patterns in plants. *In* The origin and evolution of Pacific island biotas, New Guinea to eastern Polynesia: patterns and processes. pp. 191-213. Edited by A. Keast and S. A. Miller. SPB Academic Publishing bv, Amsterdam.

Beddoes, L. R. 1980. Hydrocarbon plays in Tertiary Basins of southeast Asia. Paper presented at the Offshore SE Asia Conference, Singapore. SE Asia Petroleum Exploration Society (SEAPEX): 18.

Bemmelen, R. W. van. 1949. The Geology of Indonesia. Vol. 1a. Government. Printing Office, The Hague, Netherlands. 732 pp.

Ben-Avraham, Z. and Emery, K. O. 1973. Structural framework of the Sunda Shelf. American Association of Petroleum Geologists Bulletin 57: 2323-2366.

Bergman, S. C., Coffield, D. Q., Talbot, J. P. and Garrard, R. A. 1996. Tertiary tectonic and magmatic evolution of western Sulawesi and the Makassar Strait: evidence for a Miocene continent-continent collision. *In* Tectonic evolution of SE Asia. pp. 391-429. Edited by R. Hall and D. J. Blundell. Geological Society of London Special Publication 106.

Bishop, W. P. 1980. Structure, stratigraphy and hydrocarbons offshore southern Kalimantan, Indonesia. American Association of Petroleum Geologists Bulletin, 64: 37-58.

Bransden, P. J. E. and Matthews, S. J. 1992. Structural and stratigraphic evolution of the East Java Sea. Indonesia. Indonesian Petroleum Association, Proceedings 21st Annual Convention: 417-454.

Briggs, J. C. 1987. Biogeography and Plate Tectonics. Developments in Palaeontology and Stratigraphy 10. Elsevier, Amsterdam.

Burollet, P. F. and Salle, C. 1981. Seismic reflection profiles in the Makassar Strait. *In* The Geology and Tectonics of Eastern Indonesia. pp. 273-276. Edited by A. J. Barber and S. Wiryosujono. Geological Research and Development Centre, Bandung, Special Publication 2.

Burrett, C., Duhig, N., Berry, R. and Varne, R. 1991. Asian and south-western Pacific continental terranes derived from Gondwana, and their biogeographic significance. *In* Austral Biogeography. Edited by P. Y. Ladiges, C. J. Humphries and L. W. Martinelli, Australian Systematic Botany 4: 13-24.

Carey, S. W. 1975. Tectonic evolution of southeast Asia. Proceedings of the Indonesian Petroleum Association 4th Annual Convention: 17-48.

Carlile, J. C., Digdowirogo, S. and Darius, K. 1990. Geologic setting, characteristics and regional exploration for gold in the volcanic arcs of North Sulawesi, Indonesia. *In* Epithermal gold mineralization of the Circum Pacific: Geology, Geochemistry, Origin and Exploration. pp. 105-140. Edited by J. W. Hedenquist, N. C. White and G. Siddeley. Journal of Geochemical Exploration 35.

Carter, I. S. and Morley, R. J. 1996. Utilizing outcrop and palaeontological data to determine a detailed sequence stratigraphy of the early Miocene sediments of the Kutei Basin, East Kalimantan. *In* Proceedings of the International Symposium on Sequence Stratigraphy in SE Asia. Edited by C. A. Caughey, D. C. Carter, J. Clure, M. J. Gresko, P. Lowry, R. K. Park and A. Wonders. Indonesian Petroleum Association: 345-361.

Clennell, M. B. 1992. The melanges of Sabah, Malaysia. Ph.D. Thesis, University of London.

Cloke, I. R., Moss, S. J. and Craig, J. 1997. The influence of basement reactivation on the extensional and inversional history of the Kutai Basin, Eastern Kalimantan. Journal of the Geological Society of London 154: 157-161.

Coffield, D. Q., Bergman, S. C., Garrard, R. A., Guritno, N., Robinson, N. M. and Talbot, J. 1993. Tectonic and stratigraphic evolution of the Kalosi PSC area and associated development of a Tertiary petroleum system, south Sulawesi, Indonesia. Proceedings of the Indonesian Petroleum Association 22nd Annual Convention: 679-706.

Cranbrook, Earl of. 1981. The vertebrate faunas. *In* Wallace's line and plate tectonics. pp. 57-69. Edited by T. C. Whitmore. Clarendon Press, Oxford Monographs on Biogeography 1.

Daly, M. C., Cooper, M. A., Wilson, I., Smith, D. G. and Hooper, B. G. D. 1991. Cenozoic plate tectonics and basin evolution in Indonesia. Marine and Petroleum Geology 8: 2-21.

Davies, I. C. 1990. Geological and exploration review of the Tomori PSC, Eastern Indonesia. Proceedings of the Indonesian Petroleum Association 19th Annual Convention: 41-67.

Davies, I. 1992. The Gulf of Bone: an opening for exploration? Eastern Indonesian Symposium (abstract): 2.

Davidson, J. W. 1991. The geology and prospectivity of Buton island, S.E. Sulawesi, Indonesia. Proceedings of the Indonesian Petroleum Association 20th Annual Convention: 209-233.

de Jong, R. 1990. Some aspects of the biogeography of the Hesperiidae (Lepidoptera, Rhopalocera) of Sulawesi. *In* Insects and the rain forests of south East Asia (Wallacea). pp. 35-43. Edited by W. J. Knight and J. D. Holloway. Royal Entomological Society, London.

Dickerson, R. E. 1928. Distribution of life in the Philippines. Bureau of Printing, Manila.

Doust, H. 1981. Geology and exploration history of offshore central Sarawak. *In* Energy Resources of the Pacific Region. pp. 117-132. Edited by M. T. Halbouty. American Association of Petroleum Geologists, Studies in Geology 12.

Dransfield, J. 1981. Palms and Wallace's line. *In* Wallace's line and plate tectonics. pp. 43-56. Edited by T. C. Whitmore. Clarendon Press, Oxford Monographs on Biogeography 1.

Dransfield, J. 1987. Bicentric distribution in Malesia as exemplified by palms. *In* Biogeographical Evolution of the Malay archipelago. pp. 60-72. Edited by T. C. Whitmore. Clarendon Press, Oxford Monographs on Biogeography 4.

Duffels, J. P. 1990. Biogeography of Sulawesi cicadas (Homoptera: Cicadoidea). *In* Insects and the rain forests of south East Asia (Wallacea). pp.63-72. Edited by W. J. Knight and J. D. Holloway. Royal Entomological Society, London.

Fortuin, A. R., Smet, M. E. M. de., Hadiwasastra, S., Marle, L. J. van, Troelstra, S. R. and Tjokrosapoetro, S. 1990. Late Cenozoic sedimentary and tectonic history of south Buton, Indonesia. Journal of southeast Asian Earth Sciences 4: 107-124.

Garrard, R. A., Supandjono, J. B. and Surono. 1988. The geology of the Banggai-Sula microcontinent, eastern Indonesia. Proceedings of the Indonesian Petroleum Association 17th Annual Convention: 23-52.

George, W. 1981. Geological history of the region of Wallace's line. *In* Wallace's line and plate tectonics. pp.

3-8. Edited by T. C. Whitmore. Clarendon Press, Oxford Monographs on Biogeography 1.

Gressitt, J. L. 1961. Problems in zoogeography of Pacific and Antarctic insects. Pacific Insects Monograph 2: 1-94.

Guntoro, A. 1995. Tectonic evolution and crustal structure of the central Indonesian region from geology, gravity and other geophysical data. Ph.D. Thesis University of London 335 pp.

Hall, R. 1996. Reconstructing Cenozoic SE Asia. *In* Tectonic evolution of SE Asia. pp. 153-184. Edited by R. Hall and D. J. Blundell. Geological Society of London Special Publication 106.

Hall, R. 1998. The plate tectonics of Cenozoic SE Asia and the distribution of land and sea. *In* Biogeography and Geological Evolution of SE Asia. Edited by R. Hall and J. D. Holloway (this volume).

Haq, B. V., Hardenbol, J. and Vail, P. R. 1987. Chronology of fluctuating sea levels since the Triassic. Science 235: 1156-1167.

Haq, B. V., Hardenbol, J. and Vail, P. R. 1988. Mesozoic and Cenozoic chronostratigraphy and cycles of sea level change. *In* Sea-Level Changes — An Integrated Approach. pp. 71-108. Edited by C. K. Wilgus *et al.*, SEPM Special Publication, 42.

Hasan, K. 1991. The Upper Cretaceous flysch succession of the Balangbaru Formation, Southwest-Sulawesi. Proceedings of the Indonesian Petroleum Association 20th Annual Convention: 183-208.

Hamilton, W. 1979. Tectonics of the Indonesian region. United States Geological Survey Professional Paper 1078: 345 pp.

Heryanto, N., Williams, P. R., Harahap, B. H. and Pieters, P. E. 1993. Geology of the Sintang sheet area Kalimantan 1:250,000. Geological Research and Development Centre, Bandung, Indonesia.

Holloway, J. D. 1987. Lepidoptera patterns involving Sulawesi: what do they indicate of past geography? *In* Biogeographical Evolution of the Malay archipelago. pp. 103-118. Edited by T. C. Whitmore. Clarendon Press, Oxford Monographs on Biogeography 4.

Holloway, J. D. 1990. Patterns of moth speciation in the Indo-Australian archipelago. *In* The unity of evolutionary biology. pp. 340-370. Edited by E. C. Dudley. Dioscorides Press, Portland, Oregon.

Holloway, J. D. 1997. Sundaland, Sulawesi and eastwards: a zoogeographic perspective. Malayan Nature Journal 50: 207-227.

Holmes, D. and Phillipps, K. 1996. The birds of Sulawesi. Oxford University Press. 86 pp.

Hope, G. 1996. Quaternary change and the historical biogeography of Pacific islands *In* The origin and evolution of Pacific island biotas, New Guinea to eastern Polynesia: patterns and processes. pp. 165-190. Edited by A. Keast and S. A. Miller. SPB Academic Publishing bv, Amsterdam.

Hutchison, C. S. 1989. Geological evolution of South-East Asia. Clarendon Press Oxford. 368 pp.

Hutchison, C. S. 1996. The 'Rajang Accretionary Prism' and 'Lupar Line' problem of Borneo. *In* Tectonic evolution of SE Asia. pp. 247-261. Edited by R. Hall and D. J. Blundell. Geological Society of London Special Publication 106.

Land, D. H. and Jones, C. M. 1987. Coal geology and exploration of part of the Tertiary Kutai Basin in East Kalimantan, Indonesia. *In* Coal and Coal-bearing Strata: Recent Advances. pp. 235-255. Edited by A. C. Scott. Geological Society Special Publication 32.

Lee, T.-Y. and Lawver, L. A. 1994. Cenozoic plate reconstructions of the South China Sea region. Tectonophysics 235: 149-180.

Lee, T.-Y. and Lawver, L. A. 1995. Cenozoic plate reconstruction of Southeast Asia. Tectonophysics 251: 85-138.

Leeuwen, T. M. van. 1981. The geology of southwest Sulawesi with special reference to the Biru area. *In* The Geology and Tectonics of Eastern Indonesia. pp. 277-304. Edited by A. J. Barber and S. Wiryosujono. Geological Research and Development Centre, Bandung, Special Publication 2.

Leeuwen, T. van, Leach, T., Hawke., A. A., and Hawke, M. M. 1990. The Kelian disseminated gold deposit, East Kalimantan, Indonesia. *In* Epithermal gold mineralization of the Circum Pacific: Geology, Geochemistry, Origin and Exploration. pp. 1-62. Edited by J. W. Hedenquist, N. C. White and G. Siddeley. Journal of Geochemical Exploration 35.

Lloyd, A. R. 1978. Geological evolution of the South China Sea. SEAPEX Proceedings 4: 95-137.

Ludmadyo, E., McCabe, R., Harder, S. and Lee, T. 1993. Borneo: a stable portion of the Eurasian margin since the Eocene. Journal of Southeast Asian Earth Sciences 8: 225-231.

Kottelat, M., Whitten, A. J., Kartikasari, S. N. and Wirjoatmodjo, S. 1993. Freshwater Fishes of Western Indonesia and Sulawesi. Periplus Editions.

Kusuma, I. and Darin, T. 1989. The hydrocarbon potential of the Lower Tanjung Formation, Barito Basin, S.E. Kalimantan. Indonesian Petroleum Association Proceedings 18th Annual Convention 1: 107-138.

MacKinnon, K., Hatta, G., Halim, H. and Mangalik, A. 1996. The Ecology of Kalimantan. Periplus Editions. 802 pp.

Mayr, E. 1944. Wallace's line in the light of recent zoogeographic studies. Quarterly Review of Biology 19: 1-14.

Metcalfe, I. 1998. Palaeozoic and Mesozoic geological evolution of the SE Asian region: multi-disciplinary constraints and implications for biogeography. *In* Biogeography and Geological Evolution of SE Asia. Edited by R. Hall and J. D. Holloway (this volume).

Michaux, B. 1994. Land movements and animal distributions in east Wallacea (eastern Indonesia, Papua new Guinea and Melanesia). Palaeogeography, Palaeoclimatology, Palaeoecology 112: 323-343.

Michaux, B. 1996. The origin of southwest Sulawesi and other Indonesian terranes: a biological view. Palaeogeography, Palaeoclimatology, Palaeoecology 122: 167-183.

Monnier, C., Girardeau, J., Maury, R. C. and Cotten, J. 1995. Back-arc basin for the East Sulawesi ophiolite (eastern Indonesia). Geology 23: 851-854.

Morley, R. J. 1998. Palynological evidence for Tertiary plant dispersals in the SE Asian region in relation to plate tectonics and climate. *In* Biogeography and Geological Evolution of SE Asia. Edited by R. Hall and J. D. Holloway (this volume).

Morley, R. J. and Flenley, J. R. 1987. Late Cainozoic vegetational and environmental changes in the Malay archipelago. *In* Biogeographical Evolution of the Malay archipelago. pp. 50-59. Edited by T. C. Whitmore. Clarendon Press, Oxford Monographs on Biogeography 4.

Moss, S. J. 1998. Late Cretaceous-Palaeogene turbidites in Kalimantan/Sarawak: are they really part of an accretionary complex? Journal of the Geological Society of London, in press.

Moss, S. J. and Finch, E. M. 1998. New nannofossil and foraminiferal determinations from East and West Kalimantan and their implications. Journal of Southeast Asian Earth Sciences, in press.

Moss, S. J., Chambers, J., Cloke, I., Carter, A., Dharma Satria,

Ali, J. R., and Baker, S. 1997. New observations on the sedimentary and tectonic evolution of the Tertiary Kutai Basin, East Kalimantan. *In* Petroleum Geology of Southeast Asia. pp. 395-416. Edited by A. Fraser, S. J. Matthews and R. W. Murphy. Geological Society of London Special Publication 126.

Moss, S. J., Carter, A., Hurford, A. and Baker, S. 1998. A Late Oligocene tectono-volcanic event in East Kalimantan and the implications for tectonics and sedimentation in Borneo. Journal of the Geological Society of London 155: 177-192.

Mubroto, B., Briden, J. C., McClelland, E. and Hall, R. 1994. Palaeomagnetism of the Balantak ophiolite, Sulawesi. Earth and Planetary Science Letters 125: 193-209.

Musser, G. G. 1987. The mammals of Sulawesi. *In* Biogeographical Evolution of the Malay archipelago. pp. 73-94. Edited by T. C. Whitmore. Clarendon Press, Oxford Monographs on Biogeography 4.

Netherwood, R. and Wight, A. 1992. Structurally controlled, linear reefs in a Pliocene delta-front setting, Tarakan Basin, Northeast Kalimantan. *In* Carbonate rocks and reservoirs of Indonesia. pp. 3.1.3-3.1.36. Edited by C. T. Siemers, M. W. Longman, R. K. Park and J. G. Kaldi. A core workshop held in conjunction with 1992 Indonesian Petroleum Association Annual Convention 1.

Parkinson, C. D. 1991. The petrology, structure and geologic history of the metamorphic rocks of central Sulawesi, Indonesia. Ph.D. Thesis, University of London, 337 pp.

Peltzer, G. and Tapponnier, P. 1988. Formation and evolution of strike-slip faults, rifts, and basins during the India-Asia collision: an experimental approach. Journal of Geophysical Research 93: 15085-15117.

Pieters, P. E. and Supriatna, S. 1990. Geological Map of the West, Central and East Kalimantan Area 1:1,000,000. Geological Research and Development Centre, Bandung, Indonesia.

Pieters, P. E., Abidin, H. Z. and Sudana, D. 1993a. Geology of the Long Pahangai sheet area, Kalimantan 1:250,000. Geological Research and Development Centre, Bandung, Indonesia.

Pieters, P. E., Surono, and Noya, Y. 1993b. Geology of the Puttissibau sheet area, Kalimantan 1:250,000. Geological Research and Development Centre, Bandung, Indonesia.

Pieters, P. E., Trail, D. S. and Supriatna, S. 1987. Correlation of Early Tertiary rocks across Kalimantan. Indonesian Petroleum Association Proceedings 16th Annual Convention 1: 291-306.

Pigram, C. J. and Panggabean, H. 1984. Rifting of the northern margin of the Australian continent and the origin of some microcontinents in eastern Indonesia. Tectonophysics 107: 331-353

Polvé, M., Maury, R. C., Bellon, H., Rangin, C., Priadi, B, Yuwono, S., Joron, J. L. and Soeria Atmadja, R. 1997. Magmatic evolution of Sulawesi (Indonesia): constraints on the Cenozoic geodynamic history of the Sundaland active margin. Tectonophysics 272: 69-92.

Pupilli, M. 1973. Geological evolution of South China Sea area tentative reconstruction from borderland geology and well data. Indonesian Petroleum Association Proceedings 2nd Annual Convention 1: 223-241.

Rangin, C. and Silver, E. 1990. Geological setting of the Celebes and Sulu Seas *In* Initial Reports of the Ocean Drilling Program 124. pp. 35-42. Edited by C. Rangin, E. A. Silver and M. T. von Breymann.

Rangin, C., Jolivet, L. and Pubellier, M. 1990. A simple model for the tectonic evolution of southeast Asia and Indonesia region for the past 43 m.y. Bulletin de la Société géologique de France, 8 (VI), 889-905.

Rice-Oxley, E. D. 1991. Palaeoenvironments of the Lower Miocene to Pliocene sediments in offshore NW Sabah area. Bulletin of the Geological Society of Malaysia 28: 165-194.

Rose, R. and Hartono. P. 1978. Geological evolution of the Tertiary Kutei-Melawi Basin, Kalimantan, Indonesia. Indonesian Petroleum Association Proceedings 7th Annual Convention 1: 225-252.

Saller, A., Armin, R., La Ode Ichram and Glenn-Sullivan, C. 1992. Sequence stratigraphy of Upper Eocene and Oligocene limestones, Teweh area, central Kalimantan. Indonesian Petroleum Association Proceedings 21st Annual Convention 1: 69-92.

Saller, A., Armin, R., La Ode Ichram and Glenn-Sullivan, C. 1993. Sequence stratigraphy of aggrading and backstepping carbonate shelves, Oligocene, Central Kalimantan, Indonesia. *In* Carbonate Sequence Stratigraphy: Recent Developments and Applications. pp. 267-290. Edited by R. G. Loucks and J. F. Sarg. American Association of Petroleum Geologists Memoir 57.

Sartono, S. 1982. Genesa Danau Tempe, Sulawesi Selatan. Pertumuan Arkeologi 2: 555-560.

Siemers, C. T., Sutiyono, S. and Wiman, S. K. 1992. Description and reservoir characterization of a late Miocene, delta-front coral-reef buildup, Serang Field, Offshore East Kalimantan, Indonesia *In* Carbonate rocks and reservoirs of Indonesia. pp. 5.1.5-5.1.27. Edited by C. T. Siemers, M. W. Longman, R. K. Park and J. G. Kaldi. A core workshop held in conjunction with 1992 Indonesian Petroleum Association Annual Convention 1.

Silver, E. A., Joyodiwiryo, Y. and McCaffrey, R. 1978. Gravity results and emplacement geometry of the Sulawesi ultramafic belt, Indonesia. Geology 6: 537-531.

Simandjuntak, T. O. 1986. Sedimentology and Tectonics of the Collision Complex in the east arm of Sulawesi. Ph.D. Thesis University of London 374 pp.

Simandjuntak, T. O. 1990. Sedimentology and tectonics of the collision complex in the east arm of Sulawesi, Indonesia. Geologi Indonesia. Journal of the Indonesian Association of Geologists 13: 1-35.

Situmorang, B. 1982. The formation and evolution of the Makassar Basin as determined from subsidence curves. Indonesian Petroleum Association Proceedings 11th Annual Convention 1: 83-107.

Smith, R. B. and Silver, E. A. 1991. Geology of a Miocene collision complex, Buton, eastern Indonesia. Geological Society of America Bulletin 103: 660-678.

Steenis, C. G. G. J. van. 1979. Plant geography of east Malesia. Botanical Journal of the Linnean Society, 79: 97-178.

Stuart, C. J., Schwing, H. F., Armin, R. A., Sidik, B., Abdoerrias, R., Vijaya, S., Boer, W. D. de, Wiman, S. K., Heitman, H. L., Yusuf, S. K. and Nurhono, A. 1996. Sequence stratigraphic studies in the lower Kutai Basin, East Kalimantan, Indonesia *In* Proceedings of the International Symposium on Sequence Stratigraphy in SE Asia. pp. 363-368. Edited by C. A. Caughey *et al.* Indonesian Petroleum Association, Jakarta, Indonesia.

Sukamto, R. 1975. The structure of Sulawesi in the light of plate tectonics. Regional Conference on the Geology and mineral resources of SE Asia, Jakarta: 1-25.

Sukamto, R. and Supriatna, S. 1982. Geologi lembar Ujung Pandang, Benteng dan Sinjai quadrangles, Sulawesi. Geological Research and Development Centre, Bandung, Indonesia.

Sunaryo, R., Martodjojo, S. and Wahab, A. 1988. Detailed geological evaluation of the hydrocarbon prospects in the Bungalun area, East Kalimantan. Indonesian Petroleum Association Proceedings 17th Annual Convention 1: 423-446.

Sunyata, and Wahyono. 1991. VI. Palaeomagnetism. *In* Studies in East Asian Tectonics and Resources (SEATAR): Crustal transect VII Java-Kalimantan-Sarawak-South China Sea. Edited by C. S. Hutchison. CCOP 26: 66 pp.

Supriatna, S., Santosa, S. and Djamal, B. 1993. Geologi Lembar Muaralesan, Kalimantan Timur. Kolokium Hasil Pemetaan dan Penelitian Puslitbang Geologi, 1991/1992. Geological Research and Development Centre Special Publication 14: 20-30.

Suwarna, N., Sutrisno, Keyser, F., de, Langford, R. P. and Trail, D. S. 1993. Geology of the Singkawang sheet area, Kalimantan, 1:250,000. Geological Research and Development Centre, Bandung, Indonesia.

Takhtajan, A. 1987. Flowering plant origin and dispersal: The cradle of the angiosperms revisited. *In* Biogeographical Evolution of the Malay archipelago. pp. 26-31. Edited by T. C. Whitmore. Clarendon Press, Oxford Monographs on Biogeography 4.

Tanean, H., Paterson, D. W. and Endharto, M. 1996. Source provenance interpretation of Kutei basin Sandstones and implications for the tectono-stratigraphic evolution of Kalimantan. Indonesian Petroleum Association Proceedings 25th Annual Convention 1: 333-345.

Tate, R. B. 1991. Cross-correlation of geological formations in Sarawak and Kalimantan. Bulletin Geological Society of Malaysia 28: 63-95.

Tate, R. B. 1994. The sedimentology and tectonics of the Temburong Formation — deformation of early Cenozoic deltaic sequences in NW Borneo. Bulletin Geological Society of Malaysia 35: 97-112.

Tjia, H. D., Komoo, I., Lim, P. S. and Tungah Surat. 1990. The Maliau basin, Sabah: geology and tectonic setting. Bulletin of the Geological Society of Malaysia 27: 261-292.

Trail, D. S., John, T. V., Bird, M. C. *et al.* 1974. The general geological survey of block II, Sulawesi Utara, Indonesia. PT Tropical Endeavour Indonesia.

Truswell, E. M., Kershaw, A. P. and Sluiter, I. R. 1987. The Australian-South-east Asian connection: Evidence from the palaeobotanical record. *In* Biogeographical Evolution of the Malay archipelago. pp. 32-49. Edited by T. C. Whitmore. Clarendon Press, Oxford Monographs on Biogeography 4.

Umbgrove, J. H. F. 1938. Geological history of the East Indies. American Association of Petroleum Geologists Bulletin 22: 1-70.

Vane-Wright, R. I. 1991. Transcending the Wallace Line: do the western edges of the Australian region and the Australian plate coincide. *In* Austral Biogeography. Edited by P. Y. Ladiges, C. J. Humphries and L. W. Martinelli, Australian Systematic Botany 4: 183-197.

Wain, T. and Berod, B. 1989. The tectonic framework and paleogeographic evolution of the Upper Kutei Basin. Indonesian Petroleum Association Proceedings 18th Annual Convention 1: 55-78.

Wakita, K., Munasri, Sopaheluwakan, J., Zulkarnain, I. and Miyazaki, K. 1994. Early Cretaceous tectonic events implied in the time-lag between the age of radiolarian chert and its metamorphic basement in the Bantimala area, South Sulawesi, Indonesia. The Island Arc 3: 90-102.

Wallace, A. R. 1863. On the physical geography of the Malay archipelago. Journal Royal Geographical Society 33: 217-234.

Weerd, A. A., van de. and Armin, R. A. 1992. Origin and evolution of the Tertiary hydrocarbon bearing basins in Kalimantan (Borneo), Indonesia. American Association of Petroleum Geologists Bulletin 76: 1778-1803.

Weerd, A. A., van de., Armin, R. A., Mahadi, S. and Ware, P. L. B. 1987. Geologic setting of the Kerendan gas and condensate discovery, Tertiary sedimentary geology and paleogeography of the northwestern part of the Kutei Basin, Kalimantan, Indonesia. Indonesian Petroleum Association Proceedings 16th Annual Convention 1: 317-338.

Weissel, J. K. 1980. Evidence for Eocene oceanic crust in the Celebes Basin. *In* The Tectonic and Geologic Evolution of Southeast Asian Seas and Islands. pp. 37-48. Edited by D. E. Hayes. American Geophysical Monograph Series 23.

Whitmore, T. C. 1981a. Palaeoclimate and vegetation history. *In* Wallace's line and plate tectonics. pp. 36-42. Edited by T. C. Whitmore. Clarendon Press, Oxford Monographs on Biogeography 1.

Whitmore, T. C. 1981b. Wallace's line and some other plants. *In* Wallace's line and plate tectonics. pp. 70-80. Edited by T. C. Whitmore. Clarendon Press, Oxford Monographs on Biogeography 1.

Whitten, A. and Whitten, J. 1992. Wild Indonesia. World Wildlife Fund for Nature. New Holland, London, 208 pp.

Whitten, A. J., Mustafa, M. and Henderson, G. S. 1987. The Ecology of Sulawesi, Gadjah Mada University Press. 777 pp.

Wilson, M. E. J. 1995. The Tonasa Limestone Formation, Sulawesi, Indonesia: Development of a Tertiary Carbonate Platform. Ph.D. Thesis, University of London, 520 pp.

Wilson, M. E. J. and Bosence, D. W. J. 1996. The Tertiary evolution of South Sulawesi: a record in redeposited carbonates of the Tonasa Limestone Formation. *In* Tectonic evolution of SE Asia. pp. 365-389. Edited by R. Hall and D. J. Blundell. Geological Society of London Special Publication 106.

Wilson, M. E. J. and Bosence, D. W. J. 1997. Platform top and ramp deposits of the Tonasa carbonate platform, Sulawesi, Indonesia. *In* Petroleum Geology of SE Asia. pp. 247-279. Edited by A. J. Fraser, S. J. Matthews and R. W. Murphy. Geological Society of London Special Publication 126.

Yuwono, Y. S., Maury, R. C., Soeria-Atmadja, P. and Bellon, H. 1987. Tertiary and Quaternary geodynamic evolution of South Sulawesi: Constraints from the study of volcanic units. Geologi Indonesia 13: 32-48.

Ziegler, B. 1983. Introduction to Palaeobiology. General Palaeontology. Ellis Horwood Series in Geology, Chichester, UK.

Colour plates for:

Biogeographic implications of the Tertiary palaeogeographic evolution of Sulawesi and Borneo

Steve J. Moss[1] and Moyra E. J. Wilson
SE Asia Research Group, Department of Geology, Royal Holloway University of London
[1]*Now at School of Applied Geology, Curtin University of Technology, Perth, 6845, W.A. Australia*

Captions

Fig.4. Palaeogeographic map for 50 Ma, early Eocene. A key to the environments is shown. Note that rivers are shown schematically for all time slices.
Fig.5. Palaeogeographic map for 42 Ma, middle Eocene.
Fig.6. Palaeogeographic map for 34 Ma early Oligocene.
Fig.7. Palaeogeographic map for 21 Ma, early Miocene.
Fig.8. Palaeogeographic map for 8 Ma late Miocene.
Fig.9. Palaeogeographic map for 4 Ma, early Pliocene.

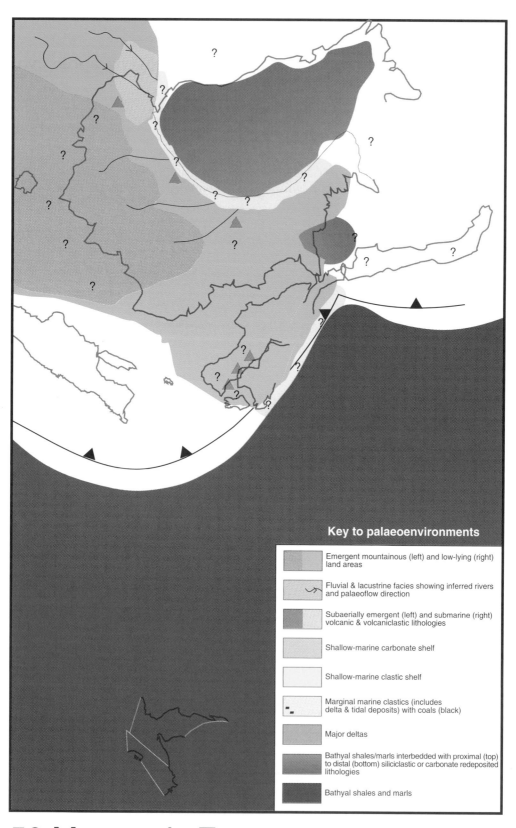

50 Ma - early Eocene

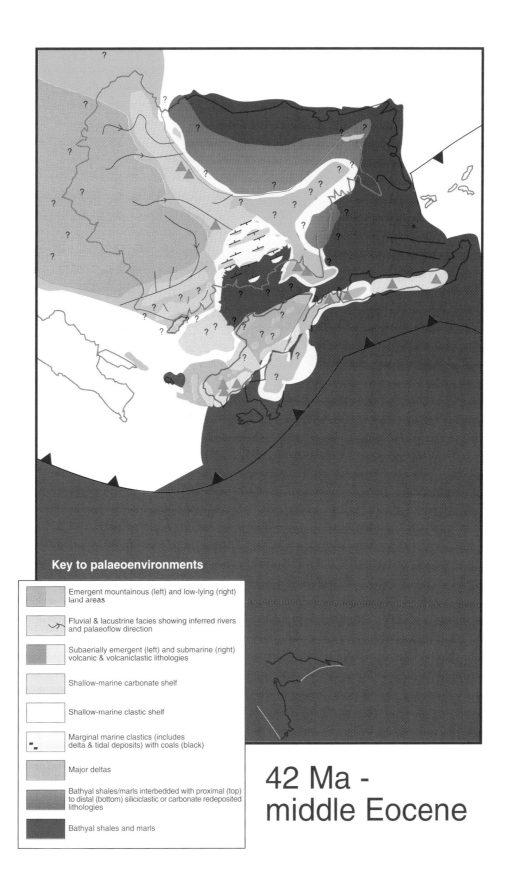

Key to palaeoenvironments

- Emergent mountainous (left) and low-lying (right) land areas
- Fluvial & lacustrine facies showing inferred rivers and palaeoflow direction
- Subaerially emergent (left) and submarine (right) volcanic & volcaniclastic lithologies
- Shallow-marine carbonate shelf
- Shallow-marine clastic shelf
- Marginal marine clastics (includes delta & tidal deposits) with coals (black)
- Major deltas
- Bathyal shales/marls interbedded with proximal (top) to distal (bottom) siliciclastic or carbonate redeposited lithologies
- Bathyal shales and marls

42 Ma - middle Eocene

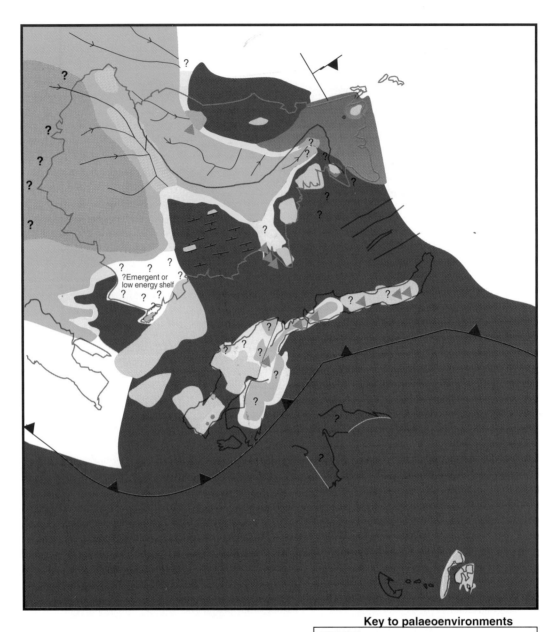

34 Ma - early Oligocene

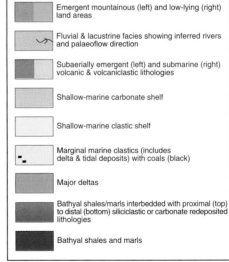

Key to palaeoenvironments

- Emergent mountainous (left) and low-lying (right) land areas
- Fluvial & lacustrine facies showing inferred rivers and palaeoflow direction
- Subaerially emergent (left) and submarine (right) volcanic & volcaniclastic lithologies
- Shallow-marine carbonate shelf
- Shallow-marine clastic shelf
- Marginal marine clastics (includes delta & tidal deposits) with coals (black)
- Major deltas
- Bathyal shales/marls interbedded with proximal (top) to distal (bottom) siliciclastic or carbonate redeposited lithologies
- Bathyal shales and marls

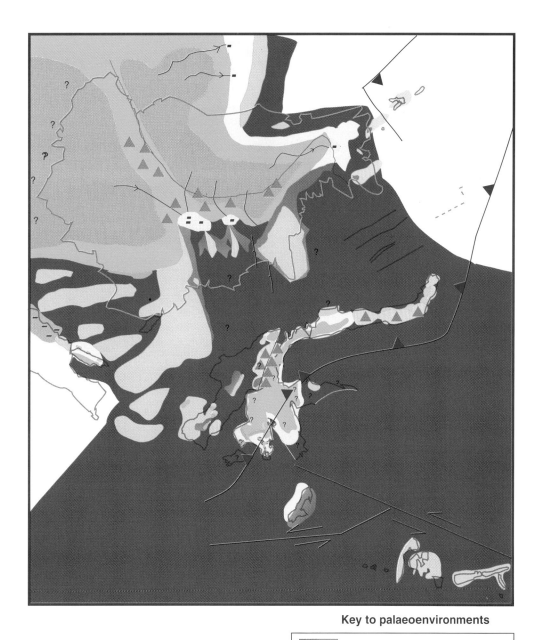

21 Ma - early Miocene

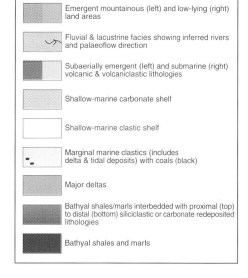

Key to palaeoenvironments

- Emergent mountainous (left) and low-lying (right) land areas
- Fluvial & lacustrine facies showing inferred rivers and palaeoflow direction
- Subaerially emergent (left) and submarine (right) volcanic & volcaniclastic lithologies
- Shallow-marine carbonate shelf
- Shallow-marine clastic shelf
- Marginal marine clastics (includes delta & tidal deposits) with coals (black)
- Major deltas
- Bathyal shales/marls interbedded with proximal (top) to distal (bottom) siliciclastic or carbonate redeposited lithologies
- Bathyal shales and marls

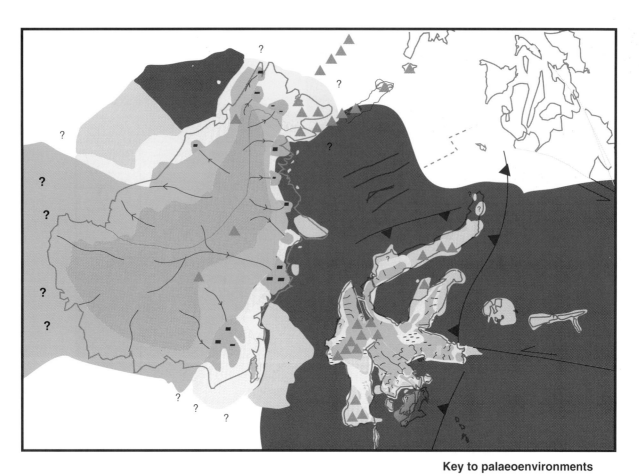

8 Ma - late Miocene

Key to palaeoenvironments

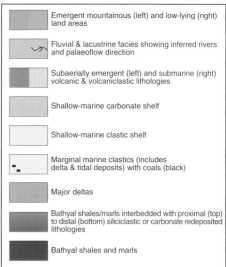

Emergent mountainous (left) and low-lying (right) land areas

Fluvial & lacustrine facies showing inferred rivers and palaeoflow direction

Subaerially emergent (left) and submarine (right) volcanic & volcaniclastic lithologies

Shallow-marine carbonate shelf

Shallow-marine clastic shelf

Marginal marine clastics (includes delta & tidal deposits) with coals (black)

Major deltas

Bathyal shales/marls interbedded with proximal (top) to distal (bottom) siliciclastic or carbonate redeposited lithologies

Bathyal shales and marls

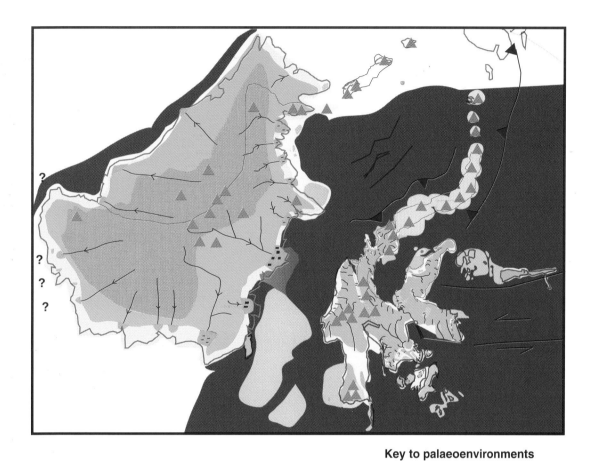

Key to palaeoenvironments

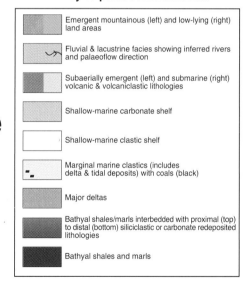

4 Ma - early Pliocene

Implications of paucity of corals in the Paleogene of SE Asia: plate tectonics or Centre of Origin?

Moyra E. J. Wilson[1] and Brian R. Rosen[2]
[1] SE Asia Research Group, Royal Holloway University of London, Egham, Surrey, TW20 0EX, UK
[2] Department of Palaeontology, The Natural History Museum, Cromwell Road, London, SW7 5BD, UK

Key words: scleractinian corals, evolution, carbonates, Indo-West Pacific, Cenozoic, Mesozoic, tectonics

Abstract

The modern Indo-West Pacific centre of marine diversity, embracing SE Asia, has the highest zooxanthellate coral diversity in the world, and is characterised by abundant coral reefs. In contrast, field and literature studies show that in Paleogene carbonates of SE Asia, corals are rare, and extensive coral reefs have not been reported. Corals and reefs of this age may have been better developed in the mid-Pacific, especially in the late Eocene, but the most diverse regions at this time were in Europe, and to a lesser degree, the Caribbean. The apparent scarcity of Paleogene corals in SE Asia is surprising in view of the long-standing theory that the Indo-West Pacific is a Centre of Origin. The question therefore arises whether this 'Paleogene gap' is a sampling problem, and if not what are its causes and implications?

A review of SE Asian Tertiary carbonates in their tectonic context shows that many shallow-water Eocene-Oligocene carbonates were dominated by larger benthic foraminifera and coralline algae. A new Eocene coral fauna (12 spp.) from Sulawesi accords with earlier views that there was little endemicity in SE Asia during the Paleogene. The general paucity of corals in carbonate platform deposits confirms that the 'Paleogene gap' is real, and not an artefact. Since SE Asia, particularly the western part, has remained in, or close to, tropical latitudes throughout the Tertiary, climatic reasons cannot account for this. Although local ecological factors would have been important, the tectonics point to a degree of geographical isolation from other coral rich regions as a key factor. These patterns changed dramatically, however, in the earliest Neogene, as Australian fragments collided with SE Asia. Collision led to decreased isolation of the region and generation of numerous shallow-water areas, with diverse and abundant zooxanthellate corals occurring in the Miocene of SE Asia, similarly to the present day. The geographical complexity of SE Asia appears to have favoured localized isolation and origination of new taxa, though the fauna also consists of older relicts and taxa which migrated into the region from elsewhere. All these processes contributed to making the Indo-West Pacific the richest region for corals from the Neogene onwards.

We compare these patterns with the three main models of high diversity in the Indo-West Pacific Centre (Centres of Origin, Accumulation and Survival). During the Paleogene the region was not a centre of any kind. From the Neogene onwards, no single model is applicable, but a combination of all three models is preferred. Comparison with Mesozoic coral data and the history of the region suggests that the patterns discussed here also apply to the Late Triassic and Late Jurassic. As with the Neogene, these were times when blocks rifted from Gondwana moved across the tropics to dock against Asia. We therefore emphasize the role of plate tectonics in controlling regional high diversity patterns of zooxanthellate corals. Since corals and coral reefs provide habitats for a myriad of marine organisms, the biogeographic history of Cenozoic reef corals has implications for the evolution of tropical reef ecosystems.

Introduction

The waters of SE Asia contain the highest marine faunal diversity in the world (Stehli, 1968; Briggs, 1974; Paulay, 1997), explanation of which has long attracted the attention of biogeographers. Reef-dwelling scleractinian corals typify this pattern, and the highest species diversity (c.400-500, Veron, 1995) and 55% of the world's coral reefs (Muller, 1995) occur in these waters (Figs.1 and 2).

The most enduring explanation of these kinds of high diversity foci is that they represent Centres of Origin (evolutionary source or 'cradle': Ekman, 1953; Stehli and Wells, 1971; Briggs, 1974, 1992; Zinsmeister and Emerson, 1981; Veron, 1995). Although detailed definitions and interpretations of Centres of Origin vary (Rosen, 1984, 1985), they all assume that taxonomic rich-

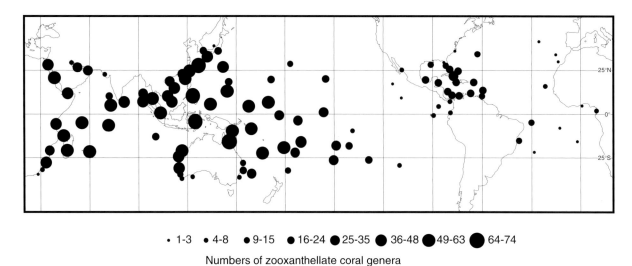

• 1-3 • 4-8 • 9-15 • 16-24 ● 25-35 ● 36-48 ● 49-63 ● 64-74

Numbers of zooxanthellate coral genera

Fig.1. Diversity (numbers of genera) of Recent zooxanthellate corals. Localities and data are mostly based on unpublished compilation by Professor John W. Wells. Map after Rosen (1984) and see this work for locality data.

ness has resulted from a concentration of evolutionary appearances over tens of millions of years (Table 1). Alternatively, the Indo-West Pacific high diversity foci may represent a Centre of Accumulation (evolutionary sinks or 'museums'), whereby species originate in other regions, but over time, their geographical ranges changed to coincide in the present high diversity focus (Table 1; Ladd, 1960; Newell, 1971; McCoy and Heck, 1976; Kay, 1984; Wallace, 1998; Rosen, 1984, 1988; Jokiel and Martinelli, 1992; Pandolfi, 1992). A third Centre of Survival

		'CRADLE' Source Centre of Origin	'MUSEUM' Sink Centre of Accumulation	'REFUGE' Centre of Survival
WITHIN THE FOCAL REGION	Speciation rates	high	low	low
	Extinction rates	low	low	low
	Immigration rates	low	high	low
OUTSIDE THE FOCAL REGION	Speciation rates	low	low	low
	Extinction rates	high	high	high
	Immigration rates	high	low	low
PREDICTED DIVERSITY PATTERNS THROUGH GEOLOGICAL TIME IN FOSSIL RECORD		Diversity conspicuously higher in the centre than in outside regions, and of greater duration than species turnover timescale (< 30 Ma)	Increasing high diversity through time in the centre, relative to outside regions.	No particular pattern of change in high diversity in the centre, through time, but diversity elsewhere diminishing through time.
OLDEST RECORDS		Mostly within centre	No particular pattern or mostly outside centre	No particular pattern

Table 1. The three main models for the origin of the Indo-West Pacific high diversity centre. See text for references. The table shows the main evolutionary characteristics within and beyond the centre, together with predicted patterns for fossil record. 'High' and 'Low' convey relative not absolute rates. The classic arguments about patterns of oldest/youngest and/or most primitive/derived taxa are reduced here to what can be reasonably easily investigated in the fossil record, especially in the context of this paper. Related ideas about competitive displacement are omitted. The dichotomy, "cradle or museum?" was posed by Stebbins (1974) for the tropics as a whole, but is here applied regionally to a feature within the tropics.

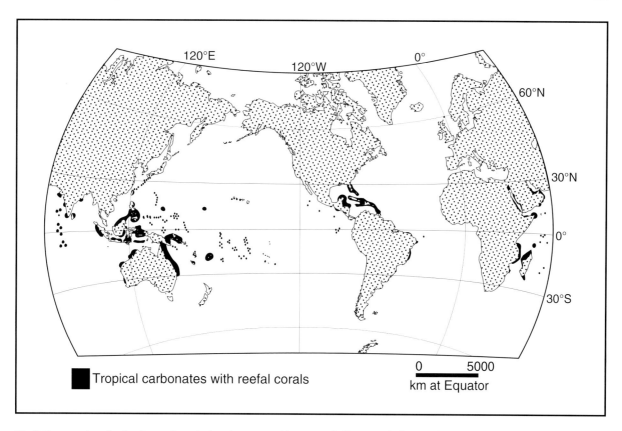

Fig.2. Present day distributions of tropical carbonates with zooxanthellate corals from Nelson (1988).

or refuge model postulates that the fauna of SE Asia is a vestige of a formerly rich, cosmopolitan distribution of broadly Cretaceous origin (Table 1; McCoy and Heck, 1976). Other ideas mostly incorporate aspects of the three main hypotheses and vary in having an ecological to a historical emphasis (Rosen 1988; Crame, 1992; Veron, 1995; Palumbi, 1996; Paulay, 1997).

This paper concentrates on the three main alternatives (Table 1). The models are essentially historical, yet paradoxically, little regional palaeontological or historical evidence has been assimilated into theories of coral evolution in SE Asia. The main way in which palaeontology has been applied to the history of the Indo-West Pacific corals is through age-and-area studies, *i.e.*, region by region analysis of global stratigraphic ages of extant taxa (Stehli and Wells, 1971; Rosen, 1984; Veron, 1995), although, as shown below, this approach is flawed. Cladistic biogeographical studies on extant corals have now also commenced (Pandolfi, 1992; Hoeksema, 1989, 1993; Wallace *et al.*, 1991) but so far, only Hoeksema (1989) has incorporated fossil evidence. He was however unable to resolve his particular fungiid patterns sufficiently to develop robust biogeographical hypotheses. Veron's (1995) theory of 'surface circulation vicariance', though drawing on fossil evidence from age-and-area patterns and evolutionary rates, is not critically dependent on the fossil record. A regional palaeontological approach is limited to pioneering endemicity work by Gerth (especially 1925, 1930), and brief observations by Umbgrove (1946) and Rosen (1988). Rosen and Smith (1988) subsequently analysed endemicity patterns of reef corals, worldwide, for the early Miocene. This modest body of work contrasts sharply with the extensive investigations of the much smaller and less diverse Caribbean fauna (*e.g.*, Budd *et al.*, 1992, 1994, 1996).

If the Indo-West Pacific focus was indeed a Centre of Origin, it should be recognizable in the fossil record as a region of persistent high diversity (Table 1) lasting significantly longer than the <30 Ma timescale of species turnover. Although the fossil record must be incomplete, it can provide constraints, particularly when compared with evidence from living species, and it provides historical clues from large scale spatio-temporal patterns (Table 1). This paper therefore examines the Cenozoic fossil coral record of this focal region, and interprets this in the light of new knowledge of the tectonics and stratigraphical distribution of carbonates and coral facies.

Background

Regional terminology

As a physical region, SE Asia broadly corresponds to Ekman's (1953) 'Indo-Malayan' biogeographical region, within which a triangular area formed by the Philippines, the Malay peninsula and New Guinea shows the richest species diversity for numerous shallow marine groups of organisms (Stehli, 1968; Briggs, 1974, 1992). Richness of corals progressively declines in all directions with increasing distance from this central triangular area (Fig.1; Rosen, 1984 – frequency plot of coral genera; Veron 1995 – interpolated 'contour' plots of coral species). In this paper, we use 'SE Asia' for the marine region of the SE Asian mainland, the Australian margins and the areas between these, and use the same term for the Cenozoic equivalents of this region (Figs.2-8). This is preferred to 'Indo-Malayan' which presumes a rigorous biogeographical definition, criteria for which may not have applied in the past.

The marine fauna of SE Asia falls within a much larger biogeographical region, the 'Indo-West Pacific' (Ekman, 1953; Paulay, 1997), which stretches from the Red Sea and western Indian Ocean to the eastern fringe of the main archipelagos of the mid-Pacific. Across the eastern Pacific is the East Pacific Barrier, with the East Pacific region beyond. The combined Indo-West Pacific and East Pacific constitute the 'Indo-Pacific'. The SE Asian area of high diversity is also referred to loosely as the 'Indo-West Pacific centre' or 'focus', though the boundaries are not rigidly defined, and to the east and southeast, the westernmost Pacific and northeast Australia are often included (Fig.1). Tectonically therefore, this centre lies across several plate boundaries, covering parts of the Indo-Australian, Eurasian, Philippine Sea and Pacific plates.

Zooxanthellate corals

Some knowledge of scleractinian coral ecology is essential to understand the significance of patterns discussed later in this paper. Scleractinians secrete hard calcareous skeletons, and those under discussion mostly belong to a broad ecological group that is confined to warm shallow waters often on, or close to, coral reefs. Today most of these scleractinian corals host symbiotic unicellular algae (zooxanthellae) within their soft parts and hence are 'zooxanthellate corals' or 'z-corals'. The algae require sunlight to photosynthesise and this restricts the corals to the euphotic zone (about 100 m depth). Scleractinians mainly occur within a low latitude circum-global band between 35-40°N and about 35°S (Fig.2). The lowest mean annual surface sea-water temperatures are about 16°C, although species tolerance may be lower (Rosen, 1984; Veron, 1995), and the richest faunas are generally found in the warmest waters (Wells, 1954a; Rosen, 1984; Veron, 1995; Fraser and Currie, 1996).

Superimposed on this latitudinal pattern are regional ecological and historical biogeographical factors, the most important of which is availability of habitable areas, itself a combination of numerous factors including temperature, nutrient flux, substrate, water depth and quality. Z-coral distributions are therefore patchy in space and time, with modern examples found especially in clear, shallow conditions along land-mass margins, and particularly in archipelagos. Although z-corals are common reefal inhabitants, the temperature limit of reefs (minimum of 18°C; Veron, 1995) is slightly warmer than that of z-corals, and the distribution of reefs and z-corals do not correspond precisely. Contrary to widespread belief, fossil z-coral faunas are often found in both reefal formations and bedded deposits, the latter being often silty, marly or even pyroclastic.

Corals mostly disperse by planktonic gametes and larvae, and the potential migration distances and biogeographic patterns of corals depend on numerous factors including duration of larval phase, life history patterns, water conditions, substrate factors, coastal and sea-floor configuration, hydrodynamics and ocean currents (Jackson, 1986; Harrison and Wallace, 1990; Done et al., 1996).

Populations of living corals tend to consist largely of either z-corals, or azooxanthellate corals (az-corals), with z-corals dominating in warm shallow conditions (Stanley and Cairns, 1988; Coates and Jackson, 1987). Mixed assemblages also occur, so simple 'either/or' assumptions cannot be made about the composition of past faunas. Hence, for present purposes, the z-coral component of a fossil fauna has to be identified and the az-coral component removed, in order to filter out the potentially conflicting ecological and biogeographical signals of az-corals.

The 'Paleogene gap' in Indo-West Pacific coral records

Preliminary evidence in the existing literature suggests that the high diversity centre of the Indo-West Pacific is relatively young, since coral diversity in the region was lower than elsewhere until the Neogene. Gerth (1925; see also Gerth, 1930) for a similar discussion in English but without his previous data compilations and map) indicated four localities with Paleogene 'reef corals' in Borneo, Java, Sulawesi, Papua New Guinea and a dubious reefal locality in New Zealand. In contrast, for the Neogene, he identified nine coral areas within the Indonesian archipelago alone, and noted a three-fold increase in generic diversity of 'reef-building corals' from 29 in the Paleogene to 76 in the Neogene-Quaternary. Umbgrove (1930), conveyed a similar picture, and in his review of the Tertiary fossil record of *extant* Indo-Pacific corals, made no note of any from Paleogene strata (Ta-Td; East India Letter Classification, cf., Adams, 1970). In comparison, for Te-Tf strata (regarded then as entirely Miocene), he mentions formations that were known to contain rich and well-described coral faunas. Umbgrove (1946) subsequently suggested that increased speciation of corals in SE Asia occurred during the late Miocene (Tg).

An Oligocene to Recent generic level comparison of the various major reef coral regions of the world (Rosen, 1988; his Table 2) showed that reef coral richness in the Oligocene was highest in the Caribbean and Mediterranean, but from the Miocene onwards the Indo-West Pacific centre became the richest region, increasing more than fourfold, from 14 genera in the Oligocene to 58 in the early Miocene (below and Fig.9).

Despite the prevalent theory that the Indo-West Pacific focus represents an evolutionary centre, it is surprising that the Paleogene of this region is apparently so poor in z-corals, thus implying that this centre is only a post-Paleogene phenomenon. Although evidence for this 'Paleogene coral gap' goes back 70 years, the numerous ideas that have been developed concerning Indo-Pacific marine biogeography (*e.g.*, Rosen, 1988; Veron, 1995; Flessa and Jablonski, 1996; Paulay, 1997), have mostly made little reference to this palaeontological pattern, preferring indirect evidence from age-and-area studies. One reason for this may be an uncertainty over how 'real' the gap is, and one of the main aims of this paper is to assess this.

Approach and Methods

Coral records

To investigate the main relevant biogeographical hypotheses (Table 1), we have compiled previously published z-coral data for SE Asia and neighbouring regions, and also included an important new fossil coral fauna found mostly in the upper Eocene part of the Tonasa Formation of Sulawesi (Wilson, 1995; Fig.10). Knowledge of fossil corals in SE Asia is derived largely from investigations made during the days of the former Dutch colonial era. More recently, the French have studied the Mesozoic of the region (see below), but no recent taxonomic accounts exist for the Tertiary. Taxonomic revision of the entire Tertiary scleractinian fauna is undoubtedly necessary. Unfortunately, the published coral collections often consist of single small samples of each species, and it is difficult to obtain any picture of species variability from them. Revisionary taxonomy is outside the present scope, and estimates of species numbers here are based on existing nomenclature, with taxonomic adjustments made only in obvious cases. Moreover, the stratigraphic, palaeoenvironmental and tectonic context of these corals in the old literature was poorly documented by modern standards. Today, since exploration for natural resources often focuses on carbonate rocks, which commonly include the remains of corals and other marine organisms, new, well-documented collections are being made. The nature, extent and history of carbonate development in SE Asia is therefore becoming clearer.

Inference of algal symbiosis in fossil corals

Zooxanthellae are not preserved in fossil coral skeletons, but algal symbiosis often needs to be inferred for palaeobiogeographical studies. A range of evidence can be used to infer symbiosis, but methods based on intrinsic features of the corals, including colonial characters, skeletal stable isotope geochemistry, palaeoautecology and phylogenetic relationships, are preferable

Table 2. Simplified scheme for designating levels of confidence in methods for inferring algal symbiosis in a fossil coral species. Note criterion 3(a) is particularly unreliable because although most living forms in such habitats are zooxanthellate, such habitats today also contain az-corals. With respect to 4, note that 'nkz' status is a neutral category, and does not signify inference of azooxanthellate status. The latter can be inferred using criteria parallel to 1-3. Inference of symbiosis in genera (*e.g.*, Fig.9) can be derived from integration of respective member species, or approximately, by applying the parallel criteria at generic level. Examples of corals in different 'z-categories' are shown in Figs.10 and 12. In this paper, we use criteria 1-2 to obtain a minimum estimate of frequency of z-corals in a given fauna, and criteria 1-4 to obtain a maximum estimate (Table 3).

'CONFIDENCE LEVELS'	CRITERIA	TERMS (abbreviated)
1.	Species is still living today and known to be zooxanthellate	**z-coral**
2.	Species is extinct, but (a) belongs to a genus whose living species are all z-corals and/or (b) has skeletal characteristics found only in living z-corals (mainly stable isotopes, morphology)	**z-like coral**
3.	Species is extinct but (a) occurs in warm shallow-water carbonate facies and/or (b) lacks skeletal characteristics consistently found in living z-corals and/or (c) there is doubt about applicability of other stronger lines of evidence (i.e. 1 and 2)	**z-like coral?**
4.	Species is extinct and there is currently no evidence for inferring algal symbiosis, but symbiosis cannot be ruled out (*i.e.*, status of algal symbiosis **n**ot **k**nown)	**nkz-like**

(Table 2; Rosen, 1998). Different methods have different merits so corals can be grouped according to a broad, albeit subjective, level of confidence according to method, strengthened where applicable by corroboration from different independent methods. Since algal symbiosis in fossil corals is an inference, it is more precise to regard fossil examples as 'z-like' or 'az-like' rather than 'z-corals' and 'az-corals' respectively (Rosen and Turnsek, 1989), although the latter pair of terms may be used for simplicity.

A simplified summary of methods, criteria and approach is shown in Table 2. Each group of criteria and its respective confidence level is denoted in the third column. If inference from first order criteria is not applicable, then second order criteria are used, and so on, working through criteria with decreasing levels of confidence. A given faunal list can then be partitioned into those for which symbiosis has been inferred (or not) according to these levels of confidence (see captions to Tables 2 and 3).

The sampling problem and reconstruction of habitable areas

The regional geographical and stratigraphic locations of all the Paleogene coral faunas have been compared with facies distributions and regional tectonics. The aim is to assess whether there are significant areas in SE Asia which are potentially rich in z-corals, but which lack coral information. Inhabitable areas for z-corals can then be assessed against other factors such as taphonomic effects, climate, and biogeographical barriers.

Tertiary carbonate rocks of SE Asia are formed mostly in shallow marine environments away from clastic input and are composed of the calcareous skeletons of marine organisms. Modern z-corals are mostly restricted to carbonate environments and through a study of these lithified sediments and their constituent components it is possible to track the availability of potential habitats for reef corals in the past. Data on the carbonate rocks of SE Asia have been extracted from the literature and are also based on current field research by the authors and other members of the London University SE Asia Research Group.

In SE Asia, local carbonate studies usually concentrate on the sedimentological rather than palaeontological aspects of the rocks, and taxonomic studies rarely include information on the sediments. However, sedimentologists usually note the presence of corals and, although taxa are not always identified, information on growth forms, together with an indication of coral abundance and diversity are commonly given. This is clearly not ideal for detailed biogeographical work, but when combined with taxonomic data, provides information on Cenozoic z-coral distributions for the region.

Reconstructions

To track the location and nature of potentially available habitats for corals during the Cenozoic, three main carbonate rock types have been plotted on to time slices of Hall's (1996) reconstructions between the late Eocene (about 37 Ma) and Recent (Figs.3-8). Paleocene and lower Eocene carbonates are rare in SE Asia and there have been no published reports of z-corals of this age, although a Paleocene locality has been reported in Java (A. Russo, J. Pignatti, pers. comms., 1997). This suggests that conditions suitable for z-coral growth were sparse, and reconstructions for these intervals are not shown. The ages of the Cenozoic carbonate formations are based on a range of nannofossil, planktonic and larger benthic foraminifera, and, in a few cases, strontium isotope dates. Although full documentation is not possible here, these have been gleaned from a large literature database and the authors' own work, and in a few cases it has been necessary to revise ages (see Table 3).

The main environments illustrated on the reconstructions of relevance to marine biogeographers are: shallow marine carbonate depositional environments, major land areas which would usually have been surrounded by clastic shelves, and deep marine areas (below 100-200 m water depth).

A wide variety of skeletal remains of shallow marine organisms are present in Cenozoic carbonate deposits, including corals, larger benthic foraminifera, coralline algae, *Halimeda*, echinoids and molluscs. However, in most SE Asian carbonates the dominant constituent grains are coral fragments and larger benthic foraminifera, although coralline algae may be locally abundant. Carbonate deposits have been divided here into three categories on the basis of these components, and the different types are illustrated on the maps (Figs.3-8). These are carbonates where z-like corals are abundant, carbonates dominated by larger benthic foraminifera and carbonates where both z-like corals and larger benthic foraminifera are abundant.

Modern larger benthic foraminifera, similar to many living corals, contain symbiotic algae, and their growth is thought to be restricted to the euphotic zone (Haynes, 1965; Hallock, 1985;

Brasier, 1995). They occur in similar tropical environments to corals and have broadly equivalent global distribution patterns (Adams, 1983; Hallock and Glenn, 1986; Adams *et al.*, 1990). A number of depth zonation schemes for larger benthic foraminifera (Ghose, 1977; Hottinger, 1983; Hallock and Glenn, 1986; Gorsel, 1988; Racey, 1994) and other shallow marine bioclasts, such as coralline algae (review in Perrin *et al.*, 1995) have been deduced for different stages during the Cenozoic. The taxa, together with their forms and the nature of the enclosing sediment, can be used to infer depositional depths for carbonate sediments (Fig.11). Therefore an ancient carbonate formation in which specific larger benthic foraminifera dominate, such as robust *Nummulites*, would have been deposited in a shallow marine area during the Eocene or early Oligocene, where one might also expect to find z-corals. For each carbonate formation illustrated on the maps (Figs.3-8), the setting (S) and the nature of corals (C) present are briefly described.

The limits of the maps correspond to 30°N and 20°S (Figs.3-8) – close to the latitudinal limits of the modern reef coral belt to the north, and somewhat within the belt to the south. The 'reefal belt' has variously widened and narrowed through geological time in response to phases of global warming and cooling. For example, the early to middle Miocene corresponds to a period of global warming, and the reef belt, together with distributions of z-like corals and larger benthic foraminifera, was as much as 10° of latitude wider (Fulthorpe and Schlanger, 1989; Adams *et al.*, 1990; James *et al.*, 1996). Davies *et al.* (1989) constructed an oceanic surface water temperature curve for NE Australia and suggested that the earliest part of the Paleocene was also a period of global warming, sufficient to sustain reefal coral growth.

Summary of Mesozoic scleractinian records in SE Asia

The earliest global records of scleractinian corals are Middle Triassic (Wells, 1956; Stanley, 1988), since when scleractinians have occurred discontinuously in SE Asia, concentrated over particular time spans. Triassic corals (largely Upper Triassic) are represented on the Asian mainland (Yunnan and around the Burma-Thailand border) and in the Indonesian archipelago, mostly east of Borneo in Sulawesi, Buru, Misool, Seram, Timor and Roti (Stanley, 1988;

Martini *et al.*, 1997). The Timor fauna appears to be the richest with over 30 nominal species. The Jurassic is represented by some Middle Jurassic, but more extensively by Upper Jurassic faunas (Beauvais, 1986), some, like those of Sarawak, being quite rich (about 20 species or more). They may continue in some places into the lowermost Cretaceous (Beauvais and Fontaine, 1990). Upper Jurassic corals are also known from Sumatra, Burma, Thailand and the Philippines (*e.g.*, Beauvais, 1983), whilst substantial Middle to Upper Jurassic faunas occur in Japan (Eguchi, 1951). In contrast, although Cretaceous carbonate platforms do occur in SE Asia (Cebu, Misool, Timor and Papua New Guinea), Cretaceous corals are rare throughout the region, except, again in Japan. A few Cretaceous corals are known from the Upper Cretaceous of western Borneo and northern Sumatra, but rudists appear to be the dominant shallow water element (P. Skelton, pers. comm., 1997). Overall, corals were most common in the Upper Triassic and Upper Jurassic, partly reflecting times when corals flourished globally.

Cenozoic plate tectonics of SE Asia and implications for coral history

SE Asia is an extremely active tectonic area, and considerable lateral and vertical movements of the Earth's crust throughout the Mesozoic and Cenozoic have affected the spatio-temporal variations in carbonate depositional environments and hence the potential habitats for corals. During this time the Indo-Australian and Philippine-Pacific plates, and a large number of smaller microcontinental and oceanic fragments, interacted and collided with Sundaland, the stable eastern margin of the SE Asian craton (Figs.3-8 and in Hall, 1998 this volume). The reconstructions of Hall (1996), were used for Figs.3-8, but other recent reconstructions (Rangin *et al.*, 1990, Daly *et al.*, 1991; Lee and Lawver, 1995) yield a similar picture for the purposes of plotting shallow marine carbonate depositional environments, although variations in timing may give slightly differing patterns (cf. Pandolfi, 1992).

At the beginning of the Tertiary, the separation of Australia from mainland SE Asia across this 'Indo-Pacific gateway' was about 3000 km. This distance, particularly if the current directions were unfavourable, is sufficient to have been a potential biogeographical barrier to the dispersal of z-coral larvae, comparable in scale with the East Pacific Barrier today (Ekman, 1953; Paulay, 1997). Until the latest Paleogene, the whole Australian craton and its coastal waters lay too far south to fall within the boundaries of the z-coral belt. It was not until between 25 to 16 Ma (late Oligocene to early Miocene) that the transition from temperate to tropical carbonates, with abundant z-coral and reef development, occurred in the northern part of what is now the Great Barrier Reef (Davies *et al.*, 1987). The central part of the Great Barrier Reef did not experience a tropical climate until 15 to 10 Ma, and the southern part only became tropical in the last few million years (Davies *et al.*, 1987).

The Indo-Pacific gateway progressively narrowed during the Cenozoic as Australia moved northwards (Figs.3-8), and closer to SE Asia. This, combined with the northward drift of various microcontinental blocks, and emergence of new islands, or development of new shallow carbonate producing areas, would have increased the potential for exchange of coral larvae with other regions and the development of z-coral communities and coral reefs (Rosen, 1988; Pandolfi, 1992). Davies *et al.* (1989) noted that the closure of this seaway would also have restricted equatorial current flow and caused the diversion of warm tropical waters to the south along NE Australia (see also Grigg, 1988).

The Indian continent drifted northwards across the zone of reef coral growth during the Late Cretaceous and early Paleogene, and is inferred to have collided with the SE Asian mainland during the Eocene (Hall, 1998 this volume). During this time suitable habitats for z-coral development may have occurred on narrow shelves with limited clastic input around the landmasses of India and mainland SE Asia. Examples include the z-like coral communities known from upper Paleocene to Miocene strata along the NW side of the Indian continent in Kaachchh and western Pakistan (Duncan, 1880; Gregory, 1930; Ghose, 1982; Rosen, 1988; revised dates in Adams 1970), belonging tectonically to India (C. Izatt, pers. comm., 1997).

Following the Eocene collision of India with Asia, and the subsequent uplift of the Himalayas and other related mountain chains, large quantities of clastic material were shed into the adjacent seas via large delta systems. These deltas, including those of the Irrawaddy and Ganges, currently inhibit reef coral growth, and form local coastal barriers to biogeographical exchange of many shallow marine organisms. Substantial clastic input along the southern margin of the Asian continent during the Paleogene may similarly have inhibited z-coral growth. The closure of the intervening seaway may also have hindered the migration of shallow marine organisms between areas such as the Middle East and SE Asia.

There are very few islands in the modern Indian Ocean between the Maldive-Chagos line and Indonesia, although volcanic islands along the Ninetyeast Ridge were emergent during the Paleogene (Kemp and Harris, 1975). Prior to the Paleogene, islands may have existed in the seaway between India and mainland Asia. However, following the juxtaposition of India with Asia the islands of the Ninetyeast Ridge would have been among the few habitable areas for corals in the eastern Indian Ocean and the Indo-Pacific gateway, and this otherwise open seaway may have formed a barrier to the survival of most coral larvae. In contrast, in the western and southern Pacific, numerous atolls, reefs and islands existed from the Cretaceous onwards (Menard, 1964; Schlanger and Premoli Silva, 1981; Grigg, 1988; Winterer, 1991), and z-coral communities existed there throughout the Cenozoic. The mid-plate foundations of most of these shallow areas of the western Pacific originated in a geographically restricted phase of volcanicity during the Mesozoic (Darwin Rise/ Superswell). Since then, they have all been moving in unison, towards Asia, so narrowing deeper water areas to their west and widening those to the east (East Pacific Barrier). This must have contributed to the development of z-coral habitats and larval exchange around SE Asia, while further diminishing potential for larval exchange across the East Pacific Barrier.

Cenozoic geological evolution and coral faunas

Overview of Paleogene corals

At the start of the Paleocene, the global scleractinian record mainly comprises az-like corals, with an apparent hiatus of up to 2-3 Ma before the first Tertiary records of z-like corals (Rosen, 1998). In the Indo-West Pacific centre the oldest recorded Tertiary corals are Eocene (Table 3), and superficially seem quite rich, although an unpublished Paleocene fauna has been found in Java (A. Russo, pers. comm., 1997). On checking the Eocene records and their associated larger benthic foraminifera (Table 3), two of the richest faunas (SE Borneo and Papua New Guinea) are probably Oligocene or younger, and a third (Tonga), although confirmed as upper Eocene, consists entirely of az-like corals. This leaves a small number of published z-like faunas of Eocene age, but even the dating of some of these is uncertain. In fact, our new coral fauna of late Eocene (Tb) age from Sulawesi (Wilson, 1995), though modest, is currently the richest known z-like fauna of definite Eocene age in SE Asia (Fig.10). A z-like fauna from the Enewetak boreholes of the Marshall Islands on the Pacific plate, has an even smaller number of species (Table 3).

Chevalier *et al.* (1971) compiled other Eocene 'reefal and parareefal' coral records including their own from New Caledonia. Most are included in Table 3 and some are probably younger than Eocene (see above). Localities mentioned by these authors, not in Table 3, are: Sumatra (no references given), Sulawesi (after Bemmelen, 1970), Saipan in the Marianas (after Yabe and Sugiyama, 1935), and Makatea in the Tuamotus (after Obellianne, 1962). We have failed to find the Sulawesi records in Bemmelen (1970). Of Makatea, the authors say that Eocene corals "should exist", as well as in "several other raised islands of the Pacific" but they cited no further work. For Saipan, Yabe and Sugiyama (1935) describe only *Saipania*, which in our opinion is a chaetitid sponge, not a coral. However, Eocene carbonates of Saipan (Matansa Limestone and Densinyama Formation), and Eocene volcaniclastic rocks on Guam (Marianas) are all reported to contain z-like corals (G. Siegrist and D. Randall, pers. comm., 1997) and require further study.

Lower Oligocene records are even sparser than for the Eocene (Table 3), but from the mid-Oligocene (Td) there are signs of increasing richness (SE Borneo). A silicified Oligocene coral reef is reported from Guam by Siegrist and Randall (above). By the late Oligocene to early Miocene (Te), richer reefal coral faunas, similar to those today, seem to have become progressively more common throughout much of the region (Fig.9).

Fig.9 shows a Cenozoic compilation of generic richness of z-corals for the four principal tropical marine regions (update of Table 2 in Rosen, 1988) incorporating new data from the Indo-West Pacific centre (Table 3) and other regions. This confirms the reality of the 'Paleogene Gap' in terms of available coral data, though not necessarily its biogeographical reality, and shows that substantially richer faunas occurred elsewhere, particularly in Europe. This pattern reversed itself in the early Miocene: as the Indo-West Pacific centre became noticeably richer,

LOCALITY	FORMATION	ORIGINAL AGE	REVISED AGE	SOURCES	TOTAL CORAL genera	TOTAL CORAL species	Z-LIKE CORAL genera	Z-LIKE CORAL species	REMARKS
NEOGENE									
SULAWESI		Oligocene	Neogene	Dollfus, 1915 Gerth, 1931 Umbgrove, 1942b					Neogene according to Umbgrove (1942b)
BUTON	Asphalt deposits			Umbgrove, 1942a					Paleogene age ruled out by author
OLIGOCENE (to Miocene)									
PAPUA NEW GUINEA (Boera Head, southern Papua)	Boera Formation		Latest Oligocene to earliest Miocene (latest Te1-5)	unpublished	c.10	c.12	>4	c.12	Z-like corals in brecciated reefal limestones (collected Rosen & Darrell, NHM Collections; work in progress)
PAPUA NEW GUINEA (Fly River, at 6°5'S and / or at 5°40'S)	(Rolled pebbles)	Eocene, probably Middle	?Oligocene - ?Miocene	Gregory & Trench, 1915 Gerth, 1931	11	13	8 (?11)	10 (?13)	Corals not found in outcrop; notwithstanding published age, the fauna looks mostly Neogene or of mixed ages
MARSHALL ISLANDS (Bikini Atoll)	Borehole: <355 m	No date given	Late Oligocene - Early Miocene (Te)	Wells, 1954b	unidentifiable coral fragments				Age revised by Adams (1970)
MARSHALL ISLANDS (Enewetak Atoll)	Boreholes: 845-355 m	Miocene		Wells, 1964	8	11	7 (?8)	10 (?11)	Age revised by Adams (1970)
SERAM (eastern part of the island near Mt. Téri)		Oligocene ('Sannoisean' = Early Oligocene)	Oligocene	Dollfus, 1908 Umbgrove, 1924 Gerth, 1931	1	1	0 (?1)	0 (?1)	Umbgrove (1924) said coral was "probably Oligocene"
KALIMANTAN, SE (Riam-Kawa and Riam-Kanan area, Meratus Complex)	Verbeek's Gamma Beds	Eocene	Mid Oligocene (Td)	Fritsch, 1878 Gerth, 1931	23	25	9 (?21)	10 (?22)	Dollfus (1908) stated that Fritsch's fauna is Oligocene ('Sannoisean' = Lower Oligocene). Bemmelen (1970) cites foraminifera that suggest mid-Oligocene (Td) (Adams 1970)
MARSHALL ISLANDS (Enewetak Atoll)	Boreholes: 931-845 m	? Miocene	? Mid Oligocene (Td)	Wells, 1964	6	7	6	7	Age revised by Adams (1970), but no direct evidence for Td
MARSHALL ISLANDS (Enewetak Atoll)	Boreholes: 976-931 m	? Miocene	? Early Oligocene (Tc)	Wells, 1964	3	3	3	3	Age revised by Adams (1970), but no direct evidence for Tc
EOCENE									
TONGA (Eua)		Late Eocene (Tb)	Late Eocene (Tb)	Wells, 1976	14	17	0	0	Z-like corals absent
MARSHALL ISLANDS (Enewetak Atoll)	Boreholes: 1220-1251 m	Late Eocene (Tb)		Wells, 1964	4	4	3 (?4)	3 (?4)	Age confirmed by Adams (1970)
SUMBA (c.80 km from western end)		Eocene		Umbgrove, 1943	1	2	0 (?1)	0 (?2)	Late Eocene (Tb) from *Pellatispira* (see Adams 1970, and A. Racey, pers. comm., 1997). See also Bemmelen (1970)
SULAWESI	Tonasa Formation (basal part)			this paper	11	12	6 (?11)	7 (?12)	
JAVA (south central)	Nanggulan Beds	Late Eocene	Mid-Eocene to Oligocene (Ta3-Td)	Gerth, 1921 Gerth, 1931 Gerth, 1933	7	8	0 (?5)	0 (?5)	Precise horizons of the corals not known
NEW CALEDONIA, SW (Baie St Vincent (îles Ducos and Saint Phalle); Oua Tom)		Mid-Late Eocene ('Eocene C')	Mid-Late Eocene (Ta3-Tb)	Chevalier et al., 1971	2	2	1	1	
PALEOCENE									
JAVA Jati Bunkus, near Karangsambung (Central Java)	reefal limestones		Late Paleocene (Thanetian) (Ta1)	Russo, Pignatti (pers. comms., 1997)	8	11	6 (?8)	8 (?11)	

Table 3. Summary of Paleogene coral faunas of the Indo-West Pacific centre including SE Asia. Ages based on foraminifera unless indicated. Revised ages show that the Eocene is even more poorly represented than previously realised. By the late Oligocene z-like corals were relatively common and the pattern is more Miocene-like, Te1-4 formations are therefore a selection only. For the period before the late Oligocene the data are as complete as possible apart from a few problematic or unstudied cases (see text). Top two rows show younger faunas which have been included because of their supposed Paleogene age. Minimum values of z-corals are in bold and maximum values are in parentheses (see caption to Table 2). Paleogene Indo-West Pacific data for Fig.9 are based on net totals of faunas in this table, except for the Oligocene (based on data from Kalimantan and the Marshall Islands).

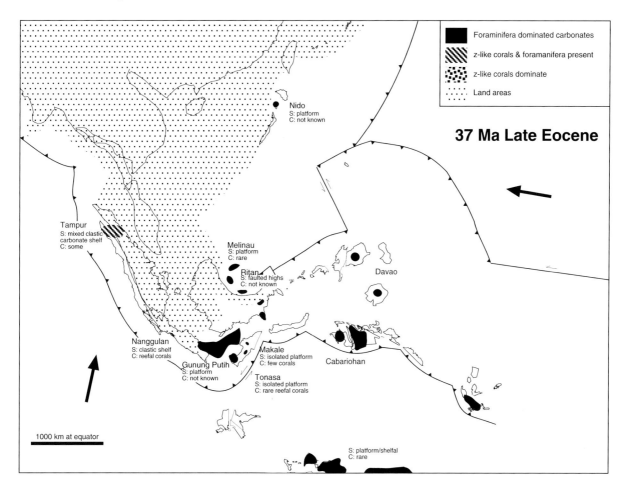

Fig.3. Late Eocene time slice from the reconstructions of Hall (1996) with carbonate deposits shown. The nature of the carbonate deposits (S) and any coral records (C) are shown. Abbreviations used in Figs.3-9 for corals records are: Br. branching, Pl. platy, M. massive, Ab. abundant, Div. diverse. Names shown are formations. Fig.8 shows present-day geography.

the Mediterranean declined. Since then, the Indo-West Pacific centre has remained the richest z-coral region. Throughout the Cenozoic, the Caribbean, by contrast, has remained relatively constant and moderately rich. The picture in SW Asia is unclear because of lack of adequate data, but Paleogene z-like faunas, such as those from western Pakistan, together with unpublished coral collections in The Natural History Museum, London, from Oman, Iran and Somalia suggest that this region was also richer than SE Asia during the Paleogene.

Late Eocene (Tb)

Carbonate deposition was common on microcontinental blocks in the eastern part of SE Asia (Fig.3) where emergent land areas were small and clastic input was low. Initiation of a number of basins occurred around Sundaland during the Paleogene (Hall, 1998 this volume), and those to the east quickly became marine and carbonates developed (Fig.3). In comparison, mainland SE Asia, Sumatra and much of Borneo (Fig.3) was a land area which was shedding considerable clastic material into the surrounding seas. Along the southern margin of this land area, the shelf would have been narrow, close to the adjacent subduction trench (Fig.3), and this, together with the amount of clastic input, would have hindered carbonate production. The distance between shallow marine areas across the Indo-Pacific gateway was greater than 2000 km in the late Eocene.

In eastern SE Asia carbonate development occurred mainly as extensive shallow-water platforms, such as in Sulawesi (Wilson, 1995), the Melinau Limestone of Sarawak (Adams, 1965) and New Guinea (Koesoemadinata, 1978). The margins of these carbonate platforms varied from steep escarpments to gently dipping ramps, but reefal buildups have not been reported (Wilson, 1995). In western SE Asia, shelfal carbonates or localised areas of carbonate deposition occurred where clastic input was insufficient to hinder carbonate production.

The Eocene carbonates are dominated by larger benthic foraminifera (Fig.11) and in some cases coralline algae, while corals are extremely rare (Adams, 1965; Kohar, 1985; Siemers et al., 1992; Wilson, 1995). Within the carbonates, there occur a wide variety of larger benthic foraminifera showing various adaptations, which suggest they occupied the full range of habitats available on the platforms within the euphotic zone (Adams, 1965; Wilson, 1995).

The first well-dated, published appearance of z-like corals in the Indo-West Pacific centre occurred in the late Eocene. There is a possibility of older (middle Eocene) z-like corals occurring in New Caledonia and Java (Table 3), but the age range of these very limited faunas also includes the late Eocene. The Nanggulan Beds are mixed clastic-carbonate deposits and show no evidence of any reefal buildups or of *in situ* growth fabrics. An enigmatic Sumba locality whose place-name we have been unable to locate on modern maps, is of late Eocene age, but contains a single genus which may not be z-like.

Following revision of the ages of Paleogene corals in SE Asia (Table 3), the only known z-like corals of definite late Eocene age in SE Asia are from the basal part of one area of the Tonasa Formation in south Sulawesi (Fig.10 and Table 3). The fauna is small, consisting of 12 species, 7 z-like and 5 z-like?, mostly known also from the Paleogene of western Pakistan (Ranikot, Khirthar and Nari series) and Europe, and therefore not strongly endemic. It includes small solitary discoidal (z-like?) forms such as *Cycloseris subcrenulata* and *Trochocyathus ? nummiformis* from the lowermost beds of the formation at the Doi doi coal quarry. Well-preserved massive colonial forms found about 2 km away, slightly higher in the succession, include *Favia pedunculata* (z-like), *Astrocoenia bistellata* (z-like?), and a striking four-lobed flabellate z-like form, *Trachyphyllia indica*. Specimens of all of these show flattened bases in response to a soft substrate. There are also colonies of *Stylocoenia* sp. (z-like?) which, with their unique spheroidal corallith habit and stout stilt-like spines (but eroded), are also soft-substrate dwellers (Darga, 1991). The corals from the basal part of the Tonasa Formation occurred where there was some clastic input, and did not form reefal buildups (Wilson, 1995).

These SE Asian z-coral faunas are paltry compared with the rich European faunas of Eocene age, such as those known from former Yugoslavia with 85 z-like species, or the 42 z-like species known from Eocene of Panama (interpretation of data from Felix, 1925, and from Budd et al., 1992, respectively). Moreover, apparent paucity of corals in SE Asia should also be seen in the context of Ladd's statement (in Wells, 1976) that upper Eocene limestones have been reported from the Pacific at Palau, the Marianas, the Marshalls, Fiji and Tonga, across a distance of 6560 km (see also Schlanger and Premoli Silva, 1981; Winterer, 1991). While some of these, as in the Marshalls (Table 3), and Marianas (G. Siegrist and D. Randall, pers. comm., 1997) contain z-like corals, this cannot be assumed in all cases, since Eocene corals from Eua, Tonga, (Table 3) are entirely az-like. It may be that a largely concealed upper Eocene sequence with z-like corals is now submerged, or lies buried beneath younger volcanics or limestones throughout much of the Pacific. Thus whereas the 'Paleogene gap' is a sampling reality for the Eocene of SE Asia, it is possibly a sampling artefact for Pacific part of the Indo-West Pacific centre.

Early (Tc) to middle (Td) Oligocene

The early and middle Oligocene were similar to the late Eocene with carbonates common on microcontinental blocks in eastern Indonesia, but rare around the margins of mainland SE Asia (Fig.4). An open seaway between the Pacific and Indian Oceans was still present and was about 1,400 km wide by the end of the middle Oligocene (Td). However, as Australia and the associated microcontinental blocks moved northwards, large carbonate platforms developed in New Guinea (Fig.4, Wilson et al., 1993), and the distance across this gateway decreased, altering the configuration of ocean currents (Grigg, 1988) and increasing the possibility of biogeographical exchange of marine biotas. Carbonates were also deposited around the margins of the developing South China Sea (Holloway, 1982). Their distribution was similar to the late Eocene, with platform and shelfal carbonates dominating to the east and west respectively. Larger benthic foraminifera again dominate in most of these deposits (Fig.11), and there is no evidence for any reefal buildups.

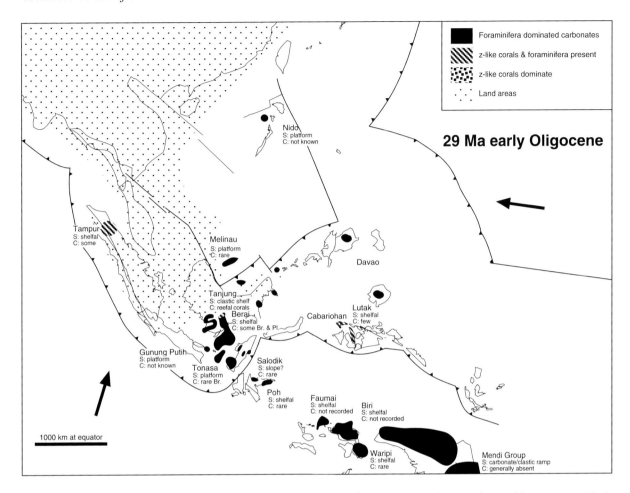

Fig.4. Early/middle Oligocene time slice from reconstructions of Hall (1996) with carbonate deposits shown; abbreviations on Fig.3.

Corals are rare in the lower Oligocene (Tc) of SE Asia, and those that do occur are described as thin branching forms, such as *Acropora* from south Sulawesi (Fig.10l). Some massive and thick branching forms have been reported from shelf margin deposits in south Kalimantan (Saller *et al.*, 1993). On the Pacific plate, a fauna possibly of this age is known from the Enewetak boreholes (Table 3). The richest Paleogene z-like fauna in SE Asia is that described by Fritsch (1878) from the Tanjung Formation of SE Borneo (Table 3). Fritsch and later workers (*e.g.*, Gerth, 1925; 1930) regarded this fauna as Eocene, although the associated foraminifera *Nummulites fichteli* and *Lepidocyclina* sp. (Bemmelen, 1970), point to an Oligocene (Td) age. The corals consist of only nine clearly z-like species, but if other corals of less certain z-status are included, the total increases to 22. This is small when compared with the rich contemporaneous faunas of Europe and the Caribbean (Fig.9). No other published z-like fauna is known from Td strata in SE Asia, but its richness relative to older Paleogene localities hints at an increase in suitable areas for z-corals at this time in parts of SE Asia. A locality in Seram with just a single coral record is dated as Oligocene (Table 3) but its exact age is not known. On the Pacific plate, a modest z-fauna of possible mid-Oligocene age is known from the Enewetak boreholes (Table 3). Grigg (1988) discussed the Cenozoic coral record of the mid- and northern tropical Pacific and emphasized that z-corals started to become more widespread in the early Oligocene, the date of their first appearance in the Hawaii-Emperor chain.

Late Oligocene (Te1-4)

Carbonates were still common in eastern Indonesia (Fig.5), and some of the basins, such as

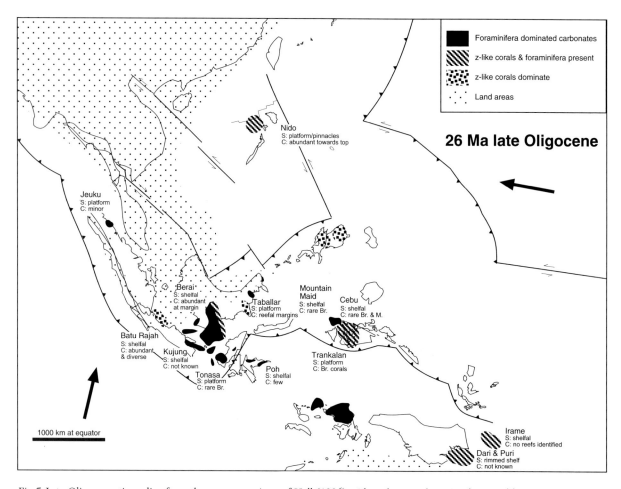

Fig.5. Late Oligocene time slice from the reconstructions of Hall (1996) with carbonate deposits shown; abbreviations on Fig.3.

Java and Sumatra, started to become marine, with the development of carbonates. Carbonates also became more extensive along the eastern side of the Sundaland craton due to continued subsidence in the region. Shelfal and platform carbonates again dominated, particularly during the earlier part of the late Oligocene, although patch reefs and some reefal buildups may have developed later in the Oligocene. Larger benthic foraminifera, such as lepidocyclinids, heterosteginids and various larger imperforate foraminifera, were still common (Fig.11). In formations which pass conformably upwards into the Miocene, such as those in Java, Sumatra or New Guinea, shelfal carbonates pass upwards into reefal buildups in the upper part of the succession (Cucci and Clark, 1993). However, it is often difficult to separate upper Oligocene from lower Miocene carbonates on the basis of information in the literature.

When corals are mentioned in the sedimentological literature, they tend to be thin delicate branching forms, or finger-like corals (such as *Acropora*), although massive framework corals occur in some localities forming patch reefs. The Boera Formation of Papua New Guinea, dated as uppermost Oligocene to lowest Miocene (Te: Rogerson *et al.*, 1981), contains a moderate fauna of z-like corals (Fig.12c,d) which is unpublished but well-represented in The Natural History Museum, London. These are the oldest z-like corals known from outcrop on or around the Australian cratonic margin, and provide a minimum age for the arrival of the craton's leading edge in tropical waters (cf., Davies *et al.*, 1987). Initial identifications suggest affinities with European Oligocene faunas. Reefal corals are also common in the upper part of the upper Oligocene in the Batu Asih limestones in Java (A. Wonders, pers. comm., 1997). Quite a rich z-like fauna is emerging from the uppermost Oligocene or lowest Miocene Gomantong Lime-

Cenozoic corals of SE Asia

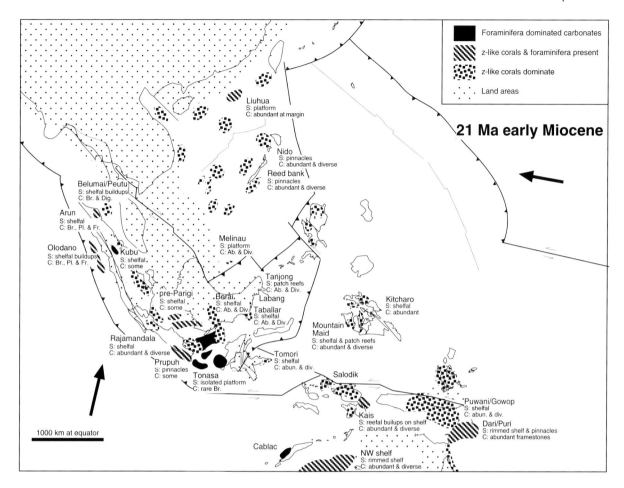

Fig.6. Early Miocene time slice from the reconstructions of Hall (1996) with carbonate deposits shown; abbreviations on Fig.3.

stone in Sabah (J. Noad, pers. comm. 1997) with 17 z-like genera identified so far (work in progress, BRR), many of them previously known from the lower Miocene elsewhere in Borneo.

Neogene

There was a striking increase in the distribution and extent of carbonates (Fig.6), and a marked change from larger foraminifera to coral dominated carbonates at the beginning of the Neogene. A distinct change also occurred in the coral faunas during the earliest Miocene, within Te time. Bemmelen's (1970) Table 15 lists some of the key coral faunas of Neogene age, many of which consist of 30 or more z-like species. They are far richer and more widespread than Tertiary faunas of pre-Te age in the same region. These patterns are clearly reflected in the generic diversity patterns of Fig.9.

The ages and taxonomy of this rich and historically important SE Asian Neogene coral fauna are much in need of modern revision, as this is the key to more rigorous studies of the global endemicity patterns and biogeographical history of corals discussed by Gerth (1925, 1930) and Rosen (1988). Nevertheless, such work is unlikely to alter radically the existing picture of substantial increase in z-like coral faunas which seem to have developed relatively suddenly throughout the region, from the early Neogene onwards. Only a few key points concerning the Neogene are made here.

Early and middle Miocene (Te5-Tf)

The early Miocene was a major phase of carbonate deposition both in SE Asia (Fig.6), and throughout much of the tropics and subtropics, with reef corals occurring in much higher lati-

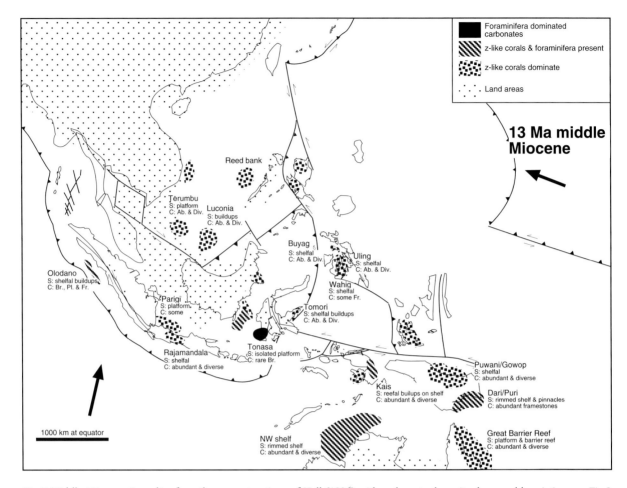

Fig. 7. Middle Miocene time slice from the reconstructions of Hall (1996) with carbonate deposits shown; abbreviations on Fig.3.

tudes than today, probably as a result of a global warm phase (Fulthorpe and Schlanger, 1989). Carbonates were common in marine basins around the margins of mainland SE Asia, although a combination of clastic input and a narrow shelf restricted carbonate development along the southeast coast (Fig.6). Carbonate production still occurred on microcontinental blocks in eastern SE Asia (Fig.6), although more islands were emerging due to collision-related uplift. During the mid-Miocene (Fig.7), the area of carbonate deposition, though still extensive and diverse, had been reduced, due to the emergence of more land areas, resulting partly from microcontinental collisions and related shedding of clastic material into adjacent marine areas.

Platform and reefal carbonates were common throughout the region and pinnacle reefs have been reported from many areas (Livingstone *et al.*, 1992). Z-like corals were widespread, abundant and diverse throughout the region, in marked contrast to the earlier Cenozoic. Many modern genera, and even species (Umbgrove, 1946; Rosen, 1984; Veron, 1995), as well as all the growth forms typical of modern reefs in the Indo-West Pacific, occur in these Miocene reefs (*e.g.*, Fig.12a,b). The accompanying marine biota is also similar to modern reef biotas, with abundant benthic foraminifera, echinoderms, molluscs, coralline and calcareous green algae.

Comparison with present day

The modern situation (Fig.8; after Wells and Sheppard, 1988) is not unlike the Miocene (Figs.6 and 7), although the extent of coral-rich carbonate is reduced, especially when compared with the lower Miocene. This is due to the continued emergence of land areas and to a corresponding influx of clastic material into adjacent coastal waters hindering carbonate produc-

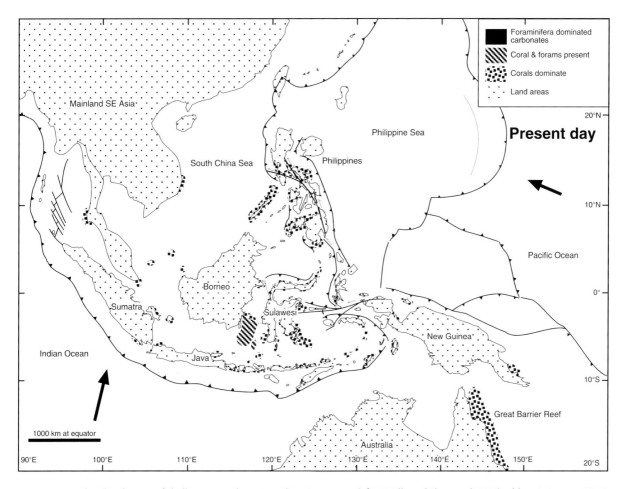

Fig.8. Present day distribution of shallow water depositional environments (after Wells and Sheppard, 1988); abbreviations on Fig.3.

tion. However, despite this slight reduction in carbonate sedimentation, reefal development with z-corals is still common in SE Asian waters, particularly in the clearer waters of eastern Indonesia (Wells and Sheppard, 1988).

As already emphasized, modern z-corals have their highest species diversity in SE Asian waters, with the whole range of morphological growth forms present. At least 262 species and 78 genera have been recorded in Sulawesi (Moll, 1983), and Veron (1995) recorded 472 species in the northern Philippines. Coral reefs are abundant, and, of the estimated 600,000 km² of these complex ecosystems worldwide, 25-30% are located in SE Asia (Bernard, 1991).

Summary of geology and coral faunas in SE Asia

Eocene z-like coral faunas appear to be even less rich than previously thought, the only certain fauna of this age being our own new finds from the Tonasa Formation. Oligocene faunas are slightly richer than previously thought. Z-coral richness is much higher in other regions in the Paleogene, and higher in SE Asia from the Neogene onwards. In the Paleogene, facies and environments in which z-like corals could have thrived (mainly shallow water carbonates) were sparse, and SE Asia was really a wide oceanic seaway in which the dense arrangement of islands of the modern region did not yet exist. There was a fundamental change at or around the beginning of the Neogene (*i.e.*, within Te time), coincident with the beginning of interaction between Australian fragments and SE Asia, leading to a notable increase in shallow water carbonates, widespread reefs, high z-coral abundance and fourfold increase in z-coral richness.

Tectonics controlled the emergence, disappearance, and movement in and out of the z-coral belt, of shallow marine areas, as well as their regional concentration. In the Paleogene however, even where shallow water carbonates did exist, they often lacked z-coral faunas, suggesting that other factors (biogeography, local ecology) were also important. In contrast to SE Asia, there are extensive Paleogene (especially upper Eocene) limestones in the tropical western Pacific, but they are mostly in the subsurface and have remained largely unsampled. The few limestones that have been sampled contain either z-like or az-like coral faunas. However, many of these limestones lay further east than now, representing perhaps a distinct region. The Cenozoic record of z-corals points to the modern Indo-West Pacific centre of diversity being a young feature, probably little older than around the beginning of the Neogene (c.24 Ma).

Discussion

Global biogeographical context

The most challenging biogeographical problem of the Indo-West Pacific centre is to understand the origins of its marked taxonomic richness, models for which are shown in Table 1. This can be partly attributed to a global pattern in which richness of numerous marine taxa increases as latitude decreases, related to available energy (Stehli, 1968; Jablonski, 1993; Fraser and Currie, 1996; Flessa and Jablonski, 1996; Paulay, 1997). This has long been known for z-corals (Yonge, 1940; Wells, 1954a), and more recently demonstrated quantitatively (Rosen, 1971, 1984; Fraser and Currie, 1996). There is also empirical evidence that more organisms originate in the tropics than elsewhere (Jablonski, 1993; Flessa and Jablonski, 1996). Temperature however cannot account for the entire global pattern. Z-corals, for example, are four times richer in the Indo-West Pacific centre than in the western Atlantic (Figs.1 and 9), but the prevailing temperatures in both are similar (Rosen, 1984), and this richness ratio has remained stable over decades of sampling and taxonomic study.

In order to consider regional anomalies, and the Indo-West Pacific centre in particular, it is useful to resolve diversity (and endemicity) patterns into two geographical components, 'latitudinal' (N-S) and 'longitudinal' (E-W) (Rosen, 1975). The strongest endemicity patterns are longitudinal. Four regions have dominated longitudinal discussion (Gerth, 1925, 1930; Paulay, 1997): the Indo-West Pacific, the eastern Pacific, western Atlantic and eastern Atlantic, and their Tertiary Tethyan counterparts (*e.g.*, Mediterranean), though the Indian Ocean (with Red Sea) is also sometimes considered separately (see Veron, 1995; Flessa and Jablonski, 1996; Budd *et al.*, 1996; Paulay, 1997; for reviews of these regions). We concentrate on the ideas shown in Table 1 for the Indo-West Pacific centre, particularly the extent to which patterns may have been affected by plate tectonics, though the influence of eustasy (not discussed here) was also important, especially in the later Cenozoic (Schlanger, 1981; Rosen, 1984, 1988; Grigg, 1988; Veron, 1995). The emphasis here on physical factors differs from models which emphasise intrinsic evolutionary processes sustained in one region over great lengths of time (Centres of Origin).

Flessa and Jablonski (1996) have remarked that Rosen's (1984,1988) physical contingency model "implies that intensive sampling of the Eocene, prior to global refrigeration and the northward migration of the Australian plate, would detect neither diversity gradient, nor age gradient, but perhaps only the legacy of the evolutionary rebound from the end-Cretaceous extinction". Their caveat that "few data are available to test (this) idea of ... young age" reflects in part the previous emphasis on age-and-area studies of extant taxa. We address Flessa and Jablonski's prediction below.

In modelling the biogeography of a particular region, three sets of factors need to be considered (Rosen, 1988): (1) originations (patterns of evolutionary turnover); (2) distributional change (often also called 'migrations', 'dispersal' etc., and meaning changes through time of biogeographical ranges of taxa, combined with changes of biogeographical barriers); and (3) maintenance (the sum of local-to-regional ecological factors). This framework is used to address the Cenozoic history of z-like corals in the Indo-West Pacific and to discuss criteria for the alternative models for the high diversity foci (Table 1). In particular, since the main conclusion of our stratigraphic review points to high diversity in the SE Asian part of the Indo-West Pacific centre being surprisingly young, we concentrate on possible reasons for this.

Maintenance factors

Although the paucity of Paleogene z-corals in SE Asia can be largely explained by sparseness of inhabitable areas (*i.e.*, shallow water carbonates) across a wide seaway, this does not account for the enigmatic absence of z-corals from most of the few shallow water carbonates that did exist. Larger benthic foraminifera are usually the dominant constituents in these carbonates, and today they inhabit very similar shallow-marine tropical environments to corals.

Temperature and latitudinal effects cannot explain the lack of corals, especially since these Paleogene habitats clearly lay within the global limits of the z-coral belt, and climatic conditions were particularly warm in the middle and late Eocene (Miller et al., 1987; Frakes et al., 1992). Rapid subsidence rates, dispersal and recruitment factors, limiting chemical conditions of the water, including salinity, oxygen, and nutrients, and unfavourable speeds and directions of currents, might all have been contributory causes, but cannot be assessed easily in an ancient context.

Substrate control is more readily inferred, since some of the foraminifera-dominated carbonates represent shifting coarse-grained shoals (Fig.11d), perhaps comparable with ooid shoals today, and hence unfavourable for coral colonisation. Water depth is also an important factor, although the taxa and growth forms of the larger benthic foraminifera and other shallow marine bioclasts, such as coralline algae, within the Paleogene limestones studied, indicate a range of water depths within the photic zone (Adams, 1970; Ghose, 1977; Hottinger, 1983; Hallock and Glenn, 1986). Some of the limestones dominated by thin flat forms of larger benthic foraminifera were probably deposited below the depth of the most prolific z-coral growth (generally < 100 m, Fig.11e). However, in other deposits, the taxa and growth forms of foraminifera, which show little signs of reworking, indicate deposition in the upper part of the photic zone (Fig.11a-d). Perhaps coral facies were once more widespread but have been largely destroyed, though there are too few deposits with abundant coral clasts to support this.

Whatever the proximal causes, secondary feedback effects would also have been important if coral communities were too scattered in time and space to sustain frequent larval exchange and recruitment between areas within the region. This would have led to high rates of local extinction, and prevented longer-term development of z-coral facies. While the Paleogene paucity of z-corals in SE Asia appears to be a real phenomenon linked to lack of habitable areas, no clear set of ecological factors yet explains the paucity of z-corals in the few areas where they might be expected.

Distributional change

The factors of distributional change most relevant to the models in Table 1 are immigration rates. Low rates may have contributed to the Paleogene paucity of z-corals in SE Asia, and higher immigration rates to the subsequent surge in their abundance and diversity in the Neogene. There are as yet few data for addressing this, but Rosen and Smith (1988) used an indirect approach, analysing the history of endemicity patterns of corals and echinoids from different time slices. In the middle to late Eocene, and in the early Miocene, Pacific and Australasian faunas were relatively isolated from Tethyan and Indian Ocean faunas, and a pan-Indo-Pacific fauna in the modern sense did not exist until these more remote elements became fully integrated during or after the early Miocene.

Further support for this comes from the tectonic evolution of SE Asia and neighbouring regions (Schlanger, 1981; Rosen and Smith, 1988; Rosen, 1984, 1988; Paulay, 1997). Paleogene paucity of z-corals in SE Asia may reflect partial isolation of its carbonate environments from z-coral areas elsewhere, though the existence of modern examples of rich reefs in isolated settings (e.g., Chagos) suggests that the factors here may be more complex. Australia was not a reefal region as it is today, and lay over several thousand kilometres away to the southeast of the Asian margins, and outside the z-coral belt. To the east, on the Pacific plate, the Paleogene forerunners of the modern mid-Pacific archipelagos lay much further to the east than now. Apart from the Hawaii-Emperor chain, most of these archipelagos are broadly confined to the limits of the Darwin Rise/Superswell. Mid-plate oceanic islands are otherwise unusual features, and few suitable areas for corals may have occurred between the Darwin Rise/Superswell and the western Pacific convergence zones in the area which has been lost due to subduction. Thus, during the Paleogene, this intervening oceanic region may have been wide enough to have caused partial or intermittent isolation of Pacific areas now represented by the Eocene coral-bearing limestones underlying these islands.

To the west of SE Asia, the collision of the Indian continent with Eurasia in the Eocene eliminated the east-west trending Tethys seaway in that region, and initiated large-scale clastic input along the surrounding coastlines. In the tropical waters of the Indian Ocean, the few oceanic island 'staging posts' (Rosen, 1983),

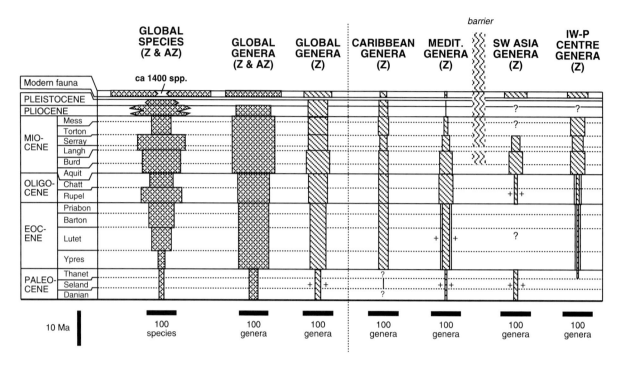

Fig.9. Taxonomic richness of zooxanthellate corals from Paleocene to present day. Z - zooxanthellate and zooxanthellate-like corals. AZ - azooxanthellate and azooxanthellate-like corals. I-WP - Indo-West Pacific. + symbols indicate interim compilation where faunas are actually richer than shown. Where data have been compiled according to Table 2, inner shaded blocks represent minimum estimate of z-like corals and unshaded outer blocks represent maximum estimates. See Rosen (1988) for discussion of time-averaging of coral data. Data sources follow. GLOBAL SPECIES; GLOBAL GENERA (Z and AZ). Rosen (1998) and sources therein. GLOBAL GENERA (Z). Eocene, Oligocene, Plio-Pleistocene and modern fauna from Veron (1995, Fig.38). These totals were independently compiled by Veron and are not summations of regional values elsewhere in this diagram. Miocene data from Rosen (1988, Table 2, from various sources); not adjusted for newer data incorporated in regional values elsewhere in this diagram. CARIBBEAN GENERA *i.e.,* Caribbean islands with adjacent mainlands of North, Central and South America. Paleocene from Bryan (1991). Eocene: compilation of middle to upper Eocene reef-coral genera from Budd *et al.* (1992, Table 1). Oligocene from Rosen (1988, Table 2, from various sources) but needs revision. Miocene to modern fauna from Budd *et al.* (1994, Fig.5). MEDITERRANEAN GENERA (*i.e.,* Mediterranean islands, and adjacent mainland mostly southern Europe and Egypt). Paleocene: partial compilation from Rosen (1998, Table 2 and sources therein). Eocene: partial compilation for a single 'sample region' of southern Europe (Bosnia, Hercegovina, Croatia and 'Dalmatia') from Felix (1925). Inclusion of other rich southern European faunas (*e.g.,* Spain, France, Italy) would show Mediterranean as the richest Eocene region. Oligocene and Plio-Pleistocene from Rosen (1988, Table 2, from various sources). Miocene from Rosen, unpublished. Modern fauna from Zibrowius (1980, and pers. comm., April 1997). SOUTHWEST ASIAN GENERA *i.e.,* Middle East, Iran, Pakistan, western India. Paleocene based on Duncan (1880) and Gregory (1930) for Ranikot fauna. Lower Miocene from McCall *et al.* (1994), and remainder from Rosen (1988, Table 2, from various sources). Eocene to Pleistocene faunas are rich in this region but await further research and compilation (the same applies to the rich adjacent areas of the eastern margins of Africa). INDO-WEST PACIFIC CENTRE GENERA *i.e.,* SE Asia, northern Great Barrier Reef and islands of western and mid-Pacific). Paleocene: this paper (Table 3); also unidentified 'coral fragments' known from boreholes. Eocene and Oligocene: this paper (Table 3). Miocene and modern fauna from Rosen (1988, Table 2, from various sources). Note that many 'lower Miocene' localities are of undifferentiated Te age and may therefore include uppermost Oligocene (Te1-4) corals, but lack adequate stratigraphical detail.

such as the Ninetyeast Ridge, may have aided only intermittent larval exchange of benthic populations between SE Asia and the rich Eocene z-like coral regions on the western side of India, the Middle East and Europe (Fig.9). These palaeogeographical factors, combined with the great width of the Indo-Pacific gateway between the Asian and Australian continents, and scarcity of shallow areas in that gateway, must have limited immigration into this gateway area during the Paleogene. This did not completely prevent colonisation by z-corals since the biogeographical affinities of the Tonasa fauna indicate that biogeographical continuity between SE Asia (Sulawesi) and the western side of the Indian continent occurred at least once, perhaps via the Ninetyeast Ridge. The faunal affinities of the contemporaneous Enewetak z-corals (Wells, 1964) suggest an analogous pattern for mid-plate Pacific areas.

The broadly contemporaneous collision of Australia with SE Asia and closure of Tethys in the Middle East during the latest Oligocene to earliest Miocene (Rosen and Smith, 1988; McCall *et al.*, 1994, for the coral history) had a profound effect on the palaeogeographic setting of z-coral faunas in SE Asia. Taxa within the Indo-West Pacific centre from this time onward must have consisted of (1) pre-Neogene, but not necessarily Indo-Pacific, relics, (2) those of local SE Asian origin and (3) taxa that originated elsewhere within the Indo-West Pacific but which immigrated into the centre (see below). It is unlikely any other taxa that had originated elsewhere could have then entered the Indo-West Pacific. Meanwhile, northern Australia and New Guinea had entered the reef coral belt, and the Pacific archipelagos of the Darwin Rise/Superswell had moved closer to Asia by over 3000 km (CGMW/UNESCO, 1990) since the late Eocene. These latter events must have greatly increased the extent and variety of potential habitats for z-coral colonisation in SE Asia. During the Neogene, SE Asia became, in effect, a biogeographical cross-roads and meeting point for faunal convergence of shallow marine organisms from numerous different directions (Rosen, 1988; Pandolfi, 1992) perhaps enhanced by changes in oceanic current configurations at this time (Grigg, 1988).

Originations

Rates and locations of originations are crucial criteria of the different models in Table 1. Rigorous treatment however requires good knowledge of the phylogenetic relationships of taxa to each other and to their times and places of occurrences. However, such work is currently confined to studies of a few extant Indo-Pacific z-corals, and until now fossil data have not been incorporated. Stratigraphic distributions also offer important historical clues, but since the fossil record is incomplete and there is insufficient taxonomic knowledge, they mainly provide indications of the minimum limits of former distributions in time and space and evidence of changing patterns of faunal affinities. A third approach, age-and-area studies of modern taxa, has until recently dominated Indo-Pacific work. Stratigraphic distributions, age-and-area studies and cladistic biogeography are discussed in this order, below.

As emphasized already, endemic Eocene z-corals have yet to be recorded from SE Asia. Their affinities (above) cannot rule out SE Asian, European or Indian Ocean origins, although given the paucity of this fauna there is therefore no evidence for high endemic origination rates in SE Asia at this time. The Oligocene picture is less clear for taxonomic reasons. The main fauna (Tanjung; Table 3), comprises a mixture of European and apparently endemic taxa. Our initial work on the Boera corals of Papua New Guinea (Fig.12), reveals strong European affinities, even though they are dated uppermost Oligocene (Table 3). For a more complete picture however, revision and/or description is urgently needed of the important Paleogene faunas of Pakistan (Duncan, 1880; Gregory, 1930), together with those of the western Indian Ocean margins from Iran and the Arabian region, to eastern Africa. Currently, however, there is no reason to change Gerth's (1925, 1930) view that relations between the corals of SE Asia during the Paleogene generally, were closer to that of the Indian and Mediterranean regions, than they were during the Neogene, and that exchange of species took place throughout these three regions during the Paleogene (see also Pandolfi, 1992). Hence palaeontological evidence of origination patterns give no support for SE Asia being a Centre of Origin (or part of one) during the Paleogene.

The fourfold surge in z-coral diversity in the Indo-West Pacific centre from around the early Neogene (Fig.9) must reflect, at least in part, the dramatic geological changes. Severance of Tethyan seaways in the Middle East at around this time would have confined subsequent originations either to the Atlantic-Mediterranean or to the Indo-Pacific, respectively, and this is reflected in their well-established post-Paleogene endemicities. Each region became an evolutionary entity, with the Indo-Pacific exhibiting the more marked taxonomic radiation (Fig.9). Independent Neogene evolution in the western Atlantic is shown by *Agaricia* species, and amongst such groups as the meandrinids, mussids and meandroid faviids (Budd *et al.*, 1994; Johnson, 1998). The important question is whether the Indo-Pacific radiation was particularly confined to the Indo-West Pacific centre.

The oldest record of the ecologically important and very diverse modern z-coral, *Acropora*, is in the Paleocene of Somalia (Carbone *et al.*, 1994). It occurs in numerous European localities from the Eocene to Miocene, but it does not appear in either the Caribbean or the Indo-West Pacific centre until the Oligocene (Fritsch, 1878; Wells, 1964; Frost, 1972; Budd *et al.*, 1992, 1994, and Fig.10l). It shows a very recent (latest Miocene or Pliocene) marked radiation in the Indo-West Pacific but not in the Caribbean (McCall et al., 1994; Budd et al., 1994). This history has parallels with the fungiids whose most primitive genus *Cycloseris*, occurs in the Paleocene of Somalia and western Pakistan, and in the Eocene (Tonasa) of SE Asia (Fig.10b,c) and Europe (Hoeksema, 1989; Carbone et al., 1994). Subsequent radiation occurred during the Neogene in SE Asia and fungiids are now richest in the Indo-West Pacific centre, though in contrast to *Acropora* they are not known from the Caribbean. These two cases show that the Indo-West Pacific centre has not been the location of their origins, although it has been important in the evolutionary history of z-corals from the Neogene onwards. Even so, it is still questionable on other grounds (see below) whether or not the region then acted as a Centre of Origin for these corals from that time.

Perpetuation of the idea that the Indo-West Pacific focus was a Centre of Origin for z-corals can be largely attributed to the landmark age-and-area study by Stehli and Wells (1971). In this particular context, however, such analyses are flawed because they give the illusion that the age of a modern taxon found in a particular area indicates that it has been in that area throughout that time. Admittedly, Stehli and Wells did not claim this, but by plotting average ages for different areas on global maps, they provided a visually powerful synthesis which compounded the illusion, leading others to interpret such analyses in this way. For example, of the 18 extant Indonesian z-genera which are of Eocene age (Rosen, 1984), only 6 actually occur in the Eocene of SE Asia, and a further one (*Astrocoenia*) is extinct (present data). None of the genera appear to have been local Eocene endemics, so the bulk of the 18 corals must have arrived in the region from elsewhere. Although this line of argument assumes that the fossil record is reliable (cf., Briggs, 1992), it is equally clear that the record does not support a Centre of Origin in SE Asia at this time. Parallel arguments apply to Eocene corals of the Pacific.

Age-and-area information alone cannot be used to infer origination patterns, but is best combined with phylogenetic work as in cladistic biogeography. Paradoxically though, Stehli and Wells (1971), used their age-and-area data to

Fig.10. Representative fossil scleractinian corals of Eocene and Oligocene age from the Tonasa Formation in South Sulawesi. Specimen numbers, prefixed AZ, are those of The Natural History Museum, London. Details of inferred zooxanthellate status are also given. Ages of Pakistan occurrences revised according to Adams (1970). So far, the upper Eocene fauna from the Tonasa Formation consists of 12 species, 7 of which are z-like and 5 are z-like? In addition to those illustrated here, are: fungiid sp. (z-like), *Stylocoenia* sp. (z-like?), and a solitary or phaceloid species (z-like?) A-C: Basal beds of the Tonasa Formation, lowest upper Eocene (Tb), Doi doi locality. A: *Trochocyathus? nummiformis* (Duncan, 1880). Transverse thin section, x1.5 (AZ 1188). Patellate to discoidal solitary coral. Modern *Trochocyathus* is azooxanthellate, but Tonasa specimens and Duncan's coral may not belong to this genus. Present species is z-like? on facies occurrence only; previously known from lower to mid-Oligocene (Tc-d) of Pakistan. B, C: *Cycloseris (Cyclolites) crenulata* (Duncan, 1880) syn?: *Funginellastraea (Cyclolites) heberti* (Tournouer, 1872). B: aboral surface, x1.5, C: polished transverse section, x1.5 (AZ 1202). Discoidal solitary 'mushroom coral'. Modern genus is a soft-substrate dweller, usually but not always zooxanthellate. Present species is z-like? on facies occurrence only; previously known from upper Paleocene of Pakistan (Ta1) and (?) various Eocene localities in southern Europe (*F. heberti*). D-K: Bulo Bunting locality, lower part of Tonasa Formation, but stratigraphically above Doi doi locality, upper Eocene (Tb). D: *Trachyphyllia (Montlivaltia) indica* (Duncan, 1880). Transverse thin section, x1.2 (AZ 1182). Flabelloid pedunculate form, consistently four-lobed in mature examples. Modern genus is a zooxanthellate soft-substrate dweller. Present species is z-like on generic and morphological grounds; previously known from middle Eocene (Ta3) of Pakistan. E: *Caulastrea (Calamophyllia) indica* (Duncan, 1880). Transverse thin section, x2 (AZ 1179). Phaceloid branching colony. Modern genus is zooxanthellate. Present species is z-like on generic grounds; previously known from middle Eocene (Ta3) of Pakistan. F: *Favites* sp. Transverse thin section, x2 (AZ 1193). Cerioid colony. Modern genus is zooxanthellate. Present form is z-like on generic grounds. G: *Diploastraea?* cf. *Rhizangia? agglomerata* Fritsch, 1878. Transverse thin section, x2 (AZ 1191). Astreoid colony. Modern genus is zooxanthellate. Present species is z-like on generic and morphological grounds; previously known from mid-Oligocene (Td) of Kalimantan (Table 3). H: *Favia (Phyllocoenia) conferta* (Duncan, 1864). Transverse thin section, x4 (AZ 1178). Plocoid colony. Modern genus is zooxanthellate. Present species is z-like on generic and morphological grounds; previously known from uncertain Tertiary horizon in Pakistan, possibly upper Oligocene to middle Miocene (Te - lower Tf). I: *Astrocoenia bistellata* (Catullo, 1856). Transverse thin section, x6 (AZ 1177). Cerioid colony. Extinct genus. Present species is z-like? on facies occurrence only; previously known from Eocene of Italy. J, K: *Favia pedunculata* Duncan, 1880. J: (AZ 1192) weathered calical surface, x0.5. K: (AZ 1187) Transverse thin section, x2. Plocoid pedunculate colony. Modern genus is zooxanthellate. Present species is z-like on generic and morphological grounds; previously known from middle Eocene to lower Oligocene (Ta3-Td) of Pakistan. L: Tonasa-II Quarry locality, upper Oligocene (Te1-4) middle part of the Tonasa Formation. *Acropora* sp. Polished longitudinal section through ramose branch, x2 (AZ 1213). Ramose branching colony with plocoid corallites. Modern genus is zooxanthellate, widespread, and an abundant reef contributor in both Indo-West Pacific and Caribbean. Present form is z-like on generic and morphological grounds.

Cenozoic corals of SE Asia

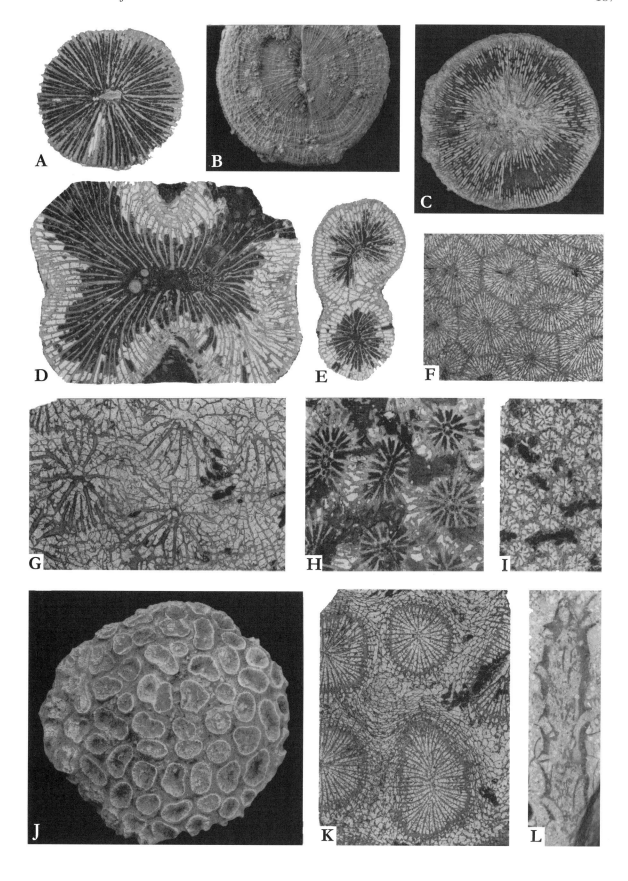

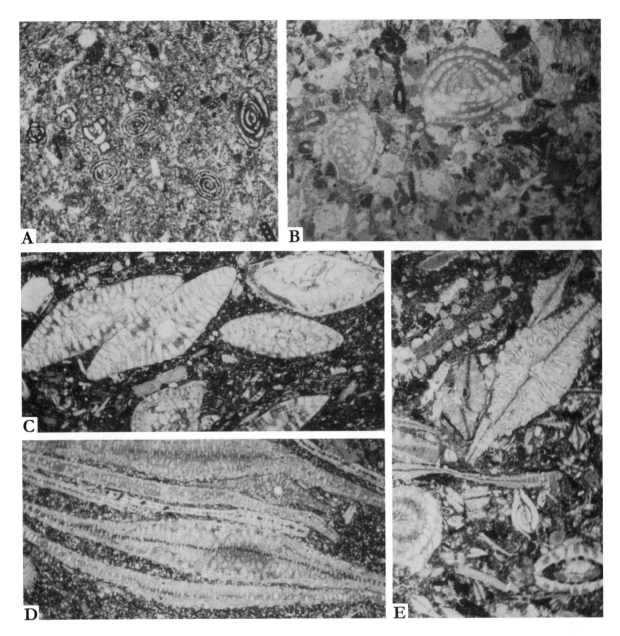

Fig.11. Thin section photomicrographs of Tertiary limestones dominated by larger benthic foraminifera and deposited in different environments in the photic zone. All were photographed in plane polarized light at x10 magnification. Samples and thin sections are held at Royal Holloway London University. A: Imperforate foraminifera bioclastic packstone (MTR114) containing soritiids, miliolids, alveolinids (*Borelis* sp.), *Austrotrillina* sp. and other fragmented bioclasts. Oligo-Miocene sample from the Taballar Limestone, Taballar River, Mangkalihat peninsula, East Kalimantan. The foraminifera are all typical of the shallower parts of the photic zone and can tolerate some fluctuations in salinity. B: Probable lower Oligocene (Tc) *Nummulites* bioclastic pack/grainstone (GU6) from Gua Unguk, Bengalon River area, north Kutai, East Kalimantan. Contains robust *Nummulites* and other fragmented and often abraded marine bioclasts and is inferred to have been deposited in moderate to high energy in the upper part of the photic zone. C: Lower Oligocene (Tc) *Nummulites* and *Heterostegina* bioclastic packstone (GB43) from Gua Bangki, Bengalon River area, north Kutai, East Kalimantan. Contains moderately robust *Nummulites* and thin forms of *Heterostegina* in a micritic matrix inferred to have been deposited under moderate to low energy conditions in moderate depths in the photic zone. D: upper Eocene (Tb) *Discocyclina* and *Biplanispira* bioclastic packstone (BR81) from the Bengalon River area, north Kutai, East Kalimantan. Contains abundant whole and sometimes fragmented larger benthic foraminifera, including *Discocyclina* sp., *Biplanispira* sp., *Pellatispira* sp., *Nummulites* sp. and other fragmented marine bioclasts such as coralline algae and echinoid plates. The species of foraminifera, their growth forms and the nature of the surrounding sediment suggests deposition in a moderate to high energy setting in shallow to moderate depths in the photic zone, possibly as a foraminiferal shoal. E: Upper Oligocene (Td/Te1-4) *Lepidocyclina* and heterosteginid packstone (MTR100) from the Taballar Limestone, Taballar River, Mangkalihat peninsula, East Kalimantan. Includes *Eulepidina* sp., *Nephrolepidina* sp., *Heterostegina* sp. and possible *Spiroclypeus* sp. Large, thin forms of these species of larger foraminifera are typical of development towards the base of the photic zone.

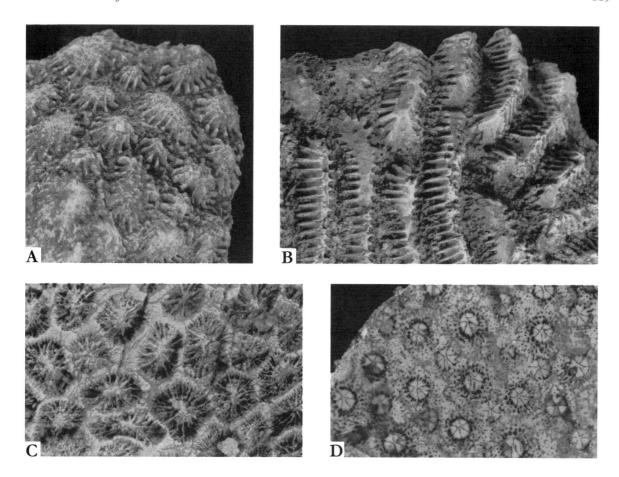

Fig. 12. Representative fossil scleractinian corals from the upper Miocene part of the Tacipi Formation, Sungai Pesing, Ujunglamuru, south Sulawesi (A-B), and from the uppermost Oligocene to lowest Miocene (Te1-5) Boera Formation, Boera Head, south Papua (C-D; see Table 3). A: *Hydnophora astraeoides* (Martin, 1880). Calical surface, x2 (AZ 1212). Continuously meandroid colony with vestigial walls reduced to isolated monticules. Modern genus is zooxanthellate. Present species is z-like on generic and morphological grounds; previously known from Miocene of Indonesia. B: *Platygyra daedalea* (Ellis and Solander, 1786). Calical surface, x2 (AZ 1211). Meandroid colony. Modern species is zooxanthellate, so same is inferred for present occurrence. Extant, widespread and important reef contributor, previously recorded from numerous fossil and modern Indo-West Pacific localities from Miocene onward. C: *Favites [Prionastraea] borneensis* (Gerth, 1925). Polished transverse section, x3 (AZ 1209). Cerioid colony. Modern genus is zooxanthellate. Present species z-like on generic grounds; previously known from Neogene of Indonesia. D: *Astreopora* cf. *meneghiniana* (D'Achiardi, 1866). Polished transverse section, x2 (AZ 1210). Plocoid colony. Modern genus is zooxanthellate. Present species z-like on generic and morphological grounds; previously known from Oligocene of Italy.

support the widely held notion that Centres of Origin have an intrinsic evolutionary characteristic in which continuous emergence of more successful taxa pushes older taxa out from the focal region. This predicts a pattern of continuous evolutionary turnover, in which (1) older faunas, prior to their displacement, should share the same central origin, and be more or less as rich, as younger faunas, and (2) a given fauna of one time slice should comprise numerous taxa that were subsequently displaced. There is as yet no evidence however that Tonasa and Enewetak Eocene genera originated in the Indo-West Pacific centre, and all but two of them still occur in the modern Indo-West Pacific centre. Therefore most of these genera were not displaced from the centre, or if they were, they evidently returned. This, as well as the paucity of the fauna is incompatible with Centre of Origin notions.

Moreover, sympatric speciation modes aside, the very commonly observed overlapping patterns of sibling species (and sister genera) amongst modern z-corals species in the Indo-Pacific, especially the central region (*e.g.*, Hoeksema, 1989, 1993; Veron, 1995; Wallace *et* al., 1991; Wallace, 1998; Pandolfi, 1992) suggest that siblings must have originated in different areas and subsequently migrated into their observed overlapping distributions (see also Paulay, 1997). Hence, whatever the evolutionary characteristics of Centres of Origin, competitive displacement, in this case, does not appear to be one of them. Since average longevity estimates for z-coral species are about 10 Ma (Veron, 1995), many of these numerous overlapping species must be post-Paleogene in origin, suggesting that distribution patterns in and around the centre are predominantly Neogene or younger. This is again incompatible with a Centre of Origin model, though it is also likely that some species originated through geographical isolation within and around the Indonesian archipelago (Pandolfi, 1992; Wallace, 1998).

Evidence for heterogeneous geographical origins of Indo-Pacific z-corals, including peripheral sources, has come not just from their overlapping distribution patterns but also from cladistic studies of the fungiids (Hoeksema, 1993), the siderastreids and mussids (Pandolfi, 1992), and some species of *Acropora* (Wallace et al., 1991). These studies do not include estimates of ages of origination from fossil evidence but Pandolfi infers these from the timing of geological events whose associated features correspond geographically to his area cladograms. Parallel studies in other groups (cf., Pandolfi, 1992; Palumbi, 1996; Paulay, 1997) give similar indications of widespread heterogeneous origins for the common marine organisms of the Indo-West Pacific centre. Reef fishes, for example, apparently evolved rapidly during the first 20 Ma or so of the Cenozoic and then remained in relative stasis (Choat and Bellwood, 1991; Bellwood 1996). Cladistic analysis of reefal parrotfish (Scaridae) by Bellwood shows that the labroid lineage which gave rise to this group, was already present by the latest early to earliest middle Eocene of Europe (*i.e.*, most recent date of Monte Bolca fish fauna from J. Pignatti, pers. comm., 1997), a pattern that has parallels with *Acropora* and the fungiids.

Plate tectonics and the implications of the Mesozoic coral record

It is instructive to view the Cenozoic coral record in the longer-term context of the Mesozoic coral record and its plate tectonic setting. During the Triassic and Jurassic, the large tropical ocean of Meso-Tethys separated blocks such as Sibumasu and Indochina in the north, from those Gondwana blocks to the south which would later form Borneo, Sulawesi, Nusa Tenggara and Australia (Metcalfe, 1996). Around the margins of Meso-Tethys, z-coral facies developed in shallow marine waters and have been preserved today in Upper Triassic (Stanley, 1988) and Upper Jurassic limestones in Burma, Thailand and eastern Indonesia. Rifting during the Jurassic and Cretaceous separated blocks, such as west Burma and west Sulawesi, from the northern part of the Australian craton by a seaway. Australia lay too far south of the z-coral belt to support coral growth. In contrast, the rifted blocks were drifting northwards across tropical latitudes to be juxtaposed against mainland SE Asia during the Jurassic to Late Cretaceous and the few known Cretaceous scleractinian corals from Borneo and Sumatra reflect this. However, as with the Paleogene, z-corals are rare in the rest of the region during the Cretaceous. Thus the 'Paleogene data gap' for z-corals is the continuation of an earlier pattern which goes back into the Cretaceous, and presumably had similar causes.

The underlying issue of interest is the relative importance of plate tectonics as an evolutionary influence compared with intrinsic evolutionary processes. In general, development and diversity of both Mesozoic and Cenozoic scleractinian z-corals in SE Asia have been ultimately controlled by the plate tectonics of the region. For the Paleogene in particular, Flessa and Jablonski (1996) gave three criteria (above) by which plate tectonics might be judged to be of greater importance than intrinsic evolutionary factors in the Paleogene history of SE Asian coral patterns. Firstly, the lack of latitudinal diversity gradients is effectively fulfilled in our study because the zooxanthellate occurrences are so few that there is no gradient to speak of. Secondly, the lack of endemics of Eocene age in particular, and the European-Indian affinities of SE Asian Paleo-gene corals in general, suggest, as predicted, that their average age cannot be much younger, if at all, than that of corals in these other regions.

Thirdly, Flessa and Jablonski predict that recovery from the K-T extinction would also have been a significant influence on SE Asian distributions if plate tectonics were important. In fa t, the paucity of both Cretaceous and Paleogene coral records within the region means that K-T extinctions and recovery were largely witnessed elsewhere. Extensive development of z-coral faunas and communities did not really occur in

SE Asia until around the earliest Neogene, whereas elsewhere in the tropics its earliest signs begin only 3-5 Ma into the Paleocene, with quite rich faunas established by the Thanetian (Fig.9; Schuster, 1996; Rosen, 1998). Thus while Flessa and Jablonski's third prediction is not fulfilled in their precise terms, the absence of K-T coral evidence in SE Asia, clearly reflects the influence of plate tectonics. This also suggests that the idea that the Indo-West Pacific centre is largely a vestige or derivation of Cretaceous patterns (McCoy and Heck, 1976; Table 1) is at best an oversimplification. More generally therefore, there is no need to suppose that SE Asia was endowed with special evolutionary processes, as Centres of Origin models imply. It has been a region in which the potential to develop rich communities has been controlled by the way in which suitable areas for them have been continually created, reorganized and lost, especially according to the timing and movements of blocks which have broken away from Gondwana, their northward passage across Tethys into tropical waters, and their eventual accretion to the Eurasian continent.

Plate tectonics or Centre of Origin?

Two contrasting major patterns can be discerned: a Paleogene one in which diversity, endemicity and origination rates of z-corals in SE Asia were low, and a post-Paleogene one which is effectively the opposite. The change coincides with tectonic events, especially the collision of Australia with SE Asia. During the Paleogene, the marine region between the continental margins of SE Asia and Australia was about 3000 km wide and, compared with the post-Neogene, there were few shallow water areas suitable for z-corals. A much more extensive, but now largely inaccessible area of shallow water carbonates existed in the mid-Pacific (on the Darwin Rise/Superswell) especially during the Eocene. However, the z-coral faunas of both regions appear to have been low diversity outposts of the very much richer faunas of Europe, eastern Tethys and the western Indian Ocean margins.

The modern high diversity pattern in SE Asia and the Indo-West Pacific centre began to emerge around the earliest Neogene with an apparent regional radiation of z-corals. However, detailed studies are revealing that relicts, and migrations of taxa into the region were both as qualitatively important as originations within the region. Local originations derived, at least in part, from these antecedent elements. The geographical complexity of the region since the Neogene favoured all three processes through increasing habitat heterogeneity and potential allopatric speciation amongst fragmented shallow water areas, further enhanced in the last 10 Ma or so by the effects of glacio-eustasy and increased climatic fluctuation.

Our fossil patterns show that a Centre of Origin model (Table 1) is completely inapplicable to Paleogene z-corals of the SE Asian region, since there was only a small non-endemic fauna here. In fact, it is misleading to think of it as a 'centre' of any kind during this time. This contrasts with the Neogene onwards, which superficially accords with a Centre of Origin model, - albeit of very short duration compared with species turnover timescales. Moreover, the combined patterns from a range of different studies show that a more appropriate model requires a combination of all three of the possibilities in Table 1. A longer-term perspective suggests that the above contrast in Paleogene and post-Paleogene patterns represents a cyclical sequence which has occurred at least twice before with strong z-coral developments in the late Triassic and late Jurassic in SE Asia, coincident with times when rifted Gondwanan blocks docked against the Asian continent in the tropics. Thus plate tectonic processes rather than intrinsic evolutionary processes (like Centres of Origin, competitive displacement, etc.) have been a major control on regional diversity patterns of z-corals, and presumably also numerous other shallow marine organisms.

Acknowledgements

The authors would like to acknowledge the financial and technical support provided by a number of groups during the course of this research: namely London University SE Asia Research Group and its consortium of funding companies, The Natural History Museum, London, and the Geological Research and Development Centre in Indonesia. We thank the following for helpful discussions or correspondence on numerous aspects of the paper: Dan Bosence, Jill Darrell, Robert Hall, Alit Ascaria, Dick Randall, Galt Siegrist and Carden Wallace. Robert Hall is gratefully acknowledged for allowing us to use time slices of his plate tectonic models as base maps. Jill Darrell also provided curatorial help with the relevant coral collections, prepared the coral figures and assisted with the field collecting of the Boera corals (with David Haig and Ava Kila). The coral photographs were taken by Phil Crabb. The comments of two reviewers, Ken Johnson and Neil Harbury, together with our editor Robert Hall, and four other readers, Johannes Pignatti, Alison Kay, Gustav Paulay and Andrew Racey greatly improved the paper.

References

Adams, C. G. 1965. The foraminifera and stratigraphy of the Melinau Limestone, Sarawak, and its importance in Tertiary correlation. Quarterly Journal of the Geological Society, London 121: 283-338.

Adams, C. G. 1970. A reconsideration of the East Indian Letter Classification of the Tertiary. Bulletin of the British Museum (Natural History) Geology 19: 87-137.

Adams, C. G. 1983. Speciation, phylogenesis, tectonism, climate and eustacy: factors in the evolution of Cenozoic larger foraminiferal bioprovinces. In Evolution, time and space: the emergence of the biosphere. pp. 255-289. Edited by R. W. Sims, J. H. Price and P. E. S. Whalley. Systematics Association Special Volume 23.

Adams, C. G., Lee, D. E. and Rosen, B. R. 1990. Conflicting isotopic and biotic evidence for tropical sea-surface temperatures during the Tertiary. Palaeogeography, Palaeoclimatology, Palaeoecology 77: 289-313.

Beauvais, L. 1983. Jurassic Cnidaria from the Philippines and Sumatra. Economic and Social Commission for Asia and the Pacific. Committee for Co-ordination of Joint Prospecting for Mineral Resources in Asian Offshore Areas (CCOP). Technical Bulletin 16: 39-67.

Beauvais, L. 1986. Evolution paléobiogéographique des formations à Scléractiniaires du Bassin téthysien au cours du Mésozoïque. Bulletin de la Société Géologique de France 8: 499-509.

Beauvais, L. and Fontaine, H. 1990. Corals from the Bau Limestone Formation, Jurassic of Sarawak, Malaysia. In Ten years of CCOP research on the pre-Tertiary of East Asia pp. 209-239. Edited by H. Fontaine. CCOP/TP (Committee for Coordination of Joint Prospecting for Mineral Resources in Asian Offshore Areas) Technical Publications 20.

Bellwood, D. R. 1996. The Eocene fishes of Monte Bolca: the earliest coral reef fish. Coral Reefs 15: 11-19.

Bemmelen, R. W., van. 1970. The geology of Indonesia and adjacent archipelagoes. Second Edition. Martinus Nijhoff, Government Printing Office, The Hague, Vol.IA, 732 pp., Vol.IB, 65 pp. and plate portfolio, Vol. II, 267 pp.

Bernard, H. U. 1991. Insight Guide: SE Asia Wildlife. APA Publications (HK) Ltd, Singapore, 430 pp.

Brasier, M. D. 1995. Fossil indicators of nutrient levels. 2: Evolution and extinction in relation to oligotrophy. In Marine Palaeoenvironmental Analysis from Fossils. pp. 133-150. Edited by D. W. J. Bosence and P. A. Allison. Geological Society Special Publication 83.

Briggs, J. C. 1974. Marine Zoogeography. McGraw-Hill, New York. 475 pp.

Briggs, J. C. 1992. The marine East Indies: centre of origin? Global Ecology and Biogeography Letters 2: 149-156.

Bryan, J. R. 1991. A Paleocene coral-algal-sponge reef from southwestern Alabama and the ecology of early Tertiary reefs. Lethaia 24: 423-438.

Budd, A. F., Johnson, K. G. and Stemann, T. A. 1996. Plio-Pleistocene turnover and extinctions in the Caribbean reef-coral fauna. In Evolution and environment in tropical America. pp. 168-204. Edited by J. B. C. Jackson, A. F. Budd and A. G. Coates, University of Chicago Press, Chicago.

Budd, A. F., Stemann, T. A. and Johnson, K. G. 1994. Stratigraphic distributions of genera and species of Neogene to Recent Caribbean reef corals. Journal of Paleontology 68: 951-977.

Budd, A. F., Stemann, T. A. and Stewart, R. H. 1992. Eocene Caribbean reef corals: a unique fauna from the Gatuncillo Formation of Panama. Journal of Paleontology 66: 570-594.

Carbone, F., Matteucci, R., Pignatti, J. S. and Russo, A. 1994. Facies analysis and biostratigraphy of the Auradu Limestone Formation in the Berbera-Sheikh area, northwestern Somalia. Geologica Romana 29: 213-235.

CGMW/UNESCO 1990. Geological Map of the World, Scale 1/25,000,000. Commission for the Geological Map of the World, France.

Chevalier, J.-P., Coudray, J. and Gonord, H. 1971. Sur la présence de coraux dans l'Eocène C de Nouvelle-Calédonie. Comptes Rendus des Séances de l'Académie des Sciences. (Série D) 272: 1972-1974

Choat, J. H. and Bellwood, D. R. 1991. Reef fishes, their history and evolution. In The Ecology of Fishes on Coral Reefs. pp. 39-65. Edited by P. F. Sale. Academic Press, New York.

Coates, A. G. and Jackson, J. B. C. 1987. Clonal growth, algal symbiosis, and reef formation by corals. Paleobiology 13: 363-378.

Crame, J. A. 1992. Evolutionary history of the polar regions. Historical Biology, 6: 37-61.

Cucci, M. A. and Clark, M. H. 1993. Sequence Stratigraphy of a Miocene Carbonate Buildup, Java Sea. In Carbonate Sequence Stratigraphy, Recent Developments and Applications. pp. 291-303. Edited by R. G. Loucks and J. F. Sarg. American Association of Petroleum Geologists Memoir 57.

Daly, M. C., Cooper, M. A., Wilson, I., Smith, D. G. and Hooper, B. G. D. 1991. Cenozoic plate tectonics and basin evolution in Indonesia. Marine and Petroleum Geology (Special Issue: South-east Asia) 8 (1): 2-21.

Darga, R. 1991. Rekonstruktion obereozäner Riffe Südostbayerns. Freunde der Bayerischen Staatssamlung für Paläontologie und Historische Geologie München e.V., Jahresberichte und Mitteilungen, (1990) 19: 31-38.

Davies, P. J., Symonds, P. A., Feary, D. A. and Pigram, C. J. 1987. Horizontal plate motion: a key allocyclic factor in the evolution of the Great Barrier Reef. Science 238: 1697-1700.

Davies, P. J., Symonds, P. A., Feary, D. A. and Pigram, C. J. 1989. The evolution of the carbonate platforms of northeast Australia. In Controls on Carbonate Platform and Basin Development. pp. 233-258. Edited by P. D. Crevello, J. L. Wilson, F. Sarg and J. F. Read. Special Publication of Economic Paleontologists and Mineralogists 44.

Dollfus, G.F. 1908. Sur quelques polypiers fossiles des Indes néerlandaises. Jaarboek Mijnwezen Nederlandsch Oost-Indië 37: 676-686.

Dollfus, G. F. 1915. Paléontologie du voyage à l'Île Célèbes de M.E.C. Abendanon. E.J. Brill, Leiden, 57 pp.

Done, T. J., Ogden, J. C., Wiebe, W. J. and Rosen, B. R. (with contributions from the BIOCORE Working Group). 1996. Biodiversity and ecosystem function of coral reefs. In Functional Roles of Biodiversity. pp. 393-429. Edited by H. A. Mooney, J. H. Cushman, E. Medina, O. E. Sala and

E.-D. Schulze. John Wiley and Sons, Chichester.
Duncan, P. M. 1880. Sind fossil corals and Alcyonaria. Memoirs of the Geological Survey of India. Palaeontologica Indica (Series 7 and 14) 1: (1), 1-110.
Eguchi, M. 1951. Mesozoic hexacorals from Japan. Scientific Reports of the Tohoku University, Sendai, Japan. Second Series (Geology) 24: 1-96
Ekman, S. 1953. Zoogeography of the sea. Sidgwick and Jackson, London, 417 pp. (previously published in German in 1935 as Tiergeographie des Meeres. Akademische Verlagsgesellschaft, 542 pp.)
Felix, J. 1925. Fossilium Catalogus. I: Animalia. Pars 28. Anthozoa eocaenica et oligocaenica. W. Junk, Berlin. 296 pp.
Flessa, K. W. and Jablonski, D. 1996. The geography of evolutionary turnover: a global analysis of extant bivalves. *In* Evolutionary paleobiology. pp. 376-397. Edited by D. Jablonski, D. H. Erwin and J. H. Lipps in honour of J. W. Valentine. University of Chicago Press, Chicago.
Frakes, L. A., Francis, J. E. and Syktus, J. I. 1992. Climate modes of the Phanerozoic. The history of the Earth's climate over the past 600 million years. Cambridge University Press, Cambridge, 274 pp.
Fraser, R. H. and Currie, D. J. 1996. The species richness-energy hypothesis in a system where historical factors are thought to prevail: coral reefs. The American Naturalist 148: 138-159.
Fritsch, K., von, 1878. Fossile Korallen der Nummulitenschichten von Borneo. Palaeontographica Supplement 3 (1): 92-135.
Frost, S. H. 1972. Evolution of Cenozoic Caribbean coral faunas. *In* Memorias VI Conferencia Geologica del Caribe - Margarita, Venezuela, pp. 461-464.
Fulthorpe, C. S. and Schlanger, S. O. 1989. Paleoceanographic and tectonic settings of early Miocene reefs and associated carbonates of offshore Southeast Asia. The American Association of Petroleum Geologists Bulletin 73 (6): 729-756.
Gerth, H. 1921. Die Fossilen von Java auf Grund einer Sammlung von Dr. R. D. M. Verbeek und von anderen bearbeitet durch Dr. K. Martin, Professor der Geologie an der Universität zu Leiden. Anthozoen von Java und die Mollusken der Njalindungschichten, erster Teil. Sammlungen des Geologischen Reichs-Museums in Leiden (Neue Folge) 1 (Zweite Abteilung) (Heft III): 387-445.
Gerth, H. 1925. Die Bedeutung der tertiären Riffkorallenfauna des malayischen Archipels für die Entwicklung der lebenden Riff-Fauna im indopazifischen und atlantischen Gebiet. Verhandelingen van het Geologisch-Mijnbouwkundig Genootschap voor Nederland en Koloniën (Geologische Serie) 8: 173-196.
Gerth, H. 1930. The evolution of reefcorals during the Cenozoic period. Proceedings of the Fourth Pacific Science Congress 2A (Physical Papers): 333-350.
Gerth, H. 1931. Coelenterata. Leidsche Geologische Mededelingen 5: 120-151.
Gerth, H. 1933. Neue Beiträge zur Kenntnis der Korallenfauna des Tertiärs von Java. Wetenschappelijke Mededeelingen Dienst van den Mijnbouw in Nederlandsch-Indië 25: 1-45
Ghose, B. K. 1977. Paleoecology of the Cenozoic reefal foraminifers and algae - a brief review. Palaeogeography, Palaeoclimatology, Palaeoecology 22: 231-256.
Ghose, B. K. 1982. Oligocene reef of Kutch and its oil potentiality. Indian Journal of Earth Sciences 9: 6-10.
Gorsel, J. T., van, 1988. Biostratigraphy in Indonesia: Methods, pitfalls and new directions. Proceedings of the Indonesian Petroleum Association, 17th Annual Convention, October, 275-300.
Gregory, J. W. 1930. The fossil fauna of the Samana Range and some neighbouring areas: Part VII. The Lower Eocene corals. Memoirs of the Geological Survey of India. Palaeontologica Indica (NS) 15: 81-128.
Gregory, J. W. and Trench, J. B. 1915. Eocene corals from the Fly River, Central New Guinea. Geological Magazine, (N.S.) (Decade 6) 3: 481-488, 529-536.
Grigg, R. W. 1988. Paleoceanography of coral reefs in the Hawaiian-Emperor chain. Science 240: 1737-1743.
Hall, R. 1996. Reconstructing Cenozoic SE Asia. *In* Tectonic Evolution of Southeast Asia. pp. 153-184. Edited by R. Hall and D. J. Blundell. Geological Society of London Special Publication 106.
Hall, R. 1998. The plate tectonics of Cenozoic SE Asia and the distribution of land and sea. *In* Biogeography and Geological Evolution of SE Asia. Edited by R. Hall and J. D. Holloway (this volume).
Hallock, P. 1985. Why are larger Foraminifera large? Paleobiology 11: (2), 195-208.
Hallock, P. and Glenn, E. C. 1986. Larger Foraminifera: a Tool for Palaeoenvironmental Analysis of Cenozoic Carbonate Depositional Facies. Palaios 1: 55-64.
Harrison, P. L. and Wallace, C. C. 1990. Reproduction, dispersal and recruitment of scleractinian corals. *In* Ecosystems of the World; 25. Coral Reefs. pp. 133-207. Edited by Z. Dubinsky. Elsevier, Oxford.
Haynes, J. 1965. Symbiosis, wall structure and habitat in foraminifera. Contributions to Cushman Foundation for Foraminifera Research 16: 40-43.
Hoeksema, B. W. 1989. Taxonomy, phylogeny and biogeography of mushroom corals (Scleractinia; Fungiidae). Zoologische Verhandelingen 254: 1-295
Hoeksema, B. W. 1993. Historical biogeography of *Fungia (Pleuractis)* spp. (Scleractinia: Fungiidae), including a new species from the Seychelles. Zoologische Mededeelingen Leiden 67: 639-654
Holloway, N. H. 1982. North Palawan Block, Philippines - Its relation to Asian mainland and role in evolution of South China Sea. American Association of Petroleum Geologists Bulletin 66: (9), 1355-1383.
Hottinger, L. 1983. Processes determining the distribution of larger foraminifera in space and time. Utrecht Micropalaeontological Bulletin 30: 239-253.
Jablonski, D. 1993. The tropics as a source of evolutionary novelty through geological time. Nature 364: 142-144.
Jackson, J. B. C. 1986. Modes of dispersal of clonal benthic invertebrates: consequences for species' distributions and genetic structure of local populations. Bulletin of Marine Science 39: 588-606.
James, N. P., Feary, D. A. and Bone, Y. 1996. Carbonate platform architecture and composition controlled by oceanic currents; Cenozoic, Southern Australia (Abstract) Proceedings of the 10th Bathurst meeting of carbonate sedimentologists, London, 30.
Jokiel, P. and Martinelli, F. J. 1992. The vortex model of coral reef biogeography. Journal of Biogeography 19: 449-458.
Johnson, K. G. 1998. A phylogenetic test of accelerated turnover in Neogene Caribbean brain corals (Scleractinia: Faviidae). Palaeontology, in press.
Kay, E. A. 1984. Patterns of speciation in the Indo-West Pacific. Special Publications of the Bernice Pauahi Bishop Museum 72: 15-31.
Kemp, E. M. and Harris, W. K. 1975. The vegetation of the Tertiary islands on the Ninetyeast Ridge. Nature 258: 303-307.
Koesoemadinata, R. P. 1978. Tertiary carbonate sedimenta-

tion in Irian Jaya with special reference to the northern part of the Bintuni Basin. Indonesian Petroleum Association Carbonate Seminar 1: 79-92.

Kohar, A. 1985. Seismic expression of late Eocene carbonate build-up features in the JS25 and P. Sepanjang trend, Kangean Block. Proceedings Indonesian Petroleum Association, 14th Annual Convention: 437-452.

Ladd, H. S. 1960. Origin of the Pacific island mollusc fauna. American Journal of Science 258A: 137-150.

Lee, T.-Y. and Lawver, L. A. 1995. Cenozoic plate reconstruction of Southeast Asia. Tectonophysics 251: 85-138.

Livingstone, H. J., Sincock, B. W., Syarief, A. M., Sriwidadi and Wilson, J. N. 1992. Comparison of Walio and Kasim reefs, Salawati Basin, Western Irian Jaya, Indonesia. *In* Carbonate rocks and reservoirs of Indonesia a core workshop. pp. 4.1-4.14. Edited by C. T. Siemers, M. W. Longman, R. K. Park and J. G. Kaldi. Indonesian Petroleum Association Core Workshop Notes 1.

Martini, R., Vachard, D., Zaninetti, L., Cirilli, S., Cornée, J.-J., Lathuilière, B. and Villeneuve, M. 1997. Sedimentology, stratigraphy, and micropalaeontology of the Upper Triassic reefal series in Eastern Sulawesi (Indonesia). Palaeogeography, Palaeoclimatology, Palaeoecology 128: 157-174.

McCall, J., Rosen, B. R. and Darrell, J. G. 1994. Carbonate deposition in accretionary prism settings: Early Miocene coral limestones and corals of the Makran mountain range in southern Iran. Facies 31: 141-178.

McCoy, E. D. and Heck, K. L., Jr. 1976. Biogeography of corals, seagrasses and mangroves: an alternative to the center of origin concept. Systematic Zoology 25: 201-210.

Menard, H. W. 1964. Marine geology of the Pacific. McGraw-Hill, New York, 271 pp.

Metcalfe, I. 1996. Pre-Cretaceous evolution of SE Asian terranes. *In* Tectonic Evolution of Southeast Asia. pp. 97-122. Edited by R. Hall and D. J. Blundell. Geological Society of London Special Publication 106.

Miller, K. G., Fairbanks, R. G. and Mountain, G. S. 1987. Tertiary oxygen isotope synthesis, sea level history, and continental margin erosion. Paleoceanography 2: 1-19.

Moll, H. 1983. Zonation and diversity of scleractinia on reefs off S.W. Sulawesi, Indonesia. Doctoral Thesis, Rijksuniversiteit Leiden. 107 pp.

Muller, K. 1995. Underwater Indonesia: A guide to the world's greatest diving. Periplus Travel Guides, Singapore, 326 pp.

Nelson, C. S. 1988. An introductory perspective on non-tropical shelf carbonates. Sedimentary Geology 60: 3-12.

Newell, N. D. 1971. An outline history of tropical organic reefs. American Museum Novitates 2465: 1-37.

Obellianne, J. M. 1962. Le gisement de phosphate tricalcique de Makatea. Sciences de la Terre, Nancy 9 (1): 60 pp.

Palumbi, S. R. 1996. What can molecular genetics contribute to marine biogeography? An urchin's tale. Journal of Experimental Marine Biology and Ecology 203: 75-92.

Pandolfi, J. M. 1992. A review of the tectonic history of New Guinea and its significance for marine biogeography. Proceedings of the Seventh International Coral Reef Symposium Guam 2: 718-728.

Paulay, G. 1997. Diversity and distribution of reef organisms. *In* Life and death of coral reefs. pp. 298-352. Edited by C. E. Birkeland. Chapman and Hall, London.

Perrin, C., Bosence, D. W. J. and Rosen, B. R. 1995. Quantitative approaches to palaeozonation and palaeobathymetry of corals and coralline algae in Cenozoic reefs. *In* Marine Palaeoenvironmental analysis from fossils. pp. 181-229. Edited by D. W. J. Bosence and P.A. Allison. Geological Society Special Publication 83.

Racey, A. 1994. Palaeoenvironmental significance of larger foraminifera biofabrics from the middle Eocene Seeb Limestone Formation of Oman: Implications for petroleum exploration. *In* Geo'94, The Middle East Petroleum Geosciences, Bahrain. pp. 793-810 Edited by M. I. Al-Husseini.

Rangin, C., Jolivet, L. and Pubellier, M. 1990. A simple model for the tectonic evolution of southeast Asia and Indonesia region for the past 43 m.y. Bulletin de la Société Géologique de France 8: (VI), 889-905.

Rogerson, R., Haig, D. W. and Nion, S. T. S. 1981. Geology of Port Moresby. Geological Survey of Papua New Guinea. Report 81/16. 56 pp.

Rosen, B. R. 1971. The distribution of reef coral genera in the Indian Ocean. *In* Regional variation in Indian Ocean coral reefs. pp. 263-299. Edited by D. R. Stoddart and C. M. Yonge. Symposium of the Zoological Society of London 28.

Rosen, B. R. 1975. The distribution of reef corals. Reports of the Underwater Association (New Series) 1: 1-16.

Rosen, B. R. 1983. Reef island staging posts and Noah's Arks. Reef Encounter 1: 5-6.

Rosen, B. R. 1984. Reef coral biogeography and climate through the Cainozoic: Just islands in the sun or a critical pattern of islands? *In* Fossils and Climate. pp. 201-264. Edited by P. J. Brenchley. Geological Journal Special Issues 11.

Rosen, B. R. 1985. Fact, hypothesis or idea? (Review of: Briggs, J. C. 1984. Centres of origin in biogeography. Biogeography Study Group, School of Geography, University of Leeds, Biogeographical Monographs 1: 1-95) Journal of Biogeography 12: 383-385.

Rosen, B. R. 1988. Progress, problems and patterns in the biogeography of reef corals and other tropical marine organisms. Helgoländer Meeresuntersuchungen 42: 269-301.

Rosen, B. R., 1998. Algal symbiosis, and the collapse and recovery of reef communities: Lazarus corals across the K-T boundary. *In* Biotic response to global change: the last 145 million years. Edited by S. J. Culver and P. F. Rawson. Chapman and Hall, London.

Rosen, B. R. and Smith, A. B. 1988. Tectonics from fossils? Analysis of reef-coral and sea-urchin distributions from late Cretaceous to Recent, using a new method. *In* Gondwana and Tethys. pp. 275-306. Edited by M. G. Audley-Charles and A. Hallam. Geological Society of London Special Publications, 37.

Rosen B. R. and Turnsek, D. 1989. Extinction patterns and biogeography of scleractinian corals across the Cretaceous/Tertiary boundary, in Fossil Cnidaria 5. Proceedings of the Fifth International Symposium on Fossil Cnidaria including Archaeocyatha and Spongiomorphs held in Brisbane, Queensland, Australia, 25-29 July 1988. pp. 355-370. Edited by P. A. Jell and J. W. Pickett. Memoirs of the Association of Australasian Palaeontologists 8.

Saller, A., Armin, R., La Ode Ichram and Glenn-Sullivan, C. 1993. Sequence stratigraphy of aggrading and backstepping carbonate shelves, Oligocene, Central Kalimantan, Indonesia. *In* Carbonate Sequence Stratigraphy: Recent Developments and Applications. pp. 267-290. Edited by R. G. Loucks and J. F. Sarg. American Association of Petroleum Geologists, Memoir 57.

Schlanger, S. O. 1981. Shallow-water limestones in ocean basins as tectonic and paleoceanic indicators. Special Publication of Economic Paleontologists and Mineralogists 32: 209-226.

Schlanger, S. O. and Premoli Silva, I. 1981. Tectonic, volcanic, and paleogeographic implications of redeposited reef faunas of Late Cretaceous and Tertiary age from the

Nauru Basin and the Line Islands. Initial Reports of the Deep Sea Drilling Project 61: 817-827.

Schuster, F. 1996. Paleoecology of Paleocene and Eocene corals from the Kharga and Farafara Oases (Western Desert, Egypt) and the depositional history of the Paleocene Abu Tartur carbonate platform, Kharga Oasis. Tübinger Geowissenshaftliche Arbeiten (A: Geologie, Paläontologie, Stratigraphie) 31: 1-96.

Siemers, C. T., Deckelman, J. A., Brown, A. A. and West, E. R. 1992. Characteristics of the fractured Ngimbang carbonate (Eocene), West Kangean-2 well, Kangean PSC, East Java Sea, Indonesia. In Carbonate rocks and reservoirs of Indonesia. pp. 10.1-10.13. Edited by C. T. Siemers, M. W. Longman, R. K. Park and J. G. Kaldi. Indonesian Petroleum Association Core Workshop Notes 1.

Stanley, G. D. 1988. The history of early Mesozoic reef communities: a three step process. Palaios 3: 170-183.

Stanley, G. D., Jr. and Cairns, S. D. 1988. Constructional azooxanthellate coral communities: an overview with implications for the fossil record. Palaios 3: 233-242.

Stebbins, G. L. 1974. Flowering plants: evolution above the species level. Harvard University Press, Cambridge, Massachusetts. 399 pp.

Stehli, F. G. 1968. Taxonomic diversity gradients in pole location: the Recent model. In Evolution and Environment. pp. 163-227. Edited by E. T. Drake. Yale University Press, New Haven.

Stehli, F. G. and Wells, J. W. 1971. Diversity and age patterns in hermatypic corals. Systematic Zoology 20: 115-126.

Umbgrove, J. H. F. 1924. Report on Plistocene and Pliocene corals from Ceram. In Geological, petrographical and palaeontological results of explorations carried out from September 1917 till June 1919 in the island of Ceram. pp. 1-22. Edited by L. Rutten and W. Hotz. Second Series: Palaeontology. Drukkerij en Uitgeverij, J. H. de Bussy, Amsterdam 1.

Umbgrove, J. H. F. 1930. Coral reefs of the East Indies. Bulletin of the American Association of Petroleum Geologists 58: 729-778.

Umbgrove, J. H. F. 1942a. Corals from asphalt deposits of the Island Buton (East Indies). Leidsche Geologische Mededeelingen 13: 29-38.

Umbgrove, J. H. F. 1942b. A revision of fossil corals from Celebes described by Dollfus. Geologie en Mijnbouw (N.S.) 5: 14-16.

Umbgrove, J. H. F. 1943. Tertiary corals from Sumba (East Indies). Verhandelingen van het Geologisch Mijnbouwkundig Genootschap voor Nederland en Koloniën (Geologische Serie) 13: 393-398.

Umbgrove, J. H. F. 1946. Evolution of reef corals in East Indies since Miocene times. Bulletin of the American Association of Petroleum Geologists 30: 23-31.

Veron, J. E. N. 1995. Corals in space in time. The biogeography and evolution of the Scleractinia. University of New South Wales Press, Sydney, 321 pp.

Wallace, C. C. 1998. The Indo-Pacific centre of coral diversity re-examined at species level. Proceedings of the 8th International Coral Reef Symposium, Panama 1996 (in press).

Wallace, C. C., Pandolfi, J. M., Young, A. and Wolstenholme, J. 1991. Indo-Pacific coral biogeography: a case study from the *Acropora selago* group. Australian Systematic Botany 4: 199-210.

Wells, J. W. 1954a. Bikini and nearby atolls, Part 2, Oceanography (Biologic). Recent corals of the Marshall Islands. U. S. Geological Survey Professional Paper 260-I: i-iv, 385-486.

Wells, J. W. 1954b. Bikini and nearby atolls. Part 4. Paleontology. Fossil corals from Bikini Atoll. U. S. Geological Survey Professional Paper 260-P: 609-617.

Wells, J. W. 1956. Scleractinia. In Treatise on invertebrate paleontology. Part F. Coelenterata. F328-F444. Edited by R. C. Moore. Geological Society of America, Boulder and University of Kansas Press, Lawrence.

Wells, J. W. 1964. Bikini and nearby atolls, Marshall Islands. Fossil corals from Eniwetok Atoll. U. S. Geological Survey Professional Paper 260-DD: 1101-1111.

Wells, J. W. 1976. Late Eocene fossils from Eua, Tonga. Eocene corals from Eua, Tonga. U. S. Geological Survey Professional Paper 640-G: G1-G13, G17-18.

Wells, S. M. and Sheppard, C. 1988. Corals reefs of the World. Volume 2; Indian Ocean, Red Sea and Gulf. United Nations Environment Programme, International Union for Conservation of Nature and Natural Resources. 389 pp.

Wilson, C., Barrett, R., Howe, R. and Lei-Kuang Leu. 1993. Occurrence and character of outcropping limestones in the Sepik Basin: Implications for hydrocarbon exploration. In Petroleum Exploration and Development in Papua New Guinea. pp. 111-124. Edited by G. J. and Z. Carman. Proceedings of the Second PNG Petroleum Convention, Port Moresby.

Wilson, M. E. J. 1995. The Tonasa Limestone Formation, Sulawesi, Indonesia: Development of a Tertiary Carbonate Platform. Ph.D. Thesis, University of London, 520 pp.

Winterer, E. L. 1991. The Tethyan Pacific during Late Jurassic and Cretaceous times. Palaeogeography, Palaeoclimatology, Palaeoecology 87: 253-265.

Yabe, H. and Sugiyama, T. 1935. Note on a new fossil coral *Saipania tayamai* n.g. n.sp. found in the Island Saipan, Mariana Group. Japanese Journal of Geology and Geography 12: 5-7.

Yonge, C. M. 1940. The biology of reef-building corals. Scientific Reports of the Great Barrier Reef Expedition 1: 353-391.

Zibrowius, H. 1980. Les Scléractiniaires de la Méditerranée et de l'Atlantique nord-oriental. Mémoires de l'Institut Océanographique 11: 1-284.

Zinsmeister, W. J. and Emerson, W. K. 1981. The role of passive dispersal in the distribution of hemipelagic invertebrates, with examples from the tropical Pacific Ocean. Veliger 22: 33-40.

Genetic structure of marine organisms and SE Asian biogeography

J. A. H. Benzie
Australian Institute of Marine Science, PMB No 3, Townsville, QLD 4810, Australia
Email: j.benzie@aims.gov.au

Key words: SE Asia, biogeography, population genetics, evolution, speciation, molecular biogeography

Abstract

SE Asia has been considered a hot spot for marine speciation, and the likely source for much of the marine biota in the Pacific. Evidence from recent genetic studies has important implications for SE Asian biogeography. Although not conclusive, these data question ideas about the origin of marine biodiversity in the region. For example, the genetic differentiation of widespread starfish species appears to be related to the separation of Indian and Pacific Oceans at times of lowered sea level. This supports data from other taxonomic work that marine biodiversity in SE Asia has resulted more from the mixing of taxa evolved in the two ocean basins than has been thought to date. Similarly, patterns of gene flow in giant clams do not parallel present-day ocean currents. Clams may disperse by other present-day mechanisms such as surface drift, or have been dispersed by palaeocurrents at times of low sea level. However, both past and present mechanisms indicate movement of material from the Pacific to SE Asia. The high marine biodiversity in the region may result more from the accumulation of species evolved in the Pacific, relative to speciation within the SE Asian region, than has been believed traditionally. The mechanisms giving rise to marine biodiversity in the SE Asian region are not well-understood, therefore, and more work is urgently required to enable effective management of these resources.

Introduction

The SE Asian region is one of the most biologically diverse in the world. The diversity of the terrestrial biota results from the meeting and mixing of the floras and faunas from two major zoogeographic regions, and the opportunities for speciation on the islands of the Indo-Malay region (Wallace, 1860, 1881; Burrett *et al.*, 1991; Michaux, 1991). Much work has focused on defining the nature of exchange between these biotas through identifying major breaks in species composition such as Wallace's line.

Knowledge of the marine biota of the region is less than that of the terrestrial sphere, and the concept of the region as a zone of overlap of two zoogeographic provinces (the Indian and the Pacific Oceans) has been considered of lesser significance. This is partly because the Indo-Pacific has generally been thought to be one major biogeographic region (Ekman, 1953; Briggs, 1974, 1987). Many taxa are found in both the Indian and Pacific Oceans, and the occurrence of widespread species in both oceans has led to the view that the floras and faunas of the two oceans are closely related.

Explanations for the high diversity of marine species in SE Asia have concentrated on the idea that most species have evolved within the region (Potts, 1983, 1985; McManus, 1985). Sea-level changes over the last few million years are thought to have created isolated seas in which marine species have had a chance to diverge and speciate. This approach has tended to dominate thinking about the origin of marine biodiversity in SE Asia and the Indo-Pacific, although a number of alternative views have been expressed with respect to the origin of species in the Pacific and Indian Oceans (Ladd, 1960; McCoy and Heck, 1976, 1983; Kay, 1980, 1984; Kay and Palumbi, 1987; Springer, 1982; Springer and Williams, 1990).

Molecular data provide the means to determine genetic structure within taxa (the spatial and temporal patterns of occurrence of gene variants, and patterns of association of different gene variants) and to assess the evolutionary relationships between species. Molecular phylogenies can be used to test whether SE Asian species were likely to have evolved in-situ, or were derivatives from the Indian or Pacific Oceans. The nature of the genetic structure of species can provide information on the principal factors influencing genetic change in populations in the region (*i.e.*, random shifts called genetic drift, strong patterns at only one gene locus suggesting natural selection at that gene, or similar patterns at several loci that are not linked and which would reflect levels of dispersal among populations).

By comparing the structure of gene products, genes or sections of DNA that evolve at different rates, at least the relative timing of speciation or dispersal events can be determined. Molecular data on marine organisms in the region are rare and it is not yet possible to examine evolutionary processes within the region in any detail. Nevertheless, new findings emerging from work throughout the Indo-Pacific is changing the way in which the dispersal and evolution of marine species is considered (Palumbi, 1992, 1994, 1997; Benzie and Williams, 1997), and this has implications for theories concerning the origin and evolution of species in the SE Asian region.

Widespread marine species have traditionally been viewed as having a high dispersal capacity that allows them to undertake long-distance movements throughout their range, presumably on major ocean currents (Briggs, 1974; Scheltema, 1977; Scheltema and Williams, 1983; Jackson, 1986). However, the notion that widespread species may not always reach their dispersal potential is increasingly recognised. A growing number of cryptic taxa within widespread species are being discovered and some widespread species have been shown to consist of a series of regionally distributed cryptic taxa (Knowlton, 1993; Knowlton and Jackson, 1994; Palumbi, 1994; Kelly-Borges and Valentine, 1995; Foltz *et al.*, 1996).

The few genetic data that have been available for widespread marine species over the last 25 years were consistent with traditional views in that little or no genetic differentiation of populations over thousands of kilometres are observed. Where structure was detected, the populations have been relatively isolated, being situated on remote islands or cut off by currents from other populations. However, new data emerging from detailed genetic studies of Indo-Pacific species over broad geographical scales demonstrates unexpected structure that is not consistent with dispersal by present-day ocean currents.

This paper reviews the new findings, and considers their implications for the extent and timing of dispersal of marine species, for the origin and maintenance of genetic variation in the Indo-Pacific, and the implications of these discoveries for SE Asian biogeography.

The nature and utility of genetic data

The aim of this section is not to give an exhaustive discussion of population genetics, but to provide a general background on key aspects of this approach for those not familiar with the topic. Details of the field of population genetics are available in a number of excellent texts such as Avise (1994) who covers different molecular techniques, their application to different problems and summarises the results of work on wild populations, Nei (1987) who provides a more theoretical approach, and Kimura (1983) who specifically addresses the issue of whether frequencies of molecular variants in wild populations reflect the processes of natural selection or are, to all intents, neutral markers. Each of these texts cites a number of reviews and key papers on the topic.

The assortment of genotypes within and among populations, and the relationships among these genotypes, provide an integrated history of the events which affect the genetic structure of populations (Avise, 1994). DNA molecules are changed by mutations which accumulate over time. Comparison of gene sequences (either directly or by examining genetically based variation in the proteins produced from the DNA) can identify the degree of relatedness of these individuals, and determine family trees (*i.e.*, gene genealogies) of the genotypes described (Zuckerkandl and Pauling, 1962; MacIntyre, 1985).

A simple case, where there is a strong association of closely related genotypes with geographical location throughout the range of a sexually reproducing species, implies that little gene exchange occurs between locations, and that each of the populations has developed from a discrete genetic base. The occurrence of asexual reproduction can lead to greater genetic divergence in the frequencies of genetic variants among populations, not because of any fundamental difference in the factors acting upon the populations, but because the ability to produce many copies of a genotype means that the effects of processes such as genetic drift (random sampling errors) or natural selection (targeted change at a locus) are magnified (Avise, 1994). A considerable reduction in population size, often referred to as a bottleneck, can result in considerable loss of variation and a marked and rapid shift in the frequency of variants (Nei, 1978). This is the result of sampling error, and can be thought of as an extreme form of genetic drift. However, bottlenecks can be recognised because the populations they affect have much reduced genetic diversity.

The time for which groups have been separated can be estimated from the degree of genetic difference between the genotypes and, all else being equal, would depend on the rate(s) of evolution of the gene(s) assayed, and factors affecting the frequencies of different variants within the groups (genetic drift, selection, dispersal among groups). While the specific rate of evolution of any gene is difficult to determine (Li and Graur, 1991), it has been established that some genes generally evolve at relatively slow rates (*e.g.*, the 18S ribosomal gene), others at a fast rate (*e.g.*, D-loop region of the mtDNA genome) and others at rates in between (*e.g.*, cytochrome oxidase I gene: the CO I gene). In general terms, genes in the mtDNA genome evolve at a faster rate than those in the nuclear genome, although the actual rate can vary considerably (Martin *et al.*, 1992). Subject to accounting for these influences, it is possible to determine the relative order of evolutionary events within and between species, and to indicate the general time scale of events, using estimates of the genetic divergence of populations derived from a variety of genetic markers (for marine examples see Cunningham and Collins, 1994).

In addition to the accumulation of mutations, genes are exchanged between populations as a result of migration and, in sexual species, between individuals through reproduction. These processes lead to the mixing of variants among populations. All else being equal, the extent to which the frequencies of variants are similar between populations provides a measure of the gene exchange between them. A survey of the spatial patterns of occurrence of gene variants can therefore provide valuable information on the degree of gene exchange between populations. Where variants are produced in relatively short time-scales (*e.g.*, mtDNA) the spatial patterns of the frequency of variants are more likely to reflect recent dispersal events than those patterns from genes that evolve at slower rates (*e.g.*, allozymes). Slower evolving markers, though, can provide information on the nature of past gene exchange, or of gene flow integrated over longer time periods.

A number of other factors can influence gene frequencies in natural populations, hence the qualification "all else being equal" being made above with reference to estimating gene exchange or dispersal among populations. These factors include molecular processes (such as meiotic drive) which might influence the frequency of particular variants, some aspects of population size (such as founder effect and other sampling effects of genetic drift) which can speed change, but probably the most important is natural selection. The extent to which selection is responsible for population genetic structure has been the subject of intense debate — the selectionist-neutralist debate (Kimura, 1983). The frequency of a gene which is favoured in particular circumstances will be likely to be maintained in the face of the influx of other variants, and would not provide an appropriate estimate of gene flow.

Although one needs to be aware of the potential influence of selection, in practice, sampling several loci provides a reasonable set of markers that can be assumed to be neutral elements (see Hillis and Moritz, 1990: pp. 5-6). This is the great advantage of many molecular markers over morphological characteristics. Selection and environmental factors are likely to have played important roles in determining the phenotypes and frequency of morphological characters. The examination of spatial patterns of gene frequencies, therefore, provides a powerful means of examining dispersal among populations over a variety of time scales. Gene genealogies are a powerful means of determining the relative order and estimating the timing of events in the evolution of natural populations. An introduction to the interpretation of such data is provided by Avise (1994) and specific examples relating to marine biogeography are illustrated by Cunningham and Collins (1994).

One of the recent interesting discoveries concerning the Indo-Pacific fauna is the number of cryptic species being revealed, and it is useful to identify criteria that can be used to determine whether the observed genetic variation reflects intra- or inter-specific differences. There is no prescribed level of genetic difference that defines species but where genetic differentiation is very great (*i.e.*, Nei's D values of 0.8 or more) it is likely that one is dealing with two taxa. More powerful evidence comes from the number of fixed gene differences between the populations (*i.e.*, no variant at a given locus found in one taxon is shared with the other taxon). The occurrence of several fixed gene differences between populations that are sympatric is strong evidence for the occurrence of two species that are reproductively isolated.

Where there are several fixed gene differences between populations that are geographically isolated from each other, the pattern may result from species differences or simply genetic differentiation as a result of isolation. The geographic pattern of genetic variation can give some clues as how to interpret such a result: if most local populations show considerable differentiation from each other, or if the differentiation of the allopatric population fits the expectations of an isolation by distance model of genetic structure, then it is less likely that the allopatric population deserves specific status.

Where genetic differentiation is within the range observed within well-defined species, abrupt shifts in gene frequency might suggest the possibility of geographically distinct taxa. In this case the default assumption would be that the populations both belonged to a more widespread taxon unless there was other biological data to indicate otherwise (*e.g.*, lack of fertility between experimental matings between members of the two populations).

Genetic data on SE Asian marine species

Genetic surveys of marine organisms from SE Asia are rare. Most focus on local areas or coastlines, or address specific questions relating to fisheries management. However, some of these are providing evidence of cryptic species in groups where it might have been expected that the taxonomy was well-based. For example, genetic studies of the northern Australian fisheries thought to be based on *Photololigo edulis* and *Photololigo chinensis* were shown to consist of four cryptic species, none of which corresponded to *P. edulis* or *P. chinensis* from type localities in the China Sea, and all of which were new to science (Yeatman and Benzie, 1994). Cryptic species in taxa of coral which occur in the region (*Montipora* spp.) have been described in Australia (Stobart and Benzie, 1994).

These data suggest that many more cryptic and regionally defined taxa may occur in the region. Studies which examine genetic variation on a variety of geographical scales and/or using a variety of genetic markers that might provide information pertinent to the evolution and biogeography of the SE Asian region are restricted to those on the butterfly fish *Chaetodon* spp. (McMillan and Palumbi, 1995), the starfish *Acanthaster planci* (Benzie, unpublished data) and *Linckia laevigata*, (Williams and Benzie, 1997) the milkfish *Chanos chanos* (Winans, 1980), and the giant clams *Tridacna gigas*, *Tridacna derasa* and *Tridacna maxima* (Macaranas *et al.*, 1992; Benzie and Williams, 1995, 1997). Within the context of the patterns of variation revealed by these taxa, information from Pacific populations of the pearl oyster *Pinctada margaritifera* (Benzie and Ballment, 1994), the sea urchin *Echinometra mathei* and related taxa (Palumbi and Metz, 1991; Palumbi, 1997), are interesting.

Patterns of gene flow in the Pacific

Early work on the genetic structure of marine species supported the view that long-distance dispersal occurred and was consistent with present day ocean circulation (Campbell *et al.*, 1975). This is well illustrated by the work on milkfish by Winans (1980). The patterns of population subdivision which emerged from his allozyme study showed populations in Hawaii were consistently and significantly differentiated from those in the Philippines, the Marshall Islands, Christmas Island and Fanning Island at all loci. The Philippines populations were differentiated from the Marshall Islands, Christmas Island and Fanning Island at some loci but not at others. He showed that there was no significant differentiation of milkfish populations over several thousand kilometres between the Marshall Islands, Christmas Island and Fanning Island. Even where gene frequencies were significantly different, genetic distances between these populations were low.

Emphasis was placed, therefore, on the extensive gene flow among all populations with the exception of the Hawaiian Islands which were considered isolated not only in respect of their geographic position but by the fact that currents flow perpendicular to the route between Hawaii, and the Christmas and Fanning Islands. The small genetic distance between the Hawaiian populations and the others (of the order of the average value among populations within species), and the lack of fixed gene differences between the Hawaiian populations and the others is evidence that the Hawaiian populations are not a cryptic species.

One of the earliest studies of genetic variation in marine organisms was on the giant clam, *Tridacna maxima*, by Ayala *et al.* (1973) and Campbell *et al.* (1975). The very small genetic distance between populations from the Great Barrier Reef and the Marshall Islands (Nei's D = 0.035) was also thought to imply that dispersal occurred throughout the species range, and over distances of several thousand kilometres. However, a more detailed survey of populations of *T. maxima* from the Coral Sea (Benzie and Williams, 1992) demonstrated as great a genetic distance between populations separated by only 400 km, as that observed by Campbell *et al.* (1975) over 5,000 km. Differences in the genes sampled in the two studies did not allow any more than a general comparison. However, the differences suggested that gene exchange was unrestricted between *T. maxima* populations up to a neighbourhood size of about 400 km, and that *T. maxima* might show a pattern of geographic variation consistent with isolation by distance.

These data provided little reason to question the traditional view that widespread marine invertebrates with a long larval phase were distributed widely over the Pacific, presumably on ocean currents, and that dispersal was contemporary. Other studies, such as those testing the origin of populations of several fish species separated by the Pacific Barrier from the American coastline (Rosenblatt and Waples, 1986), or those examining the genetic structure of tuna populations over the Pacific (Richardson, 1983; Ward *et al.*, 1994), pearl oysters in French Polynesia (Durand and Blanc, 1988), crown-of-thorns starfish over the Pacific (Nishida and Lucas, 1988), or butterfly fish over the Pacific (McMillan and Palumbi, 1995), have also provided evidence consistent with long-range dispersal.

However, Benzie and Williams (1995, 1997) noted that work has been restricted to highly vagile fish species such as tuna or coastal regions of high connectedness and that most other studies have sampled only a few widespread populations. This has restricted the capacity to evaluate any patterns of gene flow. In the case of the milkfish, the populations showing no genetic variation were all situated in the track of the strong equatorial currents that would have facilitated dispersal. Analyses of geographic variation with a reasonable density of sampling over a wide region of the Pacific has recently been achieved for giant clams, and has shown a very different picture.

The giant clams Tridacna gigas, T. derasa and T. maxima

Analysis of geographic variation in three giant clam species demonstrated much more complex genetic structures in the western Pacific (Fig.1). The best geographical coverage was achieved in *T. maxima* where nineteen populations were sampled throughout the western Pacific (Benzie and Williams, 1997). These data form a template for the interpretation of a smaller amount of data from *T. gigas* (Benzie and Williams, 1995), *T. derasa* (Macaranas *et al.*, 1992) and the pearl oyster *Pinctada margaritifera* (Benzie and Ballment, 1994). The principal patterns of gene flow in all three clam species was found to be parallel to the major island chains (following a NW-SE axis) and perpendicular to the major ocean currents flowing through the region (following a NE-SW axis) (Fig.1).

It was argued that longshore coastal currents might influence dispersal locally, but that some of the island groups are separated by several hundreds of kilometres of water and larvae entering these should be entrained in large numbers and transported in the direction of the major circulation. Relatively large gene flows were observed between the Philippines and the Marshall Islands between which there are few reefs that might act as staging posts for dispersal. In contrast, little direct gene flow was observed between the Solomon Islands and the Great Barrier Reef. This was despite the existence of many reefs in the Coral Sea that might act as staging posts, the shorter distance separating the two locations, and the fact that the South Equatorial Current flows directly from the Solomon Islands to the Great Barrier Reef.

Detailed oceanographic information is lacking for many areas of the Pacific, and the effects of local currents cannot be entirely ruled out. Similarly, wind drift, which moves thin sheets of surface water (only a few centimetres thick) and surface spray predominantly from the SE to the NW, might entrain clam larvae. This mechanism has been inferred to explain the temporal sequence of occurrence of a bacterial disease of coralline algae (Hale and Mitchell, 1995), but direct information on the nature of surface drift in the south eastern Pacific is limited. Benzie and Williams (1995, 1997) noted that currents at depths of 100 m or more were parallel to the island chains, and inferred that, at times of lowered sea level, surface currents may well have flowed from the SE to the NW. They concluded that the patterns of spatial differentiation in gi-

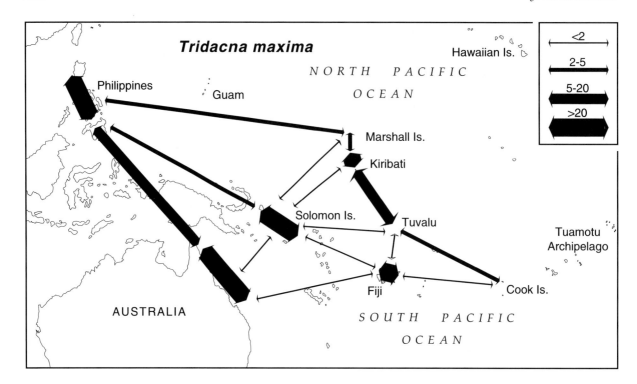
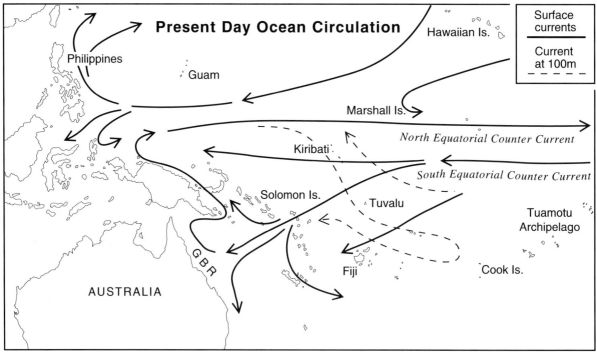

Fig.1. Patterns of gene flow among giant clam (*Tridacna maxima*) populations in the western Pacific, given as the average number of migrants per generation (a). The figure demonstrates that the major axes of gene flow are perpendicular to the major surface currents (b) in the region, but are parallel to deeper current flows (after Benzie and Williams, 1997).

ant clams were the ghosts of dispersal past, and that these had not been altered by changed oceanographic conditions since at least the early Holocene (6 Ka).

The principal finding of this work is that the general notion that widespread marine organisms with a larval phase are distributed widely throughout their range on major ocean currents is not consistent with the genetic evidence. Irrespective of whether dispersal is contemporary and occurring on surface drift, or whether major patterns of marine diversity are related to past events, it is clear we do not understand the processes driving the origin and maintenance of marine biodiversity. Given that sea-level changes are related to climatic fluctuations that may be related to periodic astronomical events (Bennet, 1990), dispersal is likely to be pulsed at periods of low sea-level when circulation might have been more intense within ocean basins, as well as different in direction. The important role these processes are likely to have played in determining connectedness of coral populations has been discussed by Rosen (1988) and Veron (1995).

These findings are of particular interest with respect to the SE Asian region. Traditional views have SE Asia as a regional centre of speciation, with species radiating from SE Asia eastwards across the Pacific. Benzie and Williams (1997) noted that the NW-SE axis of gene flow was not necessarily inconsistent with this view, since the genetic data could not determine the direction of movement along this axis. However, the mechanisms of dispersal identified (present-day surface drift or palaeocurrents) both provide for potential dispersal from the Pacific into the SE Asian region. These data therefore provide some support for a model developed by Jokiel and Martinelli (1992) which demonstrated that, when speciation was limited to islands in the Pacific, species diversity increased at the western edge of the Pacific Ocean, in the SE Asian region. The fact that there is a reduction in species diversity eastwards into the Pacific does not mean there has to be dispersal of species eastwards from a centre of origin. It is possible, for example, that the smaller size of populations on isolated islands and relatively less heterogeneous environment on a given scale may result in a greater extinction of species in the Pacific.

The sea urchins Echinometra spp.

Palumbi (1977) has summarised patterns of genetic structure based on mtDNA data for four species of closely related sea urchins. The limited mtDNA divergence between species in the central and west Pacific indicates that they have diverged in the last 1-3 myr (Palumbi, 1996) and that speciation is on-going within the Pacific basin. Strong geographic patterns of differentiation were observed within each species, but were not concordant, suggesting that different processes played a role in each case or that chance played a strong role in population differentiation. However, the wide spread of a relatively few samples across the Pacific did not provide a sound basis for discerning patterns of gene flow among these populations in the same way as the giant clam data. Nevertheless, these data do indicate regional differentiation, and evidence of speciation within ocean basins far from the SE Asian region.

Differentiation between oceans

There are very few genetic surveys which encompass both the Indian and Pacific Oceans, and those that do have only recently been acquired. All have shown marked genetic differentiation of populations in the Indian Ocean from Pacific populations. Despite the lack of significant differentiation among several populations collected throughout the western Pacific, the one population of coconut crabs collected from Christmas Island in the Indian Ocean was significantly differentiated from all others (Lavery *et al.*, 1996). Similarly, the level of genetic distinction of butterfly fish populations from different oceans was orders of magnitude greater than that between populations within oceans (McMillan and Palumbi, 1995). Major differences in gene frequencies have been observed between populations from different oceans, despite little or no genetic differentiation being observed between populations over large geographical scales within oceans, for the starfish *Linckia laevigata* (Williams and Benzie, 1996, 1997, 1998) and the starfish *Acanthaster planci* (Nishida and Lucas, 1988; Benzie, 1992, unpublished results). Most information has been obtained for *Linckia laevigata*, and it is worth looking at these findings in more detail.

The starfish Linckia laevigata

Data are available for both allozyme and mtDNA restriction fragment patterns in *L. laevigata* which allowed a deeper interpretation of the genetic data (Fig.2). Data were obtained from more than twenty sites throughout the Indian and Pacific Oceans, providing reasonable coverage of the Indo-Pacific. The complete absence of any spatial differences in gene frequencies throughout the western Pacific (Fig.2), despite the presence of a high degree of polymorphism, suggested either large scale contemporary dis-

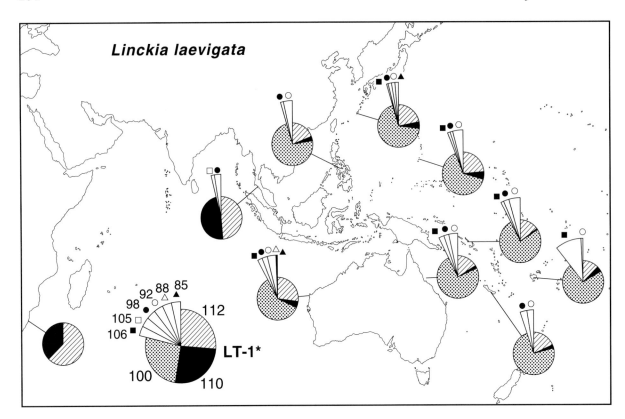

Fig.2. Spatial pattern of allele frequencies at the *LT-1* locus in the starfish *Linckia laevigata* illustrating the marked genetic differentiation of populations in the Indian and Pacific Oceans, but the lack of significant differentiation between populations within each ocean (after Williams and Benzie, 1998). The data also show clearly that the Western Australian population has a genetic constitution similar to that of the Pacific rather than the Indian Ocean.

persal, or that the populations were not at equilibrium. The four-week long larval life of the species might suggest that it could disperse considerable distances, but it was not clear that this period was sufficient to maintain homogeneity throughout the Pacific.

The discovery of some structure in the mtDNA data showed that dispersal was high over short to medium distances and regions where populations were highly connected (such as in the Great Barrier Reef). However, some distinction between the Great Barrier Reef and Fiji populations on the one hand, and Philippines and Western Australian populations on the other, indicated low present-day gene flow between these two groups. These data indicate that the allozymes are not yet at equilibrium and that some other processes (such as balancing selection) are slowing change in allozyme frequencies. The data indicate greater dispersal among populations within oceans in the past than occurs present day.

In contrast, strong and significant differences were found between the Indian and Pacific Ocean populations (Fig.2). The fact that some differences occurred at several loci (which were not linked, but were segregating independently) indicated that populations had differentiated as a result of genetic drift. The fact that the genetic diversity in each set was high suggested the differences in allele frequencies was not the result of founder effect. The differentiation was thought to occur when gene flow was restricted, probably at times of lowered sea level when land connected much of SE Asia, New Guinea and Australia, almost closing the sea connection between the Indian and Pacific Oceans. These changes are thought to be recent. Phylogenetic analysis of *Linckia* species has shown that *L. guildingi* (distributed throughout the Indo-Pacific and Caribbean) diverged from the other species about 20 Ma, and that the molecular divergence of the Indian and Pacific populations of *L. laevigata* occured during Pleistocene time (<3 Ma). It is not known whether the genetic differentiation occurred rapidly (125 Ka - 12 Ka) or

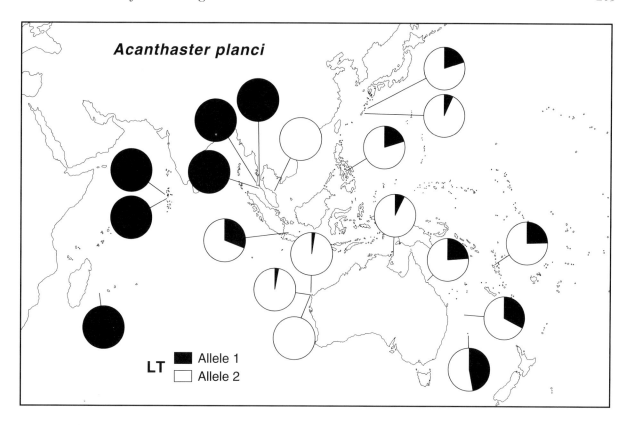

Fig.3. Spatial pattern of allele frequencies at the *LT* locus in the crown-of-thorns starfish, *Acanthaster planci,* illustrating the major genetic differentiation of populations in the Indian and Pacific Oceans, but the lack of significant differentiation between populations within each ocean (Benzie, unpublished data). The data also show clearly that the Western Australian populations have a genetic constitution similar to that of the Pacific rather than the Indian Ocean.

accumulated over several periods of lowered seas level known to have occurred in the last 3 myr.

It is important to note that the Western Australian populations of *L. laevigata* had a similar genetic constitution (allozyme and mtDNA) to the Pacific populations even though they are situated in the Indian Ocean. Oceanographic, micropalaeontological and sedimentological data provide evidence that these populations have been isolated from others in the Indian Ocean by upwelling to the west of the Australian coast (Fleminger, 1986; Wells and Wells, 1994; Wells *et al.*, 1994). It is speculated that they received (or that the populations were originally founded by) recruits from the SE Asian region through the Indonesian throughflow rather than from the east coast of Australia via the Torres Strait and the Northern Australian coast (Godfrey *et al.*, 1993; Gordon and Fine, 1996). It is pertinent to note that data from an unrelated starfish *A. planci* also shows a similar geographic pattern of allozyme variants (Fig.3).

The taxonomic composition of a number of marine faunal groups such as sponges (Berquist and Kelly-Borges, 1995) is also consistent with a connection of Western Australia with the Pacific rather than the Indian Ocean.

The distribution of mtDNA genotypes suggested that gene flow had resumed between the Indian and Pacific Ocean populations in the region of the Indonesian archipelago (Williams and Benzie, 1997). However, data within this region were limited to one population and further work will be required to determine 1) whether there is widespread introgression, 2) whether some hybrid zone has been established or 3) whether gene flow is now highly restricted. While there are a variety of colour morphs in *L. laevigata*, populations in the Pacific (and Western Australia) are predominantly blue while those in the Indian Ocean are violet or salmon pink in colour. It is pertinent to note that the crown-of-thorns starfish in the Pacific (and in Western Australia) are grey-purple to brick red in colour whereas those from the Indian Ocean are blue-purple to pink (Benzie, 1992), and the colour groups have very different allozyme frequencies (Benzie, unpublished data) (see Fig.3). No mtDNA data are yet available, but the concurrence in spatial pattern of genetic differentiation in two unrelated starfish suggests the pattern may be general to other marine invertebrates.

These findings are particularly relevant to SE Asian biogeography because they provide evidence for strong genetic differentiation, consistent with vicariant origins in oceans on either side of the SE Asian region, even within widespread marine species with long larval lives. The divergence in allopatry appears to occur in less than 3 Ma, or perhaps as little as tens of thousands of years. The lack of equilibrium in genetic structure indicates far greater dispersal among populations within oceans in past times, but less between oceans. These data provide further support for views that speciation of marine taxa did not occur within a centre of origin (SE Asia), but from successive isolation of populations outside this region (Wallace et al., 1991; Pandolfi, 1992, 1994; Wallace, 1997). Cladistic biogeographic analysis of corals has suggested species arose from successive isolation of populations consistent with major geologic events in the Indo-Pacific, rather than spreading out from the Indo-Malay region after arising within that region (Pandolfi, 1992, 1994). Wallace (1997) presents further evidence demonstrating that patterns of endemicity of *Acropora* corals are consistent with a significant level of speciation within either the Indian or Pacific Oceans followed by the spread of these taxa into the SE Asian region. Several siganid fish species distributions support this same interpretation (Woodland, 1983). The genetic structure of butterfly fish (McMillan and Palumbi, 1995) and coconut crabs (Lavery et al., 1996) also show little variation within the Indian or Pacific Oceans but marked genetic differences between oceans.

Discussion

SE Asia has long been recognised as having high levels of marine biodiversity. The many hypotheses which have been advanced to explain the biogeography of tropical marine organisms have been summarised by Rosen (1988) and each falls into one of two major models. In the first, species are thought to originate in a centre of high diversity, and then spread to peripheral areas and is known as the dispersal model. In the second, species are thought to form by divergence of populations divided by a geographic barrier, and is known as the vicariance model. In the vicariance model there is no requirement to have highly diverse centres of origin.

The principal reasons for suggesting the SE Asian region is the centre of origin of coral species is the high diversity within the region, and the reduction in diversity east and west. In the case of corals, the average age of higher taxonomic units (families, genera) also decreases moving outwards from the SE Asian region and this fact has been used to support the dispersal model (Stehli and Wells, 1971). However, Pandolfi (1992) has shown that some of the most derived coral species that show the greatest degree of endemism are at the periphery of Indo-Pacific coral distributions. This implies that many species must have originated far from the SE Asian centre of diversity. Jokiel and Martinelli (1992) have demonstrated in a simple model how speciation in the isolated peripheral areas of the Pacific still leads to an accumulation of species diversity in the western margin of the ocean at the tropics because of ocean circulation patterns. More recently the patterns of endemism of *Acropora* coral species Wallace (1997) and sponges (Kelly-Borges and Valentine, 1995) are consistent with vicariant divergence of populations in the two ocean basins, rather than within the SE Asian region.

The limited genetic data available are of interest because they suggest that patterns of genetic variation within species today are not at equilibrium, and that they are the result of historical events. The patterns of genetic diversity in giant clams indicate that dispersal from SE to NW is consistent with major ocean currents inferred at times of low sea level, and that dispersal among sites within oceans was greater in the past than now. Major genetic discontinuities between populations in the Indian and Pacific Oceans indicate divergence in allopatry at times of low sea level, and provide evidence for the origin of species in the two oceans rather than in the SE Asian region. The domination of patterns of variation explained by low sea levels should not be surprising given that sea levels have been lower than present day levels for much of the Pleistocene.

It is important to note that there is almost no data on the genetic structure of species within the SE Asian region. Potts (1985) and McManus (1985) suggested that sea level changes played an important role in creating isolated populations by cutting off local sea basins within the SE Asian region. It is likely that changing sea levels have influenced speciation of marine organisms by this means within the region, particularly those with taxa with low dispersal potential.

A review of the limited data available cannot be conclusive, therefore, but does invite speculation that present patterns of genetic variation have resulted from highly pulsed dispersal of marine taxa in the Indo-Pacific, related to global climate change and major geological events, and that regional speciation outside the Indo-Malay region has been greater than thought in the past. Movement of species originating in the Indian and Pacific Oceans into the Indo-Malay archipelago may have played a more important role in producing diversity in that region than traditionally recognised.

It will be necessary to develop specific predictions of the spatial patterns of genetic variation expected under different biogeographic hypotheses so that these can be explicitly tested. This will be a challenging task as many of the theories have identical biogeographical predictions (Pandolfi, 1992). Some approaches have been suggested by Palumbi (1997) which predict that older genotypes are likely to be found where species originate, but he also indicates that differences in population size and extinction rates in the central Pacific (lower population size and higher extinction rates) compared to the western Pacific (higher population sizes and lower extinction rates), might mask this pattern. Nevertheless, further application of molecular genetic tools to examine the structure of species within the SE Asian region will provide an important means of advancing our understanding of the origin and maintenance of biodiversity in the region.

Acknowledgements

This is contribution number 892 from the Australian Institute of Marine Science.

References

Avise, J. C. 1994. Molecular markers, natural history and evolution. Chapman and Hall, New York.

Ayala, F. J., Hedgecock, D., Zumwalt, G. S. and Valentine, J. W. 1973. Genetic variation in *Tridacna maxima*, an ecological analog of some unsuccessful evolutionary lineages. Evolution 27: 177-191.

Bennet, K. D. 1990. Milankovitch cycles and their effects on species in ecological and evolutionary time. Paleobiology 16: 11-21.

Benzie, J. A. H. 1992. Review of the genetics, dispersal and recruitment of crown-of-thorns starfish (*Acanthaster planci*). Australian Journal of Marine and Freshwater Research 43: 597-610.

Benzie, J. A. H. and Ballment, E. 1994. Genetic differences among black-lipped pearl oyster (*Pinctada margaritifera*) populations in the western Pacific. Aquaculture 127: 145-156.

Benzie, J. A. H. and Williams, S. T. 1992. Genetic structure of giant clam (*Tridacna maxima*) populations from reefs in the Western Coral Sea. Coral Reefs 11: 135-141.

Benzie, J. A. H. and Williams, S. T. 1995. Gene flow among giant clam (*Tridacna gigas*) populations in Pacific does not parallel ocean circulation. Marine Biology 123: 781-787.

Benzie, J. A. H., and Williams, S. T. 1997. Gene flow among giant clam (*Tridacna maxima*) populations in the Pacific does not parallel ocean circulation. Evolution 51: 768-783.

Berquist, P. R. and Kelly-Borges, M. 1995. Systematics and Biogeography of the genus *Ianthella* (Desmospongidae; Verongida; Ianthellidae) in the South Pacific. The Beagle, Records of the Northern Territory Museum of Arts and Sciences 12: 151-176.

Briggs, J. C. 1974. Marine zoogeography. McGraw-Hill, New York, USA.

Briggs, J. C. 1987. Biogeography and plate tectonics. Elsevier, New York, USA.

Burrett, C., Duhig, N., Berry, R. and Varne, R. 1991. Asian and South-western Pacific continental terranes derived from Gondwana, and their biogeographic significance. Australian Systematic Botany 4: 13-24.

Campbell, C. A., Valentine, J. W. and Ayala F. J. 1975. High genetic variability in a population of *Tridacna maxima* from the Great Barrier Reef. Marine Biology 33: 341-345.

Cunningham, C. W. and Collins, T. M. 1994. Developing model systems for molecular biogeography: vicariance and interchange of marine invertebrates. *In* Molecular ecology and evolution: approaches and applications. pp. 405-433. B. Schierwater, G. P. Wagner, and R. DeSalle. Birkhauser Verlag, Basel, Switzerland.

Durand, P. and Blanc, F. 1988. Divergence genetique chez un bivalve marin tropical: *Pinctada margaritifera*. Colloque Nationale de CNRS, Lyon pp. 323-330.

Ekman, S. 1953. Zoogeography of the sea. Sidgwick and Jackson, London, UK.

Fleminger, A. 1986. The pleistocene equatorial barrier between the Indian and Pacific Oceans and a likely cause for Wallace's Line. UNESCO Technical Paper in Marine Science 49: 84-97.

Foltz, D. W., Stickle, W. B., Campagnaro, E. L. and Himel, A. E. 1996. Mitochondrial DNA polymorphisms reveal additional genetic heterogeneity within the *Lepasterias hexactis* (Echinodermata: Asteroidea) species complex. Marine Biology 125: 569-578.

Godfrey, J. S., Hirst, A. C. and Wilkin, J. 1993. Why does the Indonesian throughflow appear to originate from the North Pacific? Journal of Physical Oceanography 23: 1087-1098.

Gordon, A. L. and Fine, R. A. 1996. Pathways of water between the Pacific and Indian Oceans in the Indonesian Seas. Nature 379: 146-149.

Hale, M. S. and Mitchell, J. G. 1995. CLOD spreading in the sea-surface microlayer. Science 270: 897.

Hillis, D. M. and Moritz, C. 1990. Molecular Systematics. Sinauer, Sunderland, MA.

Jackson, J. B. C. 1986. Modes of dispersal of clonal benthic invertebrates: consequences for species' distributions and genetic structure of local populations. Bulletin of Marine Science 39: 588-606.

Jokiel, P. and Martinelli, F. J. 1992. The vortex model of coral reef biogeography. Journal of Biogeography 19: 449-458.

Kay, E. A. 1980. Little worlds of the Pacific. An essay on Pacific Basin biogeography. Lyon Arboretum, University of Hawaii, USA.

Kay, E. A. 1984. Patterns of speciation in the Indo-west Pacific. Bernice P Bishop Museum Occasional Publications 2: 15-31.

Kay, E. A. and Palumbi, S. R. 1987. Endemism and evolution in Hawaiian marine invertebrates. Trends in Ecology and Evolution 2: 183-186.

Kelly-Borges, M. and Valentine, C. 1995. Taxonomy, systematics and biogeography of Porifera in the tropical Island Pacific Region (Oceania): a status review. In Marine and Coastal Biodiversity in the Tropical Island Pacific Region: 1. Species Systematics and Information Management Priorities. pp. 83-120. Edited by J. E. Maragoes, M. N. A. Peterson, L. G. Eldredge, and J. E. Bardach. East-West Centre, Honolulu, Hawaii.

Kimura, M. 1983. The Neutral Theory of Molecular Evolution. Cambridge University Press, Cambridge, USA.

Knowlton, N. 1993. Sibling species in the sea. Annual Review of Ecology and Systematics 24: 189-216.

Knowlton, N. and Jackson, J. B. C. 1994. New taxonomy and niche partitioning on coral reefs: jack of all trades or master of some? Trends in Ecology and Evolution 9: 7-9.

Ladd, H. S. 1960. Origin of the Pacific island molluscan fauna. American Journal of Science 258A: 137-150.

Lavery, S., Moritz, C. and Fielder, D. R. 1996. Indo-Pacific population structure and evolutionary history of the coconut crab *Birgus latro*. Molecular Ecology 5: 557-570.

Li, W-H. and Graur, D. 1991. Fundamentals of Molecular Evolution. Sinauer, Sunderland, MA.

Macaranas, J. M., Ablan, C. A., Pante, Ma. J. R., Benzie, J. A. H. and Williams, S. T. 1992. Genetic structure of giant clam (*Tridacna derasa*) populations from reefs in the Indo-Pacific. Marine Biology 113: 231-238.

MacIntyre, R. J. 1985. Molecular Evolutionary Genetics. Plenum, New York.

Martin, A., Naylor, G. J. P. and Palumbi, S. R. 1992. Rates of mitochondrial variation in sharks are slow compared to mammals. Science 357: 153-155.

McCoy, E. D. and Heck, K. L. 1976. Biogeography of corals, seagrasses, and mangroves: an alternative to the center of origin concept. Systematic Zoology 25: 201-210.

McCoy, E. D. and Heck K. L. 1983. Centres of origin revisited. Paleobiology 9: 17-19.

McManus, J. W. 1985. Marine speciation, tectonics and sea-level changes in southeast Asia. Proceedings of the Fifth International Coral Reef Congress, Tahiti 4: 133-138.

McMillan, W. O. and Palumbi, S. R. 1995. Concordant evolutionary patterns among Indo-west Pacific butterfly fishes. Proceedings of the Royal Society of London 260: 229-236.

Michaux, B. 1991. Distributional patterns and tectonic development in Indonesia: Wallace reinterpreted. Australian Systematic Botany 4: 25-36.

Nei, M. 1978. Estimation of average heterozygosity and genetic distance from a small number of individuals. Genetics 89: 583-590.

Nei, M. 1987. Molecular Evolutionary Genetics. Columbia University Press, New York.

Nishida, M. and Lucas, J. S. 1988. Genetic differences between geographic populations of the crown-of-thorns starfish throughout the Pacific region. Marine Biology 98: 359-368.

Palumbi, S. R. 1992. Marine speciation on a small planet. Trends in Ecology and Evolution 7: 114-118.

Palumbi, S. R. 1994. Genetic divergence, reproductive isolation, and marine speciation. Annual Review of Ecology and Systematics 25: 547-572.

Palumbi, S. R. 1996. What can molecular genetics contribute to marine biogeography? An urchin's tale. Journal of Experimental Marine Biology and Ecology 203: 75-92.

Palumbi, S. R. 1997. Molecular biogeography of the Pacific. Coral Reefs 16: s47-s52.

Palumbi, S. R. and Metz, E. 1991. Strong reproductive isolation between closely related tropical sea urchins (genus *Echinometra*). Molecular Biology and Evolution 8: 227-239.

Pandolfi, J. M. 1992. Successive isolation rather than evolutionary centres for the origination of Indo-Pacific reef corals. Journal of Biogeography 19: 593-609.

Pandolfi, J. M. 1994. A review of the tectonic history of Papua New Guinea and its significance for marine biogeography. Proceedings of the Seventh International Coral Reef Symposium, Guam 2: 718-728.

Potts, D. C. 1983. Evolutionary disequilibrium among Indo-Pacific corals. Bulletin of Marine Science 33: 619-632.

Potts, D. C. 1985. Sea-level fluctuations and speciation in Scleractinia. Proceedings of the Fifth International Coral Reef Congress, Tahiti 4: 127-132.

Richardson, B. J. 1983. Distribution and protein variation in skipjack tuna (*Katsuwonas pelamis*) from the central and south-western Pacific. Australian Journal of Marine and Freshwater Research 34: 231-251.

Rosen, B. R. 1988. Progress, problems and pattern in the biogeography of reef corals and other tropical marine organisms. Helgolander Meeresunters 42: 269-301.

Rosenblatt, R. H. and Waples, R. S. 1986. A genetic comparison of allopatric populations of shore fish species from the eastern and central Pacific Ocean: dispersal or vicariance? Copeia 1986: 275-284.

Scheltema, R. S. 1977. Dispersal of marine invertebrate organisms: Palaeobiogeography and biostratigraphic implications. In Concepts and methods of biostratigraphy. pp. 73-108. Edited by E. G. Kaufmann and J. E. Hazel. Dowden, Hutchinson and Ross, Stroudsburg, Pennsylvania, USA.

Scheltema, R. S. and Williams, I. P. 1983. Long-distance dispersal of planktonic larvae and the biogeography and evolution of some Polynesian and Pacific mollusks. Bulletin of Marine Science 33: 545-565.

Springer, V. G. 1982. Pacific plate biogeography to shore-fishes. Smithsonian Contributions to Zoology 367: 1-181.

Springer, V. G. and Williams, J. T. 1990. Widely distributed Pacific plate endemics and lowered sea-level. Bulletin of Marine Science 47: 631-640.

Stehli, F. G. and Wells, J. W. 1971. Diversity and age patterns in hermatypic corals. Systematic Zoology 20: 115-126.

Stobart, B. and Benzie, J. A. H. 1994. Allozyme electrophoresis demonstrates that the scleractinian coral *Montipora digitata* (Dana, 1864) is two species. Marine Biology 118: 183-190.

Veron, J. E. N. 1995. Corals in space and time. University of New South Wales Press, Sydney, Australia.

Wallace, A. R. 1860. On the zoological geography of the Malay Archipelago (read Nov. 3, 1859). Journal of the Linnean Society (Zoology) 4: 172-184.

Wallace, A. R. 1881. Island Life. Harper Bros, New York.

Wallace, C. 1997. The Indo-Pacific centre of coral diversity re-examined at species level. Proceedings of the Eighth International Coral Reef Congress, Panama, (in press).

Wallace, C., Pandolfi, J. M., Young, A. and Wolstenholme, J. 1991. Indo-Pacific coral biogeography: a case study from the *Acropora selago* group. Australian Systematic Botany 4: 199-210.

Ward, R. D., Elliot, N. G., Grewe, P. and Smolenski, A. J. 1994. Allozyme and mitochondrial DNA variation in yellowfin tuna (*Thunnus albacares*) from the Pacific Ocean. Marine Biology 118: 531-539.

Wells, P. E. and Wells, G. M. 1994. Large-scale reorganisation of ocean currents offshore Western Australia during the late Quaternary. Marine Micropalaeontology 24: 157-185.

Wells, P. E., Wells, G. M., Cali, J. and Chivas, A. 1994. Response of deep-sea benthic foraminifera to late Quaternary climate changes, southeast Indian Ocean, offshore Western Australia. Marine Micropalaeontology 24: 185-229.

Williams, S. T. and Benzie J. A. H. 1996. Genetic uniformity of widely separated populations of the coral reef starfish *Linckia laevigata* from the West Pacific and East Indian Oceans, revealed by allozyme electrophoresis. Marine Biology 126: 99-108.

Williams, S. T. and Benzie J. A. H. 1997. Indo-West Pacific patterns of genetic differentiation in the high-dispersal starfish *Linckia laevigata*. Molecular Ecology 6: 559-573.

Williams, S. T. and Benzie, J. A. H. 1998. Evidence of a phylogeographic break between populations of a high-dispersal starfish: congruent regions within the Indo-West Pacific defined by colour morphs, mtDNA and allozyme data. Evolution 52: 87-99.

Winans, G. 1980. Geographic variation in the milkfish *Chanos chanos*. I. Biochemical evidence. Evolution 34: 558-574.

Woodland, D. J. 1983. Zoogeography of he Siganidae (Pisces): and interpretation of distribution and richness patterns. Bulletin of Marine Science 33: 713-717.

Yeatman, J. and Benzie, J. A. H. 1994. Genetic structure and distribution of *Photololigo* in Australia. Marine Biology 118: 79-87.

Zuckerkandl, E. and Pauling, L. 1962. *In* Molecular disease, evolution and genic heterozygozity. pp.189-225. Edited by M. Kasha and B. Pullman. Horizons in Biochemistry. Academic Press, New York.

Palynological evidence for Tertiary plant dispersals in the SE Asian region in relation to plate tectonics and climate

Robert J. Morley
PALYNOVA, 1 Mow Fen Road, Littleport, nr. Ely, Cambs CB6 1PY, UK

Key words: SE Asia, Tertiary, palynology, plants, dispersal, plate tectonics, climate

Abstract

Geological evidence for plant dispersals in SE Asia is reviewed by reference to both published, and previously unpublished, evidence from the time of first appearance of angiosperms until the Quaternary. It is concluded that angiosperms did not originate in the SE Asian region, but dispersed into the area from West Gondwanaland. Many African plant species dispersed into India as the Indian plate drifted past Madagascar in the Cenomanian/Turonian, and many of their descendants subsequently dispersed into SE Asia following the collision of the Indian plate with Asia in the middle Eocene. Prior to this time, the SE Asian flora appears to have developed in some degree of isolation. There is no palynological evidence for dispersals from the Australian plate in the Cretaceous, and minimal evidence for such dispersals in the Paleocene/Eocene.

The Sundanian Eocene flora stretched as far east as the South arm of Sulawesi, and subsequent to the opening of the Makassar Straits in the late Eocene, a part of this flora became stranded to the east of Wallace's Line, and probably formed a major source for other areas to the east of Wallace's Line of taxa of Sundanian and Asiatic affinity, as islands rose above sea level during the Miocene, negating the need for wholesale Miocene dispersal eastward. A small number of plant taxa have dispersed westward across Wallace's Line since the beginning of the Miocene, at 17, 14, 9.5, 3.5 and about 1 Ma. All of the taxa involved were well adapted to dispersal, and emphasise that Wallace's Line has been a substantial barrier to plant dispersal from the Oligocene onward.

Since the Eocene, plant dispersals to and from the Sunda region have largely been controlled by climate. The Oligocene and earliest Miocene were moisture deficient over much of the region, with ever-wet rain forest climates first becoming widespread at about 20 Ma in the early Miocene, subsequent to which they have repeatedly expanded and contracted. The greatest latitudinal expansion of tropical rain forests occurred at the beginning of the middle Miocene, at which time they extended northward as far as Japan. Fluctuations between wetter and drier climates became more pronounced in the Quaternary, with 'interglacial', high sea level periods coinciding with times of rain forest expansion, and 'glacial', low sea levels coincided with periods of more strongly seasonal climates, accompanied by the expansion of forests adapted to seasonal climates (such as pine forests) and savannah.

A major montane connection existed in South and East Asia through both the Tertiary and late Cretaceous, from the equator to 60°N, allowing Laurasian mountain plants to disperse to and from the equator throughout this period. The survival of representatives of many 'primitive' northern, angiosperm families in lower montane forests within the SE Asian region is thought to be due to the continuous presence of this unbroken mountain belt, rather than an origin in SE Asia. In contrast, the New Guinea mountains were formed only in the middle Miocene, from which time many Gondwanan taxa dispersed into this area from the south. Those well adapted to dispersal, such as *Podocarpus imbricatus* and *Phyllocladus* subsequently dispersed widely into SE Asia, whereas those poorly adapted to dispersal, such as *Nothofagus*, never reached beyond New Guinea.

Introduction

Plant geographers have long recognised that the SE Asian flora has become enriched through the dispersal of taxa from other continental regions. There is clear biogeographical evidence for the dispersal of mountain plants into the region along three trackways, from the Himalayan region, East Asia, and Australasia (Steenis, 1934a,b, 1936) and also for lowland plants, especially those requiring a strong dry season, both from Asia and Australasia. Proposals have also been made for massive dispersals from the Sunda region to the east of Wallace's Line, following the mid-Miocene collision between the Australian and Sunda plates, although the scale of such migrations is debatable. Plant distributions have also been used to suggest that there was a pre-mid Miocene contact between Australia and SE Asia, with the suggestion that this contact was pre-Tertiary (Steenis, 1962). Claims that the angiosperms actually evolved in the SE Asian region also continue to be made, despite an absence of fossil evidence (Takhtajan, 1987).

Biogeographical hypotheses such as these can only find confirmation when they are based on a foundation of historical geology. This discussion attempts to review geological evidence for plant dispersals by examining the fossil pollen and spore record for the Tertiary (and Late Cretaceous) of the SE Asian region, paying particular attention to the times of appearance in the region of pollen types exhibiting clear affinities with other continental regions. The review is based on both published and unpublished data from both outcrops and boreholes, and uses the plate tectonic reconstruction of Hall (1995) for SE Asia and Daly *et al.* (1987) for the Indian plate.

Geological evidence for the dispersal of lowland plants and those of the uplands, are discussed separately, although it must be appreciated that through most of the Tertiary, it is not always easy to determine from which of these sources each pollen type is derived.

In bringing to attention names of fossil pollen, a simple convention has been followed. In cases where a fossil pollen type has been adequately described according to the botanical code, the fossil name is used, generally followed, if appropriate, by an indication of the botanical affinity of the parent plant, if this is known. In cases where a pollen type remains inadequately described according to the code, but the parent plant is known, the name of the parent plant alone is used. In cases where such a pollen type might be derived from two or more plant taxa, the taxon name is also followed by the word 'type'.

Initial stages of angiosperm radiation

The current fossil record provides no evidence to suggest that angiosperms actually originated in SE Asia, as proposed by Takhtajan (1969). It is more likely that they migrated into the region. Truswell *et al.* (1987) dismissed the suggestion of Takhtajan (1987) that they originated on an isolated Gondwanan microcontinent, which subsequently became embedded in SE Asia, on the grounds that the earliest angiosperm pollen

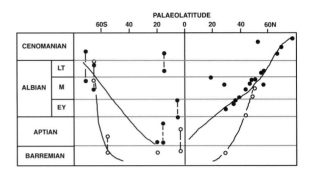

Fig.1. Poleward migration of angiosperms during the mid Cretaceous; latitude versus age for Barremian to Cenomanian monosulcate (O) and tricolp(or)ate (●) pollen records (from Hickey and Doyle, 1977).

records discovered so far for Australia post-date those from western Gondwana by 10 Ma, and also post-date the timing of separation of such microcontinents from Gondwanaland. The oldest record for Australian monosulcate pollen is from the latest Barremian or earliest Aptian (Burger, 1991), considerably later than its first appearance in the western hemisphere. Strong evidence to suggest that angiosperms originated at tropical palaeolatitudes is provided from global plots of earliest appearances against latitude (Fig.1) which indicate parallel, slow adaptation to cooler, or more seasonal, climates at higher latitudes in both hemispheres during the course of the mid-Cretaceous. The parallel diversification of both pollen and macrofossils in the mid-Cretaceous (Hickey and Doyle, 1977) suggests that this pattern reflects the radiation of the early angiosperm flora, and not simply the development of those angiosperm groups with recognisable pollen.

Fossil evidence of the representation of the earliest angiosperms in the SE Asian region is still very meagre. In the past, this was due to the lack of studies from the region. Recent palynological analyses of thick fluvial Lower Cretaceous sediments from Thailand, ranging in age from Neocomian to Aptian, have, however, failed to yield a single angiosperm pollen grain (Racey *et al.*, 1994), and probable Barremian sediments from the Malay peninsula have produced but a single tentative record of the chloranthaceous *Clavatipollenites* (Shamsuddin and Morley, 1994), among an assemblage dominated by pollen of *Classopollis* spp. and fern spores. Current evidence, therefore, suggests that angiosperms were much less well repre-

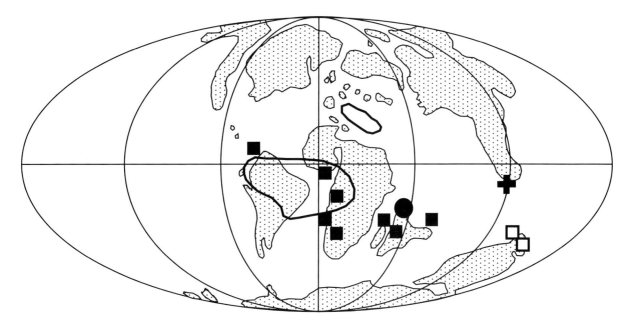

Fig.2. Mid-Cretaceous migration routes along the southern shore of Tethys. *Elateropollenites africaensis* (□) is recorded from the Turonian of Irian Jaya (Bates, unpublished) and Papua New Guinea (Lowe, pers. comm., 1987), and has a late Albian to Turonian centre of abundance in Africa and South America (outlined area); *Afropollis jardinus* exhibits an identical distribution pattern in the Cenomanian of Pakistan (●) to that seen in West Africa (IEDS, unpublished); *Constantinisporis, Victorisporis* and *Andreisporis* (■) are very well represented in the Turonian to Campanian of sub(palaeo)equatorial Africa, but appear in the Senonian of Madagascar (Chen 1982) and India (Venkatachala, 1974; Nandi, 1991; Morley, unpublished), also recorded rarely from the mid-Tertiary (earliest Miocene) of Java (+). Turonian palaeogeographic reconstruction by Smith *et al.* (1994).

sented during the time of their initial radiation in the eastern tropics, compared to the western.

Some evidence is now emerging to suggest the direction of plant dispersal routes into the eastern Tethyan region in the mid and later Cretaceous. Aptian shales from the Tarim basin of China contain very rare pollen of taxa characteristic of Africa and South America, such as *Afropollis zonatus* (which has been suggested by Doyle *et al.* (1990) to be derived from probable Winteraceae). In addition, low diversities of presumed ephedroid elater-bearing pollen in the Albian of Tibet (Herngreen and Duenas Jimenez, 1990), demonstrate mid-Cretaceous dispersal along the northern Tethyan coast.

Evidence is also emerging for dispersal along the southern shore of Tethys. Unpublished records by IEDS (1995) demonstrate that some typical African/South American palynomorphs, such as *Afropollis jardinus* show a similar temporal distribution pattern within the Cenomanian of Pakistan to that seen in West Africa, suggesting dispersal via Madagascar (Fig.2). Also, the presence of specimens of the Turonian elater-bearing pollen type *Elateroplicites africaensis* in Irian Jaya and Papua New Guinea can perhaps be explained by dispersal along the southern Tethyan shoreline, or along an island arc stretching from the India to Australia, along the leading edge of the Indian plate. Such a route would have closely paralleled the southern margin of the mid-Cretaceous equatorial low pressure zone (Barron and Washington, 1985), and warrants further consideration with respect to the dispersal of megathermal angiosperms into Australia during the mid-Cretaceous.

Late Cretaceous and early Tertiary dispersal paths

A somewhat later, very clear dispersal path eastward from Africa is shown by the distribution of the subequatorially triporate *Constantinisporis* group (Fig.2), thought by Srivastava (1977) to reflect ancestral Palmae pollen with affinity to *Sclerosperma* (although this affinity is considered unlikely, as their exine structure is different). The *Constantinisporis* group ranges from the Turonian to Campanian in West Africa, and shows an abundance maximum in Gabon during the Turonian.

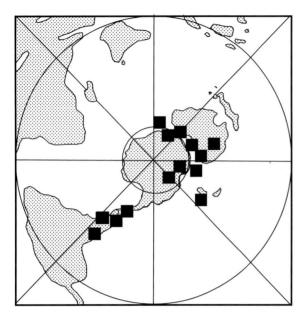

Fig.3. Late Cretaceous and early Tertiary records for pollen of *Nothofagus*; early Tertiary palaeogeographic reconstruction.

Constantinisporis has also been reported from the Senonian of Madagascar (Chen, 1978) and India (Venkatachala, 1974; Nandi, 1991; Morley, unpublished), and rarely from the mid-Tertiary of Java reflecting an important dispersal path, possibly associated with low Turonian sea levels (Haq *et al.*, 1988), which was followed, with a high degree of probability, by many groups of tropical angiosperms, such as members of the Sapindaceae (Ham, 1990) and Myrtaceae, as well as *Ctenolophon* (Ctenolophonaceae), and many Palmae, including the *Monocolpopollenites Palmaepollenites* complex, and the *Longapertites* group (probably ancestral to *Eugeissona*), even some *Normapolles* (Kar and Singh, 1986; Nandi, 1991) and mammals (Krause and Maas, 1990). This dispersal path probably became severed in the later Cretaceous, during which time both flora and fauna evolved in isolation.

An examination of the Late Cretaceous distribution of *Nothofagus* (Nothofagaceae) pollen also helps to clarify the nature and timing of south-

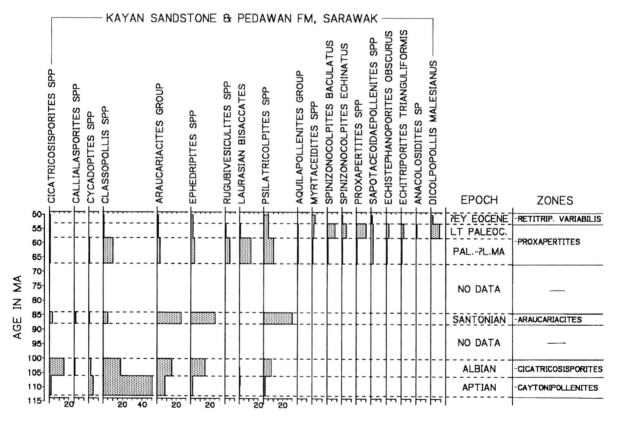

Fig.4. Proposed reinterpretation of the ages of the palynological zones of Muller (1968); Pal. = Paleocene, ?L.Ma = ?late Maastrichtian, Retitrip. variabilis = *Retitriporites variabilis*.

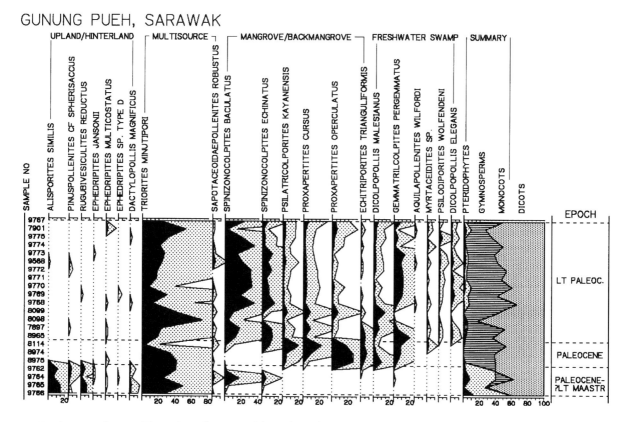

Fig.5. Palynomorph assemblages from the Kayan (Plateau) Sandstone Formation assemblages, constructed from raw data from Muller (1968). Pollen sum total miospores, selected taxa shown only.

ern hemisphere dispersal routes (Fig.3). The first *Nothofagus* pollen is recorded from the Santonian onward in Australia and Antarctica and from younger Cretaceous rocks of South America and New Zealand, but is absent from South Africa and India, demonstrating that the latter were well separated from Gondwanaland at the time of its initial radiation.

These data show that by the earliest Tertiary, as the Indian subcontinent was drifting close to Southern Asia and Sundaland, it bore a vegetation which contained three distinct elements: an ancient, eastern Gondwanan element, comprising gymnosperms, and perhaps some angiosperms; an African allochthonous element consisting predominantly of angiosperms of tropical west Gondwanan aspect; and, as a result of its isolation during the latest Cretaceous and earliest Tertiary, a distinct endemic element. Members of each of these groups were released into Asia following docking in the middle Eocene.

Turning now to SE Asia, the best database for the earliest Tertiary remains that of Muller (1968) from Sarawak. I have reconsidered the ages assigned to the Plateau (now Kayan) Sandstone and Pedawan Formation assemblages recorded by Muller and believe that these need to be revised (Fig.4). His *Cicatricosisporites* zone, from the Pedawan Formation, is more likely to be Albian, and the *Araucariacites* zone, from the upper part of the Pedawan Formation, is most likely of Santonian, or possibly Turonian, age. Assemblages from the *Araucariacites* zone contain abundant *Araucariacites*, common *Ephedripites* and rare spores, and suggest a dry, tropical climate. Angiosperms are common, but little diversified, consisting of poorly differentiated tricolpates (*Retitricolpites vulgaris*), tricolporates (*Psilatricolporites acuticostatus, P. prolatus*) and triporate pollen (*Triorites minutipori, T. festatus*). It has become customary to extend the range of *Myrtaceidites* to the *Araucariacites* zone, but only one specimen of this tiny pollen type was recorded by Muller, hardly the basis for such a range extension. None of the typical West African, Indian or Australian lower Senonian taxa was noted, suggesting that at this time, Sundaland was isolated from these regions.

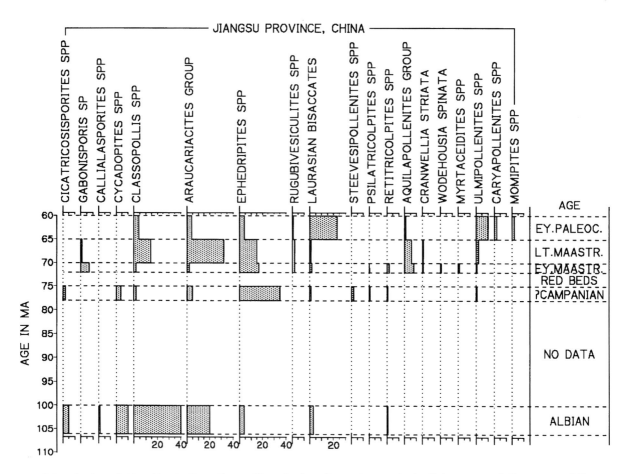

Fig.6. Distribution of *Rugubivesiculites, Classopollis* and other diagnostic taxa in the Paleogene and Cretaceous of Jiangsu Province, China (Song Zhichen et al., 1981).

Reconsideration of the assemblages from the Kayan Sandstone (Fig.5) suggests that there is no positive evidence for a Late Cretaceous age for the lower part of this formation. It is more likely that the formation is no older than Paleocene in age, although a late Maastrichtian age cannot be wholly ruled out. Muller's main reason for extending the age of this group into the Late Cretaceous was the common occurrence of the gymnosperm pollen type *Rugubivesiculites reductus* with regular *Classopollis* spp. in the basal section of the formation. *Rugubivesiculites reductus*, which at the time of his publication, was recorded only from the mid-Cretaceous of North America, has now been demonstrated by Song Zhichen *et al.* (1981) to comprise a common element in the Paleocene of China, where *Classopollis* is also common (Fig.6), opening up the possibility of a Paleocene age for the *Rugubivesiculites* zone of Muller. This conclusion is important, since it demonstrates that there is no firm evidence for the late Cretaceous ages proposed from this region for a number of critical pollen types, such as *Nypa, Proxapertites,* Sapotaceae, *Ilex* and Myrtaceae. The Sarawak Paleocene pollen flora is of somewhat lower diversity than contemporaneous floras in Africa, India (Frederiksen, 1994) and Australia (Harris, 1965) and lacks both Gondwanan and African elements.

A pollen record is also available for about the same time period from Irian Jaya (Fig.7). Assemblages from equivalents of the Lower Eocene Waripi Formation (which contains evaporites, consistent with accumulation on the northern coast of Australia, within the southern sub-tropical high pressure zone), are essentially of Australian aspect, with common *Casuarina* (Casuarinaceae) pollen, *Malvacipollis* (similar to *Austrobuxus* and *Dissiliaria*, Euphorbiaceae), Myrtaceae, some proteaceous pollen, and rare *Nothofagus*, together with a tropical shoreline element, provided by the presence of ancestral

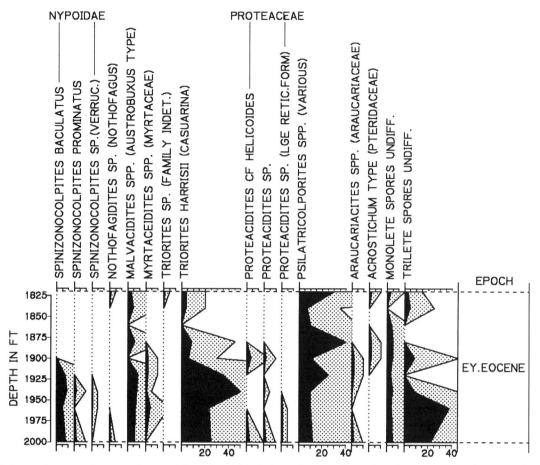

Fig. 7. Lower Eocene palynomorph assemblages from the Birds Neck area of Irian Jaya, from a stratigraphic equivalent of the Waripi Formation; pollen sum total miospores, selected taxa shown only.

Nypa pollen (*Spinizonocolpites baculatus* and *S. prominatus*), but, other than the widespread *Nypa* and Myrtaceae, with no affinities to the Kayan assemblages. Evidence therefore suggests that in the Paleocene and early Eocene, Sunda and Irian Jaya were widely separated.

SE Asian floras following India's collision with Asia

The Indian plate collided with the Asian plate in the middle Eocene (Daly *et al.*, 1987). At this time, both the northern margin of the Indian plate, and the Sunda region experienced an ever-wet, equatorial climate, as reflected by the common occurrence of coals in both areas. With neither oceanic, nor climatic barriers to dispersal, many plant taxa were able to disperse into the Sunda region (Fig.8). The palynological succession through the middle and upper Eocene of Sundaland can be observed in the very rich and well preserved assemblages recorded from the Nanggulan Formation in Central Java (Takahashi, 1982; Morley, 1982; Harley and Morley, 1995; Morley *et al.*, 1996) and Malawa Formation in South Sulawesi (see below), and an uppermost Eocene succession of coals from Mangkalihat peninsula, Kalimantan (Morley, unpublished). The 'oldest' middle Eocene sediments from Nanggulan contain much more diverse palynofloras than the Paleocene and ?lower Eocene assemblages from Sarawak, and there is good reason to believe that this increased diversity is due in considerable part to wholesale dispersal from the Indian plate, as indicated by the taxa listed in Table 1 and shown in Fig.9, all of which are recorded from Paleocene or older sediments from India. Obvious Gondwanan elements in this assemblage are aff. *Beaupreadites matsuokae* (Plate 2.4), *Palmaepollenites kutchensis* (Plate 3.1-3.2) and

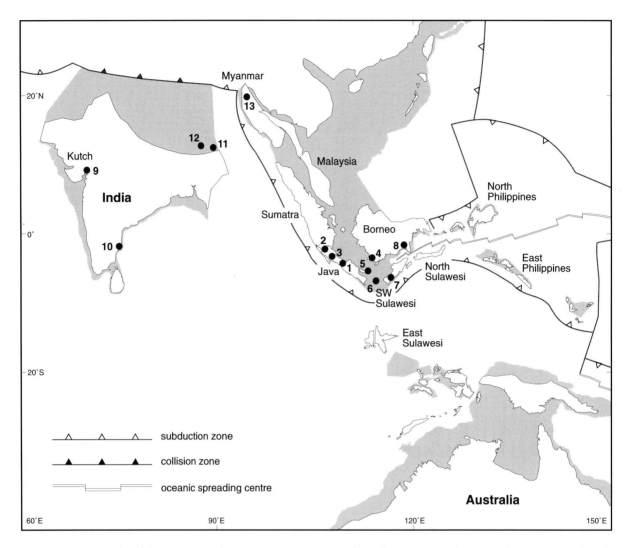

Fig.8. Palaeogeography of the SE Asian and Australian region at about 40 Ma (reconstruction based on Hall, 1998 this volume), with geographical distribution of *Palmaepollenites kutchensis* (numbered circles) from Harley and Morley (1995). Grey shaded areas are present Australian, Eurasian and Indian continental margins shallower than 200 m and inferred area of Greater India.

Polygalacidites clarus (Plate 2.6).

Close to the middle/upper Eocene boundary, two Gondwanan gymnosperm taxa, *Dacrydium* and *Podocarpus*, appear in the Sunda region for the first time. It might be expected that these also dispersed via the Indian plate. This may have been the case for *Podocarpus*, but the oldest records of *Dacrydium* post-date the separation of India from Gondwana by 20 Ma; its oldest record is from the Santonian/Coniacian of South Australia and the Antarctic peninsula (Dettmann and Thomson, 1987). Paleocene records of *Dacrydium* are known from the Ninetyeast Ridge (Kemp and Harris, 1975), which are interpreted as due to long distance dispersal, but records from India are very rare, and dubious (*e.g.*, Mathur (1984) from northern India). Perhaps its occurrence here was also a result of long distance dispersal, for Australia was positioned far to the south.

Two additional taxa with Indian connections appear within the latest Eocene and Oligocene. *Meyeripollis nayarkotensis* (Plate 3.3) of unknown affinity, appears in the latest Eocene of Assam (Baksi, 1962; Handique, 1992) where it first appears at about 37 Ma (Morley, unpublished). Its first appearance in East Kalimantan is at about the same time. Striate spores referred to *Magnastriatites grandiosus*, derived from the adiantaceous aquatic fern *Ceratopteris*, appear shortly after the earliest *Meyeripollis* specimens in the Sunda region, close to the Eocene-

Oligocene boundary, but in Assam just predate the oldest *Meyeripollis*, which led Kar (1982) to suggest that *Ceratopteris* evolved in India before dispersing across the tropics at the beginning of the Oligocene. The rare occurrence of the otherwise upper Cretaceous pollen type *Constantinisporis* cf. *jacquei* from the lowest Miocene of the Talang Akar Formation in the West Java Sea (Fig.2; Plate 3.9) indicates that the now extinct parent plant of this taxon followed the same dispersal path.

Note that several of the immigrant taxa, such as *Gonystylus* (*Cryptopolyporites cryptus*), *Ixonanthes* (Plate 1.3-1.4), *Eugeissona* (*Quilonipollenites* sp., Plate 2.1) and *Durio* (*Lakiapollis ovatus*, Plate 2.3) are considered 'typical' Malesian taxa today, and are rare, or absent from India, having been wiped out by Neogene and Quaternary climatic changes. A similar history would explain the present distribution of Dipterocarpaceae, with rafting from Africa, and subsequent range reduction in both Africa and India. Dipterocarpaceae also have two genera in Africa (*Monotes* and *Marquesia*, subfamily Monotoideae) as well as fossils of *Dipterocarpus* (Ashton, 1982), representatives in South America (*Pakaraimaea* and *Pseudomonotes*), Seychelles (*Vateriopsis*) and Sri Lanka, and a centre of diversity in Borneo (Ashton, 1969; Ashton and Gunatilleke, 1987).

Although today, *Eugeissona* is endemic to Borneo and Malaysia, its pollen shows some similarities to members of the fossil genus *Longapertites* (see Frederiksen, 1994), which is recorded widely in the uppermost Cretaceous and lower Tertiary of South America, West Africa and India, raising the possibility that *Eugeissona* may be derived from the parent taxon of this group, which is of very ancient origin, with a former pantropical distribution.

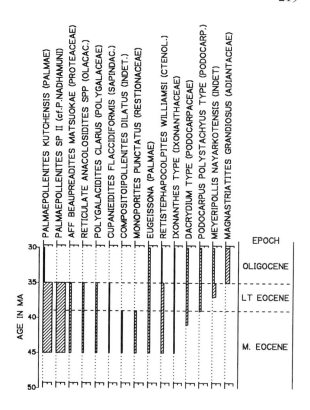

Fig.9. Ranges of taxa appearing in the Eocene of Java with either Gondwanan affinities, or recorded from the lower Eocene and Paleocene of India.

Table 1. Pollen taxa recorded from the pre-collision Tertiary of India (lower Eocene and Paleocene), but from the middle Eocene of the Sunda region.

Family/Tribe	Extant taxon	Fossil taxon
Proteaceae	aff. *Beauprea*	aff. *Beaupreadites matsuokae*
Palmae/Calamoidae	*Eugeissona*	*Quillonipollenites* spp.
Palmae/Iguanurinae	*Lepidorrachis*	*Palmaepollenites kutchensis*
Bombacaceae/Durioneae	*Durio* type	*Lakiapollis ovatus*
Linaceae	*Ixonanthes*	*Spiniulotriporites spinosus*
Gonystylaceae	*Gonystylus*	*Cryptopolyporites cryptus*
Ctenolophonaceae	*Ctenolophon parvifolius*	*Retistephanocolpites williamsi*
Ctenolophonaceae	*Ctenolophon*	*Polycolpites* spp.
Polygalaceae	*Polygala* or *Xanthophyllum*	*Polygalacidites clarus*
Sapindaceae	Indet.	*Cupanieidites flaccidiformis*
Alangiaceae	*Alangium* (sect *Conostigma*)	*Lanagiopollis* spp.
Indet	Indet.	*Compositoipollenites dilatus*
Adiantaceae	*Ceratopteris*	*Magnastriatites grandiosus*
Indet	Indet.	*Dandotiaspora laevigata*
Indet.	Indet.	*Constantinisporis* sp.

Mid and late Tertiary climates

The Oligocene and earliest Miocene were periods of much drier and cooler climates in the Sunda region, although obvious dry climate palynomorph indicators are few. Dry climates are inferred more from the limited representation of taxa characteristic of wet climates and the character of lithologies. The driest intervals are thought to have been in the early Oligocene and early Miocene, during which periods Gramineae pollen is also relatively common. Coniferous pollen is also a common component in many areas during this time, of which *Pinus* is most frequently represented, and although a background presence of montane conifer pollen clearly reflects the presence of upland areas, it is probable that maxima of *Pinus* reflect the widespread occurrence of seasonally dry vegetation (Ashton, 1972a) within lowland areas. Today *Pinus* is a common element of seasonally dry vegetation in Thailand, North Sumatra and Luzon (Whitmore, 1975), and has probably been similarly associated with such climates throughout the Tertiary.

There were few new immigrants into the Sunda region during the Oligocene, which is characterised by reduced diversities compared to the Eocene. Muller (1972) notes the earliest records of *Casuarina* at this time in Sarawak. Since *Casuarina* is absent from India, and there is no positive Indian pollen record (it first appears in the Australian fossil record well after the separation of the Indian plate from Gondwana), rafting on the Indian plate is unlikely. However, Hall (1995) suggests docking of Halmahera with New Guinea in the latest Oligocene, and the possibility of dispersal via Halmahera and the eastern Philippines should be given consideration. Such a dispersal path might also explain the recent discovery of *Dacrydium guillauminii* pollen within the uppermost Oligocene and lower Miocene of the West Java Sea; *Dacrydium* spp. producing the *D. guillauminii* pollen type occur today only in New Caledonia.

A major climatic change occurred in the Early Miocene (Morley and Flenley, 1987), at about 21 Ma, subsequent to which time coals were formed in many areas of the Sunda region, and taxa characteristic of peat-swamps became widespread, such as *Blumeodendron, Calophyllum, Cephalomappa, Durio* and *Stemonurus* (Fig.10).

The latest part of the early Miocene, and initial part of the Middle Miocene, coinciding with nannofossil zones NN4 and NN5, is well established as a period of globally high sea levels, based on $O^{16/18}$ data (*e.g.,* Miller *et al.,* 1987), and the maximum degree of Neogene coastal onlap based on sequence stratigraphic studies (Haq *et al.,* 1988). This interval was also a period of markedly warm and moist climatic conditions through a large part of SE and East Asia (Fig.10). During this short period, warm and moist paratropical conditions allowed the proliferation of diverse mangroves, and mixed warm temperate and paratropical forests, as far north as Japan, with *Dacrydium,* Sapotaceae, and *Alangium,* preserved within the Japanese Daijima Flora (Tsuda *et al.,* 1984; Yamanoi, 1974; Yamanoi *et al.,* 1980). The effect of this warming event was seen as far north as Korea, where pollen of warm temperate taxa, such as *Alangium,* are commonly recorded in the same time interval. Clearly, during this time of markedly equable climate, many species migrations may have occurred. A number of additional moist and warm episodes can be noted in East Asia, principally coinciding with periods of high sea level, but none was so pronounced as the warming phase at the early-middle Miocene boundary.

Whereas a few years ago evidence for the former expansion of seasonally dry climates in the younger Neogene and Quaternary of the Sunda region was purely conjectural, evidence for such climates is now becoming more widespread. Intermittent periods of drier climates,

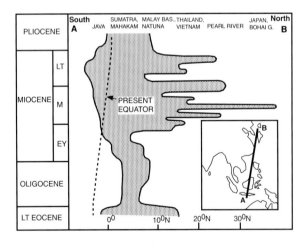

Fig.10. Schematic, and simplified distribution of tropical rain forest climates in SE and east Asia during the Tertiary. Note that Eocene climates were moist, whereas Oligocene ones were dry or seasonal. The greatest northward extension of rain forests occurred in the earliest middle Miocene. Position of palaeoequator according to Smith *et al.* (1994).

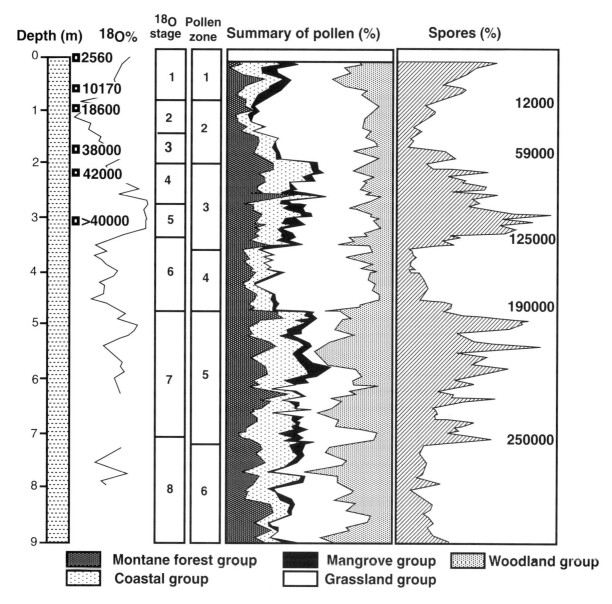

Fig.11. Summary pollen diagram from Lombok Ridge, modified from Kaas (1991). 'Glacial' climates, coinciding with oxygen isotope stages 2+3, 6 and 8 are characterised by abundant Gramineae pollen, but the low representation of pollen of coastal plants and mangroves, and fern spores, reflecting periods of widespread savannah vegetation. 'Interglacial' climates, coinciding with oxygen isotope stages 4+5 and 7, are characterised by abundant pteridophyte spores, and increased pollen of coastal plants and mangroves, but greatly reduced frequencies of Gramineae pollen, reflecting periods of forest and mangrove swamp expansion during periods of wetter climates. Dates marked (□) are radiocarbon dates, other dates refer to oxygen isotope stage boundaries.

reflected by maxima of Gramineae pollen, can be observed within the uppermost middle and upper Miocene, and the Plio-Pleistocene of the South China Sea, which sometimes contain pollen of the vicariad *Aegialitis* (Plumbaginaceae), a mangrove genus now confined to India and northern Australia, and pollen of Leguminosae. Seasonal climates in Java developed from the start of the Pliocene, indicated by the common occurrence of Gramineae (both pollen and charred cuticle) and *Casuarina* pollen since that time (Morley *et al.*, 1996). Dry 'glacial' climates within the Quaternary are clearly illustrated in pollen diagrams from cores from south of Nusa Tenggara (Fig.11) by Kaas (1991a). There is also evidence to suggest that *Pinus* savannah was widespread across the Malay peninsula during dry intervals of the Pleistocene. One sample,

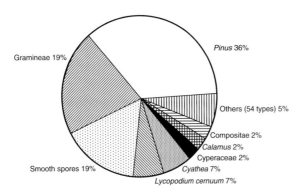

Fig.12. Percentage representation of the main palynomorph types from sample 3/15, from a probable mid-Pleistocene site close to Subang Airport, near Kuala Lumpur, Malay peninsula.

Table 2. Palynomorphs recovered from middle Pleistocene sample 3/15 from Subang, near Kuala Lumpur, Malaysia.

Pinaceae, *Pinus*	34%
Gramineae	18%
Smooth fern spores	18%
Lycopodiaceae, *Lycopodium cernuum*	7%
Cyatheaceae, *Cyathea*	7%
Cyperaceae	2%
Pandanaceae, *Pandanus*	1%
Aquifoliaceae, *Ilex*	2%
Palmae, *Calamus* type	2%
Compositae	1%
Pteridaceae, *Pteridium*	1%
Rubiaceae, *Nauclea* type	tr
Myrtaceae	tr
Sapotaceae, *Palaquium* type	tr
Blechnaceae, *Stenochlaena palustris*	tr
Hamamelidaceae, *Altingia excelsa*	tr
Guttiferae, *Calophyllum*	tr
Podocarpaceae, *Dacrydium*	tr
Dilleniaceae, *Dillenia*	tr
Dipterocarpaceae, *Shorea*	tr
+ 42 others	

collected from a probable mid-Pleistocene locality from near Subang Airport (Table 2, Fig.12), yielded abundant *Pinus* pollen (Plate 3.7-3.8) and Gramineae (Plate 3.6), together with Compositae (Plate 3.5) (Morley and Flenley, 1987). Steenis (1961) explained disjunct distributions of leguminous (papilionaceous) taxa, which require a marked dry season in Nusa Tenggara and Indochina in terms of the more extensive occurrence of drier climates during glacial periods. This hypothesis is fully borne out by the fossil record, but whether the migrations of these species occurred in the Quaternary, as suggested by Steenis, or at an earlier time, remains unanswered; Ashton (1972b) thought that Steenis' disjunct distributions predated the evolution of the 'dry dipterocarps', of the Asian mainland, since these are absent from Nusa Tenggara.

Miocene migrations

All plate tectonic reconstructions indicate that the main phase of collision between the Sunda and Australian plates was during the middle Miocene; the most clear instances of westward plant migration within the region also date from about this time.

Muller (1972) remarked on a distinct increase in abundance of Myrtaceae pollen in Sarawak within the lower part of the Miocene. He tentatively interpreted this to reflect dispersal of Myrtaceae from the east, but Martin (1982) subsequently remarked on this event, and suggested that it was due to deteriorating soil conditions. The event has now been independently calibrated in Sarawak, the Malay basin (Azmi *et al.*, 1996) and Mahakam delta, Kalimantan (Carter and Morley, 1996) to have occurred at about 17 Ma. Such a widespread, synchronous event is unlikely to relate to changing soil factors, which will differ from area to area, and a migration from the east appears to be the most probable explanation. A number of myrtaceous genera of Australasian origin, such as *Baeckia, Melaleuca, Leptospermum, Rhodamnia* and *Tristaniopsis,* occur in the Sunda region, predominantly in coastal settings and on poor soils, and the increase of Myrtaceous pollen at 17 Ma may reflect immigration of these taxa when Sundaland and Australia were closely juxtaposed.

Stephanoporate echinate pollen referable to the mangrove genus *Camptostemon* (Bombacaceae) shows sudden appearances in Sarawak (Muller, 1972) and East Kalimantan (Morley, unpublished) in the middle Miocene, at about 14 Ma, and a somewhat later appearance in the Malay basin, at 10, or 9.5 Ma (Azmi *et al.*, 1996). The sudden appearance of this pollen type coincides with times of pronounced sea level lowstand, and suggests that its parent plant immigrated from elsewhere, and, since the pollen type is not recorded to the west, derivation from the east, where the pollen record is incomplete, is most likely. *Camptostemon* is currently a common mangrove tree in New Guinea, and is now virtually extinct west of Wallace's Line,

Tertiary plant dispersals in SE Asia

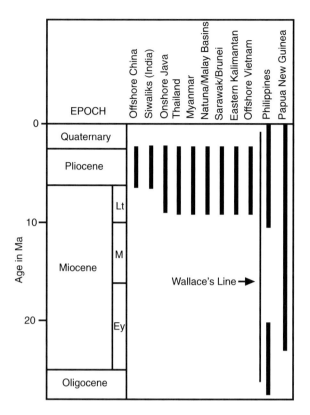

Fig.13. Stratigraphic range of *Stenochlaenidites papuanus* in SE Asia.

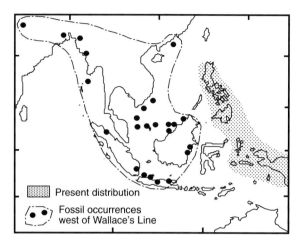

Fig.14. Modern distribution of *Stenochlaena milnei* and *S. cumingii* in the Philippines and Irian Jaya (shaded), and fossil distribution of *Stenochlaenidites papuanus* west of Wallace's Line (●, dashed line).

with rare occurrences in SE Kalimantan.

A further clear migration occurred in the earliest late Miocene, indicated by the widespread occurrence west of Wallace's Line of spores referred to *Stenochlaenidites papuanus* (Fig.13, 14) derived from a climbing fern allied to the East Malesian *Stenochlaena milnei* (synonym, *S. laurifolia*) or *S. cumingii*. This species rapidly dispersed across Borneo into Java, Sumatra, and into Indochina, China (Barre de Cruz, 1982) and India (Mathur, 1984), and then suddenly disappeared from the record from the western area during the earliest Pleistocene (Morley, 1978, 1991; Caratini and Tissot, 1985). Its arrival in the Sunda region has been independently dated at about 9 Ma, in Sarawak, the Malay basin (Azmi *et al.*, 1996) and Java (Rahardjo *et al.*, 1994). Within the Malay basin, the sedimentary succession demonstrates that this spore type first appears following the most pronounced sea level lowstand of the Neogene (Azmi *et al.*, 1996), and it is proposed that migration took place during this phase of low sea level. The spore type ranges from the middle or lower Miocene in Papua New Guinea, and the parent plant is common today in New Guinea, the Moluccas and the Philippines.

Truswell *et al.* (1987) suggested that a number of additional Australian taxa migrated into SE Asia in the Miocene. The pollen record in SE Asia, however, cannot provide support for any of their additional suggestions. This may be due to the fact that the pollen of some of their taxa are difficult to differentiate bearing in mind the high diversity of assemblages from SE Asia, but for those which have been recorded (but are as yet unpublished), *Gardenia* (Rubiaceae) and Loranthaceae are noted from the middle Eocene of Java, and thus predate Australian records, whereas pollen referable to Malvaceae, of the *Thespesia* type, shows a clear base at about 21 Ma in the lower Miocene offshore Vietnam, wholly consistent with the lower Miocene record from Australia (Morley, unpublished).

In summarising westward dispersals across Wallace's Line, it is noteworthy that all of the taxa considered are well adapted to dispersal (ferns, small-seeded Myrtaceae and mangroves), and emphasise that Wallace's Line remained a major barrier to plant dispersal throughout the Neogene.

Origin of floras to the east of Wallace's Line

The flora to the east of Wallace's Line as far east as Fiji is essentially 'Malesian', with many taxa exhibiting poor dispersal by the possession of large, heavy fruits, suggesting a continental ori-

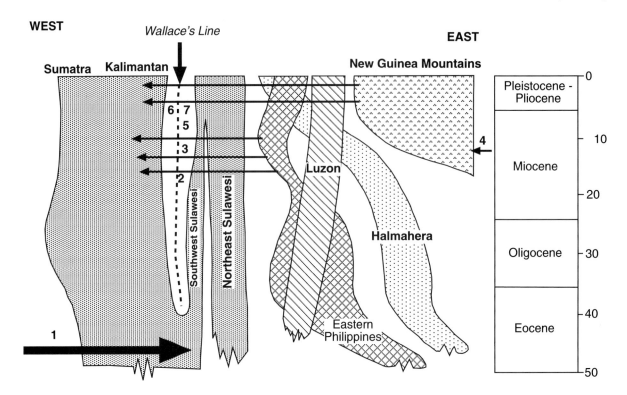

Fig.15. Proximity of some Sundanian and East Malesian land masses through time (data source, Hall, 1995); South Sulawesi and Sunda were joined in the middle Eocene, and shared the same flora; whereas most islands of Eastern Indonesia were not formed until the middle Miocene, Halmahera and the Philippines have a much older history, and would have borne a flora of tropical aspect throughout the Tertiary. The islands of the proto-Philippines, and Halmahera were in close proximity at a number of times during the Tertiary, especially in the late Eocene/Oligocene, and late Miocene, and the possibility of dispersal between these islands and Sunda at these times is strong. ———> main migrations; 1) dispersals from Indian plate, c.45 Ma; 2) Myrtaceae, c.17 Ma; 3) *Camptostemon*, c.14 Ma; 4) *Nothofagus*, middle Miocene; 5) *Stenochlaena milnei/cumingii*, c.9.5 Ma; 6) *Podocarpus (Dacrycarpus) imbricatus*, 3.5 Ma; 7) *Phyllocladus*, c.1 Ma.

gin (Whitmore, 1973), and in trying to reconcile this with the theories of continental drift (Wegener, 1929) and plate tectonics, Diels (1934) and Steenis (1979) predicted massive migrations of Sundanian taxa eastward across Wallace's Line at the time of collision of the Australasian and Sunda plates during the middle Miocene. Recently, Dransfield (1981) has reviewed the biogeography of palms with respect to this line, and concluded that only a few palm genera made the crossing, and that numbers were of the same order in each direction. He probably over estimated the number of dispersals, since it is likely that *Calamus* was already present to the east prior to the collision, based on the occurrence of *Dicolpopollis* spp. of the *Calamus* type from the Eocene of Sulawesi (Plate 3.4) and the Australian Eocene (Truswell et al., 1987) and abundant occurrences in the middle Miocene of Mindoro, in the Philippines (Morley, unpublished). Also, *Oncosperma* appears to be a relict genus, with a number of Neogene records east of its present area of distribution. Similarly, Ham (1990) concludes that the only genus of the tribe Nephelieae (Sapindaceae) to cross Wallacea from the west is *Pometia*, other genera having dispersed into Wallacea either from the Sunda region or from the Australian plate.

Truswell et al. (1987) could not find any evidence for such a major post-collision invasion from the Sunda region from examining the Australian pollen record. However, the sudden and widespread latest Oligocene (Truswell et al., 1985; Pocknall, 1982) or earliest Miocene (Stover and Partridge, 1973, Martin, 1978), appearance of *Acacia* pollen (*Acaciapollenites myriosporites*), and the early Miocene appearances of pollen of *Merremia* (*Perfotricolpites digitatus*) *Caesalpinia* type (*Margocolporites vanwijhei*) and spores of *Stenochlaena palustris* (*Verrucatosporites usmensis*) in Australia

(Truswell et al., 1985, Hekel, 1972) may reflect such dispersals from the Asian plate, but may not necessarily reflect dispersals across Wallace's Line.

Consideration of the geological history of islands east of Wallace's Line may help explain the limited evidence for Neogene crossings, but the 'Malesian' aspect of the eastern flora (Fig.15). Firstly, whereas many of the islands of Eastern Indonesia may be very young, such as Seram, Irian Jaya and eastern Sulawesi, the island chains of the Philippines, and Halmahera probably included emergent areas through much of the Tertiary, and these islands are likely to have borne a tropical aspect, rather than Australian flora. Secondly, although South Sulawesi is east of Wallace's Line, its geological affinity is with the Sunda plate; it became separated from Kalimantan in the late Eocene following the opening of the Makassar Straits (Situmorang, 1982; Hall, 1995), but at the time of separation, already bore a flora with affinities to the Eocene flora of India, Java and SE Kalimantan. Middle and upper Eocene paralic sediments from the Malawa Formation of South Sulawesi contain abundant pollen with affinities to the Sunda region, including aff. *Beaupreadites matsuokae* (Proteaceae, Plate 2.4), *Cupanieidites flaccidiformis* (Plate 2.4) of sapindaceous affinity, *Dicolpopollis* spp., from Palmae, Calamoideae (Plate 3.4), *Ixonanthes* pollen (Plate 1.3-1.4), *Lakiapollis ovatus*, produced by *Durio* (Plate 2.3), *Lanagiopollis*, from *Alangium* (Plate 1.1-1.2) *Palmaepollenites kutchensis*, from the palm subtribe Iguanurinae (Plate 3.1), *Quillonipollenites* spp., indicating *Eugeissona* (Plate 2.1) and *Retistephanocolpites williamsi* (from *Ctenolophon*, Plate 1.5-1.6), together with the taxa of indeterminate affinity *Meyeripollis nayarkotensis* (Plate 3.3) and *Compositoipollenites dilatus* (Plate 2.2).

Reconstructions by Hall (1995) for 45 to 40 Ma suggest that at these times, dispersal paths were possibly present which may have allowed elements of this flora to migrate eastward to the eastern Philippines and possibly beyond. Thirdly, recent studies by the Indonesian Riset Unggulan Terpadu (RUT) demonstrate that there is a virtually continuous pollen record through the Oligocene and Miocene in this area (N. Polhaupessy and S. Sugeng, pers. comm., 1997), suggesting that the Sundanian Paleogene flora may have persisted in South Sulawesi until the time of the Miocene collision.

The pre-collision flora to the east of Wallace's Line, with all probability, originated from the Paleogene flora of South Sulawesi, and the island arcs of the Philippines. As the islands of the Banda arc and New Guinea formed during the Middle Miocene, they probably became vegetated by floras comprising elements of both the South Sulawesi and Philippine floras, mixed with Australian rain forest elements. Such an origin is fully compatible with results of a biogeographical appraisal of the flora of Sulawesi, which suggests generic affinities firstly with New Guinea, and secondly with the Philippines and Moluccas (Balgooy, 1987).

Dispersal of mountain plants

The pollen record reveals the presence of two broad dispersal paths which have allowed mountain plants to disperse into the Sunda region during the Tertiary. The first is a long established route, along which both Laurasian conifers and north temperate broad-leaved trees have dispersed, whereas the second, much later path from Australasia, is characterised by the migration into the area of southern hemisphere podocarps. Both of these dispersal paths were first noted by Muller (1966).

Typical Laurasian elements include pollen of *Pinus, Picea, Abies, Rugubivesiculites, Tsuga canadensis* type, *Tsuga diversifolia* type, *Taxodium* type, and pollen of the temperate trees, *Alnus, Betula, Carya, Juglans* and *Pterocarya*, whereas the southern podocarps comprise *Podocarpus (Dacrycarpus) imbricatus* type and *Phyllocladus hypophyllus* type pollen (Fig.16). The typical abundances of pollen from each of these sources differ in that the Laurasian elements may be extremely abundant, and in some settings may actually dominate palynomorph assemblages, whereas the southern element is invariably a minor, background component of the assemblages.

Dispersals from Asia

The Laurasian connection of Sunda with East Asia is very ancient. The presence of common *Rugubivesiculites reductus* and *Pinus* type pollen (*Alisporites similis* and *Pinuspollenites* cf. *spherisaccus*) in the probable Paleocene of Sarawak shows that the connection was already well established at the beginning of the Tertiary (Fig.5). There is little evidence for the representation of montane elements in the area during the Eocene, since the pollen record for this period is poor, but Laurasian montane elements,

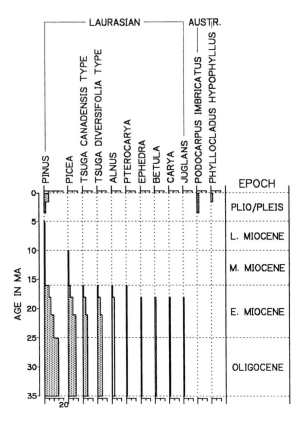

Fig.16. Stratigraphic distribution of temperate taxa in the south China Sea region. The diagram brings together records from NW Borneo to offshore Vietnam. The taxa recorded need not all occur in the same location. Horizontal axis percentage total miospores.

consisting of *Pinus, Picea, Abies, Tsuga* and temperate angiosperms are abundant in the Oligocene and earliest Miocene, after which time they exhibit a gradual decline, with marked reductions in representation at about 21 Ma and 18 Ma in the early Miocene, although they occasionally return in small pulses, especially in the Plio-Pleistocene.

The Laurasian element is not abundant throughout the Sunda region, but is characteristically present in sediments within the Malay, Penyu, Vietnamese and Natuna basins, offshore Sarawak, the Gulf of Thailand and to a lesser extent, central Sumatra (Fig.17). It occurs as a background element in the lower Miocene of Mahakam Delta, but occurs only sporadically from the Java Sea basins. *Pinus* is the most abundant and widespread pollen type, followed by *Picea, Alnus* and *Tsuga*. The remaining types become progressively more common toward the Gulf of Thailand, and offshore Vietnam. Pollen of temperate taxa, including *Alnus*, is extremely common in the early Miocene and Oligocene of the intermontane basins of Thailand (Watanasak, 1988, 1990) where it occurs in association with abundant *Quercus* pollen, suggesting a vegetation similar to some modern temperate forests.

The approximate extent of the upland topography through the late Cretaceous and Tertiary of the Sunda region which bore this upland flora can be obtained through the comparison of the relative representation of elevated and lowland terrain proposed on the global palaeogeographic maps of Smith *et al.* (1994). The widespread representation of elevated terrain was a major feature throughout the late Cretaceous and Early Tertiary, but became reduced in the Neogene, complying closely with (but probably based partly on) the distribution of the Laurasian gymnosperm and temperate pollen (Fig.18).

The presence of a major elevated area, joining Sunda with East Asia, as indicated by Smith *et al.* (1994) throughout the period of radiation of the angiosperms, and also through the period of global climatic deterioration since the end of the Eocene, is very important from the point of view of survival of certain primitive angiosperms in SE Asia. One of the reasons for Takhtajan (1969) suggesting that SE Asia was the birthplace of angiosperms was the occurrence of many more archaic angiosperms in the sub-montane forests of this region compared to South America or Africa. The presence of a continuous elevated area in SE Asia, stretching from the equator to over 60°N, throughout this period, is certainly responsible in a large part for the retention of archaic angiosperms in the Far Eastern tropics. Such upland latitudinal connections in the South American and African tropics are either absent, or have been intermittent.

Steenis (1936) defined two migration tracks from Asia, the Sumatra track, characterised mainly by herbaceous taxa with connections in the Himalayas, and the Luzon track, with connections with East Asia. Migrations discussed here do not appear to relate to either of these directly, but suggest that the Malay peninsula provided the main dispersal route from Asia during the Tertiary. The Sumatran track probably became established following the uplift of the Barisan mountains in the mid Miocene, whereas the Luzon track may relate more to the southerly drift of Palawan (although there is no palaeobotanical evidence available to test this suggestion), or to dispersal via Taiwan during

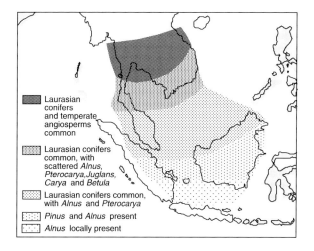

Fig.17. Areal distribution of Laurasian montane elements in SE Asia during the early Miocene.

phases of low Pleistocene sea levels, since these areas became juxtaposed only in the latest Tertiary (Hall, 1995).

Dispersals from Australasia

Podocarpus sect *Dacrycarpus* (now elevated to generic level by Laubenfels, 1988), which produces trisaccate pollen, is more or less restricted to moist, upland regions of SE Asia. It has a good pollen record in the Sunda region, which extends only over the last 3.5 Ma, since the latest phase of the Early Pliocene (Fig.16). *P. imbricatus* is presently widespread in SE Asia, occurring throughout Sundaland, and into Indochina, and is the classic example of westward migration across Wallace's Line. New data from Irian Jaya, and the southern Philippines, show that Wallace's Line itself was no barrier to this species; it appears to have arrived in the Bird's Head area of Irian Jaya only in the earliest Pliocene. Kaas (1991b) subsequently dates its earliest appearance in the Southern Philippines in the mid Pliocene, attributing its arrival there to the establishment of suitable habitats as a result of local tectonic uplift. This pollen type is particularly common in the uppermost lower Pliocene of Java (Rahardjo *et al.*, 1994), and is recorded at identical times from offshore Sarawak/Sabah. It appears to have taken longer to reach the Malay peninsula, for its pollen is absent from the Pliocene of the Malay basin, but is present within Malay peninsula Quaternary sediments.

The pattern of migration displayed by *Podocarpus imbricatus* is followed, although to a lesser degree, by that of *Phyllocladus hypophyllus* (Fig.16), which dispersed only as far as Borneo, where it occurs today, and where its pollen can be seen only in sediments of Quaternary age (Muller, 1966, 1972; Morley, 1978; Caratini and Tissot, 1985). It is also recorded in the southern Philippines at about the same time (Kaas, 1991b).

In Papua New Guinea and Irian Jaya, montane podocarp pollen is accompanied by that of *Nothofagus* (Khan, 1976), whose arrival in this area also appears to have been strongly controlled by the uplift of the New Guinea mountains. It first appears in the Birds Head of Irian Jaya in the late Miocene, but was never able to disperse further to the west, presumably because of its inability to disperse across water barriers. As previously noted by Muller (1972), the fossil record for southern podocarps reflects the New Guinea track of Steenis (1936).

Anomalies

Consideration of the palynological record in relation to plate tectonics leaves remarkably few unexplained anomalies. The main anomalous distributions were previously in the Oligocene, but the late Oligocene collision of Hall (1995) may explain these. Only the record of *Dacrydium* from the upper Eocene of Java cannot be readily explained from geological evidence for past continental positions. The Paleocene occurrence of *Dacrydium* pollen on the Ninetyeast Ridge was explained by Kemp and Harris (1975), as due to long distance dispersal, and a similar explanation may apply here. The major 'anomaly' seen in the pollen flora is its uniformity over long periods of time (cf. Martin, 1982, Truswell *et al.*, 1987). It is remarkable to realise that, despite such tectonic upheavals as witnessed in SE Asia, the pollen flora of the region reveals so few changes, and regional migrations appear to be so few. Perhaps we are not looking carefully enough to see the real picture?

Conclusions

The current palynological record provides no evidence to suggest that the SE Asian region was an area of initial radiation of angiosperms. Angiosperm diversities were probably lower than in other tropical areas until the middle

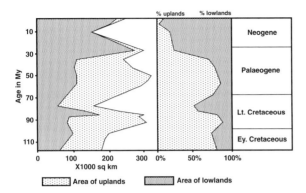

Fig. 18. Representation of upland and lowland topography through time in SE Asia, source of data: palaeogeographic maps of Smith *et al.* (1994).

Eocene, when floras were enriched by widespread migrations from the Indian plate. Since Gondwanan elements were brought to SE Asia via the Indian plate, there is less need invoke an earlier, Cretaceous, connection between Sunda and Gondwana in order to explain present day plant distributions.

The Sundanian Eocene flora stretched as far east as the south arm of Sulawesi. Subsequent to the opening of the Makassar Straits in the late Eocene, a part of the Sundanian flora became stranded in South Sulawesi to the east of Wallace's Line. This area probably provided a major source of taxa of Sundanian, and Asian affinity, to areas to the east of Wallace's Line, throughout the remainder of the Tertiary, negating the need for wholesale eastward dispersal across the Makassar Straits during the Miocene.

A small number of taxa have migrated westward across Wallace's Line since the main phase of collision of the Australian and Sunda plates during the Miocene. Such events occurred at about 17 Ma, with the immigration of Myrtaceae, *Camptostemon* at 14 Ma, and at about 9.5 Ma, at which time a fern, ancestral to *Stenochlaena milnei*, dispersed from the East, and spread widely across the region. All of the above are well adapted to dispersal, and emphasise that Wallace's Line remained a substantial barrier to dispersal throughout the Oligocene and Neogene. *Acacia* was among the taxa which probably dispersed from the Sunda plate (though not necessarily across Wallace's Line) to Australia in the latest Oligocene or earliest Miocene.

The pre-collision East Malesian flora probably developed on island arcs of the Palaeo-Philippines, and from the Paleogene Sundanian flora which became stranded in South Sulawesi after the late Eocene opening of the Makassar Straits. It is speculated that the New Guinea flora formed as a result of the intermixing of these floras with the Australian flora in the Miocene.

SE Asian rain forest floras successively migrated northward during periods of Miocene high sea level and moist, warm climates. Their greatest northward migration was at the early/middle Miocene boundary, when several mangroves, and other tropical taxa, dispersed northward as far as Central Japan. Some migrations from Australia probably took place in the Oligocene. These include *Casuarina*, and *Dacrydium* spp. producing the *D. guillauminii* pollen type. There is a significant number of pollen types from the mid-Tertiary of Java which compare closely with pollen of endemic, or characteristic extant taxa from New Caledonia. These include *Beaupreadites matsuokae* (*Beauprea*, Proteaceae), *Palmaepollenites kutchensis* (*Burretiokentia* and *Basselinia*; Palmae) and the *Dacrydium guillauminii* type.

A major montane connection existed throughout the Tertiary, and probably late Cretaceous, in SE Asia, from the equator, to 60°N. This allowed Laurasian mountain plants to disperse freely to and from the equator throughout this period. This mountain range became of much reduced extent during the later Tertiary, especially after about 20 Ma. The proliferation of primitive 'northern' angiosperm families within lower montane forests of SE Asia, such as Magnoliaceae, is thought to be due largely to the continuous presence of this mountain belt, which probably provided an unbroken succession of niches for such groups irrespective of climatic changes at lower altitudes. Both Magnoliaceae and Fagaceae would fall into this category. In contrast, the Neogene convergence of the Sunda and Australian plates, and the development of substantial upland areas in New Guinea, has provided niches for related southern families (Nothofagaceae) or archaic previously cosmopolitan families (*e.g.*, Winteraceae) which, during the later Tertiary, found refuge in the southern hemisphere. In this manner, some bihemispheric family pairs have been brought close together in SE Asia, without requiring an origin in this area, thus knocking a few more nails in the coffin of the theory that angiosperms evolved in some isolated area 'somewhere between Assam and Fiji'.

The dispersal of Australian mountain plants, such as *Podocarpus imbricatus* and *Phyllocladus hypophyllus*, became possible as a result of the development of suitable upland niches in both East and West Malesia, rather than the formation of continuous terrestrial connections. The SE Asian region is a vast area, and still remains very poorly studied using palynology. One safe prediction is that further detailed studies will undoubtedly bring to attention many new records, which will certainly modify, and improve, the sketchy picture of migration and dispersal outlined above.

Acknowledgements

The author is grateful to Eko Budi Lelono and Lucila Nugrahaningsih of the Indonesian Riset Unggalan Terpadu, or RUT (an Integrated Research Team from ITB, LEMIGAS, PERTAMINA, PPPG, PPGL and UGM) formalising biostratigraphic zonations for the different provinces of Indonesia) for discussions about Eastern Indonesian palynology, and Petronas Research and Scientific Services (PRSS) for the opportunity to undertake a detailed appraisal of Malay basin palynostratigraphy. Moyra Wilson kindly provided the samples from the Malawa Formation from South Sulawesi. Tim Whitmore kindly provided many useful comments on the first draft of the manuscript, and Mary Dettmann helped in naming proteaceous pollen and some taxa of Australian affinity.

References

Ashton, P. S. 1969. Speciation among tropical forest trees: some deductions in the light of recent evidence. Biological Journal of the Linnean Society 1: 155-196.

Ashton, P. S. 1972a. Discussion in Muller, J. Palynological evidence for change in geomorphology, climate and vegetation in the Mio-Pliocene of Malesia. In The Quaternary Era in Malesia pp. 1-34. Edited by P. S. Ashton and M. Ashton. Geography Department, University of Hull, Miscellaneous Series 13.

Ashton, P. S. 1972b. Discussion in Medway, Lord. The Quaternary mammals of Malesia: a review. In The Quaternary Era in Malesia pp. 63-98. Edited by P. S. Ashton and M. Ashton. Geography Department, University of Hull, Miscellaneous Series 13.

Ashton, P. S. 1982. Dipterocarpaceae. Flora Malesiana 9 (2): 283-285.

Ashton, P. S. and Gunatilleke, C. V. S. 1987. New light on the plant geography of Ceylon 1. Historical plant geography. Journal of Biogeography 14: 249-285.

Azmi, M. Y., Awalludin, H., Bahari, M. N. and Morley, R. J. 1996. Integrated biostratigraphic zonation for the Malay Basin. Bulletin Geological Society of Malaysia, in press.

Baksi, S. K. 1962. Palynological investigation of Simsang River Tertiaries, South Shillong Front, Assam. Bulletin of Geology, Mining, Metallurgy Society of India 26: 1-21.

Barre de Cruz, C. 1982. Étude palynologique du Tertiare du Sud Ouest asiatique (Kalimantan, delta de la Mahakam, Mer de Chine, Permis du Beibu). Ph.D. thesis, Université de Bordeaux, 2 vols, 161 pp. and 61 pp.

Barron, E. J. and Washington, W. M. 1985. Cretaceous climate: a comparison of atmospheric simulations with the geologic record. Palaeogeography, Palaeoclimatology, Palaeoecology 40: 103-133.

Balgooy, M. M. J. van. 1987. A plant geographical analysis of Sulawesi. In Biogeographical Evolution of the Malay Archipelago. pp. 94-102. Edited by T. C. Whitmore. Oxford Monographs on Biogeography 4, Oxford Scientific Publications.

Burger, D. 1991. Early Cretaceous angiosperms from Queensland, Australia. Reviews of Palaeobotany and Palynology 65: 153-163.

Carter, I. S. and Morley, R. J. 1996. Utilising outcrop and palaeontological data to determine detailed sequence stratigraphy of the Early Miocene sediments of the Kutai Basin, East Kalimantan. International Symposium on Sequence Stratigraphy in Southeast Asia Proceedings. IPA Jakarta, May 1995.

Caratini, C. and Tissot, C. 1985. Le sondage misedor, études palynologique. Etude Géographie Tropiques CNRS 3: 49 pp.

Chen, Y. Y. 1978. Jurassic and Cretaceous palynostratigraphy of a Madagascar well. Ph.D. Thesis, University of Arizona, 264 pp.

Daly, M. C., Hooper, B. G. D. and Smith, D. G., 1987. Tertiary plate tectonics and basin evolution in Indonesia. Proceedings of the Indonesian Petroleum Association, 16th Annual Convention 1: 399-428.

Dettmann, M. E. and Thomson, M. R. A. 1987. Cretaceous palynomorphs from the James Ross Island area, Antarctic – a pilot study. British Antarctic Survey Bulletin 77: 13-59.

Diels, L. 1934. Die Flora Australiens und Wegener's Verschiebungstheorie. Sitzungsberichte der Preussischen Akademie der Wissenschaften zu Berlin. 33: 533-545.

Doyle, J. A., Hotton, C. A. and Ward, J. A. 1990. Early Cretaceous tetrads, zonasulcate pollen and Winteraceae. I. Taxonomy, Morphology and ultrastructure. American Journal of Botany, 77: 1544-1567.

Dransfield, J. 1981. Palms and Wallace's Line. In Wallace's Line and Plate Tectonics pp. 43-56. Edited by T. C. Whitmore. Oxford Monographs on Biogeography 1, Oxford Scientific Publications.

Frederiksen, N. 1994. Middle and Late Paleocene angiosperm pollen from Pakistan. Palynology 18: 91-197.

Good, R. 1962. On the geographical relationships of the angiosperm flora of New Guinea. Bulletin British Museum (Natural History) Department of Botany 12: 205-226.

Hall, R. 1995. Plate tectonic reconstructions of the Indonesian region. Proceedings Indonesian Petroleum Association 24th Annual Convention: 71-84.

Ham, R.W. J. M., van der. 1990. Nephelieae pollen (Sapindaceae): form function and evolution. Rijksherbarium/Hortus Botanicus Leiden 255 pp.

Handique, G. K. 1992. Stratigraphy, depositional environment and hydrocarbon potential of Upper Assam Basin, India. Symposium on Biostratigraphy of Mainland Southeast Asia: Facies and Palaeontology, Chiang Mai, Thailand. 1: 151-169.

Haq, B. U., Hardenbol, J. and Vail, P. R. 1988. Mesozoic and Cenozoic chronostratigraphy and cycles of sea level change. In Sea Level Changes: an Integrated Approach. Edited by Wilgus et al., Society of Economic Palaeontologists and Mineralogists Special Publication 42: 71-108.

Harley, M. M. and Morley, R. J., 1995. Ultrastructural studies of some fossil and extant palm pollen, and the reconstruction of the biogeographical history of the subtribes Iguanurinae and Calaminae. Reviews of Palaeobotany and Palynology 85: 153-182.

Harris, W. K. 1965. Basal Tertiary microfloras from the Princetown area, Victoria, Australia. Palaeontographica Abt. B. 113, Leifg 4-5: 75-106.

Hekel, H. 1972. Pollen and spore assemblages from Queensland Tertiary sediments. Geological Survey of Queensland, Palaeontological paper 30: 34 pp.

Herngreen, G. F. W. and Duenas Jimenez, H. 1990. Dating of the Cretaceous Une Formation, Colombia and the relationship with the Albian-Cenomanian African-South American microfloral province. Reviews of Palaeobotany and Palynology 66: 345-359.

Hickey, L. J. and Doyle, J. A. 1977. Early Cretaceous fossil evidence for angiosperm evolution. The Botanical Review 43: 3-104.

IEDS (Integrated Exploration and Development Services Ltd.). 1995. Poster display on hydrocarbon potential of Pakistan. International Symposium on Sequence Stratigraphy in Southeast Asia, Jakarta, May 1995.

Kaas, W. A. van der. 1991a. Palynology of eastern Indonesian marine piston-cores: A Late Quaternary vegetational and climatic history for Australasia. Palaeogeography, Palaeoclimatology, Palaeoecology 85: 239-302.

Kaas, W. A. van der. 1991b. Palynological aspects of site 767 in the Celebes Sea. Proceedings of the Ocean Drilling Program, Scientific Results, 124: 369-374.

Kar, R. K. 1982. On the original homeland of *Ceratopteris* Brong and its palaeogeographical province. Geophytology 12 (2): 340-341.

Kar, R. K. and Singh, R. S. 1986. Palynology of the Cretaceous sediments of Meghalaya, India. Palaeontographica B. 202: 83-153.

Kemp, E. M. and Harris, W. K. 1975. The vegetation of Tertiary islands on the Ninetyeast Ridge. Nature 258: 303-307.

Khan, A. H. 1976. Palynology of Neogene sediments from Papua (New Guinea) stratigraphic boundaries. Pollen et Spores 16: 265-284.

Krause, D. W. and Maas, M. C. 1990. The biogeographic origins of late Paleocene-early Eocene mammalian immigrants to the Western Interior of North America. Geological Society of America Special Paper 243: 71-105.

Laubenfels, D. J. de 1988. Coniferales. Flora Malesiana. Series 1 — spermatophyta (flowering plants). Sijthoff-Noordhoff, Alphen a/d Rijn 10(3): 337-453.

Martin H. A. 1978. Evolution of the Australian flora and vegetation through the Tertiary: evidence from pollen. Alcheringia 2: 181-202.

Martin, H. A. 1982. Changing Cenozoic barriers and the Australian palaeobotanical record. Annals of the Missouri Botanical Garden 69: 625-67.

Mathur, Y. K. 1984. Cenozoic palynofossils, vegetation, ecology and climate of the north and northwestern sub-Himalayan region, India. *In* The evolution of the East Asian environment. Volume II. pp. 433-551. Edited by R. O. Whyte. Centre of Asian Studies, University of Hong Kong.

Miller, K. G., Fairbanks, R. G. and Mountain, G. S. 1987. Tertiary oxygen isotope synthesis, sea level history and continental margin erosion. Paleoceanography 2: 1-19.

Morley, R. J. 1978. Palynology of Tertiary and Quaternary sediments in southeast Asia. Proceedings Indonesian Petroleum Association 6th Annual Convention: 255-276.

Morley, R. J. 1982. Development and vegetation dynamics of a lowland ombrogenous swamp in Kalimantan Tengah, Indonesia. Journal of Biogeography 8: 383-404.

Morley, R. J. 1991. Tertiary stratigraphic palynology in southeast Asia: current status and new directions. Geological Society of Malaysia Bulletin 28: 1-36.

Morley, R. J. and Flenley, J. R. 1987. Late Cainozoic vegetational and environmental changes in the Malay Archipelago. *In* Biogeographical Evolution of the Malay Archipelago. pp. 50-59. Edited by T. C. Whitmore. Oxford Monographs on Biogeography 4, Oxford Scientific Publications.

Morley, R. J., Lelono, E. B., Nugrahaningsih L. and Nur Hasjim. 1996. LEMIGAS Tertiary palynology project: aims, progress and preliminary results from the Middle Eocene to Pliocene of Sumatra and Java. GRDC Palaeontology Series, Bandung Indonesia (in press).

Muller, J. 1966. Montane pollen from the Tertiary of NW Borneo. Blumea, 14: 231-5.

Muller, J. 1968. Palynology of the Pedawan and Plateau Sandstone Formations (Cretaceous-Eocene) in Sarawak. Micropalaeontology 14: 1-37.

Muller, J. 1972. Palynological evidence for change in geomorphology, climate and vegetation in the Mio-Pliocene of Malesia. *In* The Quaternary Era in Malesia pp. 6-34. Edited by P. S. Ashton and M. Ashton. Geography Department, University of Hull, Miscellaneous Series 13.

Nandi, B. 1991. Palynostratigraphy of Upper Cretaceous sediments, Meghalaya, northeastern India. Reviews of Palaeobotany and Palynology 65: 119-129.

Pocknall, D. T. 1982. Palynology of the Late Oligocene Pomahaka Estuarine Bed sediments, Waikoikoi, Southland, New Zealand. New Zealand Journal of Botany 20: 7-15.

Racey, A., Goodall, J. G. S., Love, M. A., Polachan, S. and Jones, P. D. 1994. New age data for the Mesozoic Khorat Group of Northeast Thailand. Proceedings International Symposium on Stratigraphic Correlation of Southeast Asia, Bangkok, November 1994: 245-252.

Rahardjo, A. T., Polhaupessy. T. T., Sugeng Wiyono, Nugrahaningsih, H. and Eko Budi Lelono. 1994. Zonasi Polen Tersier Pulau Jawa. Makalah Ikatan Ahli Geologi Indonesia, December 1994: 77-84.

Shamsuddin, J. and Morley, R. J. 1994. Palynology of the Tembeling Group, Malay peninsula. Proceedings International Symposium on Stratighigraphic Correlation of Southeast Asia, Bangkok, November 1994: 208.

Situmorang, B. 1982. The formation and evolution of the Makassar Basin, Indonesia. PhD Thesis, University of London 313 pp.

Smith, A. G. and Briden, J. C. 1977. Mesozoic and Cenozoic Palaeocontinental maps. Cambridge University Press, 63 pp.

Smith, A. G., Smith, D. G. and Funnell, B. M. 1994. Atlas of Mesozoic and Cenozoic coastlines. Cambridge University Press, 99 pp.

Song Zhichen, Zheng Yahui, Liu Jinling, Ye Pingyi, Wang Cong Feng and Zhou Shan Fu. 1981. Cretaceous-Tertiary palynological assemblages from Jiangsu. Geological Publishing House, Peking, China. 268 pp.

Srivastava, S.K. 1977. *Ctenolophon* and *Sclerosperma* palaeogeography and Senonian Indian plate position. Journal of Palynology 23-24, pp 239-253.

Steenis, C. G. G. J. van. 1934a. On the origin of the Malaysian mountain flora, Part 1 Bulletin du Jardin Botanique de Buitenzorg, III, 13: 135-262.

Steenis, C. G. G. J. van. 1934b. On the origin of the Malaysian mountain flora, Part 2. Bulletin du Jardin Botanique de Buitenzorg, III, 13: 289-417.

Steenis, C. G. G. J. van. 1936. On the origin of the Malaysian mountain flora, Part 3, Analysis of floristic relationships (1st instalment). Bulletin du Jardin Botanique de Buitenzorg, III, 14: 36-72.

Steenis, C. G. G. J. van. 1961. Introduction: The pathway for drought plants from Asia to Australia. Reinwardtia 5: 420-429.

Steenis, C. G. G. J. van. 1962. The land-bridge theory in botany. Blumea, 11: 235-372.

Steenis, C. G. G. J. van. 1979. Plant geography of East Malesia. Botanical Journal of the Linnean Society 79: 97-178.

Stover, L. E. and Partridge, A. D. 1973. Tertiary and Late Cretaceous spores and pollen from the Gippsland Basin, Australia. Royal Society of Victoria Proceedings 85(2): 237-286.

Takahashi, K. 1982. Miospores from the Eocene Nanggulan Formation in the Yogyakarta region, Central Java. Transactions Proceedings Palaeontological Society Japan N.S.126: 303-326.

Takhtajan, A. 1969. Flowering plants, origin and dispersal. (Translated by C. Jeffrey) Oliver and Boyd, Edinburgh: Smithsonian Institution, Washington D. C.

Takhtajan, A 1987. Flowering plant origin and dispersal: the cradle of the angiosperms revisited. In Biogeographical Evolution of the Malay Archipelago. pp. 26-31. Edited by T. C. Whitmore. Oxford Monographs on Biogeography 4, Oxford Scientific Publications.

Truswell, E. M., Sluiter, I. M. and Harris, W. K. 1985. Palynology of the Oligocene-Miocene sequence in the Oakvale-1 corehole, western Murray Basin, South Australia. BMR Journal of Australian Geology and Geophysics 9: 267-295.

Truswell, E. M., Kershaw, P.A. and Sluiter, I. R. 1987. The Australian-South-east Asian connection: evidence from the palaeobotanical record. In Biogeographical Evolution of the Malay Archipelago. pp. 32-49. Edited by T. C. Whitmore. Oxford Monographs on Biogeography 4, Oxford Scientific Publications.

Tsuda, K., Itoigawa, J. and Yamanoi, T. 1984. On the Middle Miocene palaeoenvironment of Japan with special reference to the ancient mangrove swamps. In The evolution of the East Asian environment. Volume II. pp. 388-396. Edited by R. O. Whyte. Centre of Asian Studies, University of Hong Kong.

Venkatachala, B. S. 1974. Palynological zonation of the Mesozoic and Tertiary subsurface sediments in the Cauvery Basin. In Aspects and Appraisal of Indian Palaeobotany. pp. 476-494. Edited by K. R. Surange et al. Birbal Sahni Institute of Palaeobotany, Lucknow, India.

Watanasak, M. 1988. Palaeoecological reconstruction of Nong Ya Plong Tertiary Basin (Central Thailand). Journal of Ecology (Thailand) 15: 61-70.

Watanasak, M. 1990. Mid Tertiary palynostratigraphy of Thailand. Journal of South-East Asian Earth Sciences 4: 203-218.

Wegener, A. 1929. The origin of continents and oceans. Freidrich Vieweg und Sohne, Braunschweig. 246 pp.

Whitmore, T. C. 1973. Plate tectonics and some aspects of Pacific plant geography. New Phytologist 72: 1185-1190.

Whitmore, T. C. 1975. Tropical rain forests of the Far East. Clarendon Press, Oxford. 282 pp.

Yamanoi, T. 1974. Note on the first fossil record of genus *Dacrydium* from the Japanese Tertiary. Journal Geological Society Japan 80: 421-423.

Yamanoi, T., Tsuda, K., Itoigawa, J. and Taguchi, E. 1980. On the mangrove community discovered from the Middle Miocene formations of southwest Japan. Journal Geological Society Japan 86: 635-638.

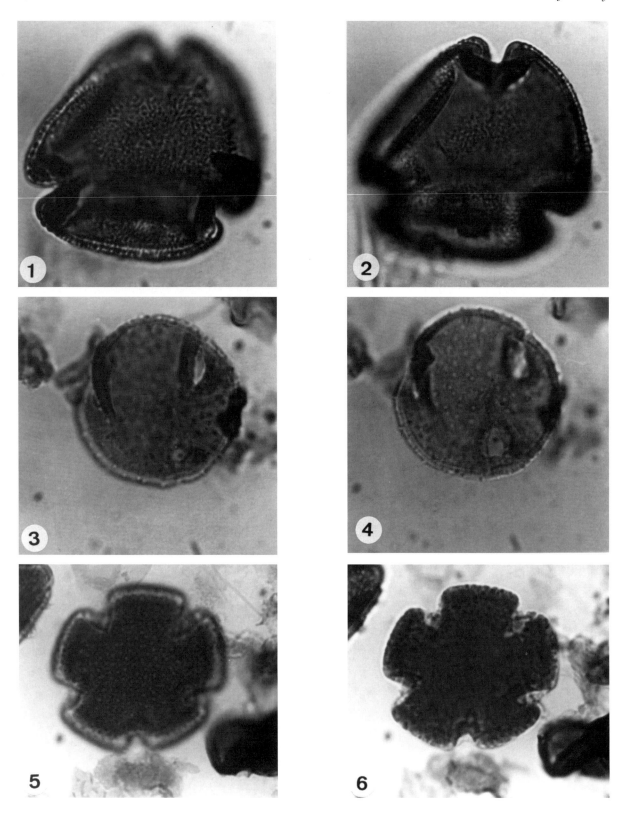

Plate 1. Middle Eocene palynomorphs from Central Java and S Sulawesi, all x1000. 1-2: *Lanagiopollis* cf. *regularis* (*Alangium* sect. Conostigma, Alangiaceae), S Sulawesi, Malawa Formation, sample P-04. 3-4: *Ixonanthes* pollen (Ixonanthaceae), Central Java, Nanggulan Formation, sample XIVF. 5-6: *Retistephanocolpites williamsi* (*Ctenolophon parvifolius*, Ctenolophonaceae), S Sulawesi, Malawa Formation, sample P-04.

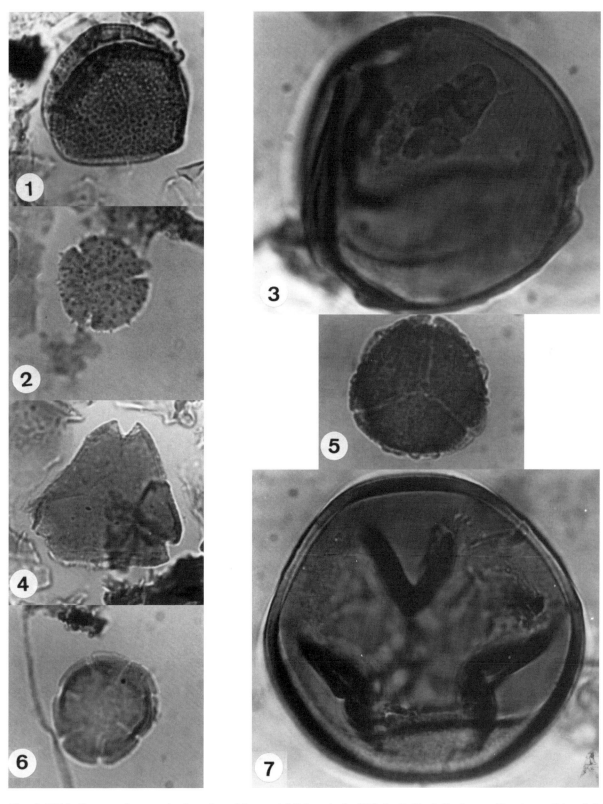

Plate 2. Middle Eocene palynomorphs from Central Java and S Sulawesi, all x1000. 1: *Quillonipollenites* sp. (*Eugeissona*, Palmae), S Sulawesi, Malawa Formation, sample P-04. 2: *Compositoipollenites dilatus* (family indet.), Central Java, Nanggulan Formation, sample XIVF. 3: *Lakiapollis ovatus* (*Durio* type, Bombacaceae, sect Durioneae). S Sulawesi, Malawa Formation, sample P-04. 4: aff. *Beaupreadites matsuokae* (*Beauprea*, Proteaceae), S Sulawesi, Malawa Formation, sample P-04. 5: *Cupanieidites flaccidiformis* (Sapindaceae), Central Java, Nanggulan Formation, sample XIVF. 6: *Polygalacidites clarus* (*Polygala* type, Polygalaceae). S Sulawesi, Malawa Formation, sample P-04. 7: *Dandotiaspora laevigata* (Pteridophyta, family indet.), S Sulawesi, Malawa Formation, sample P-04.

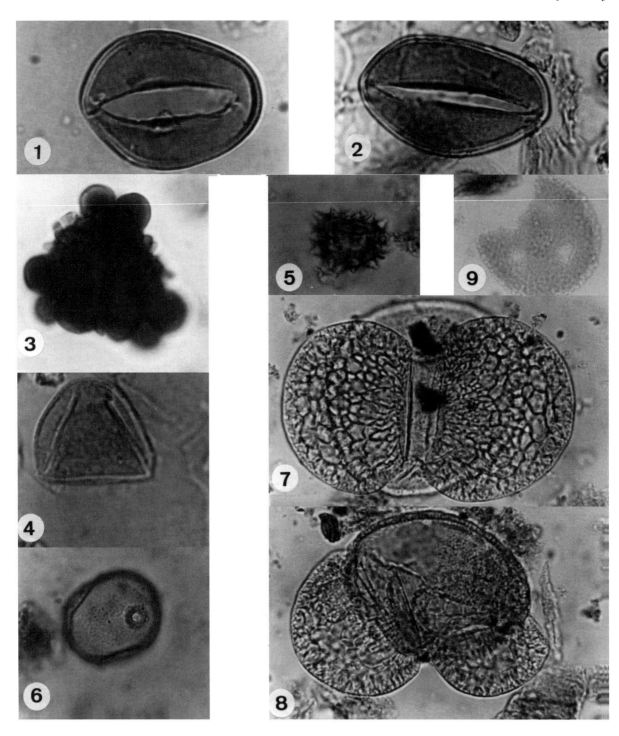

Plate 3. Pollen from the upper and middle Eocene of Sulawesi, the lowest Miocene of the Java Sea and from the mid-Pleistocene of the Malay peninsula, all x1000. 1: *Palmaepollenites kutchensis* (Palmae, subtribe Iguanurinae) S Sulawesi, Malawa Formation, middle Eocene, sample P-04. 2: *Palmaepollenites* sp. II of Harley and Morley (1993) (Palmae, subtribe Iguanurinae) S Sulawesi, Malawa Formation, middle Eocene, sample P-04. 3: *Meyeripollis nayarkotensis* (family indet.) S Sulawesi, Malawa Formation, upper Eocene, sample UL16. 4: *Dicolpopollis* sp. (*Calamus* type, Palmae), S Sulawesi, middle Eocene, sample P-04. 5: *Echitricolporites spinosus* (Compositae, sect Tubiflorae) West Malaysia, near Subang Airport, Old Alluvium, mid-Pleistocene, sample 3/15. 6: *Monoporites annulatus* (Gramineae), West Malaysia, near Subang Airport, Old Alluvium, mid-Pleistocene, sample 3/15. 7-8: *Pinuspollenites* sp. (*Pinus*, cf. *P. merkusii,* Pinaceae), West Malaysia, near Subang Airport, Old Alluvium, mid-Pleistocene, sample 3/15. 9: *Constantinisporis* cf. *jacquei,* Arjuna Basin, West Java Sea, Talang Akar Formation, lowest Miocene.

Noteworthy disjunctive patterns of Malesian mosses

Benito C. Tan
Farlow Herbarium, HUH, Harvard University, 22 Divinity Avenue, Cambridge, MA 02138, USA

Key words: Malesian mosses, disjunction, plate tectonics, dispersal

Abstract

A review of some noteworthy patterns of disjunction of Malesian moss distributions is presented. Correlation with new information from the plate tectonic history of the Malesian region offers a new interpretation of the origin and migration of Malesian mosses. Although long distance and chance dispersal remain important explanations for local plant distribution patterns, it appears that the climatic and vegetation changes in Malesia during the Cenozoic are two important factors contributing to the present-day patterns of moss distribution in Malesia.

Introduction

Since the distribution of many Malesian mosses is intimately tied to the rain forest habitat, the study of Malesian moss diversity and distribution has become more urgent today because of the increasing rate of destruction of rain forests across the entire region. The complex patterns of the distribution of mosses in the region is matched by the diverse composition of the Malesian moss taxa.

The impact and the biogeographical implication of the knowledge of moss distribution in Malesia must not be underestimated. Based on published information, the distribution of Malesian mosses, like their seed plant counterparts, supports recognition of distinct eastern, western and southern floristic provinces (Steenis, 1950; Tan, 1984, 1992; Touw, 1992a, 1992b). The proposed scenario of a Central Malesian province (see Johns, 1995) in place of the South Malesian province is not well supported by moss distribution patterns. On the other hand, the distributions of moss taxa across the Makassar Strait support the concept of a Wallacean region over Wallace's Line as a demarcation separating the Asiatic and Australasian floras. Nonetheless, the filtering effect of this biogeographical barrier on the spore-producing mosses is not as dramatic as is the case with seed plants (Tan, 1984, 1992; Hyvönen, 1989). Additionally, some areas in Borneo and Palawan that harbour an aggregation of uncommon moss taxa sharing similar ecological preference and also the same pattern of distribution have been taken to be refugia of species belonging to the Tertiary moist and the Quaternary dry forests (Meijer, 1982; Tan, 1996).

An appreciation of the importance of Malesian moss biogeography can be better achieved with an overview of the composition and affinities of Malesian mosses, which total c.330 genera and 1,755 species. It is clear from Table 1 that the majority of Malesian species of mosses (c.60%) have evolved *in situ* during the geological formation of the various island groups, notably Borneo and New Guinea. Taxa of clearly Laurasian and Gondwanan origins are nearly equal in number. With the recent refinements in the plate tectonic history of the Malesian region (see Hamilton, 1979; Audley-Charles, 1987; Hall, 1996), a review of the regional patterns of moss distribution, especially those of the disjunctive taxa, is both instructive and timely. Several of these patterns can now be better explained by relating them to local plate tectonic movements than by a long distance dis-

Table 1. Phytogeographical groupings/affinities of Malesian mosses (330 genera and 1,755 species).

Cosmopolitan taxa	3.0%
N Hemisphere/Laurasia	4.0%
Pantropical	2.3%
Palaeotropical	1.0%
Palaeotropical and Oceania	4.8%
Tropical Asia (including Eastern Himalayas) and Oceania	5.2%
Sri Lanka, S India, E Himalayas, Indochina, China, Taiwan, Japan and Malesia	9.2%
Indochina, China, Taiwan, Japan and W Malesia	2.2%
Malesia	22.9%
Widespread	13.2%
W Malesia	6.4%
E Malesia	1.9%
S Malesia	1.4%
Narrow Endemics (restricted to 1-2 islands)	30.9%
W Malesia	11.3%
E Malesia	15.7%
S Malesia	3.9%
Malesia and Oceania	4.0%
S Hemisphere/Gondwana	4.5%
Disjunctive	3.0%
Not Certain	3.0%

persal hypothesis. This is especially true in the bryogeographical interpretation of the floras between Luzon and Mindanao, and also between the floras of North Borneo and Kalimantan Borneo. New geological evidence indicates an independent plate tectonic history for each of these areas.

A word of caution is necessary here. In spite of the large amount of new taxonomic information on Malesian mosses published in the present decade, the moss flora of some pivotal islands, such as Mindanao, Sulawesi and Halmahera, remains insufficiently known at present to make definitive conclusions. It is probable that some of the disjunction patterns outlined below represent under-collection in intervening areas. A good example to illustrate how improved taxonomy and new collections alter the distributional status of a species is found in *Luisierella barbula* (Schwaegr.) Steere. This species was long thought to be a neotropical taxon until two collections from Java and New Caledonia, which were erroneously named as *Didymodon brevicaulis* (Hampe ex C. Muell.) Fleisch., were correctly identified in 1977 (Touw, 1992a). To date, the species is known additionally from Japan, China, Borneo, Sulawesi, Java, the Lesser Sunda Islands and New Caledonia. Its presence in Asia is probably incompletely documented because the species is a tiny moss of disturbed sites and can be easily overlooked.

Another noteworthy revelation shown by the distribution patterns of mosses is the relatively slow rate of speciation in many groups. Studies of fossilized mosses dated from Palaeozoic, Mesozoic and the Tertiary, reveal a remarkable similarity between the ancient and extant moss floras supporting the alleged evolutionary conservatism of the plant group (Frahm, 1994). Many of the fossil specimens can be identified to families and genera that still exist today (Miller, 1984; Krassilov and Schuster, 1984; Ignatov, 1990). It is therefore not surprising that some patterns of moss disjunction parallel those seen in flowering plants and vertebrates at a higher taxonomic level.

Noteworthy disjunctive patterns

Chameleion peguense (Besch.) Ellis & Eddy and *Horikawaea redfearnii* B. C. Tan & P.-J. Lin

These two taxa (Fig.1) represent the continental Asia and Philippine disjunction which was earlier called the Indochina-Philippines disjunction by Tan (1984). Members of this group (c.2.2% of the total Malesian species) are widespread in continental Asia, becoming rare in wet parts of archipelagic Malesia, and are often known only from the Philippines (Luzon and Palawan, see Fig.1). Some extend southward to the Malay peninsula, but they are absent from Borneo, Sumatra and New Guinea. A broader version of this pattern is seen in *Pogonatum microstomum* (Dozy & Molk.) Dozy & Molk. (see Touw, 1992b, p.149, Fig.2), with additional populations present also in Java and the Lesser Sunda Islands. Together, they probably represent denizens of monsoonal, semi-deciduous forest and savannah communities that had a broad range across tropical Asia during the Oligocene (Morley, 1991) and Pleistocene (Morley and Flenley, 1987; Heaney, 1991). The subsequent climatic changes, beginning in the Miocene, may have caused the fragmentation of the once expansive, seasonally dry vegetation of SE Asia, and its replacement by ever-wet rainforest such as exists today in Borneo, Sumatra and New Guinea. Consequently, mosses requiring seasonal drought disappeared.

In the case of *Horikawaea redfearnii*, which is known only from the seasonally dry forests of Hainan and Palawan Islands (see Fig.1), the disjunctive pattern might have been brought about by the drifting of the North Palawan-Mindoro

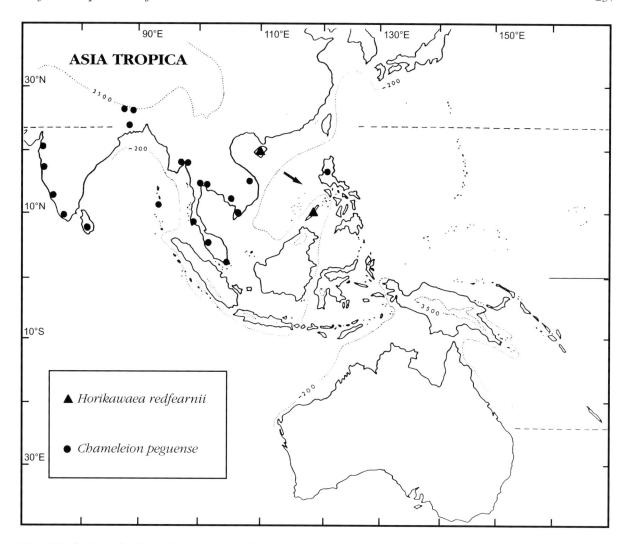

Fig.1. Distributions of *Chaemeleion peguense* (after Ellis, 1992) and *Horikawaea redfearnii* showing the continental Asia-Philippines disjunction. Arrow indicates the possible direction of dispersal.

terrane from the vicinity of South China toward its present position in the Philippines during the Oligocene-Miocene (Hall, 1996). Additional examples of a similar disjunction across the South China Sea are *Sphagnum robinsonii* Warnst. (Indochina and northern Luzon) and *S. luzonense* Warnst. (Indochina, Yunnan and Luzon).

A superficially similar pattern of disjunction, but attributable to a different ecological factor, is shown by *Trachycladiella aurea* (Mitt.) Menzel. The species is a moss of humid montane forest; its total range was mapped by Menzel and Schultze-Motel (1994, p. 79, Fig.4). It is common in the eastern Himalaya, scattered through the mountains of S China, Taiwan, southern Japan and northern Luzon, and reappears in the mountains of Sulawesi, Seram and W Java. Its present distribution appears to follow a hopping migration along the Himalayan-Philippine mountain track proposed by Steenis (1964). Its presence on Mt. Kinabalu in Borneo and high mountains in Mindanao can be predicted.

Dawsonia superba Grev., *Bescherellia elegantissima* Duby, *Mittenia plumula* (Mitt.) Lindb., *Thuidium sparsum* (Hook. f. & Wils.) Reichdt. and *Orthorrhynchium elegans* (Hook. f. & Wils.) Reichdt.

Bescherellia elegantissima Duby (Fig.3) and *Dawsonia superba* Grev. are two mosses that exhibit an Australasian/Oceania or South Malesia/Philippine disjunction. Members of this group (about 4.5% of the total Malesian species) are mostly widespread in rain forests in Australia, New Zealand and New Guinea, spreading

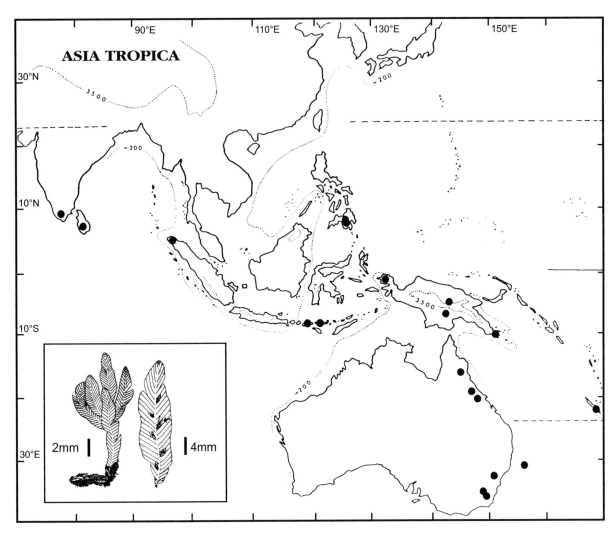

Fig.2. Distribution of *Orthorrhynchium elegans* showing a basically Gondwanan range. New Zealand populations are not shown on the map. Locality information from S.-H. Lin (1984a, 1984b).

westward into Sulawesi or Halmahera, and reaching northward to Mindanao Island in the Philippines. They represent the Gondwana elements in the Malesian moss flora. On the basis of Cenozoic plate tectonic history of SE Asia (Hall, 1996), their arrival in Malesia probably followed the collision of SE Asia and the Australian margin in the mid-Miocene.

A good example of this disjunction is seen in *Thuidium sparsum* which is common in eastern Australia, Tasmania and New Zealand, and reappears in S Malesia, touching the seasonally dry, southeastern margin of Borneo and southern Sulawesi (see Touw, 1992b, p.152, Fig. 5). Because of its absence in continental SE Asia, Touw (1992b) attributed this peculiar pattern to post-glacial colonization in Malesia after extensive everwet rainforests had re-established in Borneo and other wet parts of W Malesia. Another example of an Australasian moss that probably became established post-glacially in W Malesia is *Mittenia plumula* (Mitt.) Lindb. An isolated population of this widespread luminescent moss occurring in Papua New Guinea, Australia, Tasmania and New Zealand, was recently discovered on Mt. Kinabalu in North Borneo (Tan, 1990).

In the case of *Dawsonia superba*, outlier populations have penetrated the wet rain forest in north Borneo. Since north Borneo has a different geological history from Kalimantan Borneo (Michaux, 1991; Hall, 1996), the north Bornean populations of *D. superba* most likely originated from Mindanao and reached Sabah,

Brunei and Sarawak via either the Palawan or Sulu archipelagic corridor.

As to the Philippine island groups, new geological evidence has shown the island of Mindanao to have an origin south of the equator (Hall, 1996). The movement of Mindanao since the mid-Eocene from 140°E, 10°S to its present position in the Philippines could have provided an opportunity for exchange of plant taxa between the W and E Malesia provinces before the arrival of the New Guinean block. Compared to Luzon Island, Mindanao has a moss flora with an apparent Australasian influence. Good examples of disjunction between Mindanao (absent from Luzon and the Visayas Islands) and E Malesia (mainly New Guinea) are *Ectropotheciopsis novo-guineensis* (Geh.) Fleisch. and *Plagiotheciopsis oblonga* (Broth.) Broth. (Tan, 1984).

Among the many Malesian Gondwanan taxa, *Orthorrhynchium elegans* has a broad, discontinuous range reaching the Lesser Sunda Islands and Mindanao (Fig.2). West of the Wallace Line, this species is known additionally from Sri Lanka, S India and northern Sumatra, but is absent from Borneo and the Malay peninsula. The species probably reached Malesia by two separate events of introduction: the arrival of the Shan-Thai-Sumatra terranes in SE Asia in late Cretaceous or early Paleocene (Michaux, 1991) and the joining of the Australia-New Guinean plate in the late Miocene (Burrett *et al.*, 1991).

Aongstroemia orientalis Mitt., *Tortula caroliniana* Andrews and *Diphyscium chiapense* Norris

These are three species representing East Asia and Tropical American disjunction. They can also be considered examples of the amphi-Pacific disjunction. They constitute about 1% of the Malesian mosses (see Table 1). *Tortula caroliniana* is common in the SE United States, and Central and South Americas, with a disjunct population in Papua New Guinea. In contrast, *Aongstroemia orientalis* is widespread in the mountains in Asia, being reported from the E Himalaya, India, Indochina, China, Japan, Philippines (northern Luzon), Borneo (Mt. Kinabalu), Java (G. Sumbing), Lombok (G. Rinjani), and is found disjunctively in high mountains of Mexico and Guatemala. Their absence in tropical Africa seems to preclude a common Tethyan origin. The unique pattern of distribution of *Diphyscium chiapense* needs special mention. The species is known from Japan where it is locally widespread from Honshu to Shikoku, Mindanao (Mt. Kitanglad) and Mexico (Chiapas). It possibly represents an example of long distance and chance dispersal, although the ecology of this mainly terrestrial moss, as well as its sessile capsules and seemingly large and green spores, would argue against a long distance dispersability.

Entodon concinnus (De Not.) Par., *Rhytidium rugosum* (Hedw.) Kindb. and *Hageniella micans* (Mitt.) B. C. Tan and Y. Jia

These three species have a wide, albeit discontinuous, circum-boreal range, with scattered outlier populations in tropical Asia/Malesia and/or tropical America. Outside the northern hemisphere, *Entodon concinnus* is known from Papua New Guinea (Mt. Sarawaket, Mt. Wilhelm and Huon peninsula) and Ecuador (Enroth, 1991a). Interestingly, the circum-boreal and calcicole *Rhytidium rugosum* also has a single Malesian station in Irian Jaya of New Guinea (Schultze-Motel, 1963). In *Hageniella micans* (Mitt.) Tan & Jia, a recent revision (unpublished data, 1997) demonstrates a broad and discontinuous range which includes Britain, W Europe, SW China, Taiwan, Philippines (Luzon), Borneo (Mt. Kinabalu), Java (Mt. Gedeh), Hawaii, Canada (British Columbia), E United States and Mexico (Oaxaca). On a smaller scale, *Brachymenium bryoides* Hook. ex Schwaegr. has a range from the Himalayas to S India and is disjunctly present in Papua New Guinea. This type of disjunctive pattern is best explained by long distance and chance dispersal. In the case of *Rhytidium rugosum* and *Hageniella micans*, populations with sporophytes are infrequent throughout their ranges, therefore, long distance dispersibility must be inefficient and chancy.

Brachymenium capitulatum (Mitt.) Par. and *Caduciella mariei* (Besch.) Enroth

Brachymenium capitulatum has a reported range of continental Africa, Madagascar, Taiwan and Papua New Guinea. It exhibits a tropical Asian and tropical African disjunction which has been summarized and discussed by Pócs (1992) who provided additional examples of liverworts showing the same disjunction pattern. *Caduciella mariei*, which has been reported from Tanzania, the Comoro Islands, India (Assam), SW China, Indochina, Malesia, Oceania and Australia (Queensland), is another example (Fig.3). Many members of this group, though, are not

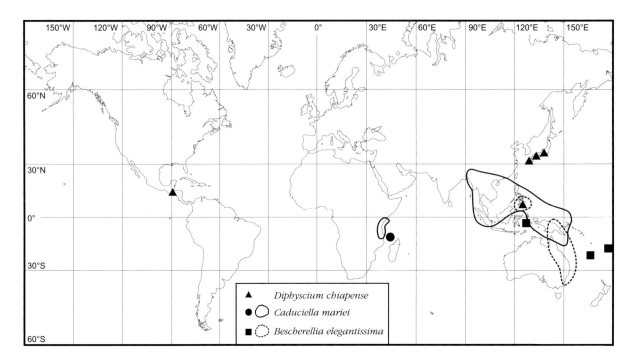

Fig.3. Distributions of *Diphyscium chiapense, Caduciella mariei* and *Bescherellia elegantissima*. Distribution of *Diphyscium chiapense* shows an E Asiatic and Tropical American or amphi-Pacific disjunction. Locality information from Deguchi (1997). Distribution of *Caduciella mariei* shows a Tropical Africa and Tropical Asia disjunction. Locality information from Pócs (1992) and Enroth (1991a, b). Distribution of *Bescherellia elegantissima* includes eastern Australia and Papua New Guinea, with outlier populations in Mindanao, Buru, New Caledonia and Fiji. Locality information from Sastre-Ines (1987).

known from the African continent, but only from Madagascar, in addition to SE Asia. Others have a more or less continuous range across the Old World tropics and can be considered palaeo-tropical taxa. According to Pócs (1992), they represent either the fragmented parts of a former continuous Gondwana range or the result of chance dispersal between extant populations on the two continents.

Takakia lepidozioides Hatt. & Inoue

This is probably the most widely disjunctive moss species in the world. Individual populations have been collected from small areas in coastal British Columbia and Alexander archipelago of Alaska, the Japanese Alps in Honshu, E Nepal, SE Tibet of China, and Mt. Kinabalu in north Borneo. The widely separated populations in Asia and North America and the restrictive habitat of this species seem to imply a long history of survival probably dating back to the time of the Pangaea supercontinent, with subsequent extinction of intervening populations. However, Schuster (1976) considered the distribution of *T. lepidozioides* to have resulted from the northward drifting of the Indian plate, followed by dispersal events across the land to reach North America.

Syrrhopodon strictus Thwaites & Mitt. and *Fissidens subbryoides* Hampe ex Gang

These are species of Malesian mosses with an odd or seemingly erratic disjunctive distribution. For example, *Syrrhopodon strictus* is known only from Sri Lanka and Irian Jaya of Indonesia, whereas *Fissidens subbryoides* has a few sporadic reports from E Nepal, Assam (India), the Andaman Islands and Seram (Indonesia). Apparently, they represent species of insufficiently known distribution awaiting more collections from the vast intervening and neighbouring regions. They could also be artefacts of inaccurate taxonomy.

Conclusions

As our knowledge of plate tectonic history of the Malesian region progresses, there is a corresponding need to learn more about the detailed distribution of various plant species, including mosses, across this vast region, as well as their ecology, reproductive and dispersal biology. Information from these sources can elucidate and corroborate hypotheses that lead us to a deeper understanding of the co-evolution of the geological environs and the flora it has come to support.

Acknowledgements

I am grateful to Prof. R. Hall and colleagues at the Natural History Museum for making it possible for me to attend the symposium. Drs. W. B. Schofield and D. Boufford kindly read and criticized the draft of this paper. The comments of the two reviewers and the editor of this volume are equally appreciated. Travel support to attend the symposium was provided by HUH through the Director M. Donoghue's Office and is acknowledged.

References

Audley-Charles, M. G. 1987. Dispersal of Gondwanaland: relevance to evolution of the angiosperms. In Biogeographical Evolution of the Malay Archipelago. pp.5-25. Edited by T. C. Whitmore, Clarendon Press, Oxford.

Burrett, C., Duhig, N., Berry R. and Varne, R. 1991. Asian and southwestern Pacific continental terranes derived from Gondwana, and their biogeographic significance. Australian Systematic Botany 4: 13-24.

Deguchi, H., Ueno, J. and Yamaguchi, R. 1997. Taxonomic notes on *Diphyscium* species with unipapillose leaf cells. Journal of Hattori Botanical Laboratory 82: 99-104.

Ellis, L. 1992. Towards a moss flora of southern India. Bryobrothera 1: 133-136.

Enroth, J. 1991a. Bryophyte flora of the Huon peninsula, Papua New Guinea. XLII. Entodontaceae (Musci). Acta Botanica Fennica 143: 43-55.

Enroth, J. 1991b. Notes on the Neckeraceae (Musci). 10. The taxonomic relationships of *Pinnatella mariei*, with the description of *Caduciella* (Leptodontaceae). Journal of Bryology 16: 611-618.

Frahm, J.-P. 1994. Moose-lebende Fossilien. Biologie in unseren Zeit 24: 120-124.

Hall, R. 1996. Reconstructing Cenozoic SE Asia. In Tectonic Evolution of Southeast Asia. Edited by R. Hall & D. J. Blundell, Geological Society Special Publication 106: 153-184.

Hamilton, W. 1979. Tectonics of the Indonesian region. United States Geological Survey Professional Paper 1078: 345pp.

Heaney, L. R. 1991. A synopsis of climatic and vegetational changes in Southeast Asia. Climatic Change 19: 53-61.

Hyvönen, J. 1989. On the bryogeography of Western Melanesia. Journal of Hattori Botanical Laboratory 66: 231-254.

Ignatov, M. S. 1990. Upper Permian mosses from the Russian platform. Paleontographica 217: 147-189.

Johns, R. J. 1995. Malesia – an introduction. Curtis's Botanical Magazine 12: 52-62.

Krassilov, V. A. and R. M. Schuster. 1984. Paleozoic and Mesozoic fossils. In New Manual of Bryology, v.2. Edited by R. M. Schuster, Hattori Botanical Laboratory, Nichinan.

Lin, S.-H. 1984a. A taxonomic revision of Phyllogoniaceae (Bryopsida). Part II. Journal of Taiwan Museum 37: 1-54.

Lin, S.-H. 1984b. Contribution to the knowledge of Orthorrhynchiaceae (2) *Orthorrhynchium* specimens in the Rijksherbarium, Leiden, The Netherlands. Yushania 1(2): 31-44.

Meijer, W. 1982. Plant refuges in the Indo-Malesian region, pp. 576-584. In Biological Diversification in the Tropics. Edited by G. T. Prance, Columbia University Press, New York.

Menzel, M. and Schultze-Motel, W. 1994. Taxonomische notizen zur Gattung *Trachycladiella* (Fleisch.), stat. nov. (Meteoriaceae, Leucodontales). Journal of Hattori Botanical Laboratory 75: 73-83.

Michaux, B. 1991. Distributional patterns and tectonic development in Indonesia: Wallace reinterpreted. Australian Systematic Botany 4: 25-36.

Miller, N. G. 1984. Tertiary and Quaternary fossils, pp. 1194-1232. In New Manual of Bryology, vol. 2. Edited by R. M. Schuster, Hattori Botanical Laboratory, Nichinan.

Morley, R. 1991. Tertiary stratigraphic palynology in Southeast Asia: current status and new directions. Bulletin of Geological Society of Malaysia 28: 1-36.

Morley, R. and Flenley, J. R. 1987. Late Cainozoic vegetational and environmental changes in the Malay Archipelago, pp. 50-59. In Biogeographical Evolution of the Malay Archipelago. Edited by T. C. Whitmore, Clarendon Press, Oxford.

Pócs, T. 1992. Correlation between the tropical African and Asian bryofloras II. Bryobrothera 1: 35-47.

Sastre-De Jesus, I. 1987. Revision of the Cyrtopodaceae and transfer of *Cyrtopodendron* to the Pterobryaceae. Memoirs of the New York Botanical Gardens 45: 709-721.

Schultze-Motel, W. 1963. Vorläufiges Verzeichnis der Laubmoose von Neuguinea. Willdenovia 3: 399-549.

Schuster, R. M. 1976. Plate tectonics and its bearing on the geographical origin and dispersal of angiosperms, pp. 48-138. In Origin and Early Evolution of Angiosperms. Edited by C. B. Beck, Columbia University Press, New York.

Steenis, C. G. G. J. van. 1950. The delimitation of Malaysia and its main plant geographic regions. Flora Malesiana Series 1, 1: lxxvi-cvi.

Steenis, C. G. G. J. van. 1964. Plant geography of the mountain flora of Mt. Kinabalu. Proceedings of Royal Society of London, Series B, Biological Sciences 161: 7-38.

Tan, B. C. 1984. A reconsideration of the affinity of Philippine moss flora. Journal of Hattori Botanical Laboratory 55: 13-22.

Tan, B. C. 1990. Six new taxa of Malesian mosses. Bryologist 93: 429-437.

Tan, B. C. 1992. Philippine Muscology (1979-1989). Bryobrothera 1: 137-141.

Tan, B. C. 1996. Biogeography of Palawan mosses. Australian Systematic Botany 9: 193-203.

Touw, A. 1992a. A survey of the mosses of the Lesser Sunda Islands (Nusa Tenggara), Indonesia. Journal of the Hattori Botanical Laboratory 71: 289-366.

Touw, A. 1992b. Biogeographical notes on the Musci of South Malesia, and of the Lesser Sunda Islands in particular. Bryobrothera 1: 143-155.

Patterns of distribution of Malesian vascular plants

W. J. Baker[1], M. J. E. Coode, J. Dransfield, S. Dransfield, M. M. Harley, P. Hoffmann and R. J. Johns
The Herbarium, Royal Botanic Gardens, Kew, Richmond, Surrey, TW9 3AE, UK
[1]*Department of Botany, Plant Science Laboratories, University of Reading, Whiteknights, Reading, Berkshire, RG6 6AS, UK*

Key words: biogeography, phytogeography, palynology, SE Asia, Malesia, Palmae, Gramineae, Euphorbiaceae, Elaeocarpaceae, *Antidesma*, *Elaeocarpus*, *Nypa*, *Spinizonocolpites*

Abstract

A miscellaneous selection of Malesian plant distributions is presented, including examples from the Palmae, Gramineae, Euphorbiaceae, Elaeocarpaceae, and various fern genera. Hypotheses of the tectonic evolution of the area may be required to explain many of the observed patterns that are described. Two major distribution types are identified repeatedly, the first displaying a strongly Sundaic bias and the second focusing on E Malesia. Patterns involving New Guinea are complex as they tend to include a variable combination of other islands such as Sulawesi, Maluku, the Bismarck archipelago and the islands of the W Pacific. A number of striking disjunctions exist, some of which have relatively narrow overall ranges, such as those of the palm genera *Cyrtostachys* and *Rhopaloblaste* which occur across Malesia excluding central parts of the region. In other examples, however, the separation is more profound including disjunctions between parts of Malesia and Madagascar in *Orania* (Palmae) and *Nastus* (Gramineae), and between Fiji, Vanuatu and Palawan in *Veitchia* (Palmae). At this stage, the significance of these distributions for the understanding of the geological history of SE Asia remains unclear. It is noted that distributions of species from the genus *Antidesma* (Euphorbiaceae) are more easily explained in terms of dispersal and environmental factors. Formal cladistic biogeographic analyses of these and other groups should aid interpretation of the region's history.

Introduction

Over the last four decades, there has been a radical change in the methods used by systematists to classify organisms. This change has had a concurrent effect on biogeographical methods. Ball (1975) describes three phases in biogeographical studies which he calls the descriptive or empirical phase, the narrative phase and the analytical phase. Biogeographical work concerned with the analytical phase has appeared increasingly in the systematic literature and it is here that modern methods are most evident. Previously, most classifications have been based on intuition and overall similarity which, though they may stand the test of time, are nevertheless subjective. Despite the introduction of statistical techniques which aimed to make similarity-based or phenetic classifications more testable, theoretical objections have led to the decline of phenetics in favour of cladistic methods. Cladistics uses character patterns of extant organisms to construct a hypothesis of their evolutionary or phylogenetic relationships. These patterns are generally visualised in the form of branching diagrams or cladograms which can be interpreted appropriately.

This paper is concerned mainly with the necessary precursors to the analytical phase, the descriptive and narrative phases of biogeography. In selecting taxa for discussion we have chosen only those which we believe will prove to be sound when subjected to the rigours of cladistics. The only groups which are acceptable in the cladist's eyes are those described as monophyletic. A monophyletic (natural) group contains all the descendants of a common ancestor and is defined by shared characters, or synapomorphies, that support a specific node on a cladogram. A group which contains some, but not all descendants of a common ancestor may appear to be coherent, but cannot be defined in the same way and, indeed, it can be argued that it cannot be defined at all. These paraphyletic groups are a partial and arbitrary expression of the product of common ancestry and make no sense in an evolutionary context. The inclusion of non-monophyletic groups in biogeographical studies is flawed for these reasons. For basic introductions to cladistic theory, general overviews may be found in Patterson (1982), Wiley et al. (1991) and Scotland (1992).

There is a substantial literature describing plant distributions in Malesia which we do not summarise here. Instead we present distributions from a miscellany of vascular plant families. Firstly, two monocotyledonous families are discussed; from the palm family, generic distributions within the three subfamilies Nypoideae, Calamoideae and Arecoideae, and from the grass family, distributions within the two bamboo genera, *Dinochloa* and *Nastus*. Secondly, in the dicotyledons, the genera *Antidesma* (Euphorbiaceae) and *Elaeocarpus* (Elaeocarpaceae) are considered. The fern genus *Christensenia* (Marattiaceae) and a variety of other fern and angiosperm examples from the diverse flora of New Guinea complete the assortment. The term Malesia is used in this paper to identify the area bounded to the north by the Thai-Malaysian border and to the south by the Torres Straits (Steenis, 1950; Johns, 1995). It includes the following political entities: Brunei Darussalam, Indonesia, Malaysia, Papua New Guinea, Philippines and Singapore. It is an area of exceptional biological diversity.

Palmae

The Palmae (Arecaceae) are a diverse, largely tropical and subtropical family of about 200 genera and 2700 species. In the latest classification of the family (Uhl and Dransfield, 1987) six subfamilies are recognised. The delimitation of these subfamilies is currently under rigorous phylogenetic analysis, and preliminary evidence suggests that four of the six (Coryphoideae, Calamoideae, Nypoideae and Phytelephantoideae) are monophyletic whereas two, Ceroxyloideae and Arecoideae, may be paraphyletic (Uhl et al., 1995). However, the relationships of many of the tribes, subtribes and genera within the subfamilies need further study.

The Malesian and W Pacific region represents one of the richest areas of palm diversity in the world. An estimated 1000 species in 93 genera are found in the area from Indochina to Australia and the W Pacific islands of Fiji and New Caledonia. All subfamilies mentioned above are represented in the region except for the Phytelephantoideae.

Palm distributions in the region have been discussed previously in an anecdotal fashion (Dransfield, 1981, 1987; Uhl and Dransfield, 1987) but these discussions lack a firm cladistic basis. However, with the explicit analytical framework that we are developing, we have been able to select a number of genera or groups of genera that we believe to be monophyletic and display distribution patterns of considerable biogeographic interest.

We present examples from three subfamilies: Nypoideae, Calamoideae and Arecoideae. As mentioned above, the last subfamily is probably paraphyletic but the examples we have chosen from within it are, we believe, monophyletic.

Nypoideae

This is the only subfamily in the Palmae represented by a single extant species. *Nypa fruticans* is distinguished from all other palms by a number of unique features including the prostrate dichotomously branched stem, the erect inflorescence bearing a terminal head of pistillate flowers, the lateral spikes of staminate flowers and unusual features of the flowers themselves (Uhl and Dransfield, 1987). Furthermore, the spiny zonosulcate pollen grains are highly distinctive. With the exception of similar but much smaller and incompletely zonosulcate pollen produced by many species of the genus *Salacca* (Calamoideae), such pollen grains are not encountered elsewhere in the family, and are unknown outside the Palmae. The pollen grains are notably uniform in their morphology throughout the geographic range of the genus. In a recent phylogenetic study of the whole family, *Nypa* is resolved as sister taxon to all other palms (Uhl et al., 1995).

Nypa fruticans is a mangrove palm, often growing in vast natural stands in a variety of estuarine conditions. Its present day distribution extends throughout Malesia and also includes Sri Lanka, the Ganges delta, Indochina, NW Australia, the Solomon Islands and Ryukyu Islands (Uhl and Dransfield, 1987). It has also been introduced to W Africa and to Panama where it is now well established.

Unlike its modern counterpart, fossil *Nypa* pollen, *Spinizonocolpites* (Muller, 1968), has been shown to possess a range of spine lengths and spine distributions. This variation has been demonstrated by a number of authors to be very

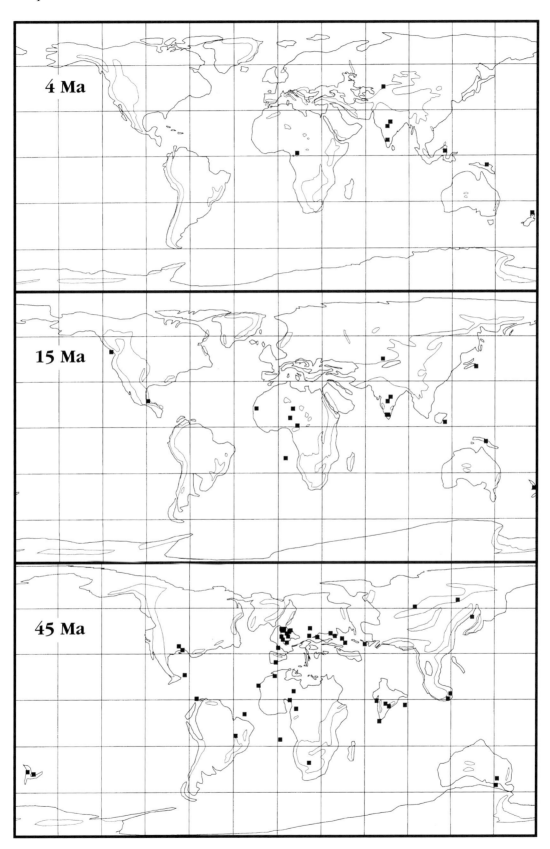

Fig. 1. Distribution maps for *Spinizonocolpites* generated using Atlas palaeomapping program (Cambridge Paleomap Services, 1992), incorporated into Plant Fossil Record 2.2 (Lhotak and Boulter, 1995).

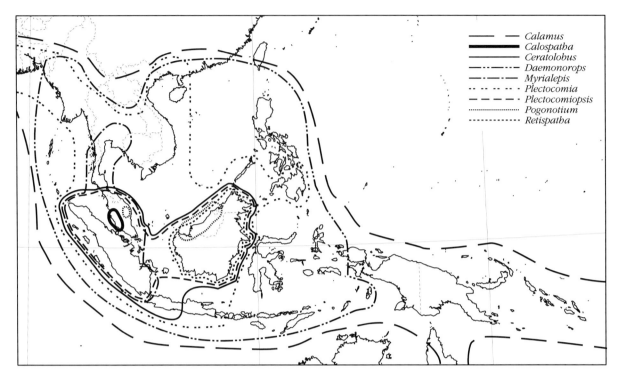

Fig.2. Distributions of genera in subfamily Calamoideae (Palmae), rattans.

localised, often occurring in quite small matrix samples (*e.g.,* Harley *et al.,* 1991). The distribution of *Spinizonocolpites* in the fossil record is remarkable (Fig.1). It is known from the Upper Cretaceous (Senonian) of Borneo (Muller, 1968) although there is some doubt about the age of this record (Morley, 1998 this volume). Upper Cretaceous records are widespread throughout the palaeotropics, including Meso and northern S America, W and N Africa, the Middle East and India. From the Paleocene, northern palaeolatitude records extend to c.65°N, and include examples from northern USA, Europe and Pakistan (Frederiksen, 1994). In the southern hemisphere, there are records of *Spinizonocolpites* in S Australia (Stover and Evans, 1973) and N Island, New Zealand (McIntyre, 1965). Eocene records are more frequent and by the Eocene the widespread pantropical distribution was stable (Fig.1C). There are additional records from NW Australia (Hekel, 1972) and the Lower Eocene of Tasmania, which is the most southerly occurrence of *Spinizonocolpites* at palaeolatitude 65°S (Cookson and Eisenack, 1967; Pole and McPhail, 1996). For the Miocene (Fig.1B), records are substantially reduced. There are records from Africa, India and Malaysia, including numerous and varied *Spinizonocolpites* grains from offshore cores in the West Java Sea (R. J. Morley, pers. comm., 1997). By the end of the Miocene (Fig.1A), the distribution did not extend far outside the area now occupied by *Nypa fruticans*. There are few African records, one or two unsubstantiated records in Europe, and there is a Pliocene record from New Zealand (Couper, 1953). Records for India, Papua New Guinea and SE Asia indicate a similar distribution to the early Miocene.

Calamoideae

This large subfamily with 22 genera and 650 species is pantropic. Three highly distinctive genera with palmate leaves are confined to northern S America. Currently, they are regarded as comprising a separate tribe, Lepidocaryeae, but continuing studies have proposed stronger links with genera within the other tribe, Calameae, than had been suspected.

In Africa, there are three endemic genera of climbing palms (rattans). A fourth rattan genus, *Calamus*, is represented in Africa by a single species, but is extraordinarily diverse in Asia. The massive tree palm genus *Raphia* is very diverse in Africa. One species extends to S

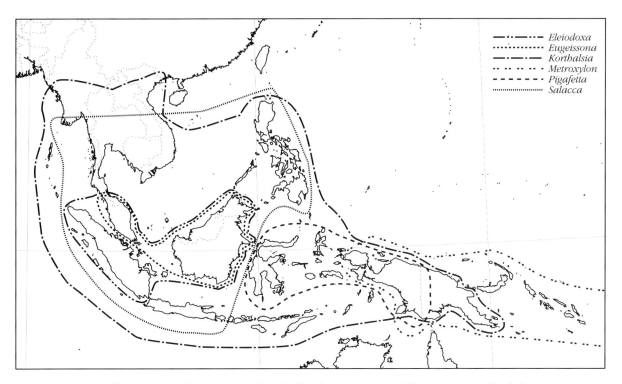

Fig.3. Distributions of genera in subfamily Calamoideae (Palmae), non-rattans and the rattan genus *Korthalsia*.

America and another to Madagascar, but both these distributions are thought to be man-made.

All other calamoid genera are restricted to the Asian and W Pacific region where they display varied distribution patterns (Figs.2 and 3). However varied these patterns may be, there is one conspicuous trend. The greatest diversity in terms of species and genera is overwhelmingly Sundaic rather than Papuasian. The largest genus *Calamus*, distributed from the W Ghats and China to Fiji and Australia and in Africa (1 species) and with a total of 370 species, is almost certainly paraphyletic (Kramadibrata, 1992, Baker unpublished), but even when the distributions of individual, potentially monophyletic groupings within the genus are examined, there is a similar bias of diversity in Sundaland. The genera *Korthalsia* with 26 species of climbing palms and *Daemonorops* with about 115 species are two other examples of genera with major diversity in Sundaland and with a decrease in number of species eastwards. The distribution of these genera strongly suggests a south-eastwards invasion from Sundaland to Papuasia that could have occurred after the Miocene juxtaposition of the two ends of Malesia (but see also Morley, 1998 this volume).

A further type of distribution can be found in certain genera which are almost exclusively Sundaic, but with local variations. For example, *Ceratolobus* (6 species) is found in Sumatra, Java, Malay peninsula and Borneo. *Salacca* (c.20 species) is more widely distributed reaching China in the north and Palawan and Mindanao in the east. *Plectocomiopsis* (6 species) is found in Sumatra, the Malay peninsula (including S Thailand) and Borneo, while closely related *Myrialepis* (1 species) occurs in Indochina southwards to the Malay peninsula and Sumatra, but not Borneo or Java. *Eleiodoxa* (1 species) is restricted to peat-swamp forest in the Malay peninsula (including S Thailand), Sumatra and Borneo (see also *Cyrtostachys* below). *Pogonotium* (3 species) is confined to peninsular Malaysia and Borneo, while *Calospatha* (1 species) is endemic to peninsular Malaysia and *Retispatha* (1 species) to Borneo. These local distribution patterns within Sundaland emphasise the close relationship of the floras of these land masses. Differences may be explicable in terms of vegetation changes in relation to climatic fluctuations and chance dispersal and extinction events.

Eugeissona (6 species), distinguished by a large range of unique characters but still clearly a member of the subfamily, is distributed in the

Malay peninsula and Borneo. Two fossil palynomorphs of lower and middle Miocene age in Borneo have been referred to *Eugeissona* suggesting the presence of the genus in Borneo from at least the early Miocene. Fossil pollen, *Quilonipollenites*, of Lower to Middle Miocene (21-14 Ma) age from India (Muller, 1972, 1979; Morley, 1977; Phadtare and Kulkhani, 1984) has also been referred to *Eugeissona* (see also Morley, 1998 this volume).

There are two highly distinctive genera which do not display the Sundaic bias. *Metroxylon*, the sago palm, with five species, is found in the Caroline Islands, Fiji, Vanuatu, the Solomon Islands, the Bismarck archipelago and New Guinea. *M. sagu*, now widely dispersed by man across Malesia as a starch-producing crop, is thought to be native to New Guinea. The other genus, *Pigafetta*, with two species of massive pioneer tree palms, is represented in Sulawesi by *P. elata* and in Maluku and W New Guinea by *P. filaris*. Both species of *Pigafetta*, possessing relatively very small seeds, produced in large quantities, are apparently efficient colonisers of open habitats within their range so it seems surprising that the genus is not more widespread than it is. That these two genera are so distinct from other Asian calamoids both in their morphology and their distribution is clearly suggestive of a very different biogeographic history from the other calamoid genera.

Current molecular and morphological work is providing exciting new insights into the generic relationships within the Calamoideae (Baker *et al.*, 1998; Baker and Dransfield, unpublished work). A robust phylogeny of the Calamoideae will be a powerful tool in biogeographic studies of SE Asia.

Arecoideae

As stated above, this very large subfamily may prove to be paraphyletic. We have chosen four unusual examples from tribe Areceae, a tribe which, while being pantropic, is most diverse in the W Pacific. In marked contrast to the Calamoideae, the tribe displays greatest diversity at the Papuasian end of the Malesian region, with only a few genera (*e.g., Pinanga, Nenga* and *Iguanura*) displaying a Sundaic bias.

Orania with about twenty species is the sole member of subtribe Oraniinae. There are three species in Madagascar, one in the western part of Sundaland, three or four in the Philippines, one in Maluku, one in Aru Islands and nine in New Guinea (Fig.4A). This remarkable genus deserves a modern phylogenetic study.

Veitchia with about eighteen species belongs to subtribe Ptychospermatinae. This subtribe is defined by a suite of presumed apomorphies unusual within the Areceae. Intergeneric relationships are currently being reassessed by Scott Zona at Fairchild Botanic Garden; while the limits of *Veitchia* may be uncertain, there seems no doubt that the subtribe itself represents a monophyletic group. *Veitchia* is present in Fiji (10 species) and Vanuatu (5 species). A single species (*V. merrillii*) is restricted to karst limestone in Palawan and neighbouring islands (Fig.4A). This is an extraordinary disjunction. Even if the limits of *Veitchia* are changed, the presence of this species in Palawan is still remarkable, as the rest of the members of the subtribe are found only in Maluku, New Guinea, Australia and W Pacific islands.

Rhopaloblaste (subtribe Iguanurinae), a morphologically remarkably uniform genus that we regard as being natural, also displays unusual disjunction. One species is present in the Nicobar Islands and one in peninsular Malaysia while the other four species are found in Maluku, New Guinea and the Solomon Islands (Fig.4B).

Cyrtostachys (sole genus of subtribe Cyrtostachydinae) has a single species, *C. renda*, widespread in peat-swamp forest in Sumatra, Malay peninsula and Borneo, while the seven remaining species are restricted to New Guinea, the Bismarck archipelago and the Solomon Islands (Fig.4C). Pollen referred to *Cyrtostachys* occurs in Upper Miocene deposits in Borneo (Muller, 1972).

Gramineae

The genera *Dinochloa* and *Nastus* are members of the tribe Bambuseae, the woody bamboos (subfamily Bambusoideae, family Gramineae). Phylogenetic relationships within the tribe are not yet understood and are currently being investigated, but at present it is divided into 8-9 subtribes (Soderstrom and Ellis, 1987, Dransfield and Widjaja, 1995). These two genera have been chosen to illustrate the widely differing distributions that occur within the Bambuseae.

Dinochloa is a natural, well-defined genus of 23-39 species, each with a limited distribution. It is placed in the Bambusinae, a subtribe with a largely tropical Asian distribution which is most diverse in tropical mainland Asia. The genus it-

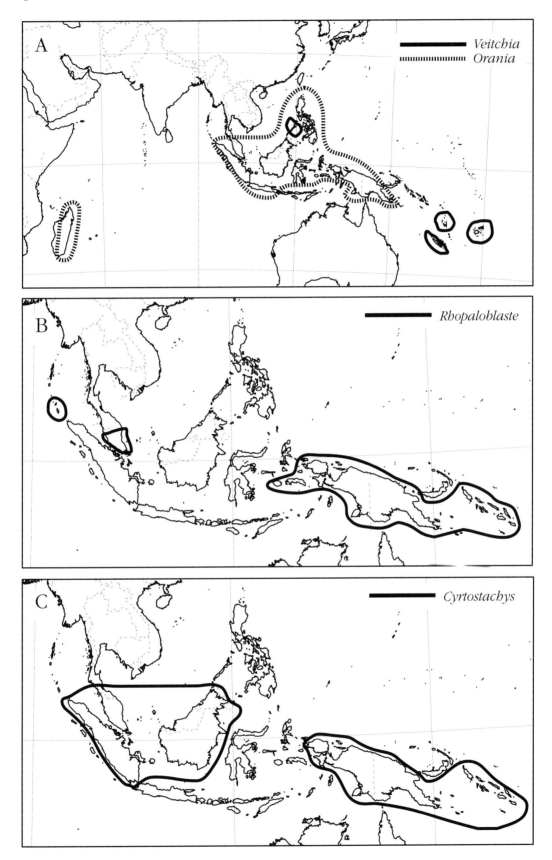

Fig.4. Distributions of genera in subfamily Arecoideae (Palmae). A: *Orania* and *Veitchia*, B: *Rhopaloblaste*, C: *Cyrtostachys*.

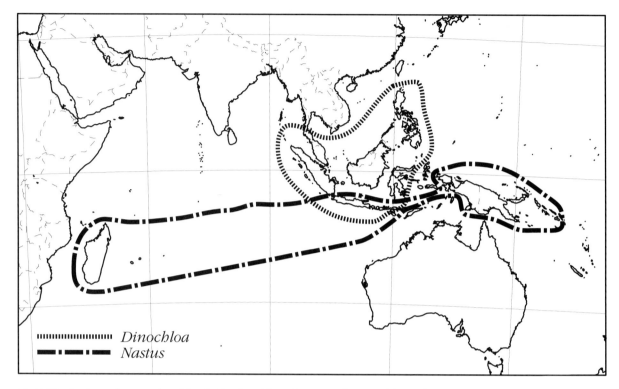

Fig.5. Distributions of genera in subfamily Bambusoideae (Gramineae), *Dinochloa* and *Nastus*.

self is found mainly in Malesia where it occurs in Sundaland, the Philippines, Sulawesi and Flores, but it also extends to the Andaman Islands and peninsular Thailand (Fig.5). It is absent from Maluku eastwards. Clearly, *Dinochloa* is a genus with a strongly Sundaic distribution. Given the limited distribution of the majority of its members, a phylogenetic analysis could be biogeographically informative.

The genus *Nastus* is included in subtribe Nastinae, which contains genera distributed in the southern hemisphere of the Old World tropics. The genus comprises about 19 species, distributed from Reunion and Madagascar to Java, Sumba through New Guinea and Solomon Islands, and possibly in Sumatra (Fig.5). This disjunct distribution is remarkable and parallels to some degree that of the palm genus *Orania*. However, *Nastus* remains a poorly known genus.

Antidesma (Euphorbiaceae-Phyllanthoideae)

Antidesma, a genus of dioecious shrubs and trees, is commonly found in the understorey of the tropical rain forest. It is distributed throughout the palaeotropics from W Africa to the Pacific Islands and from the Himalayas to N Australia. The highest diversity, both in species number and character variation, exists in Malesia, where about 90 of the 150 currently recognised species occur, c.70 of which are endemic to the region (Fig.6A). Despite the large number of species, no subgenera or sections can be recognised within *Antidesma*, and the great variability in some species complexes suggests that speciation is actively taking place.

The genus *Antidesma* belongs to the subtribe Antidesminae, the most notable character of which is the enlarged, U-shaped connective of the anthers that is found nowhere else in the family. The unilocular fruits and highly reduced flowers of *Antidesma* suggest that the genus is monophyletic. Furthermore, the genus is the only representative of the subtribe in Asia. Its closest relative *Hyeronima* occurs in S America and the Caribbean, while the genus with the largest number of ancestral character states in the subtribe, *Thecacoris*, is distributed in Africa and Madagascar. This strongly suggests that the subtribe and the genus *Antidesma* have originated in Gondwana, although no confirmatory fossil evidence has yet been presented.

The distribution data presented below have

been taken from a forthcoming revision of the genus in western and central Malesia (Hoffmann, unpublished work).

W and S Malesian elements

The majority of species fall into this category. Examples of species endemic to one island include *A. brachybotrys* (E Sarawak and Brunei Darussalam), *A. montis-silam* (Sabah), *A. orthogyne* (Malay peninsula) and *A. pleuricum* (Philippines). Other species are common to more than one island, e.g., *A. tetrandrum* (Sumatra, Java and Bali), *A. pendulum* (Sumatra, Borneo and Malay peninsula) and *A. leucopodum* (Sumatra, Borneo, Malay peninsula and Mindanao: Fig.6B). The Philippines show the highest degree of endemism, followed by Borneo and the Malay peninsula.

In *Antidesma* there is no sharp boundary between the W and the S Malesian province of Steenis (1950). *A. tetrandrum* for example occurs in Sumatra, Java and Bali, while its presumed sister species *A. venenosum* is common throughout Borneo. *A. minus*, a species common to the submontane regions of Java and Sumatra, illustrates Steenis' statement about the similarity of the mountain flora of Java and that of most of Sumatra. Furthermore, the two species extending across the Makassar Straits, *A. stipulare* and *A. tomentosum* (see below), occur in W Malesia and Java but not in the Lesser Sunda Islands.

E Malesian elements

In his account of the Euphorbiaceae of New Guinea, Airy Shaw (1980) recognised 32 species on the island, 27 of which he reported to be endemic. As most of them are poorly known at the moment, these figures might change with a critical revision. *A. excavatum* (syn. *A. moluccanum*) is the only E Malesian species with a wide distribution. It is common to N Sulawesi, Maluku, New Guinea, the Admiralty and Solomon Islands (Fig.6B).

Species extending across Makassar Straits

The distribution of *A. stipulare* bridges the phytogeographic demarcation line of the Makassar Straits (Steenis, 1950), including comparably dry Maluku and S Sulawesi (Fig.6B). *A. tomentosum* has a similar but slightly more northwestern distribution (Malay peninsula, Sumatra, W Java, Borneo, Philippines and N Sulawesi) with more humid conditions.

Species extending across the Isthmus of Kra

Several species do not respect the demarcation line between Malaysia and SE Asia slightly north of the Thai-Malay border (Steenis, 1950), e.g., *A. puncticulatum* (Fig.6C) and *A. velutinosum* (Burma, Thailand, Malay peninsula, Anambas and Natuna Islands, Sumatra, Java). *A. japonicum* (peninsular Malaysia, Thailand, Burma, Indochina, SE China, S Japan, Taiwan, Luzon) is an example of the similarity of the upland flora of N Luzon and E Asia (Steenis, 1950).

Monsoon forest elements

A. acidum (syn. *A. diandrum*) is the only species in the genus with a disjunct distribution (Fig.6D). This disjunction is known from many other taxa including the teak tree, *Tectona grandis* (Verbenaceae). *A. acidum* is found in the teak forests of Java and listed among the drought indicating plants (Steenis and Schippers-Lammertse, 1965). The species is also somewhat isolated morphologically.

The taxa sharing this distribution pattern require a two-seasonal climate with an annual drought period. They are present in the monsoon forests of Burma, Thailand and Indochina, avoid the everwet central part of W Malesia from the Isthmus of Kra southwards, but reappear in Java, parts of the Philippines, Sulawesi and the Lesser Sunda Islands. Because of the high number and the composition of taxa concerned, this cannot be explained by anthropogenic dispersal only. Instead, it dates back to the extension of areas with periodical drought during the Pleistocene, which then disappeared again in the post-glacial period leaving many taxa with disjunct distribution areas. This has been discussed in detail by Steenis (Meeuwen *et al.*, 1960).

Widespread species with a broad ecological spectrum

The three most widespread Asian species hardly show any ecological preferences. This applies particularly to the very common *A. montanum*,

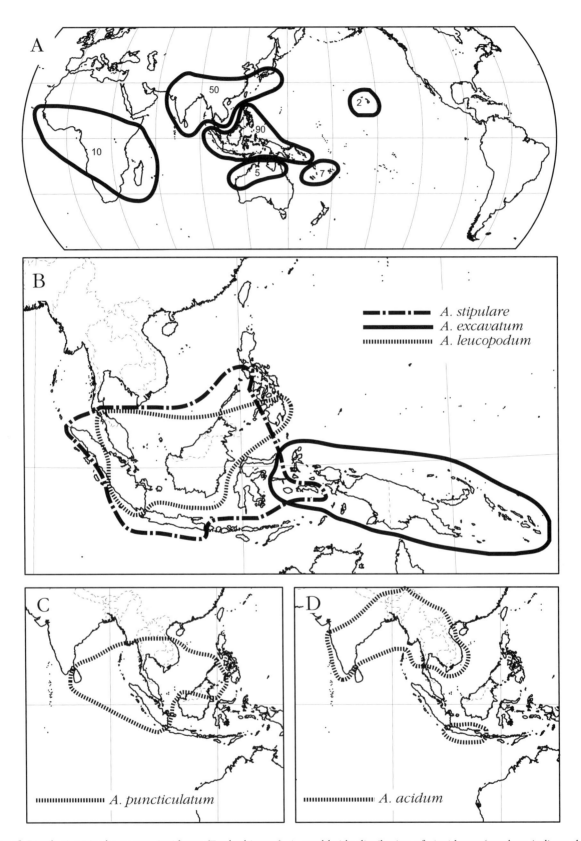

Fig.6. Distributions in the genus *Antidesma* (Euphorbiaceae). A: worldwide distribution of *Antidesma* (numbers indicate the number of species in the outlined areas); B: *A. stipulare* and *A. excavatum, A. leucopodum*; C: *A. puncticulatum*; D: *A. acidum*.

which occurs in most habitats from India (excluding Sri Lanka) to S Japan, Indochina, the Philippines, W Malesia, Java and Sulawesi. *A. ghaesembilla* is found throughout the region including India, S China, New Guinea and N Australia. It is the only species in the genus that prefers open vegetation such as secondary scrub and savannah. *A. bunius* is widely cultivated as a fruit tree, which makes it impossible to establish its original distribution. The species is known from Nepal and Sri Lanka to the Philippines and Maluku, although no collections from the Malay peninsula have been seen.

It has been shown for many plant taxa that the ecological amplitude and geographic distribution of polyploids may exceed that of related diploid taxa (Briggs and Walters, 1984, Skalinska, 1946). According to the chromosome numbers compiled by Hans (1970, 1973), three African species and the two Asian species *A. ghaesembilla* and *A. acidum* are diploid with n = 13, which is the base number in most Phyllanthoideae. On the other hand, *A. acuminatum* (*A. montanum*-complex) and *A. bunius* have been found to be polyploid. In the former species both diploid (n = 13) and hexaploid (n = 39) races have been observed, whereas *A. bunius* has been shown to be 18-ploid (n = 117). All polyploid records are based on collections from India, where the genus reaches its northern limit, and both taxa are among the most widespread, variable and ecologically tolerant taxa.

A possible evolutionary scenario based on the facts presented here would suggest the origin of the genus in wet tropical eastern Gondwana, followed by migration to and extensive radiation in tropical Asia. Climatic changes during the Pleistocene affected the disjunct distribution of *A. acidum* (see above). Unfortunately, we have no means yet to date any of the other events. The most promising systematic approach to render this outline more precise would involve an extensive survey of chromosome numbers in the genus and analysis of molecular data.

Elaeocarpaceae

The Elaeocarpaceae are a mostly tropical family of some 500 species, the majority of which are rainforest trees. Six of the family's nine genera (*Aceratium, Aristotelia, Dubouzetia, Elaeocarpus, Peripentadenia* and *Sloanea*) occur in Australia which therefore has more genera of Elaeocarpaceae than any other land mass. Furthermore, several species of *Elaeocarpus* in Australia do not fit into the otherwise effective classification within the genus. Preliminary analysis (Coode, 1987) supports the suggestion that the family has undergone a major radiation in the Australian region, certainly in Gondwana although, of course, the number of taxa in an area does not necessarily indicate that the group has evolved there. Furthermore, the family "is represented in the Tertiary fossil record in Australia by pollen, leaves and fruit" while fruits identifiable as belonging to *Elaeocarpus* have "been recognised as a ubiquitous element of the Tertiary flora of Australia" (quotes from Rozefelds and Christophel, 1996). Rozefelds and Christophel also quote references reporting fossil wood from the Tertiary in India and the Paleocene in Patagonia.

Elaeocarpus is a genus of some 360 species distributed widely in the Old World tropics, but not in Africa. It is currently known from the area bounded by Madagascar, eastern India including Ceylon, the warmer parts of China and Japan, New Caledonia, eastern Australia, New Zealand, the western Pacific and Hawaii. The majority of the species occur in SE Asia, Malaysia, the Philippines, Indonesia and New Guinea. Many species appear to be successional while others are restricted to the forest depths. They occur from sea-level to 3000 m or more.

Although formal analysis is lacking, several groupings within the genus *Elaeocarpus* are almost certainly monophyletic. The distribution patterns of four of these groupings are presented. Unfortunately, the large and widespread groups (currently known as section *Coilopetalum* and section *Monocera*) cannot be discussed as it is doubtful that they are monophyletic.

Section *Elaeocarpus* is most speciose in the western part of its range, tailing off eastwards after Borneo (Fig.7A). There is evidence of further speciation in New Guinea, with two small groups of species, one endemic, the other not found west of Sulawesi, in addition to the widespread *E. solomonensis*. In Fig.7B the *Polystachyus* group presents a different pattern. Also included in Fig.7B are the hypothesised nearest relatives to the *Polystachyus* group, which are in Sulawesi and Australia (Coode, 1997).

The *Acronodia* group has a rather similar pattern of current distribution (Fig.7C) but apparently its closest links are with China and Madagascar (Coode, 1996). Section *Oreocarpus* includes a single species (*Elaeocarpus culminicola*) found in various forms from the Philippines, Sulawesi, Maluku, New Guinea and the Bismarck archipelago, Aru Island, Melville Island

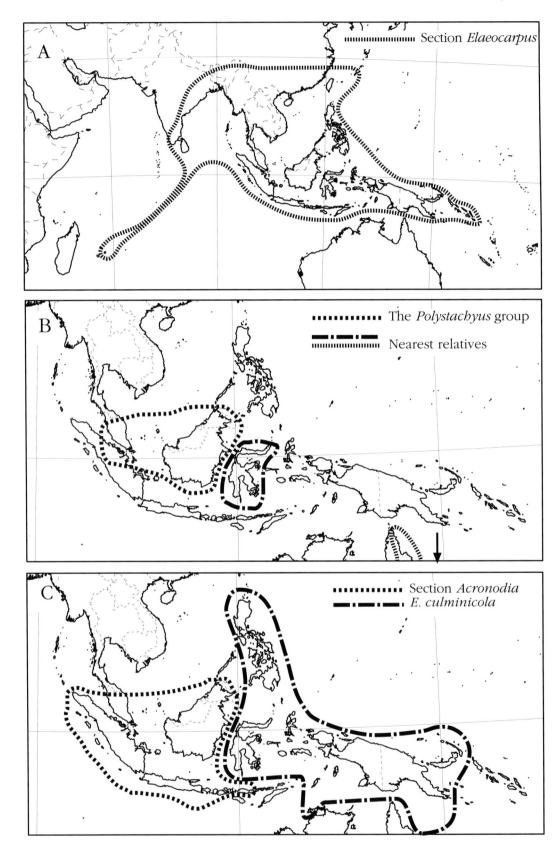

Fig. 7. Distributions in the genus *Elaeocarpus* (Elaeocarpaceae). A: section *Elaeocarpus*; B: the *Polystachyus* group and its hypothetical nearest relatives; C: section *Acronodia* and *E. culminicola*.

and NE Queensland (Fig.7C). The six remaining species of the section are all found in Australia alone.

Curious plant distributions in Papuasia

Some curious distributions of ferns and angiosperms have been selected as evidence for relationships between certain land masses and parts of Papuasia. Such evidence carries a caveat that the plant collection density for New Guinea is particularly low (Johns, 1995).

Christensenia, a fern genus in the Marattiaceae, comprises a single species, *C. aesculifolia*, distributed from India and SE Yunnan to Malesia and the Solomon Islands (Camus, 1990). With the exception of a single collection from the Vogelkop peninsula in Irian Jaya, this species has not been recorded from the mainland of New Guinea. Within Papuasia the genus occurs at several sites on New Ireland and in the Solomon Islands.

The Vogelkop peninsula includes several other floristic elements which it shares with central and western Malesia, but not with the remainder of New Guinea. The dipterocarp *Hopea inexpecta* (*Hopea* Section *Dryobalanoides*), the only representative of this section east of Wallace's Line, is restricted to the Kebar Valley in the Vogelkop where it is locally frequent in lowland rain forest. The rubiaceous ant plant genus *Myrmephytum* is represented in the Philippines, Sulawesi, Maluku and by five species in the Vogelkop peninsula.

Distribution patterns also suggest a relationship between western and central Malesia, the Bismarck archipelago and the Solomon Islands, excluding New Guinea. Several other distributions support this suggestion. The genus *Sararanga* (Pandanaceae) occurs in the Philippines (*S. philippinensis*), and is represented by a second species (*S. sinuosa*) along the island arc to the north of New Guinea. It is found also on Manus Island and New Ireland in the Bismarck archipelago and throughout the Solomon Islands with three small localised populations known from the north coast of New Guinea on Japen Island, near Jayapura and Vanimo. A similar pattern is also shown by the ferns *Pneumatopteris rodgersiana* and *Christella harveyi*.

Some species which reflect this pattern of distribution also occur on the New Guinea mainland in SE Papua, suggesting an additional relationship between the Solomon Islands and the Papuan peninsula. For example, the fern *Cephalomanes boryanum* occurs on Manus, New Ireland, the Solomon Islands, Woodlark Island, the Louisiade archipelago and on the mainland of SE Papua.

Discussion

At this stage, it would be inappropriate to speculate on the geological significance of the data presented here. Any one explanation of these very preliminary data could easily be replaced by another interpretation while formal analysis is lacking. A few generalisations can be made. Firstly, we believe that the distributions that we have discussed probably require a tectonic explanation. This seems to be particularly true for the curious disjunctions that have been discussed, such as those found in *Orania* and *Nastus* between Malesia and Madagascar, and in *Rhopaloblaste* and *Cyrtostachys* between E Malesia and W Malesia. However, it is important to be aware of the potential impact of extinction in generating apparently disjunct distributions which at one time may have been continuous. Furthermore, we find it far easier to explain W Malesian plant distributions than those of E Malesia as the geological evolution of the former region is better understood. We anxiously await the elucidation of the geological evolution of E Malesia, especially of the Papuasian region.

Although dispersal events are of no interest to vicariance biogeographers, they are a biological reality. It is important to take dispersal into account when one is considering biogeographical patterns of a single group in isolation from other groups so as to avoid mistaking ecological noise for biogeographical signal. One of our examples, the distribution of *Antidesma*, is thought to be heavily influenced by long distance dispersal, as the small bright red or black fruits are attractive to birds. Therefore, it is likely that the distribution of many *Antidesma* species is a reflection of their climatic, altitudinal and edaphic preferences rather than of tectonic movements.

The efficiency of birds in long-distance plant dispersal is difficult to assess, as a number of factors must be taken into account, such as the retention time of the seed in the digestive system and the viability of the transported seed after defecation or regurgitation. Both depend on the characteristics of the seed (*e.g.*, attractiveness and nutrient content of the fruit pulp or aril, seed size and robustness) as well as the bird (*e.g.*, size, presence of grit in the muscular gizzard, flight speed, territorial behaviour). For example, some birds may avoid open spaces to avoid potential attacks by predators, thereby failing to deposit seeds in these habitats, while other birds may cross open areas flying at high altitude and act as effective dispersal agents for seed. The major difficulty lies in the fact that most studies necessarily concentrate on only one aspect of this complex process, usually either the behavioural patterns and physiology of a particular bird taxon or the dispersal type of the plants found in a certain ecosystem.

A recent project on forest regeneration on Krakatau investigates the re-colonisation of the once sterilised islands by bird-dispersed plants (Whittaker and Jones, 1994). Three or four species of *Antidesma* have arrived on Krakatau so far. *A. montanum*, the most common and most widely distributed species of the genus, has become the dominant element of the forest understorey (Whittaker and Schmitt, pers. comm., 1997). However, as Krakatau is only 40 km away from the nearest mainland, the dispersal of seeds across greater distances may follow different rules.

Robust biogeographic hypotheses could be constructed from the data presented in this paper if the examples used were taken further into the analytical phase, historical biogeography. Historical biogeography can be practised on different levels. A simple approach is to discuss the biogeographical implications of a single group, with reference to a phylogeny. However, this is, in a sense, still narrative biogeography as it lacks independent support. The alternative, cladistic biogeography in the strict sense, uses phylogenetic hypotheses derived from any number of organisms relevant to a region to assess relationships between areas of endemism within that region. Under the assumption that the Earth and its biota share a common history, it uses biological data to identify area relationships that occur repeatedly and that are congruent with each other, and then relates them to Earth history, principally geological events. Despite the reality of influences such as climate, dispersal and pollination, it is very probable that a geological theme underpins all current plant distribution patterns. The methods of cladistic biogeography aim to reveal the theme by using multiple hypotheses of phylogenetic relationship to identify the repeating signal and dismiss the ecological noise. The theory and techniques used in cladistic biogeography are well summarised in Humphries and Parenti (1986), Humphries *et al.* (1988) and Humphries (1992), and some excellent examples of these techniques being applied in Malesian plant groups may be found in Linder and Crisp (1995), Turner (1995) and Ridder-Numan (1996).

SE Asia contains a diverse flora which undoubtedly reflects its complex geological history. When rigorous systematic methods are applied to the members of this flora, further patterns may be revealed and related to tectonic evolution. Preliminary discussions such as ours are necessary to assess the potential value of groups in biogeographical studies.

Acknowledgements

The authors would like to thank Terry Hedderson, Anne Bruneau, Estrela Figueiredo, Sasha Barrow, Freek Bakker, Joe Mullins, Peter Stevens and an anonymous reviewer for helpful comments on the manuscript and Chris Humphries for valuable advice.

References

Airy Shaw, H. K. 1980. The Euphorbiaceae of New Guinea. Kew Bulletin Additional Series 8: 1-243.

Baker, W.J., Dransfield, J., Harley, M.M. and Bruneau, A. 1998. Morphology and cladistic analysis of the subfamily Calamoideae (Palmae). Memoirs of New York Botanical Garden, in press.

Ball, I. R. 1975. Nature and formulation of biogeographical hypotheses. Systematic Zoology 24: 407-430.

Briggs, D. and Walters, S. M. 1984. Plant Variation and Evolution. Cambridge and New York.

Cambridge Paleomap Services. 1992. ATLAS version 3.2. Cambridge Paleomap Services, P.O. Box 246, Cambridge, U.K.

Camus, J. 1990. Marattiaceae. *In* The Families and Genera of Vascular Plants, I. Pteridophytes and Gymnosperms. pp. 174-180. Edited by K. U. Kramer and P. S. Green. Springer Verlag, Berlin.

Coode, M. J. E. 1987. *Crinodendron, Dubouzetia* and *Peripentadenia*, closely related in Elaeocarpaceae. Kew Bulletin 42: 777-814.

Coode, M. J. E. 1996. *Elaeocarpus* for Flora Malesiana: notes, new taxa and combinations in the *Acronodia* group. Kew Bulletin 51: 267-300.

Coode, M. J. E. 1997. *Elaeocarpus* for Flora Malesiana: notes, new taxa and combinations in the *Polystachyus* group. Kew Bulletin, in press.

Cookson, I. C. and Eisenack, A. 1967. Some early Tertiary microplankton and pollen grains from a deposit near Strahan, western Tasmania. Proceedings of the Royal Society Victoria 80: 131-140.

Couper, R.A. 1953. Upper Mesozoic and Cainozoic spores and pollen grains from New Zealand. New Zealand Geological Survey Palaeontological Bulletin 22: 1-77.

Dransfield, J. 1981. Palms and Wallace's Line. *In* Wallace's Line and Plate Tectonics. pp. 43-56. Edited by T. C. Whitmore. Oxford Monographs on Biogeography 1, Oxford Scientific Publications.

Dransfield J. 1987. Bicentric distributions in Malesia as exemplified by palms. In Biogeographical Evolution of the Malay Archipelago. pp. 60-72. Edited by T. C. Whitmore. Oxford Monographs on Biogeography 4, Oxford Scientific Publications.

Dransfield, S. and Widjaja, E. A. (Editors) 1995. Plant Resources of South East Asia No. 7. Bamboos. Backhuys Publishers, Leiden, 34-35.

Frederiksen, N. O. 1994. Middle and late Paleocene angiosperm pollen from Pakistan. Palynology (Dallas) 18: 91-137.

Hans, A. S. 1970. Polyploidy in Antidesma. Caryologia 23: 322-327.

Hans, A. S. 1973. Chromosomal Conspectus of the Euphorbiaceae. Taxon 22: 591-636.

Harley, M. M., Kurmann, M. H. and Ferguson, I. K. 1991. Systematic implications of comparative morphology in selected fossil and extant pollen from the Palmae and the Sapotaceae. In Pollen and Spores: Patterns of Diversification. pp. 225-238. Edited by S. Blackmore and S. Barnes. Clarendon Press, Oxford.

Hekel, H. 1972. Pollen and spore assemblages from Queensland Tertiary sediments. Geological Survey of Queensland Palaeontological Papers 30: 1-33.

Humphries, C. J. 1992. Cladistic Biogeography. In Forey, P. L., Humphries, C. J., Kitching, I. L., Scotland, R. W., Siebert, D. J. and Williams, D. M. Cladistics, A Practical Course in Systematics, Systematics Association Publication 10, Oxford University Press: 137-159.

Humphries, C. J. and Parenti, L. R. 1986. Cladistic Biogeography. Oxford University Press.

Humphries, C. J., Ladiges, P. Y., Roos, M. and Zandee, M. 1988. Cladistic biogeography. In Analytical Biogeography, An integrated approach to the study of animal and plant distributions. pp. 371-404. Edited by A. A. Myers and P. S. Giller, Chapman and Hall.

Johns, R. J. 1995. Malesia — an introduction. Curtis's Botanical Magazine 12: 52-62.

Kramadibrata, P. 1992. A revision of the genus Calamus (Palmae) section Macropodus sensu Furtado. Ph.D. Thesis, University of Reading.

Lhotak, M. and Boulter, M.C. 1995. Towards the creation of an international database of palaeontology. In Geological Data Management. Edited by J. R. A. Giles. Geological Society Special Publication 97: 55-64.

Linder, H. P. and Crisp, M. D. 1995. *Nothofagus* and Pacific Biogeography. Cladistics 11: 5-32.

McIntyre, D. J. 1965. Some new pollen species from New Zealand Tertiary deposits. New Zealand Journal of Botany 3: 204-214.

Meeuwen, M. S. van, Nooteboom, H. P. and Steenis, C. G. G. J. van 1960. Preliminary revisions of some genera of Malaysian Papilionaceae I. Reinwardtia 5: 419-456.

Morley, R. J. 1977. Palynology of Tertiary and Quaternary sediments of Southeast Asia. Proceedings of the Indonesian Petroleum Association, 16th Annual Convention 1: 255-275.

Morley, R. J. 1998. Palynological evidence for Tertiary plant dispersals in the SE Asian region in relation to plate tectonics and climate. In Biogeography and Geological Evolution of SE Asia. Edited by R. Hall and J. D. Holloway (this volume).

Muller, J. 1968. Palynology of the Pedawan and Plateau sandstone formations (Cretaceous - Eocene) in Sarawak, Malaysia. Micropalaeontology 14: 1-37.

Muller, J. 1972. Palynological evidence for change in geomorphology, climate and vegetation in the Mio-Pliocene of Malesia. In Transactions II Aberdeen-Hull Symposium of Malesian Ecology. University of Hull, Department of Geography, Miscellaneous Series 13: 6-16.

Muller, J. 1979. Reflections on fossil palm pollen. In Proceedings of IV Palynolological Conference Lucknow I: 568-579.

Patterson, C. 1982. Cladistics and Classification. New Scientist 94: 303-306.

Phadtare, N. R. and Kulkarni, A. R. 1984. Affinity of the genus *Quilonipollenites* with the Malaysian palm *Eugeissona* Griffith. Pollen and Spores 26: 217-226.

Pole M. S. and McPhail, M. K. 1996. Eocene *Nypa* from Regatta Point, Tasmania. Review of Palaeobotany and Palynology 92: 55-67.

Ridder-Numan, J. 1996. Historical Biogeography of the Southeast Asian genus *Spatholobus* (Legum.-Papilionoideae) and its allies. Ph.D. Thesis, Rijksherbarium/Hortus Botanicus, Leiden University, The Netherlands. Blumea Supplement 10: 1-144.

Rozefelds, A. C. and Christophel, D. C. 1996. *Elaeocarpus* (Elaeocarpaceae) endocarps from the Oligo-Miocene of eastern Australia. Papers and Proceedings of the Royal Society of Tasmania 130: 41-48.

Scotland, R. W. 1992. Cladistic Theory. In Forey, P. L., Humphries, C. J., Kitching, I. L., Scotland, R. W., Siebert, D. J. and Williams, D. M. Cladistics, A Practical Course in Systematics, Systematics Association Publication 10, Oxford University Press: 3-13.

Skalinska, M. 1946. Polyploidy in *Valeriana officinalis* Linn. in relation to its ecology and distribution. Botanical Journal of the Linnean Society 53: 159-184.

Soderstrom, T. R. and Ellis, R. P. 1987. The position of bamboo genera and allies in a system of grass classification. In Grass Systematics and Evolution. pp. 225-238. Edited by T. R. Soderstrom, K. H. Hilu, C. S. Campbell and M. I. Barkworth. Proceedings of the International Symposium on Grass Systematics and Evolution, Washington, 27-31 July 1986. Smithsonian Institute Press, Washington, D.C.

Steenis, C. G. G. J. van 1950. The delimitation of Malaysia and its main plant geographical divisions. Flora Malesiana. Noordhoff-Kolff N. V. (Djakarta) 1(1): LXX-LXXV.

Steenis, C. G. G. J. van and Schippers-Lammertse, A. F. 1965. Concise plant-geography of Java. In Flora of Java 2. pp. 3-72. Edited by C. A. Backer and R. C. Bakhuizen van den Brink. N. V. P. Noordhoff, Groningen, The Netherlands.

Stover, L. E. and Evans, P. R. 1973. Upper Cretaceous-Eocene spore-pollen zonation, offshore Gippsland Basin, Australia. Special Publications of the Geological Society of Australia 44: 55-72.

Turner, H. 1995. Cladistic and biogeographic analyses of *Arytera* Blume and *Mischarytera* Gen. Nov. (Sapindaceae) with notes on methodology and a full taxonomic revision. Ph.D. Thesis, Rijksherbarium/Hortus Botanicus, Leiden. Blumea Supplement 9: 1-230.

Uhl, N. W. and Dransfield, J. 1987. Genera *Palmarum*. A classification of the palms based on the work of H. E. Moore Jr. The L. H. Bailey Hortorium and International Palm Society, Allen Press, Lawrence, Kansas.

Uhl, N. W., Dransfield, J., Davis, J. I., Luckow, M. A., Hansen, K. S. and Doyle, J. J. 1995. Phylogenetic relationships among palms: cladistic analyses of morphological and chloroplast DNA restriction site variation. In Monocotyledons: Systematics and Evolution, Volume 2. pp. 623-661. Edited by P. J. Rudall, P. J. Cribb, D. F. Cutler and C. J. Humphries. Royal Botanic Gardens, Kew.

Whittaker, R. J. and Jones, S. H. 1994. The role of frugivorous bats and birds in the rebuilding of a tropical forest ecosystem, Krakatau, Indonesia. Journal of Biogeography 21: 245-258.

Wiley, E. O., Siegel-Causey, D., Brooks, D. R. and Funk, V. A. 1991. The Compleat Cladist, a Primer of Phylogenetic Procedures. Special Publication 19, The University of Kansas, Museum of Natural History, Lawrence, Kansas.

Historical biogeography of *Spatholobus* (Leguminosae-Papilionoideae) and allies in SE Asia

J. W. A. Ridder-Numan
Rijksherbarium/Hortus Botanicus, University of Leiden, PO Box 9514, 2300 RA Leiden, The Netherlands
Email: ridder@rulrhb.leidenuniv.nl

Key words: historical biogeography, phylogeny, Leguminosae, SE Asia

Abstract

A phylogenetic analysis of *Spatholobus* and its allies (Leguminosae-Papilionoideae) is presented. The resulting phylogeny is used for a historical biogeographical analysis of SE Asia together with the phylogenies of three other genera: *Fordia* (Leguminosae-Papilionoideae), *Genianthus* (Asclepiadaceae), *Xanthophytum* (Rubiaceae). Areas of distribution, based on all species, are defined. The distribution areas coincide in large part with the tectonic terranes of SE Asia, and they are briefly discussed in connection with their possible geological history.

The result of the biogeographic analysis, a general area-cladogram of SE Asia, is compared with the available geological information. The general area-cladogram shows the existence of four major groups of areas: areas on continental SE Asia, areas around the Isthmus of Kra, areas on the Malay peninsula and areas on Borneo.

The first split in the cladogram is between the continental SE Asia group of areas (including the Isthmus of Kra) and the Sundaland groups. This can be interpreted in terms of varying sea levels between the mid-Eocene and the Pliocene.

The areas on the Malay peninsula do not show any tectonic event, because the areas were welded together long ago, before the development of the analyzed genera. The separation of areas on the peninsula are perhaps due to ecological factors.

The first four splits in the Borneo clade probably reflect uplift in the late Miocene-Pliocene of the areas of Meratus, SW Borneo, Semitau and NE Borneo. For other areas such as the Philippines and Sulawesi, not enough data were available.

The general area-cladogram is also used to reconstruct the biogeographical history of the genus *Spatholobus*. It is suggested that the genus came from around India, invaded the dry Sunda shelf during a period of low sea level, and speciated due to the isolation of the area after sea level rises. There were renewed invasions from the mainland into the Malesian archipelago during later periods of low sea level.

Introduction

Although biogeographers have always had an interest in geology, this has been renewed since the broad acceptance of plate tectonic theory in geology and the introduction of cladistic methods in taxonomy. Taxonomists are not only interested in describing the flora and fauna, but also in discovering historical relationships, the phylogeny, within a group of taxa under study. When the phylogeny of a group is known, it is also possible to set up a hypothesis about the history of the group in space and time. In many cases speciation will be related to events concerning the area defined by the taxa, e.g., isolation. On the cladogram, distribution data of the taxa within the group can be plotted. By comparing and analysing the area-cladograms of different non-related groups, a general pattern will appear. This pattern, a general area-cladogram, may give information about the history of the areas, and thus be of interest to geologists.

Spatholobus consists of 29 species of lianas in SE Asia (Ridder-Numan and Wiriadinata, 1985; Ridder-Numan, 1992). The closely allied genera *Butea* and *Meizotropis* occur only on the mainland of SE Asia. Both genera are comprised of two species (Sanjappa, 1987). *Spatholobus* and its allies, *Butea* and *Meizotropis*, belong to the Leguminosae, subfamily Papilionoideae. As an outgroup for the phylogenetic analysis the genus *Kunstleria* has been chosen (Ridder-Numan and Kornet, 1994). This genus resembles *Spatholobus*, but differs in the amount of leaflets, the shape of the pod and the flower.

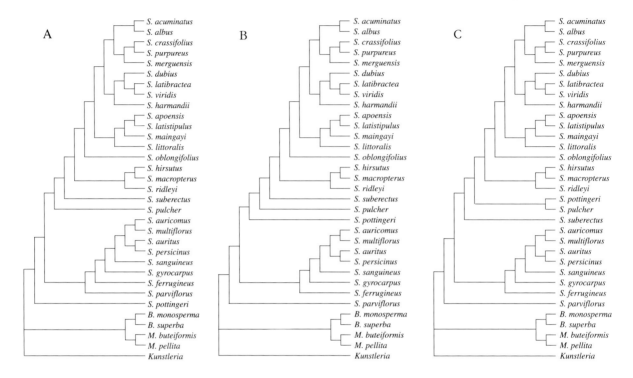

Fig.1. Three most parsimonious trees found by using PAUP under the heuristic search option, 'stepwise addition sequence random', tree-bisection-reconnections (TBR); all characters analysed unordered, uninformative characters ignored and characters with multiple states treated as polymorphisms; length 589, CI 0.46.

Phylogeny

The phylogenetic method is based primarily on the phylogenetic principles of Hennig (1966); the relationship between species can be found by looking for shared derived (apomorphic, homologous) characters, and not by overall similarities, which can arise through the sharing of common inherited 'primitive' characters (plesiomorphy). In this study, which is described in more detail in Ridder-Numan (1996), the phylogeny of the genus *Spatholobus* and allies and the cladistic biogeography of SE Asia are worked out. I will not add to the discussion on the methods and ideas behind phylogenetics, as there is a wealth of information on methodology and already many practical studies have been published (Brooks, 1990; Brooks and McLennan, 1991; Cracraft, 1983, 1988; Forey *et al.*, 1993; Nelson and Platnick, 1981; Page, 1988, 1990; Wiley, 1981, 1987; Zandee and Roos, 1987; for a recent overview: Morrison, 1996). Several computer programs have been developed to analyse the morphological or molecular data, *e.g.*, PAUP (Swofford, 1991, 1993), Hennig86 (Farris, 1988), PeeWee (Goloboff, 1993), and CAFCA (Zandee, 1995).

For a phylogenetic treatment an existing group is used that is supposed to be monophyletic, although monophyly can never necessarily be assumed on beforehand. One of the purposes of cladistics is to find out if groups are monophyletic. In addition, it is necessary to include an outgroup, to polarise the character states used in the phylogenetic analysis.

Data

A data matrix has been made for all species of *Spatholobus*, *Butea* and *Meizotropis*, which consists of 97 characters (relating to pollen morphology, leaf anatomy, and macromorphology). For details see Ridder-Numan (1996). The outgroup *Kunstleria* was considered a single taxon by coding the information on species as a combined column of data. Autapomorphic character states occurring in only one species were ignored, and any state occurring in the rest of the genus was taken as representative. Data for the matrix, as well as samples of pollen and leaves, were taken from herbarium specimens.

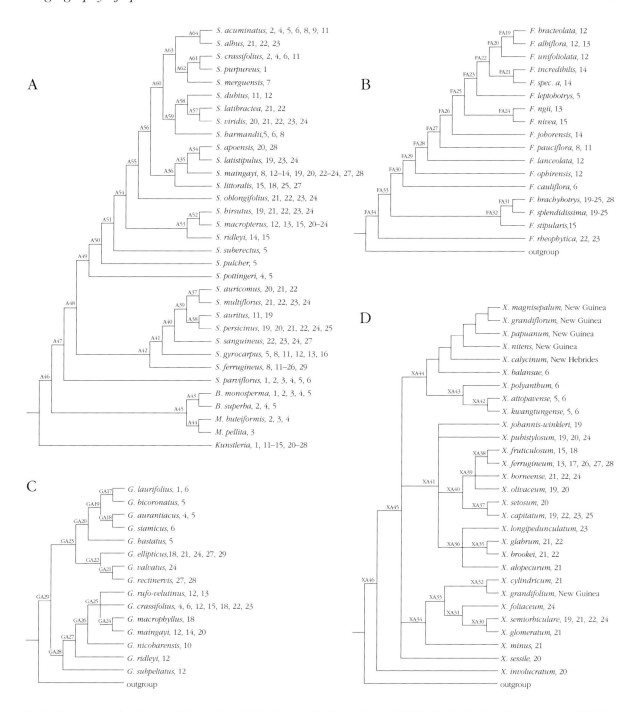

Fig.2. Cladograms of A: *Butea, Meizotropis* and *Spatholobus;* B: *Fordia* (Schot, 1991); C: *Genianthus* (Klackenberg, 1995); D: *Xanthophytum* (Axelius, 1990). Numbers in the cladograms refer to the numbering of the ancestors used for compiling the area-taxon datamatrix. Numbers by species names refer to the areas shown in Fig.5.

Method of analysis

PAUP (Swofford, 1993) was used for the phylogenetic analysis. This program compares the data of all taxa in the data matrix with each other, and calculates cladograms. Out of the possible cladograms the Most Parsimonious Tree (MPT) was chosen, *i.e.,* with fewest character state changes. The heuristic search option was used with a stepwise addition sequence set at random, and as branch-swapping method tree-bisection-reconnection (TBR). All characters were unordered, and uninformative characters, *e.g.,* autapomorphies, were ignored. Characters with multiple states were treated as polymorphic. To avoid homoplasy due to the increased variability by polymorphy, I kept the number of states for each character low, by re-evaluating each character state using MacClade (Maddison and Maddison, 1992), and recoding the characters where possible. In addition Pee-Wee (Goloboff, 1993) was used.

Results

Following the phases of preliminary analyses, checking and rechecking data, coding and recoding characters, an eventual phylogenetic analysis was performed. The analysis with PAUP resulted in three most parsimonious trees (MPT) with a length of 589 steps, and a consistency index of 0.46 (Fig.1). The consistency index (CI) is, according to Sanderson and Donoghue (1989), within the expected ranges for analysis with this number of taxa involved. An evaluation of 10,000 random trees from the set of all possible trees with the same options as above was carried out to see if it was likely that the MPT's found were equal or within the range of randomly found trees. The three MPT's (length = 589 steps) were much shorter than the 805.6 steps of the random tree set, and it is very likely that the data set has more structure than would be expected by chance alone. The analysis with Pee-Wee resulted in a tree which made the genera paraphyletic, and with a length within the range of the random trees.

The choice of the tree

The three MPT's show similar topologies. There are several ways to choose between equally parsimonious trees, including successive approximation weighting (Farris, 1969; Carpenter, 1988). Successive approximation weighting resulted in one most parsimonious tree slightly different from the set of three MPT's found as result of the initial analysis. This tree is 1 step longer than the original set of most parsimonious trees, and is nearly identical to the second of the three MPT's found without weighting (Fig.1B).

When taking into account the amount of support for the branches in the three different trees, differences are minor. Except for some other slight changes, the second tree has more support for the first branch leading to the genus *Spatholobus*. As two of the three equally MPT's are little different it may also be possible to give preference to one of the trees on morphological arguments, although implying subjective weighting criteria. *S. parviflorus* differs from the other species in *Spatholobus*, and more closely resembles *Butea*. Although it certainly belongs to *Spatholobus*, a place in the cladogram as close to *Butea* as possible would be expected. In the first cladogram *S. parviflorus* is included in the *S. ferrugineus* clade, in the other two it is placed at the base of the whole *Spatholobus* clade, which is preferable in my view. In addition, there are similar arguments for other species possible (Ridder-Numan, 1996). As a result of all these considerations tree 1 was discarded. There are no convincing arguments to choose between the remaining two, but I accepted tree 2 as a working hypothesis (Fig.1B).

Cladistic biogeography

Considerable research has been undertaken to find appropriate methods for analysis of data provided by phylogenies of taxa and their distribution, *e.g.,* Bremer (1992, 1995), Brooks (1990), Brooks and McLennan (1991), Cracraft (1983, 1988), Morrone and Carpenter (1994), Nelson and Ladiges (1991), Page (1988, 1990), Turner (1995), Turner and Zandee (1995), Welzen (1992), Wiley (1981, 1987), and Zandee and Roos (1987). Many computer programs are helpful for analysing or comparing taxon-area cladograms, such as: PAUP (Swofford, 1991, 1993) and Hennig86 (Farris, 1988), CAFCA (Zandee, 1995), COMPONENT (Page, 1993), and TAS (Nelson and Ladiges, 1992).

Apart from limitations in the computer implementation, problems may occur when choosing taxa for the historical biogeographical analysis. In order to avoid comparing cladograms that reflect different time spans and different vicariance events, it is best to use groups of about the same age, *e.g.,* not an 'old' group known from the Cretaceous with a more 'recent' group. Different groups of taxa can also have a 'conflicting' distribution pattern. For instance, some taxa have migrated from SE Asia into the Malesian archipelago eastward, whereas others migrated from the east towards SE Asia. In comparing such patterns it is possible that in the analysis the patterns will cancel each other and reflect nothing. With these limitations in mind I used the results of the cladistic analysis of *Spatholobus* and allies, and those of the other groups. As geological background I used Hamilton (1973, 1979, 1988), Hutchison (1989a, 1989b, 1992, 1996), Metcalfe (1988, 1990, 1991, 1994a, 1994b, 1996), Audley-Charles (1987), Dercourt *et al.* (1993), Rangin *et al.* (1990), Hall (1995, 1996).

The data

The data used for the biogeographical analysis were the phylogenies of the genera *Fordia* (Schot, 1991), *Genianthus* (Klackenberg, 1995), *Xanthophytum* (Axelius, 1990) and *Spatholobus* and allies (Fig.2). The genera occur in continental SE Asia and the West Malesian archipelago (Figs.3 and 4). A small monophyletic group in *Xanthophytum* occurs in New Guinea.

The delimitation of the areas

To facilitate comparison of the different area cladograms, areas were recognised from consideration of the distribution of all species (Fig.5). These are not exclusively areas of endemism, but areas of distribution as described by Axelius (1991). Some areas contain endemics (= only occurring in the chosen area) and would be considered areas of endemism, but others are composed of overlapping distributions of more widespread species. Some authors (Sosef, 1994), state that each area should at least contain one endemic, and that areas without endemics ('remnant areas') are not informative. In my opinion, however, species occurring in an area – endemic or not – will reflect their relation to species sets in other areas, and thus a solution for placement of 'remnant areas' is also possible. As a result of using these areas, however, it is possible that some end-branches in the general area cladogram will be empty, because the species present in that particular area, becomes optimised on a node lower in the cladogram, when it is occurring in more than one area in the clade.

The delimitation of the areas also depends on the number of species present in an area. Problems may arise during analysis when there are too few species in an area. Due to lack of data (many zeros in the area-taxon matrix), these areas often turn up at the base of a cladogram, especially when the program requires all zero outgroups. Furthermore, the size of the area can be important. When the size chosen is too large results concerning several parts of the area will remain uninformative. It is clear that by adding other genera to the analysis, delimitation of areas has to be reviewed and changed wherever necessary.

An area-taxon matrix was based on the available cladograms and the distributions (Figs.2, 3, 4, 5). The area-taxon matrix gives the presence (or absence) in an area of each terminal node (= taxon) in the cladogram, as well as for all other (internal) nodes (= ancestors). In this way the structure of the cladogram is added to the matrix. The basic working hypothesis is that speciation results from passive allopatric speciation (vicariance), which implies no taxon-specific mechanisms. Therefore, it is assumed that the ancestor was present in the same areas as its descendants, thus excluding, *e.g.*, dispersal events. The coding of the ancestor nodes is thus not independent, and is in fact an imperfect and unlikely assumption. In the case of incongruencies, dispersal is one of the possibilities to explain these patterns. Dispersal events may be recognised by parallelisms. This factor, often described as 'ad hoc' explanation, has to be kept in mind, because dispersal is an important strategy for plants, often leading to widespread species. In the case of extinction or primitive absence in part of the area, this way of coding leads to reversals (= species from state present to state absent). And, as mentioned above, if only one taxon (or a few) is present in an area (a column with many zeros), the area may be placed artificially low in the area cladogram after analysis.

When one or more genera are absent in part of the total area, these missing data for areas (usually called 'missing areas') can be treated under different assumptions. Assumption 0 (Zandee and Roos, 1987) considers missing data as primitively absent (= the genus was never present). In this case missing data are coded as zero. However, under assumptions 1 and 2 (Nelson and Platnick, 1981) these missing data are treated as uninformative (unknown optimisation state), and coded as '?' (Nicobar Islands for *Spatholobus*). The latter two assumptions are less restrictive than the former method, but require some manipulation for areas. In this analysis missing taxa were treated both under assumption 0 and assumption 1. Widespread species were only treated under assumption 0, *i.e.*, the areas inhabited by the widespread species formed one single area in the past (monophyletic origin). The area-taxon matrices of the genera combined with an artificial (all-zero) outgroup into one large matrix were analysed with PAUP

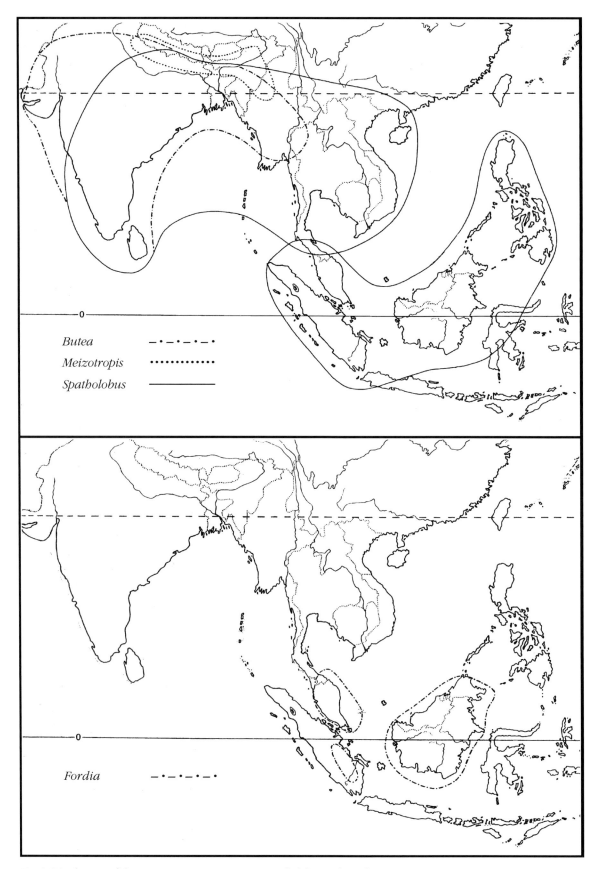

Fig. 3. Distribution of the genera *Butea, Meizotropis, Spatholobus* and *Fordia*.

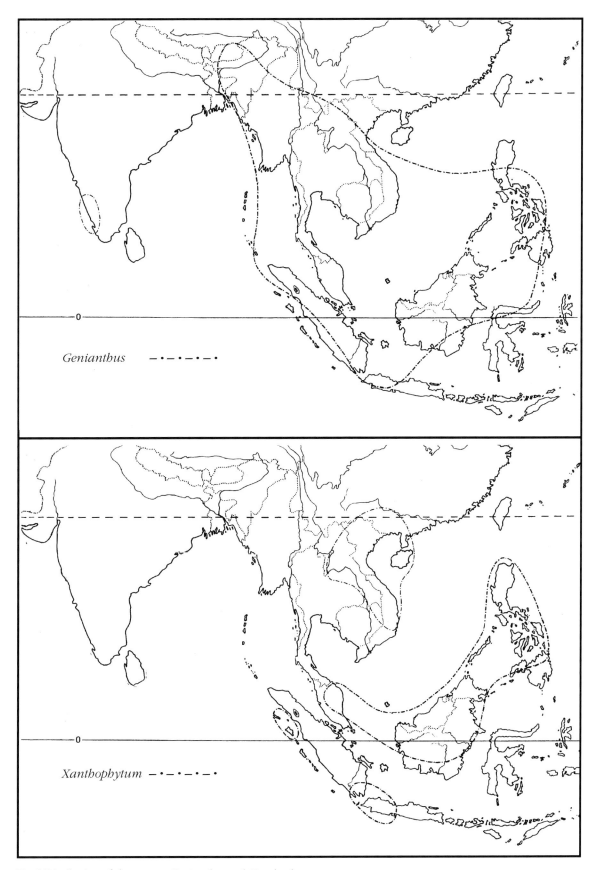

Fig.4. Distribution of the genera *Genianthus* and *Xanthophytum*.

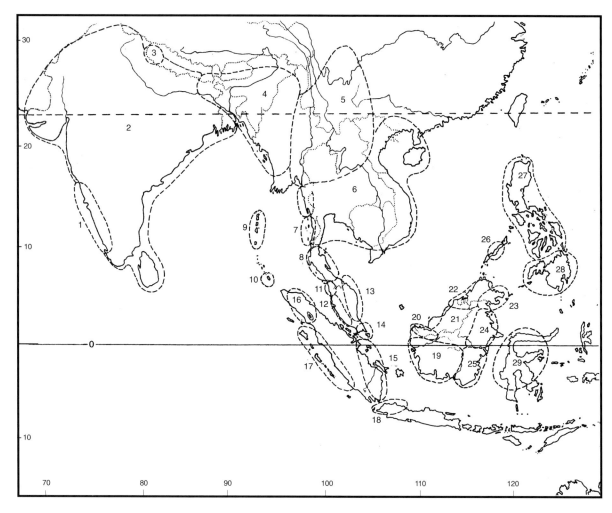

Fig.5. Principal areas of distribution. 1: Kerala; 2: India; 3: Kumaon and W Nepal; 4: Bangladesh and eastern India up to Burma; 5: E Burma, northern Thailand and northern Laos, Yunnan; 6: Indochina, Hainan, S Guangxi, central Thailand and part of Burma; 7: Mergui Archipelago/Tenasserim; 8: S Thailand; 9: Andaman Islands; 10: Nicobar Islands; 11: Penang and Kedah; 12: W Malay peninsula; 13: E Malay peninsula; 14: Singapore; 15: SE Sumatra; 16: N Sumatra; 17: Sumatra coast and islands; 18: W Java; 19: SW Borneo; 20: Semitau; 21: Sarawak and central mountain range of Borneo; 22: NW Borneo; 23: NE Borneo; 24: Central E Borneo; 25: Meratus; 26: Palawan; 27: Northern Philippines; 28: Southern Philippines; 29: Sulawesi.

3.1.1 (Swofford, 1993), both under assumptions 0 and 1 for missing areas. It is important to check for clades solely based on the absence of taxa, as mentioned above. For the analysis with CAFCA (Zandee, 1995) no outgroup was needed. The four matrices (for each group its own matrix) were analysed under both assumptions 0 and 1.

The areas

The distribution areas shown in Fig.5 are based only on the distribution of the species. A large part of these areas coincide with the tectonic terranes of Fig.6. In the text below, geological (and other historical) information on each of the separate areas is summarised. Abbreviations of area names (as used in the computer analyses) are given in parentheses. These abbreviations (instead of full geographical names) are also used further on in the text of the present chapter. In addition to the historical information of the areas, the presence of endemic species is given.

Kerala (1) is part of the Indian continent. This part of India has retained tropical rain forest during all of the last 18,000 years. In other parts of India, there have been extensive grasslands during dry periods, but this seems to have been one of the rain forest refuges. There are four species, one endemic (*S. purpureus*).

India (2) is the continental part of the Indian plate that separated as a large fragment from Australia in the Jurassic (c.160 Ma). It started colliding with Eurasia in the Early Eocene (c.50 Ma), causing the Himalayas to uplift. There are six species, no endemics.

W Nepal (3). West Nepal and Kumaon are part of the Himalayan region, which was uplifted after the Late Eocene. This area changed significantly after the uplift and provided a geographical barrier to the north. There are four species, no endemics.

W Burma/E India (4). Bangladesh and eastern India up to Burma almost entirely overlaps with the W Burma plate, which was uplifted in the Early Eocene by the subducting Indian plate, and during the Cenozoic moved northwards to its present position. This area is composed of several parts, because Bangladesh forms part of the Indian plate. There are nine species, no endemics.

Yunnan/N Thailand (5). East Burma, northern Thailand, northern Laos and Yunnan amalgamated from different continental fragments. East Burma and the north of Thailand (west part) are part of Sibumasu; N Laos and the northeast of Thailand are part of Indochina; Yunnan is the southwest part of S China. S China and Indochina rifted away from Gondwanaland in the Palaeozoic, and sutured along the Song Ma line in the Early Carboniferous. Sibumasu rifted off the Gondwana margin, during the Early to Middle Permian. Indochina and Sibumasu collided in the Late Permian/Early Triassic, and since that time have behaved as a single geological entity. There are thirteen species, five endemics (*F. leptobotrys, G. bicoronatus, G. hastatus, S. pulcher, S. suberectus*).

Indochina (6). Indochina, Hainan, South Guangxi, Central Thailand and part of Burma are parts of Indochina, South China, and Sibumasu respectively. This composite area has a history comparable to area 5. There are eleven species, four endemic (*F. cauliflora, G. siamicus, X. balansae, X. polyanthum*).

Mergui/Tenasserim (7). The Mergui archipelago and Tenasserim are part of Sibumasu. Tenasserim is part of the continent itself. The Mergui archipelago, however, consists largely of much younger material which did not exist before c.30-25 Ma. There are two species, one endemic (*S. merguensis*). Due to incomplete material, the other species (*S. bracteolatus*) was excluded from the analysis.

S Thailand (8). South Thailand is the narrow part of Thailand, near the Isthmus of Kra. It mainly represents the transition between the more seasonal and ever-wet climates during the present. This part has been inundated extensively in the past, during periods of high sea level. Tropical rain forest became established here at least 18,000 years ago. At that time sea level was low and a savannah corridor from Indochina to Java was present, bordered by monsoon forest (Adams, 1995; Steenis, 1961). Only the largest parts of Sumatra and Malaya, extending up to the Isthmus of Kra, were carrying rain forest. Other tropical rain forest refugia were present in Borneo, the Philippines and Sulawesi. There are six species, no endemics.

Andaman Islands and Nicobar Islands (9 and 10) represent an outer arc ridge, produced during subduction of the Indian plate. The Andaman Sea began to open in the Middle Miocene (13 Ma). The Andaman Islands have one species. On the Nicobar Islands there is one endemic (*G. nicobarensis*).

Penang/Kedah, W Malaya and N Sumatra (11, 12, 16). Penang and Kedah, the West Malay peninsula and North Sumatra are part of Sibumasu (see also area 5). For Penang/Kedah there are seven species, no endemics. *Spatholobus auritus* is possibly endemic here. For W Malaya there are sixteen species, three endemics (*F. lanceolata, F. ophirensis, F. unifoliolata*). For N Sumatra there are two species, no endemics.

E Malaya, Singapore, and SE Sumatra (13, 14, 15). The East Malay peninsula, Singapore, and SE Sumatra are part of the same fragment as Indochina. The Bentong-Raub line is the suture between East Malaya and West Malaya (Late Triassic). E Malaya has seven species, one endemic (*F. ngii*). Singapore has seven species, three endemic (*F. incredibilis, F. johorensis, F. spec. A*). SE Sumatra has eight species, two endemic (*F. nivea, F. stipularis*).

Sumatra coast (17) The west coast of Sumatra and its islands are largely the same as the Woyla terranes of Metcalfe (1996). These terranes rifted from the Gondwana margin and were accreted to the Sundaic margin in the Cretaceous. All islands are forearc material and emerged for the first time in the Miocene (R. Hall, pers.comm., 1995). There are two species, no endemics.

W Java (18). West Java is underlain by Cenozoic volcanic rocks formed by subduction of the Indian plate resting on pre-Tertiary continental crust. There are six species, one endemic (*Genianthus macrophyllus*).

SW Borneo (19) is the oldest part of Borneo,

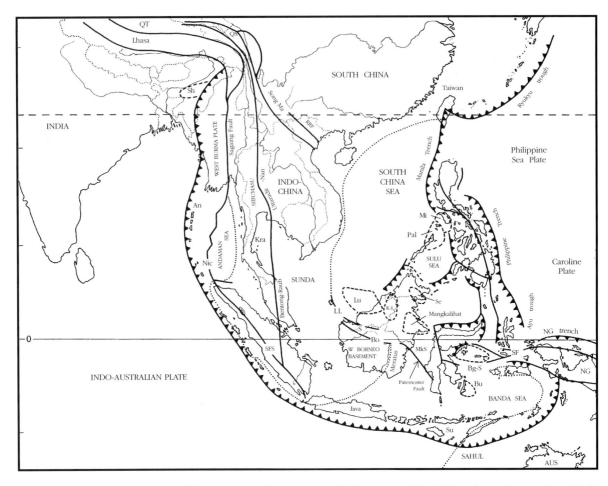

Fig.6. Principal geological features of SE Asia. Heavy lines represent fault systems, indented lines show plate margins with the indentation on the overriding plate, and dotted (bold) lines represent the border of Sahul and Sunda platform. An: Andaman Islands, AUS: Australia, Bg-S: Banggai-Sula, K-L: Kelabit-Longbowan, Kra: Isthmus of Kra, LL: Lupar Line, Lu: Luconia, Mg: Mangkalihat, Mi: Mindoro, MkS: Makassar Strait, NG: New Guinea, Nic: Nicobar Islands, Pal: Palawan, QS: Qamdo Simao terrane, QT: Qiangtang terrane, RRF: Red River Fault, S: Semitau, Se: Segama, SF: Sorong Fault, SFS: Sumatra Fault System, Sh: Shillong Plateau, Su: Sumba.

parts of which may have been emergent since the late Mesozoic. There are thirteen species, one endemic (*X. johannis-winkleri*).

Semitau (20), including the region around Kuching, includes Mesozoic and Lower Tertiary rocks and has been an area close to sea level or emergent since the early Tertiary. There are fifteen species, three endemics (*X. involucratum, X. sessile, X. setosum*).

Sarawak/C Borneo and NW Borneo (21 and 22). Sarawak, the central mountain range of Borneo, and NW Borneo are part of the Rajang-Crocker accretionary complex that was uplifted in the Early Miocene. Parts of the central mountain range have probably been emergent since the Paleocene. There are nineteen species in both areas; in addition, six in Sarawak/C Borneo, of which four are endemics (*X. alopecurum, X. glomeratum, X. grandifolium, X. minus*). No endemics in NW Borneo.

NE Borneo (23) is underlain by Mesozoic and early Tertiary accretionary complexes, and younger Tertiary sediments. Parts of central NE Borneo probably became emergent in the Late Miocene-Pliocene. There are ten species, one endemic (*X. longipediculatum*).

CE Borneo (24). Central East Borneo includes the Mangkahilat peninsula, and the Kutei and Tarakan basins. The Makassar Strait opened in the early Tertiary. The Tarakan and Kutei deltas began to fill the deep basins from the early Miocene. There are seventeen species, two endemics (*G. valvatus* and *X. foliaceum*).

Meratus (25). The Meratus mountains were up-

lifted rather late, probably in the Late Miocene, before which most of SE Borneo was a shallow marine area. There are six species, no endemics.
Palawan (26) is part of the South China margin that rifted away after the opening of the South China Sea in the mid-Oligocene (32 Ma). Part of the Philippines (Mindoro) has the same origin. Rifting stopped at 17 Ma. There are two species, no endemics.
N Philippines and S Philippines (27, 28). The northern and southern Philippines are part of a geologically complex and active region. Most of the southern part formed before the Miocene as an arc at the southern edge of the Philippine Sea plate, whereas Luzon formed part of an arc at the north side of the Celebes Sea-West Philippine Sea basin in the same plate. The N Philippines has five species. The S Philippines has five species. There are no endemics.
Sulawesi (29) is geologically very complex. The groups studied are not well represented in this area and, consequently, no subdivision of Sulawesi can be made using my data. There are two species, no endemics.

Results

The results of the analysis with PAUP under assumption 0 (the most restricted option: information taken as it is) are: 19 equally most parsimonious area cladograms with a length of 366 steps and a consistency index of 0.45 (homoplasy index (HI) = 0.55, retention index (RI) = 0.66, rescaled consistency index (RC) = 0.3). The strict consensus tree and the 50% majority rule consensus tree of the 19 most parsimonious area cladograms are presented in Fig.7. The consensus trees are shown to give an indication of the support to particular groups in the 19 most parsimonious cladograms. The percentages of the trees that support a particular branch are indicated in the 50% majority rule cladogram. Differences within the 19 trees are caused by the changing position of the following areas, more or less within their clade: 1) India/W Burma, 2) Mergui/Tenasserim (and Andaman Islands), 3) the clade of Sumatra coast and Palawan, and N Sumatra. The summary results (for details on all analyses see Ridder-Numan, 1996) show that in the area cladograms obtained there are parts with similar topologies: a continental SE Asia group, a Borneo group, an Isthmus of Kra group and a Malay peninsula–SE Sumatra group.

The areas Sulawesi, N Sumatra, Sumatra coast, and Palawan are not very informative due to a lack of species occurrences and endemics, and appear low at the base of the cladogram. On the Nicobar Islands an endemic is present. This one species, however, is outnumbered in its clade to define a node in the area cladogram by the other six species with a distribution around the Sundaland plateau, and the area is placed basal in the area cladogram. West Nepal is probably at this basal position because the species present in this area (*Butea monosperma, Meizotropis buteiformis, M. pellita,* and *Spatholobus parviflorus*) are all basal in the fundamental cladogram. As working hypothesis the summarized area-cladogram in Fig.7A is used in the rest of this study, because it best shows the four groups of areas mentioned above.

Comparison with the geological information

In the case of splitting areas (vicariance events), it is possible to construct a geological 'cladogram' in which branches represent geographical entities. The splitting up in the case of SE Asia, however, took place at the Gondwanaland margin, and the continental fragments accumulated in different areas in SE Asia. This geological information dominated by accretion makes it difficult to construct a cladogram-like structure (even when reticulations are taken into consideration). Instead, these events are summarised below according to the information found in the literature mentioned before; the different continental fragments and sutures are shown in Fig.6.

Before the Eocene no reliable fossils are known for Leguminosae (Raven and Polhill, 1981; Herendeen *et al.* 1982), and one cannot expect to see geographical vicariance events before that time reflected in the distribution of the group. Only Miocene and younger fossils can be attributed to specific genera within the Leguminosae. This is probably the reason why the areas of distribution on the continent and Indochina (*e.g.,* areas 4, 5, and 6 in Fig.5) are less similar to the geological entities than those in, *e.g.,* Borneo, because they had already welded in the Triassic and formed one area at the time the Leguminosae evolved.

Before the Eocene the Tethys Ocean was subducting beneath the Asian margin forming a volcanic arc from Lhasa, through W Burma, Sumatra and part of Java to W Sulawesi. In the Early Eocene India started to collide with Asia. Between the Oligocene to the mid-Miocene the South China Sea opened, and areas such as Palawan-Mindoro moved southwards, separat-

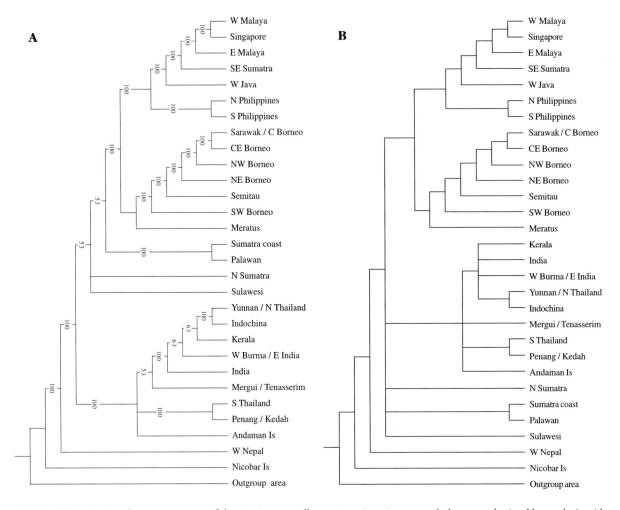

Fig. 7. A. 50% majority rule consensus tree of the nineteen equally most parsimonious area cladograms obtained by analysis with PAUP of the area-taxon matrix of *Spatholobus, Butea, Meizotropis, Fordia, Genianthus,* and *Xanthophytum,* under assumption 0 (missing areas coded as 0). Values on the branches indicate the percentage of trees supporting a specific clade. B. Strict consensus tree. Differences between the 19 MPT's are mainly caused by the changing position of the following areas, more or less within their clade: 1) India/W Burma, 2) Mergui/Tenasserim (and Andaman Islands), 3) the clade of Sumatra coast and Palawan, and N Sumatra.

ing from the Chinese continental margin. In the Palaeocene large parts of Borneo were not yet uplifted, but there was land south of what is now Sarawak and Sabah. Sarawak and Sabah were part of the Rajang-Crocker accretionary complex receiving sediment from the uplifted parts of Borneo as well as from the Indochinese part of the Sunda shelf. The Makassar Strait opened in the Late Eocene/Oligocene resulting in the separation of W Sulawesi. The Meratus mountains were uplifted rapidly in the Late Miocene-Pliocene. In the Miocene the Philippines were formed. The Andaman Sea began to open at the end of the Middle Miocene.

In the Late Oligocene, the sea level was as low as 250 m below the present level, thus the Sunda platform was for the largest part above sea level (Hutchison, 1989a). The climate at that time was probably tropical monsoon, and before that, in the Oligocene, there was a wet tropical monsoon or a tropical rain forest climate. In the mid-Miocene the sea level changed again in three stages up to 220 m above present sea level, after which there was a lowering again of up to 220 m below the present sea level in the Upper Miocene. At that time the Leguminosae became dominant among the phanerogams in India. According to Awasthi (1992) this was after migration from Africa and SE Asia to India. In the Pliocene the sea level rose up to 140 m above the present level. After the Pliocene there were many glacial periods and interglacials which coincided with lower and higher sea levels, although by then the spectacular sea-level rises and falls were over (Hutchison, 1992). Hall (1998 this volume) indicates that the Sunda shelf remained above sea level during most of the Neogene, and that areas below sea level, e.g., the Malay basin and the S China Sea, did so for tectonic reasons and not because of global sea level change. If the groups studied were present in the region at the time these events took place and if they did react to these events, one may expect that this will be reflected in a general area cladogram.

The first major split in the general area cladogram is between the continental SE Asia group (including the Isthmus of Kra group and the Andaman Islands) and the Sundaland groups (Malay peninsula/Sumatra and the Borneo areas). This may be due to one of the high sea levels, which occurred between the mid-Eocene and the Pliocene as indicated above. The reconstructions of Rangin *et al.* (1990) show, that during high sea level there was a separation between the Malay peninsula and Indochina. It is impossible to indicate more exactly in what period this splitting off took place. If the place of the Andaman Islands is correct, *i.e.*, below the split between continental SE Asia and the Sundaland groups, it may be possible to date the split of the Andaman Islands to the time of the opening of the Andaman Sea (13 Ma). This is very speculative, however, because on the Andaman Islands only the widespread, but continental, species *Spatholobus acuminatus* is found. It would be better for future research to analyse taxa with more endemics on the Andaman Islands.

Nearly all area cladograms are resolved for the areas in Borneo. In the general area cladogram (Fig.7), the first area to split off is the Meratus. It is possible that this reflects the uplift of the Meratus mountains in the Late Miocene-Pliocene. The second area is SW Borneo, which is the oldest part of Borneo. The third area is the Semitau, which is in fact the extended part of the Semitau ridge, which was accreted to SW Borneo in the Cretaceous. According to Hutchison (1992), this part of Borneo is probably uplifted in the same period as the Meratus. The next area, NE Borneo, was, also part of the extensive landmass, and uplifted in the same period as the Meratus. This implies that the first four areas in the Borneo group were not yet uplifted prior to the Late Miocene-Pliocene, and were probably submerged during some of the sea level rises. The first four splits may indicate these uplift events and the related changes in climatological/ecological factors. The other parts of Borneo in these times consisted of a large Rajang basin and the Crocker Range (Sarawak/C Borneo and NW Borneo). The last split in the Borneo group is formed by the central parts of Borneo: Sarawak/C Borneo and CE Borneo. The mountains of the Crocker Range form a natural delimitation between CE Borneo and the west central part of Borneo (Sarawak/C Borneo).

In the Malay peninsula group, W Malaya and E Malaya are in most cases placed as sister areas in the cladogram, together with Singapore. SE Sumatra is at the base. In the general area cladogram W Java is basal in the same clade, and the N and S Philippines are also connected to the base. The general area cladogram shows a sister relationship between the Singapore and the W Malaya part of the Sibumasu block; these two have a sister relation to the E Malaya part of the Indochina block. SE Sumatra belongs to the Sibumasu block. However, the connections between these blocks date back to the Triassic, long before the Leguminosae existed there. Although it is possible to recognise different distribution areas, more or less according to the delimitations of Sibumasu and E Malaya, it is not possible to distinguish in the area cladogram between these areas. The delimitation of the distribution areas may also be due to ecological or climatological factors: the areas around Singapore and SE Sumatra are both lowland areas. The E and W Malaya parts of the Malay peninsula consist mainly of a large, mountainous central area. If there is a relation between Malaya and Sumatra, one would expect, on account of distance, that there would be as many species occurring in W Malaya and mountains of NW Sumatra as there are species in the lowlands around Singapore and SE Sumatra. From the distribution of the species in this study the former is not evident. No species occurs exclusively in both W Malaya and N Sumatra. The relation between the Malay peninsula (West, East and Singapore), SE Sumatra and W Java may be understood in geological terms also as part of the border of Sundaland below which the Indian plate was subducting. Only after the rotation of the Malay peninsula, the northward movement of Burma and the opening of the Andaman Sea (13 Ma) were these parts slightly rearranged. The reconstructions of Rangin *et al.* (1990) indicate that during certain periods of high sea level, be-

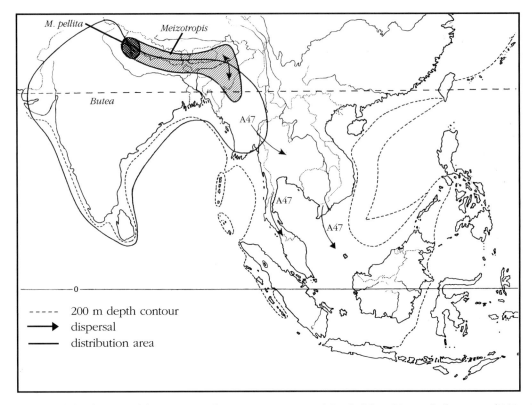

Fig.8. Hypothetical distributions of the ancestors of *Butea, Meizotropis* and *Spatholobus*. Dispersal of ancestor 47 (*Spatholobus*) into the Malay Archipelago.

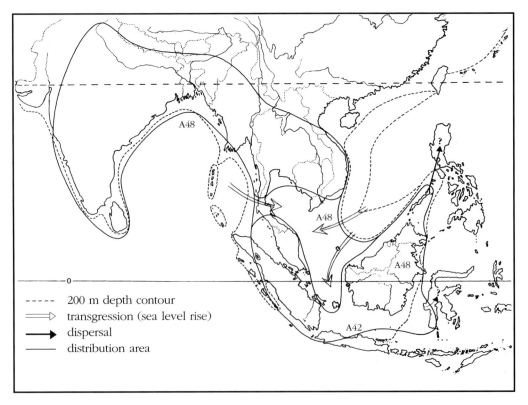

Fig.9. After a sea level rise part of the ancestor 48 became isolated and developed into the ancestor of the *Spatholobus ferrugineus* clade A42.

Biogeography of Spatholobus

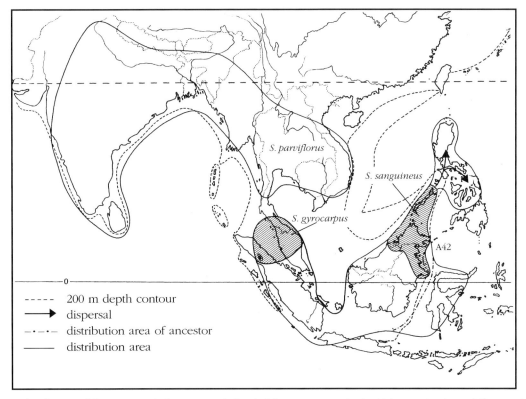

Fig.10. In isolated parts of the area speciation occurred, *Spatholobus gyrocarpus* in the Malay peninsula, and *S. sanguineus* in northern Borneo.

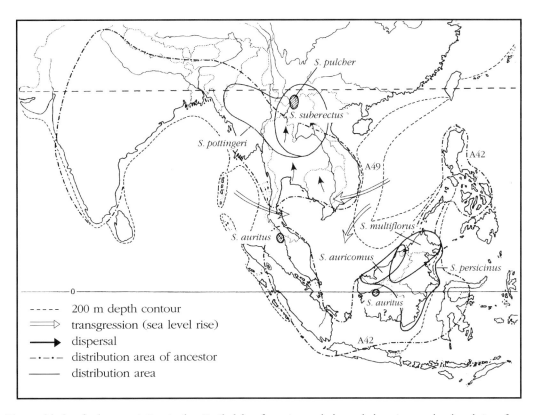

Fig.11. Dispersal led to further speciation in the *Spatholobus ferrugineus* clade, and changing sea level and rise of mountains caused speciation throughout Sundaland.

tween the Middle Eocene and the Pliocene, there were emergent areas in Borneo, the Malay peninsula, large parts of Sumatra, and West Java. Other parts are shown as submerged, although based on continental crust, *e.g.*, the areas of S Thailand and Penang/Kedah. The S Thailand and Penang/Kedah areas are placed very close to each other in all area cladograms. In the general area cladogram this clade is in polytomy to the continental SE Asian areas and the Andaman Islands. S Thailand and Penang/Kedah form the region around the Isthmus of Kra, which was easily inundated during periods of high sea level. Floristically this is the place where the distribution of the species from ever-wet and seasonal areas overlap. This small group of two areas does not contain any endemic species. It is possible that rain forest species do not have as much potential for dispersal than monsoon forest ones, and it is then probable that the invading species came mainly from the north and less from the south (with its rain forest climate).

The history of Spatholobus

Although historical biogeography is about areas, it is also possible to reconstruct the history of speciation within *Spatholobus* in the area. The speciation events indicated by the cladogram of *Spatholobus* and allies (Fig.1) were used to relate the general area cladogram (Fig.7) to the phylogenetic history of the genus as reconstructed in the *Spatholobus* cladogram (see for more details Ridder-Numan, 1996). It is assumed that the phylogenetic cladogram and the general area cladogram correctly reflect the history of speciation of the genus and the historical relation of the areas respectively.

It is rather speculative to superimpose geological events older than Pleistocene on the cladogram of *Spatholobus*. Although some events can be reconstructed, it is not possible to designate the time these events happened, especially because most events, *e.g.*, changes in sea levels, have occurred several times during geological history. Later events, such as Pleistocene ice ages, could easily have disrupted part of the distribution area of some of the species and their ancestors, resulting in disjunct distribution patterns, which in their turn may have resulted in speciation events. As not all species respond with speciation, some disjunct patterns remain.

The ancestor of the genus *Spatholobus* probably originated on the SE Asian continent. The genera *Butea* and *Meizotropis* developed in that part of the area where also the ancestral species of *Spatholobus* occurred; this species spread out over the whole area in a period when the sea level was low and the land emergent: during other periods these parts (the Sunda shelf) were below sea level (Fig.8). On the continent the ancestor of *S. parviflorus* had already developed.

For the *S. ferrugineus* clade, the first split in the cladogram after *S. parviflorus*, it can be concluded that its ancestor speciated in a period of isolation (high sea level) from the continent of SE Asia, spreading out over the whole area of the Malay peninsula, Sumatra, W Java and the oldest parts of Borneo (Fig.9). At the margins of its distribution area species developed, perhaps during a later period of isolation by high sea level that isolated Borneo from the Malay peninsula and Sumatra (Late Oligocene?): firstly *S. gyrocarpus* in the Malay peninsula and N Sumatra; secondly *S. sanguineus* more to the north, on all of the newly uplifted parts of northern Borneo (Fig.10). During later periods of lower sea level, the whole area was inhabited by the predecessor of *S. ferrugineus*, probably an adaptable species that was able to occupy larger regions at all latitudes, and frequently dispersed to parts outside the original range of its ancestor. In Borneo, at lower altitudes, *S. persicinus* developed, and at the same time or later on the island of Penang *S. auritus* (Fig.11). On Borneo, but only in the north, a small-flowered species evolved, later splitting into two species due to the separation by mountainous areas: *S. auricomus* and *S. multiflorus*. In the meantime speciation events occurred on the SE Asian continent as well (Fig.11). For the other species a similar history of invasion, isolation and speciation may be concluded from the results above (Ridder-Numan, 1996).

Conclusions

There have been many events that have led to isolation of parts of the Sundaland plateau from the Asian continent, as is also evident from the history of *Spatholobus*. The most obvious events are changes in sea level, which occurred during all periods. It would be useful if more was known about these sea level fluctuations in the SE Asian region. Studies like that of Kaars and Dam (1994), describing a rather recent but long period (135,000 years) on Java, are helpful.

Spatholobus probably originated on the continent of SE Asia, together with *Butea* and *Meizotropis*. After several invasions of *Spatholobus* into the West Malesian region, and after an equal number of isolation events, the present distribution pattern can be surmised. In this history, vicariance and dispersal events both played a role. The genera *Butea* and *Meizotropis*, and the first species of *Spatholobus*, *S. parviflorus*, remained on the continent. The ancestor of the rest of *Spatholobus* invaded the West Malesian archipelago, after which the history of invasion and isolation began.

A major problem is timing. In a large area it is possible that more than one species develops at the same time of the same widespread ancestor. For example, an ancestor occurring in the entire area from Indochina to Borneo, may split off species in each of the peripheral parts of the area: *e.g.*, the north of Indochina and the northern part of Borneo. The species occupying the majority of the ancestral range after some time may have re-invaded the peripheral areas. This latter species may in turn again be split up into other species in the same peripheral areas. In this way it is evident that we recognise a split in the fundamental phylogenetic cladogram, but not in the general area cladogram, as an area can only appear in one position. It is possible that two species are placed as most related to each other because they are in fact the first splits from the ancestor; these two species are present in areas only related to each other by the ancestral distribution. In the N Borneo example this area is placed as sister to N Indochina, and these two together are placed as a sister group to the area in which the ancestor continued to exist.

For biogeographical analysis, it is not enough that an area contains one endemic species. The endemic should have relations with taxa in 'related' areas as well. It is best to use at least two groups with endemics in other areas (C. J. Humphries, pers. comm., 1997). If there are not enough relations, the endemic – with its distribution area – will stand on its own, *e.g.*, in a basal position in the area cladogram, and will not be very informative. Similarly, it is unnecessary to use only distribution areas that are areas of endemism; areas containing overlapping widespread species can be informative as well. It depends on how informative the relatives are.

Similarly, some areas are not very informative due to a lack of species in the area. It would be worthwhile to expand this analysis with groups with a similar distribution pattern, but with more representatives in India, E India and W Burma, Sumatra, Sulawesi, Palawan and the Philippines.

The groups added should preferably be of about the same geological age – in this case not older than the Eocene – because more ancient groups may reflect a different pattern, and the results will be mixed and end up with an uninformative area cladogram.

On the other hand it would be worthwhile to extend the analysis also with biogeographical runs on larger basic areas, in that way avoiding too many 'absent areas'. In a way the outcome will be expected to be the same, but by different optimisations it is possible that different cladograms will turn out to be the most parsimonious ones.

Another interesting option is the possibility to use the geological entities as found in the geological literature instead of the areas of distribution. With these geological entities it may be possible to say something about the relationships of the areas of distribution if the phylogenetic relationships of the species are reflected in their distribution. If other factors are dominant in the distribution pattern and disturb the historical pattern it will not be possible to make sensible hypotheses on these relationships. In this case it will be even more difficult to score the presence/absence of a species, because the limitations are in this case independent from the distribution of the species. In some cases they may occur to some extent outside the chosen area. It may then be necessary to set a limit to the percentage of occurrences outside the scored area, in that way correcting for an occasional dispersal event.

I believe, however, that most will be gained by adding more genera to the analysis, thus probably giving a better supported generalised area cladogram and distribution areas that are better delimited.

References

Adams, J. M. 1995. Influence of terrestrial ecosystems on glacial-interglacial changes in the carbon cycle. Ph.D. Thesis, University of Marseille II.

Audley-Charles, M. G. 1987. Dispersal of Gondwanaland: relevance to evolution of the angiosperms. *In* Biogeographical evolution of the Malay Archipelago. pp. 5–25. Edited by T. C. Whitmore. Clarendon Press, Oxford.

Awasthi, N. 1992. Indian fossil Legumes. *In* Advances in Legume Systematics, part 4. The Fossil Record. pp. 225–250. Edited by P. S. Herendeen and D. L. Dilcher. Royal Botanic Gardens, Kew.

Axelius, B. 1990. The genus Xanthophytum (Rubiaceae). Taxonomy, phylogeny and biogeography. Blumea 34: 425–497.

Axelius, B. 1991. Areas of distribution and areas of endemism. Cladistics 7: 197–199.

Bremer, K. 1992. Ancestral areas: a cladistic reinterpretation of the centre of origin concept. Systematic Biology 41: 436–445.

Bremer, K. 1995. Ancestral areas: optimization and probability. Systematic Biology 44: 255–259.

Brooks, D. R. 1990. Parsimony analysis in historical biogeography and coevolution: Methodological and theoretical update. Systematic Zoology 39: 14–30.

Brooks, D. R. and McLennan, D. A. 1991. Phylogeny, ecology, and behavior. University of Chicago Press, Chicago.

Carpenter, J.M. 1988. Choosing among multiple equally parsimonious cladograms. Cladistics 4: 291-296.

Cracraft, J. 1983. Cladistic analysis and vicariance biogeography. American Science 71: 273–281.

Cracraft, J. 1988. Deep-history biogeography: Retrieving the historical patterns of evolving continental biotas. Systematic Zoology 3: 221–236.

Dercourt, J., Ricou, L. E. and Vrielynck, B. (Editors.) 1993. Atlas Tethys Palaeoenvironmental Maps. Gauthier-Villars, Paris, 307 pp., 14 maps, 1 plate.

Farris, J. S. 1969. A successive approximations approach to character weighting. Systematic Zoology 18: 374-385.

Farris, J. S. 1988. Hennig86, version 1.5. Computer program and manual. University of Stony Brook, New York.

Forey, P. L., Humphries, C. J., Kitching, I. L., Scotland, R. W., Siebert, D. J. and Williams, D. M. 1993. Cladistics. A practical course in systematics. Clarendon Press, Oxford.

Goloboff, P.A. 1993. Pee-Wee, version 2.0. Computer program and manual. Published by the author, New York.

Hall, R. 1995. Plate tectonic reconstructions of the Indonesian region. Proceedings of the Indonesian Petroleum Association 24th Annual Convention 71-84.

Hall, R. 1996. Reconstructing Cenozoic SE Asia. In Tectonic Evolution of Southeast Asia. Edited by R. Hall and D. J. Blundell. Geological Society Special Publication 106: 153-184

Hall, R. 1998. The plate tectonics of Cenozoic SE Asia and the distribution of land and sea. In Biogeography and Geological Evolution of SE Asia. Edited by R. Hall and J. D. Holloway (this volume).

Hamilton, W. B. 1973. Tectonics of the Indonesian region. Geological Society of Malaysia Bulletin 6: 3–10.

Hamilton, W. B. 1979. Tectonics of the Indonesian region. U. S. Geological Survey Professional Paper 1078.

Hamilton, W. B. 1988. Plate tectonics and island arcs. Geological Society of America Bull. 100: 1503–1527.

Hennig, W. 1950. Grundzüge einer Theorie der phylogenetischen Systematik. Berlin.

Hennig, W. 1966. Phylogenetic Systematics. Urbana.

Herendeen, P. S., Crepet, W. L. and Dilcher, D. L. 1992. The fossil history of the Leguminosae: phylogenetic and biogeographic implications. In Advances in Legume Systematics, part 4. The Fossil Record. pp. 303–316. Edited by P. S. Herendeen and D. L. Dilcher. Royal Botanic Gardens, Kew.

Hutchison, C. S. 1989a. Geological evolution of South-east Asia. Oxford Monographs on Geology and Geophysics 13, Clarendon Press, Oxford.

Hutchison, C. S. 1989b. The Palaeo-Tethyan Realm and Indosinian Orogenic System of Southeast Asia. In Tectonic evolution of the Tethyan Region. Edited by A. M. C. Sengör, Kluwer Academic Publishers, Dordrecht.

Hutchison, C. S. 1992. The Eocene unconformity on Southeast and East Sundaland. Geological Society of Malaysia Bulletin 32: 69–88.

Hutchison, C. S. 1996. The 'Rajang accretionary prism' and 'Lupar Line' problem of Borneo. In Tectonic Evolution of Southeast Asia. Edited by R. Hall and D. J. Blundell. Geological Society Special Publication 106: 247–262.

Kaars, W. A. van der and Dam, M. A. C. 1994. A 135,000-year record of vegetational and climatic change from the Bandung area, West-Java, Indonesia. Palaeogeography, Palaeoclimatology, Palaeoecology 117: 55–72.

Klackenberg, J. 1995. Taxonomy and phylogeny of the SE Asian genus Genianthus (Asclepiadaceae). Botanisches Jahrbuch 117: 401–467.

Maddison, W. P. and Maddison, D. R. 1992. Macclade: Analysis of phylogeny and character evolution. Version 3.0. Sinauer Associates, Sunderland, Massachusetts.

Metcalfe, I. 1988. Origin and assembly of south-east Asian continental terranes. In Gondwana and Tethys. Edited by M. G. Audley-Charles and A. Hallam. Geological Society of London Special Publication 37: 101–118.

Metcalfe, I. 1990. Allochthonous terrane processes in Southeast Asia. Philosophical Transactions of the Royal Society of London A331: 625–640.

Metcalfe, I. 1991. Late Palaeozoic and Mesozoic palaeogeography of Southeast Asia. Palaeogeography, Palaeoclimatology, Palaeoecology 87: 211–221.

Metcalfe, I. 1994a. Gondwanaland origin, dispersion, and accretion of East and Southeast Asian continental terranes. Journal of South American Earth Sciences 7: 333–347.

Metcalfe, I. 1994b. Late Palaeozoic and Mesozoic palaeogeography of eastern Pangea and Tethys. In Pangea: Global environments and resources. Edited by A. F. Embry, B. Beauchamp and D. J. Glass. Canadian Society of Petroleum Geologists Memoir 17: 97–111.

Metcalfe, I. 1996. Pre-Cretaceous evolution of SE Asian terranes. In Tectonic Evolution of Southeast Asia. Edited by R. Hall and D. J. Blundell. Geological Society Special Publication 106: 97–122.

Morrison, D. A. 1996. Phylogenetic tree-building. International Journal for Parasitology 26 (6): 589-617.

Morrone, J. J. and Carpenter, J. M. 1994. In search of a method for cladistic biogeography: an empirical comparison of Component analysis, Brooks Parsimony analysis, and Three-Area Statements. Cladistics 10: 99–153.

Nelson, G. and Ladiges, P. 1991. Standard assumptions for biogeographic analysis. Australian Systematic Botany 4: 41–58.

Nelson, G. and Ladiges, P. 1992. TAS (MS-DOS computer program). Published by the authors, New York and Melbourne.

Nelson, G. and Platnick, N. I. 1981. Systematics and biogeography: cladistics and vicariance. Columbia University Press, New York.

Page, R. D. M. 1988. Quantitative cladistic biogeography: constructing and comparing area cladograms. Systematic Zoology 37: 254–270.

Page, R. D. M. 1990. Component analysis: a valiant failure? Cladistics 6: 119–136.

Page, R. D. M. 1993. COMPONENT, version 2.0. Computer program and manual. Natural History Museum, London.

Rangin, C., Pubellier, M., Azéma, J. et al. 1990. The quest for Tethys in the western Pacific. 8 paleogeodynamic maps for Cenozoic time. Bulletin de la Société Géologique de France, 8, tome 6: 909–913 + 8 colour maps.

Raven, P. H. and Polhill, R. M. 1981. Biogeography of the Leguminosae. In Advances in Legume Systematics, part 1. pp. 27–34. Edited by R. M. Polhill and P. H. Raven, Royal Botanic Gardens, Kew.

Ridder-Numan, J. W. A. 1992. Spatholobus (Leguminosae–Papilionoideae): a new species and some notes. Blumea 37: 63–71.

Ridder-Numan, J. W. A. 1996. Historical Biogeography of the

Southeast Asian Genus Spatholobus (Legum.-Papilionoideae) and its allies. Ph.D. Thesis, Rijksherbarium/Hortus Botanicus, University of Leiden.

Ridder-Numan, J. W. A. and Kornet, D. J. 1994. A revision of the genus Kunstleria (Leguminosae–Papilionoideae). Blumea 38: 465–485.

Ridder-Numan, J. W. A. and Wiriadinata, H. 1985. A revision of the genus Spatholobus (Leguminosae–Papilionoideae). Reinwardtia 10: 139–205.

Sanderson, M. J. and Donoghue, M. J. 1989. Patterns of variation in levels of homoplasy. Evolution 43: 1781–1795.

Sanjappa, M. 1987. Revision of the genera *Butea* Roxb. Ex Willd. and *Meizotropis* Voigt (Fabaceae). Bulletin of Botanical Survey in India 29: 199–225.

Schot, A. M. 1991. Phylogenetic relations and historical biogeography of *Fordia* and *Imbralyx* (Papilionaceae: Millettieae). Blumea 36: 205–234.

Sosef, M. S. M. 1994. Refuge begonias. Taxonomy, phylogeny and historical biogeography of Begonia sect. Loasibegonia and sect. Scutobegonia in relation to glacial rain forest refuges in Africa. Wageningen Agricultural University Papers 94-1: 1–306.

Steenis, C. G. G. J. van 1961. Introduction. *In* M. S. van Meeuwen, H. P. Nooteboom and C. G. G. J. van Steenis, Preliminary revision of some genera of Malaysian Papilionaceae I. Reinwardtia 5: 420-429.

Swofford, D. L. 1991. PAUP: Phylogenetic analysis using parsimony, version 3.0. Computer program and manual. Illinois Natural History Survey, Champaign.

Swofford, D. L. 1993. PAUP: Phylogenetic analysis using parsimony, version 3.1.1. Computer program and manual. Illinois Natural History Survey, Champaign.

Turner, H. 1995. Cladistic and biogeographic analysis of Arytera Blume and Mischarytera gen. Nov. (Sapindaceae), with notes on methodology and a full taxonomic revision. Blumea Supplement 9: 1–230. Rijksherbarium/Hortus Botanicus, Leiden.

Turner, H. and Zandee, M. 1995. The behaviour of Goloboff's tree fitness measure F. Cladistics 11: 57–72.

Welzen, P. C. van 1992. Interpretation of historical biogeographical results. Acta Bot. Neerl. 41: 75–87.

Wiley, E. O. 1981. Phylogenetics: the theory and practice of phylogenetic systematics. Wiley-Interscience, New York.

Wiley, E. O. 1987. Methods in vicariance biogeography. *In* Systematics and evolution: a matter of diversity: pp. 283–306. Edited by P. Hovenkamp *et al.*, Utrecht University.

Zandee, M. 1995. CAFCA – a Collection of APL Functions for Cladistic Analysis. Macintosh and PowerPC version 1.5d. Computer program and manual. Institute of Evolutionary and Ecological Sciences, Leiden University, Leiden.

Zandee, M. and Roos, M. C. 1987. Component-compatibility in historical biogeography. Cladistics 3: 305–332.

Biogeography of *Aporosa* (Euphorbiaceae): testing a phylogenetic hypothesis using geology and distribution patterns

Anne M. Schot
Rijksherbarium/Hortus Botanicus, PO Box 9514, NL-2300 RA Leiden, The Netherlands
Email: schot@rulrhb.leidenuniv.nl

Key words: *Aporosa*, hybrids, phylogeny, biogeography

Abstract

A cladistic analysis of the SE Asian angiosperm genus *Aporosa* (Euphorbiaceae) leads to a hypothesis about the possible formation of hybrid species on New Guinea. It is suggested that members of two different Sundanian lineages reached New Guinea independently and hybridized on contact. Distributional patterns and current interpretations of the geological evolution of the area confirm the possibility of two different pathways to New Guinea from Sundaland: one along the Outer Melanesian arc, and a second via the Banda arc.

Introduction

Aporosa is a genus of about 80 species of small trees which inhabit the tropical rain forests of SE Asia. Six species are found in the remnant forests of South India and Sri Lanka; the others are found from the Himalaya southwards into the Malay archipelago and onwards to the Solomon Islands. Borneo, with 30 species, of which 10 are endemic, and New Guinea, with probably up to 20 endemic species, many of which are still poorly known, are the main centres of diversity for *Aporosa* (Fig.1).

Aporosa has been revised and cladistically analysed (Schot, 1998). Shaping the cladogram for *Aporosa* has met with some difficulties. Some problems are caused by the size of the genus. There are 110 taxa, including 82 species, 8 varieties, 6 forms, 7 poorly known species, and a multiple outgroup. The number and quality of the available macromorphological characters which were used to build the data set presented another problem. *Aporosa* belongs to the subfamily Phyllanthoideae in the family Euphorbiaceae. The Phyllanthoideae comprise the paraphyletic part of the family as a whole, in which the more derived 'eu-Euphorbiaceae' are thought to nest. The main characteristic of the Phyllanthoideae is the lack of specialized characters. *Aporosa* conforms completely with this characterization; it is non-descript. The species are minimally distinct from each other and defined on combinations of characters. There are very few unique apomorphies. 83 multistate macromorphological characters can be scored for the 110 taxa. Of these characters, most turned out to be phylogenetically uninformative because of the recognition of combinations instead of unique characters.

Due to these two factors, cladistic analysis of *Aporosa* has yielded thousands of equally parsimonious solutions. It is impossible to make biologically meaningful choices among these; the accuracy of any one cladogram is probably very low. However, it was found that many of the ambiguities of the various solutions resulted from the placements of one category of taxa; part of the New Guinean species. These particular New Guinean species possess anomalous character combinations in comparison with the Sundanian species and the remainder of the New Guinean species. The character combinations that are found in the majority of species do not apply. The species vary interspecifically in characters that are stable in West Malesia and are stable for characters that vary interspe-

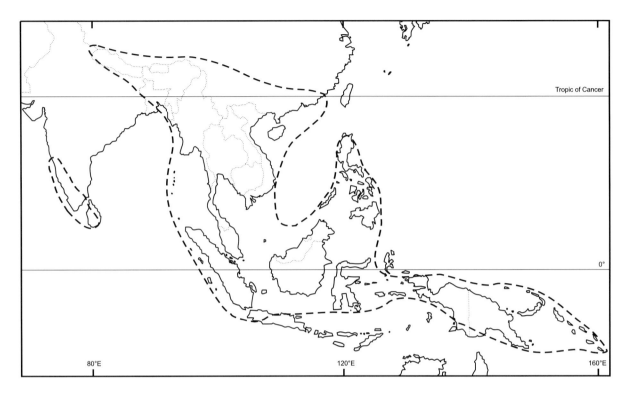

Fig.1. The distribution of *Aporosa*.

cifically in West Malesia. In an attempt to clarify some of the cladistic patterns, an analysis was conducted excluding these anomalous New Guinean taxa. This resulted in a strong reduction of the number of solutions. In consequence, it was hypothesized that the species that possess the anomalous character combinations and disrupt the cladogram might be the products of extensive hybrid speciation between unrelated lineages. The remaining New Guinean taxa, which were placed in two lineages in the cladogram when excluding the anomalous ones, are interpreted to represent the two parental clades that independently reached New Guinea and hybridized on contact.

In this paper I investigate whether historical biogeographical patterns support this hypothesis. First, I give a summary of the main phylogenetic patterns found in the cladistic analysis. Next, I describe the distributions of selected groups of *Aporosa* and try to identify those events in the geological evolutionary history of the Malay archipelago that could be relevant in the evolution of *Aporosa*. Comparison of the distributional pathways as suggested by the geology of the area with the phylogenetic patterns in *Aporosa* may then be used as independent test of the hypothesis of hybrid speciation.

Phylogenetic patterns in *Aporosa*

Cladistic analysis was undertaken with the computer program HENNIG86 (Farris, 1988). This program searches for the most parsimonious trees. These are calculated by looking for trees with the fewest character changes, or the highest consistency index (CI). The retention index (RI) is a measure of the quality of the characters.

The data matrix for *Aporosa* contained 110 taxa, of which 103 were taxa of *Aporosa* and seven were outgroup taxa, and 83 macromorphological characters. Analysis with the tree search commands 'mhennig*' and 'bb*' produced 875 trees of length 1428, CI 0.11, and RI 0.57. The selection of the characters and outgroup and the results of the analyses are treated in detail by Schot (1998). Here I give a brief outline of the procedures followed and the results. As can be seen, the CI and RI in the analysis are low. Since a CI and RI of 1.00 indicate perfect fit, the low CI and RI indicate that the found cladograms of *Aporosa* are untrustworthy. In addition, because an overflow of computer-memory was encountered during the search, more equally parsimonious trees are present. The possibility that the trees found are not the most parsimonious can not be ruled out.

To search for more equally parsimonious or more parsimonious trees various options of the phylogenetic analysis were changed. For instance, characters were weighted differently or only subsets were taken, the taxa were given in a different order, or other tree calculating commands were used. All various options resulted in even more different solutions. It was found that all trees shared a pattern in which 8 clusters of species could be identified, 3 of which were monophyletic groups. The other 5 components were in some analyses placed as paraphyletic tails to the monophyletic groups, and in other analyses as a monophyletic sister groups. Part of the species had no constant placement among the cladograms. Together a general pattern was formed consisting of three lineages, each containing one monophyletic group with one or more of the other components as its sister or paraphyletic tail and a changing compound of other taxa. An Adams consensus tree is given in Fig.2 showing these 8 clusters and the switching remainder. A ninth cluster in the lower half of Fig.2 is in my view a completely arbitrary clustering of incompletely known anomalous taxa and does not deserve recognition. I have simplified the tree by not naming the species in the clusters. They are named after the most common taxon in the cluster. Taxa not placed in any of the clusters are taxa that swap positions in the various solutions. Accolades with numbers indicate how the three lineages are formed. Note that some clusters and species switch between the lineages.

The Adams consensus tree of Fig.2 was calculated using PAUP 3.1.1 (Swofford, 1991) and all the cladograms found in the various analyses using the complete and unweighted set of characters. Adams consensus trees summarize the patterns found by checking the clades for identical elements. In contrast with the more often used strict consensus tree, an Adams consensus tree will thus not collapse clades in which one or a few taxa are missing in part of the solutions. For *Aporosa*, where some taxa are switching and some clusters of taxa are either paraphyletic or monophyletic, a strict consensus gives no solution at all. The Adams consensus tree moves the swapping taxa to a polytomy at the base and keeps the unchanging part of the clade intact. In this way, the consensus tree can be used for identifying the identical elements between the thousands of found solutions. However, it will sometimes show paraphyletic elements in a monophyletic configuration. Thus the 5 non-monophyletic clusters of species of Fig.2 are

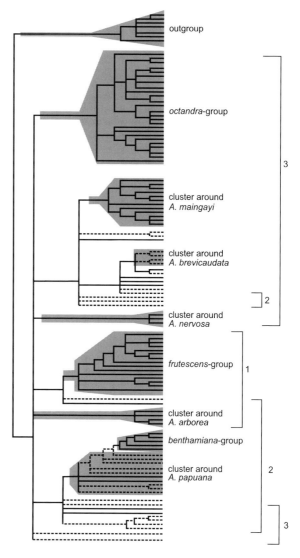

Fig.2. Adams consensus tree of the 4374 trees found in various phylogenetic analyses of *Aporosa*. For simplicity the names of taxa have been left out. Eight clusters of taxa can be identified: three named monophyletic groups and five paraphyletic components. Species not part of any of these components represent the ambiguous paraphyletic tail. The three lineages that were present are indicated by accolades on the right with numbers. Some taxa or clusters were variously placed. Asiatic and West Malesian species are in black, New Guinean species in dashed lines. Note the scattering of the New Guinean species.

depicted monophyletically only as a result of the Adams consensus technique. The three lineages do not show in the consensus tree, because their identical elements are the 8 clusters; the lineages themselves are formed of these clusters and a varying component of other taxa. They can be seen by studying the cladograms.

The three monophyletic groups are the *frutes-*

cens group, the *benthamiana* group, and the *octandra* group (Fig.2). The species of the first two groups occur on Sundaland, in the Malay peninsula, Sumatra, West Java, and Borneo, and some species reach the Philippines and NE Sulawesi. The *octandra* group is mainly found on the mainland of Asia. The six endemic species of Sri Lanka and South India all belong in this clade, and two species are widespread in West Malesia.

Two of the five paraphyletic components are rather large: the species around *A. papuana* and the species around *A. maingayi*. It is possible that these two components only show up in a paraphyletic configuration in part of the analyses because of the way in which the phylogenetic algorithm works. They may represent the local sister group to the last two monophyletic clades respectively. Because they do not possess a synapomorphy, parsimony analysis optimizes them in a pectinate diagram, but this is not necessarily accurate (Baum and Estabrook, 1996). *Aporosa maingayi* and its allies are Sundanian, *A. papuana* and allies New Guinean.

The remaining three smaller recognizable paraphyletic elements are the Sundanian species around *A. nervosa* and those around *A. arborea*, and the New Guinean species around *A. brevicaudata*. Together with the taxa without consistent placement they form the paraphyletic tail. Their configuration is ambiguous, and nothing can be said at this point about the taxa in it.

Biogeographically speaking, the pattern of the consensus tree shown in Fig.2 is disturbing: the distribution of the Sundanian and the New Guinean taxa in the paraphyletic tail is completely mixed. As seen in Fig.2, many of the New Guinean species (dashed lines) are interspersed among their Sundanian relatives (in black). Interpreted as biogeographical pathways, this would mean a constant coming and going between Sundaland and New Guinea. An evolutionary scenario with fewer dispersals might be more plausible.

To seek a less ambiguous pattern, and one with, preferably, the New Guinean taxa less scattered, further cladistic analyses were undertaken using only subsets of taxa. Taxa placed in the polytomies in the Adams consensus tree in Fig.2 are taxa that have ambiguous placements. It may clarify part of the phylogenetic relationships if such taxa are omitted from the analysis. Thus subanalyses were undertaken with various taxa of the basal polytomies excluded. One subanalysis (the others are treated by Schot, 1998) was an analysis without the New Guinean taxa of the basal polytomy. These particular taxa are different from all West Malesian taxa and the remaining New Guinean taxa around *A. papuana* and *A. brevicaudata* in having an anomalous combination of the characters they are described on. The species can be identified by characters that vary within a species in Malesia, such as size differences, whereas characters that are used in West Malesia to delimit the species, such as inflorescence, are variable. It was found that excluding these taxa resulted in a strong reduction of the number of trees found. 24 trees were found with length 977, CI 0.17 and RI 0.53. Besides a reduction in the number of solutions, the ambiguity between the solutions decreased. Differences were only found in minor branches, resulting in a better resolved consensus among them. The other subanalyses with different taxa of the basal polytomy excluded did, in contrast to the analysis without the anomalous New Guinean taxa, not give fewer trees or a better solution of the consensus pattern.

This obvious improvement of the cladistic analysis when excluding part of the taxa led to the idea that these particular taxa were the main causes of disturbance of the phylogenetic patterns. The question then arises of what to do with the excluded New Guinean taxa. How should they be placed in the phylogeny of *Aporosa*? Where did they come from and why do they disrupt the phylogenetic pattern so much? The macromorphological characters of these species give an answer to this last question. The species are characterized by anomalous combinations of characters. The unchanging combinations as shown by the Sundanian species are disrupted and recombined. Additionally, more primitive characters are retained by these various New Guinean taxa. And finally, exceptional characters emerge that have no parallel within the genus. Such characteristics might indicate hybrid species (Rieseberg, 1995). Furthermore, when formed between distantly related lineages, hybrid species are expected to disrupt phylogenetic patterns (McDade, 1992). With this idea in mind the data set was analysed again, but now with only subsets of taxa and recoded characters. A new consensus tree, built by combining the various results of the subanalyses after excluding the anomalous New Guinean taxa, is given in Fig.3. The basic pattern is an Adams consensus tree, with polytomies solved as far as possible based on cladistic analyses of smaller groups. Not all polytomies could be solved in a trustworthy manner and

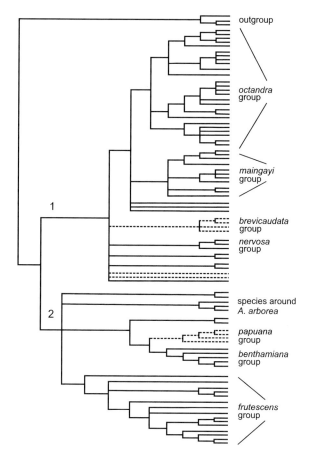

Fig.3. Accepted combined consensus tree for *Aporosa*. The names of the taxa are left out. The consensus tree was shaped on several subanalyses. The New Guinean taxa with anomalous character combinations are excluded. The remaining taxa settle into two lineages, indicated by numbers, each with a New Guinean component, shown in dashed lines. The eight components of Fig.2 are also indicated.

these were left as polytomies. However, only the broad picture is considered here. The separation of the remaining New Guinean species into two lineages was evident in other analyses as well and can be viewed with confidence.

Looking at the consensus tree of Fig.3 the broad picture that was found before with three intermingling lineages (Fig.2), has changed to consist of two distinctly separate lineages. Both contain one component of New Guinean species (Fig.3, dashed lines). One of the lineages contains the now monophyletic cluster with *A. papuana*, the other the cluster with *A. brevicaudata* and two further species, *A. nigropunctata* and *A. longicaudata*. Macromorphological characters and results of other subanalyses showed that these two species belonged to the ambiguously placed West Malesian species, rather than to the anomalous New Guinean species. The clade with *A. papuana* shows a close relationship with the monophyletic *frutescens* and *benthamiana* groups and the cluster around *A. arborea*; the clade with *A. brevicaudata* is more related to the *octandra* group and the clusters around *A. maingayi* and *A. nervosa* and the remainder of the paraphyletic tail. In a biogeographic sense, only two colonization events from West Malesia to New Guinea now need to be proposed, which is a more parsimonious scheme than that mentioned above.

The phylogenetic pattern found with two independent lineages that occur on New Guinea supports the hypothesis that two distinct lineages colonized New Guinea. The fact that the pattern is disrupted by including the anomalous species is consistent with the hypothesis that these originated by hybridization events that followed the double colonization. Fig.3 shows the two possible parental lines to the hybrid species and their nearest West Malesian relatives. Next, distribution data can be studied to examine if the two New Guinean clades inhabit different areas and if the supposed hybrid species occur in their contact zones.

Distributions of *Aporosa*

The distributions of the New Guinean species are shown in Fig.4. The New Guinean species are divided into three groups: the two separate lineages in the cladogram of Fig.3 and the excluded putative hybrids. The species of the first clade consisted of the four species around *A. brevicaudata*, together with *A. nigropunctata* and *A. longicaudata*. Their distribution is depicted in Fig.4A. The species of the second lineage, the *papuana* group, are in Fig.4B. Fig.4C shows the distribution of the putative hybrid species.

The four species around *A. brevicaudata* are montane species of the central mountains ranging from the Arfak Mountains in the Bird's Head to Markham in the east. The two lowland species *A. longicaudata* and *A. nigropunctata* are found in the easterly and westerly adjacent lower parts of this range respectively. *Aporosa nigropunctata* also occurs sporadically in the Papuan peninsula (Fig.4A).

The *papuana* group consists of species of the lowland rain forests and seems mostly concentrated in north New Guinea, with one species, *A. brassii*, near the Vailala River on the southern side of the central mountains. *A. papuana* itself

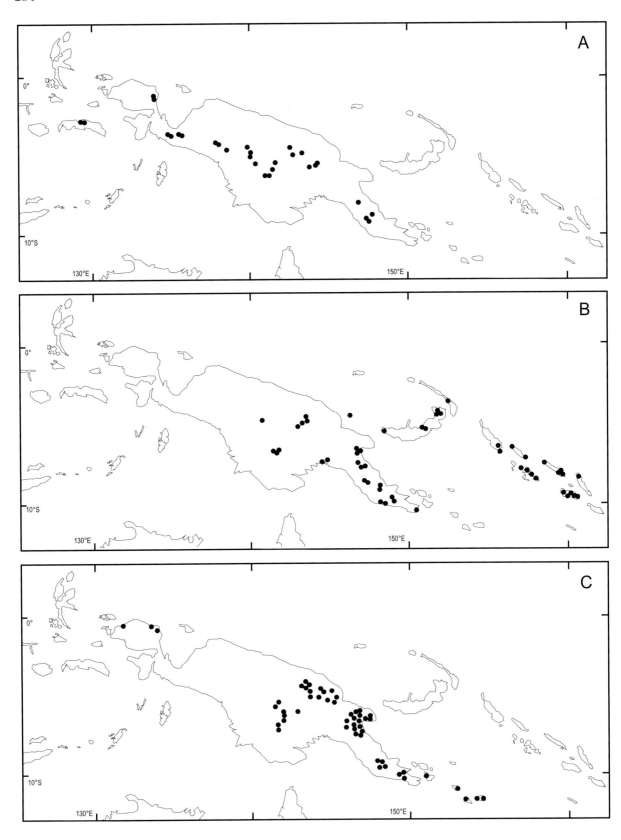

Fig.4. Distribution of the three groups of New Guinean species of *Aporosa*: A: the four species around *A. brevicaudata* with *A. nigropunctata* and *A. longicaudata*; B: the species around *A. papuana*; and C: the putative hybrids.

is widespread over north New Guinea, the Bismarck archipelago, and reaches the Solomon Islands. A few specimens have been found in the Papuan peninsula (Fig.4B).

The putative hybrids are found mostly in the Sepik area, the western part of the central mountain range and adjacent lowlands, and in the Owen Stanley range of the Papuan peninsula southwards into the Papuan Islands (Fig.4C). The places where this last group is found coincide largely with places where the first two groups co-occur. Thus, the hypothesis of hybridization is consistent with the distributional data. A third confirmation would be if historical geographical patterns allow at least two different ways and/or times to disperse from Sundaland to New Guinea. This is discussed next.

Relevant geological history

The Euphorbiaceae are assumed to have originated in the Cretaceous in the Old World tropics, since the largest number of primitive taxa of the basal subfamily Phyllanthoideae are found in Africa/Madagascar. From here they spread over America and the Malay archipelago (Webster, 1994). This may mean that *Aporosa*, whose greatest diversity is found in the tropical forests of Sundaland, originated in a time when the Sibumasu, East Malayan, and the SW Borneo block had already accreted to the Asian mainland (see Ridder-Numan, 1996, or Metcalfe, 1998 this volume, for a summary of the drifting sequences of the various terranes). This sets the starting point of this story at its earliest in the Cenozoic (Metcalfe, 1994). The onset of the collision of India in the early Eocene probably caused the clockwise rotations of the various islands on the Sunda shelf. This might have given *Aporosa* the chance to establish itself here over the next few million years, as is indicated by the mainly Sundaland distribution of the more primitive species of the genus (Fig.3).

The consensus tree shows that at least the species around *A. papuana* originated from West Malesian relatives; for the second New Guinean component this is hard to tell with the base of the lineage unresolved. My interpretation is that, after originating, the genus dispersed over the archipelago, including part of New Guinea. These species are those that are found in the polytomies. A later speciation event gave rise to the more derived resolved components, such as the three monophyletic groups. This included a second invasion of New Guinea by the ancestor to the species around *A. papuana*.

An origin of the primitive *Aporosa* in New Guinea, which then spread to West Malesia and returned to New Guinea in the form of *A. papuana*, is less plausible, since the outgroup occurs in West Malesia. However, this would still require two separate pathways. Dispersal events to New Guinea became possible with the forming of the Tertiary island arc complex later in the Eocene and the arrival of Australia in Malesia in the Miocene (Boer, 1995). Since I am here mainly concerned with the possible distributional pathways for *Aporosa* to reach New Guinea from Sundaland, I will concentrate on the geological events in the period from the Eocene to recent that allowed dispersal from Asia to Australia (Fig.5).

Boer (1995) gives a summary of the most important geological events in Indo-Melanesia. The Indo-Melanesian part of the Tertiary island arc system started to develop about 40 million years ago. Remnants of this arc are found in parts of the central Philippines, northwest and east Sulawesi, and north and east New Guinea. This part has been termed the Outer Melanesian arc (OMA) in biogeographic studies and served as an important pathway for dispersal between SE Asia and Australia (Boer, 1995). At 30 Ma the OMA connected with SE Asia, possibly somewhere south of the Philippines and started breaking. At 20 Ma the Australian craton had arrived in the area and collided with the fragments of the OMA. Little is known of the amount of emergent land on this OMA; dispersal patterns of biota suggest that at times some dry land must have been present to allow colonization.

In southeast Indonesia the amalgamation of the various fragments of the Philippines and Sulawesi and the continuous movement of Australia was also felt (see Hall, 1998 this volume). According to Daly *et al.* (1991) at about 10 Ma the Banda arc bent and Buru-Seram rotated, initiating accretion of the northwest margin of Australia to south Indonesia (see also Burrett *et al.*, 1991). This might have opened connections of the Sunda island chain to Buru-Seram and onto the Bird's Head of New Guinea.

In the meantime, New Guinea itself was growing. Pigram and Davies (1987) reconstructed the accretion history of New Guinea (Fig.6). According to them fragments of the Tertiary arc chain were accreted onto the passive margin of the Australian craton in various steps. The Sepik terrane was the first to dock at 25 Ma, followed at 15 Ma and 10 Ma by the Composite East Papuan block and north New Guinea respectively.

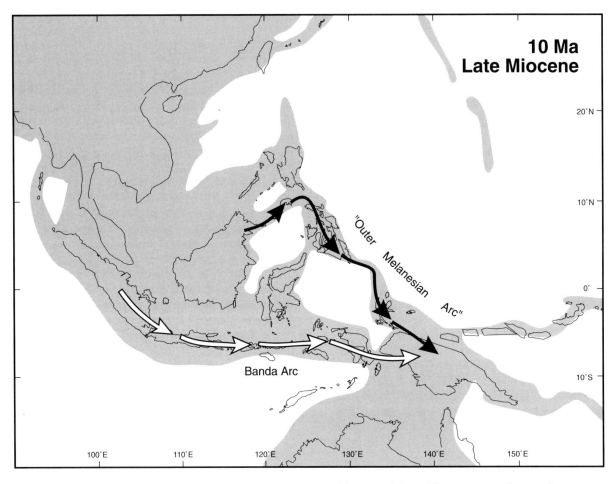

Fig.5. A geological reconstruction of SE Asia at 10 Ma. The two possible routes followed by *Aporosa* to disperse from Asia to Australia are indicated by arrows. Modified from Hall (1998 this volume).

In the west, the south Bird's Head was accreted at 10 Ma following the bending of the Banda arc and the rotation of Buru-Seram. At 2 Ma the Arfak Mountains in the north Bird's Head and the Solomon Island chain followed suit.

Two possible pathways can be supposed on the basis of these geological events. The first goes in the north along the fragments of the OMA between the Philippines, Moluccas, northern New Guinea, and New Guinea. A second dispersal route in the south could have followed the Sunda islands, Banda arc, Buru-Seram, to the south Bird's Head and, after docking at 10 Ma, to the rest of New Guinea (Fig.5).

Comparison of phylogenetic, distribution and geological patterns

With the phylogenetic patterns, the distributions, and the geological events we can try to complete the picture of the putative colonization events of *Aporosa* from Sundaland to New Guinea. The cladogram had two lineages (Fig.3). The New Guinean clade of the first lineage, the *brevicaudata* group with *A. nigropunctata* and *A. longicaudata* was concentrated mostly in the central mountain range, the Arfak Mountains, and some specimens in the Papuan peninsula (Fig.4A). This clade showed close relationships with the Asiatic *octandra* group, the Sundanian *maingayi* group, the Sindanian species around *A. nervosa,* and the unresolved part of the cladogram (Fig.3). The distribution of these relatives is shown in Fig.7A. The *octandra* group, which is the only clade found mainly on the mainland of Asia, is indicated by its derived position in the cladogram not to be relevant to the dispersal to New Guinea of the more primitive species. This

Biogeography of Aporosa

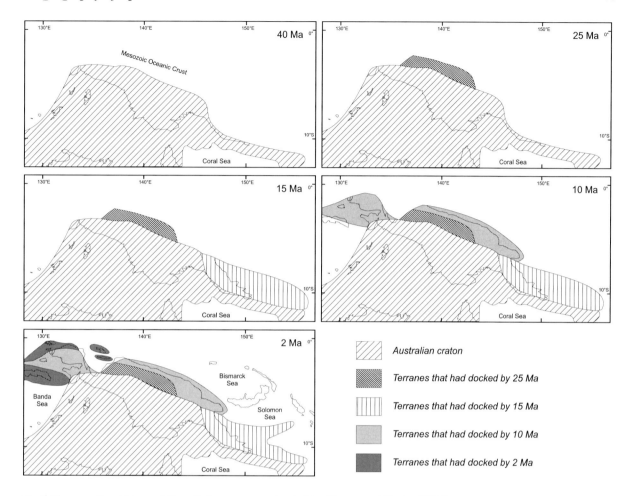

Fig.6. The accretion history of New Guinea as reconstructed by Pigram and Davies (1987).

group also contains the species nowadays found on Sulawesi. The *maingayi*, the *nervosa* clade, and the remainder contain 18 species which are found on peninsular Malaysia, Sumatra, Java, Borneo, the Philippines, and Buru-Ambon (Fig.7A). The main diversity is found in peninsular Malaysia and Borneo with 9 species in both, of which 3 are widespread in Sundaland. One species is found in the Philippines with a different variety on West Java. One species occurs in Buru and Ambon, and the last is a fragmentary specimen from the Philippines.

Coupled to the fact that some of the New Guinean relatives of this lineage occur in the Bird's Head, it is likely that this group reached New Guinea along the Banda arc, and spread over New Guinea when the Bird's Head came in close contact at approximately 10 Ma (Figs.5 and 6). The central mountain range of New Guinea, where the clade is still best represented, was almost certainly emergent at this time and onwards (Rangin *et al.*, 1990).

According to such a distributional pathway it might be expected that the lineage would also occur on the Lesser Sunda Islands. But they do not. This absence is explicable. *Aporosa* is mostly restricted to an ever-wet climate. However, nowadays the Lesser Sunda Islands and parts of Sulawesi and the Moluccas have a more or less seasonal climate (Steenis, 1979). Because of changing climates or ecological circumstances the range where the lineage once occurred may now be shrinking. The Buru-Ambonese *A. dendroidea*, the West Javanese *A. sphaeridiophora* var. *campanulata*, and the most northerly distributed *A. duthieana* are all little known species that might represent the boundaries of the distribution of this lineage in former times. Supporting this interpretation is the fact that *A. sphaeridiophora* var. *campa-*

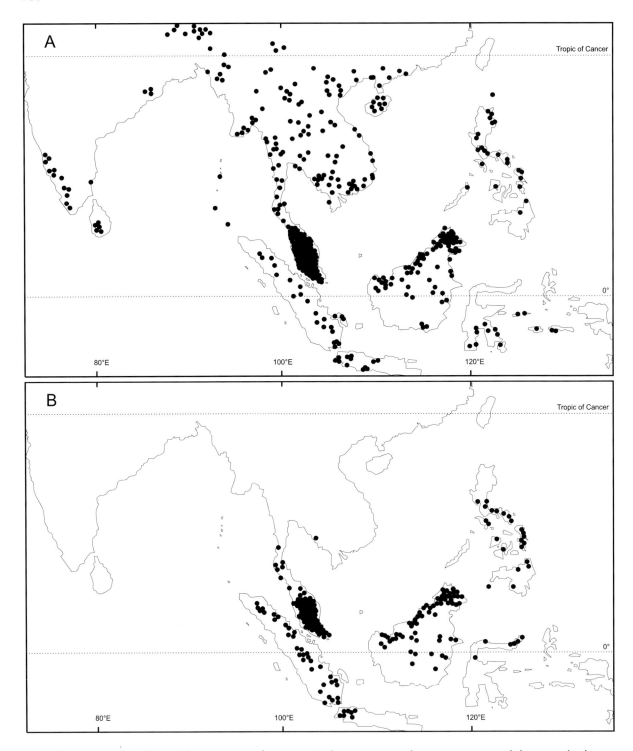

Fig. 7. Distribution of the West Malesian species of *Aporosa*. A: the *maingayi* and *nervosa* groups and the unresolved paraphyletic tail; and B: the *benthamiana*-group and the species allied to *A. arborea*.

nulata has never been collected again after the first few times in 1890 and is now almost certainly extinct. Considering this possible extinction of *A. sphaeridiophora* var. *campanulata* from Java, other extinct species may once have occupied the Lesser Sunda Islands, bridging the gap between Sumatra and Buru-Seram.

The New Guinean clade of the second lineage

is distributed in north New Guinea and the Solomon Islands (Fig.4B). The Sundanian relatives of this lineage, the *benthamiana* group, the *frutescens* group, and the species allied to *A. arborea* (Fig.3) are found in Sundaland, south Philippines, and the Minahassa peninsula of Sulawesi (Fig.7B). This pathway coincides with a dispersal along the OMA. The south Philippines were colonized from Borneo and from there to Minahassa. The lineage is absent from east Sulawesi, the Moluccas, and part of the Sepik district, but abundant again in north New Guinea, the Bismarck and Solomon Islands. Two aspects might play a role to effect this. The dispersal along the OMA may have been taking place only after the OMA became partly incorporated in the Indonesian archipelago, at about 10 Ma. Halmahera or other islands of the Moluccas now subjected to a seasonal climate may have functioned as stepping stones from Minahassa directly to north New Guinea. It is also possible that more *Aporosa* species occur on Sulawesi, which is one of the under-collected areas of Indonesia. But this would only strengthen the pathway. The possibility that such fragments may have served as stepping stones is also supported by the occurrence of two species of putative hybrid origin in the most northern part of the Bird's Head, which is thought to have docked to New Guinea at 2 Ma (Fig.6). This area might represent another more recent point of contact between the first lineage, which was present in south Bird's Head, and the second lineage, coming in along the OMA.

The larger part of the second lineage may have reached New Guinea with the North New Guinea terrane that docked at 10 Ma (Fig.6). Spreading from there over New Guinea it met in the central mountain range with the first lineage. Hybrids or introgression yielded species with various adaptations that were able to colonize different parts of New Guinea, like *A. ledermanniana* that established itself in the swamps along the Sepik River, or *A. petiolaris* and *A. hermaphrodita* that became locally thriving populations in the striction zones caused by the subsequent docking of Finisterre. Others spread over the Papuan peninsula and reached places neither lineage was able to colonize before, like *A. misimana* on the Papuan Islands. Next to the newly inhabited areas the non-hybrid species had established themselves in other parts before contact and were able to persist there, such as the major part of Irian Jaya for the first lineage and the Solomon Islands for the second lineage.

These distributions of the three groups also raise the possibility that the first lineage was earlier to reach central New Guinea, because the mountain ranges contain both first and hybrid taxa, but none of the second lineage. This could mean that the Bird's Head was earlier in docking than the North New Guinea terrane, or that climatological, oceanic currents, or the presence of appropriate land extensions were more favourable to promote a dispersal from the south. It remains an area that needs further study.

Conclusions

Geological events and distributional patterns support the phylogenetic hypothesis that two different lineages of *Aporosa* have reached New Guinea from Sundaland independently at approximately 10 Ma. One clade might have undertaken the journey along the Lesser Sunda Islands to the Banda arc and the Bird's Head, reaching the central mountain ranges of New Guinea when the Bird's Head came in close contact with the rest of New Guinea. The other lineage might have travelled along the Outer Melanesian arc from Mindanao, through the Moluccas, to north New Guinea and the Solomons. New Guinea was colonized with the arrival of the North New Guinea terrane, which possibly occurred slightly later than the colonization from the south.

References

Baum, B. R. and Estabrook, G. F. 1996. The influence of including outgroup taxa on topology. Taxon 45: 243-257.

Boer, A. J. de 1995. Islands and cicadas adrift in the West-Pacific. Biogeographic patterns related to plate tectonics. Tijdschrift voor Entomologie 138: 169-244.

Burrett, C., Duhig, N., Berry, R. and Varne, R. 1991. Asian and South-western Pacific continental terranes derived from Gondwana, and their biogeographic significance. In Austral Biogeography. Edited by P. Y. Ladiges, C. J. Humphries and L. W. Martinelli, Australian Systematic Botany 4: 13-24.

Daly, M. C., Cooper, M. A., Wilson, I., Smith, D. G. and Hooper, B. G. D. 1991. Cenozoic plate tectonics and basin evolution in Indonesia. Marine and Petroleum Geology 8: 2-21.

Farris, J. S. 1988. Hennig86, version 1.5. Computer program and manual. University of Stony Brook, New York.

Hall, R. 1998. The plate tectonics of Cenozoic SE Asia and the distribution of land and sea. In Biogeography and Geological Evolution of SE Asia. Edited by R. Hall and J. D. Holloway (this volume).

McDade, L. A., 1992. Hybrids and phylogenetic systematics II. the impact of hybrids on cladistic analysis. Evolution 46: 1329-1346.

Metcalfe, I. 1994. Gondwanaland origin, dispersion, and accretion of East and Southeast Asian continental terranes. Journal of South American Earth Sciences 7: 333–347.

Metcalfe, I. 1998. Palaeozoic and Mesozoic geological evolution of the SE Asian region: multi-disciplinary constraints and implications for biogeography. *In* Biogeography and Geological Evolution of SE Asia. Edited by R. Hall and J. D. Holloway (this volume).

Pigram, C. J. and Davies, H. L. 1987. Terranes and the accretion history of the New Guinea orogen. Bureau of Mineral Resources, Journal of Australian Geology and Geophysics 10: 193–211.

Rangin, C., Pubellier, M., Azéma, J. *et al.* 1990. The quest for Tethys in the western Pacific. 8 paleogeodynamic maps for Cenozoic time. Bulletin de la Société Géologique de France, 8, tome 6: 909-913 + 8 colour maps.

Ridder-Numan, J. W. A. 1996. The historical biogeography of the Southeast Asian genus *Spatholobus* (Legum.-Papilionoideae) and its allies. Blumea Suppl. 10: 1-144. Rijksherbarium/Hortus Botanicus, Leiden.

Rieseberg, L. H. 1995. The role of hybridization in evolution: old wine in new skins? American Journal of Botany 82: 944-953.

Schot, A. M. 1998. Taxonomy, phylogeny, and biogeography of *Aporosa* Blume (Euphorbiaceae). Blumea suppl. Ph.D. Thesis, Rijksherbarium/Hortus Botanicus, University of Leiden.

Steenis, C. G. G. J. van 1979. Plant geography of East Malesia. Botanical Journal of the Linnean Society 79: 97-178.

Swofford, D. L. 1991. PAUP: Phylogenetic analysis using parsimony, version 3.0. Computer program and manual. Illinois Natural History Survey, Champaign.

Webster, G. 1994. Classification of the Euphorbiaceae. Annals of the Missouri Botanical Garden 81: 3-32.

Geological signal and dispersal noise in two contrasting insect groups in the Indo-Australian tropics: R-mode analysis of pattern in Lepidoptera and cicadas

Jeremy D. Holloway
Department of Entomology, The Natural History Museum, Cromwell Road, London, SW7 5BD, UK

Key words: phenetic biogeography, cladistic biogeography, panbiogeography, cluster analysis, R-mode pattern analysis, Lepidoptera, cicadas, Sundaland, Sulawesi, Melanesian archipelagos

Abstract

Biogeographic methodology and philosophy are reviewed within the context of obtaining an objective means of identifying geological signal in a complex archipelagic system where dispersal has played a major role in the development of biogeographic pattern. A method of R-mode analysis is explored that associates phylogenetically related groups of species (clades) which share common features in terms of geographical representation of endemics and more widespread species. Such associations of clades, perhaps compatible with the panbiogeographic concept of generalised tracks, can provide the basis for an independent analysis of area relationships, facilitating the identification of contrasting patterns in the associations which may reflect spatially or temporally different episodes in the past geography of the archipelago. The method is applied to data for Lepidoptera and cicadas on the basis of areas of endemism recognised for the former and illustrates both the differences in pattern between the two, due to much more restricted powers of dispersal of cicadas, and common features that probably have a geological basis. A second analysis for cicada data, based on their much smaller areas of endemism, reveals a lack of clear groupings similar to that of the Lepidoptera at a grosser geographical scale, indicating that similar stochastic pattern generation through dispersal over the geological template may occur, but in a more localised fashion. The results are assessed in relation to geological events, in particular, the independent evolution of groups in the regions of Sundaland, the Sulawesi area and inner and outer Melanesian archipelagos. Evidence is presented for association of such groups on broader northern (Philippines and outer Melanesian archipelagos) and southern (Sundanian, Banda arc, inner Melanesian archipelago and Australia) axes. Sulawesi (or its components) appears to have interacted with both axes in a complex fashion.

Introduction

Geologists investigating the tectonic history of the Indo-Australian tropics would probably be aware of the interest shown by biogeographers in the results of their work and the possibility for some mutual benefit to be gained by the sharing of ideas and data. They might also be puzzled at the frequent lack of agreement amongst biogeographers on the best way to approach data on plant and animal distributions, particularly where the objective was to derive from them information on Earth history.

Indeed there has been a danger in the not too distant past that biogeographers would polarise into two or three extreme groups. One rather traditionalist group was accused by the other, more iconoclastic groups of embracing biological dispersal to explain all biogeographic pattern. Each dispersal event was unique and therefore did not generate scientifically testable hypotheses. Such scenarios were condemned as story-telling, and dispersal was dismissed as random and therefore uninformative.

In attempts to secure biogeography on a sound methodological basis of hypothesis testing and falsification, there developed a countertendency to reject any notion of biological dispersal and veer to another extreme where modern biogeographic pattern was considered to have developed by a process of fragmentation, or vicariance, of both Earth and life together, with the potential to discover, using the new methodology, one unique set of area relationships from biological data that would encapsulate earth history. Conflicting biogeographic patterns in even very small archipelagos led some biogeographers, who might be termed geological dispersalists, to postulate complex scenarios of island or terrane integration that would, if taken to their limits, rival the story-telling of the biological dispersalists. Page and Lydeard (1994) have reviewed the debate in relation to biogeography in the Caribbean archipelago.

The philosophies of three different biogeographic schools: cladistic/vicariant, dispersalist/migrationist and panbiogeographic, have been contrasted by Wilson (1991) in relation to a number of criteria, such as methods of analysis used, assumptions about dispersal, evolution and other process factors, attitudes to fossil evidence, non-structural physical factors such as climate, and the explanatory and predictive powers of the models. From these philosophies has emerged a general consensus on the need to separate as far as possible the analysis of biogeographic pattern from considerations of the processes that have led to development of such pattern. But disagreement has remained on the means whereby this may best be done, eliminating the likelihood of circularity of argument, yet not introducing such constraints in the cause of methodological rigour that the results foreclose on some process option in subsequent interpretation. Crisci *et al.* (1991) stated, "Although it is valid to investigate the existence of a unique pattern of interrelationships among areas of endemism, it cannot be accepted as an *a priori* assumption of the analysis".

Arguments with much of the middle ground excluded are never ultimately very productive in a real and messy world, and there are signs of a welcome return to a more pragmatic approach to analysis of biogeographic pattern in relation to earth history (*e.g.*, Page, 1989; Page and Lydeard, 1994; Wagner and Funk, 1995). There are still many potential pitfalls in the interpreting of biogeographic pattern: for example, the types of distribution patterns in the New Zealand biota that have been related to tectonic arc structures in the islands (*e.g.*, Craw, 1989) are also manifest in the distributions of plants introduced by Europeans, and suggested for these to have a climatic basis by Wilson *et al.* (1992).

There are a number of problems in attempting to investigate biogeographic pattern in relation to geological history in the Indo-Australian tropics. If the geological hypotheses current today for the history of the archipelago are taken at face value, then biological dispersal is likely to be the primary means whereby ancestral ranges of higher taxa (natural groupings of species) are established. Thereafter, fragmentation and vicariance can lead to biogeographic pattern informative about the juxtaposition of land at the time when that dispersal occurred. But subsequent dispersal events equally will obscure such pattern, possible overlaying it with pattern reflecting a more recent juxtaposition of lands. Any methodology of pattern analysis must therefore optimise the possibility of recognising sequential, overlaid patterns. Also, terrane accretion in areas such as the Sunda shelf and New Guinea may lead to biotic assemblages that are disharmonic and of diverse affinity (Polhemus, 1996).

Dispersal, though random, stochastic, need not be uninformative (see also Jong, 1998 this volume). Its frequency will relate to factors such as distance and area among the islands existing at any point in geological time. This was modelled for island systems by MacArthur and Wilson (1967), shown to hold in the establishment of the biota of Norfolk Island where the source areas are unambiguous (Holloway, 1977, 1996), and thence applied to help understand the biological enrichment of Sulawesi (Holloway, 1991) and to highlight anomalies in the biogeographical affinities of elements of the biota of Lord Howe Island (Holloway, 1977, 1979).

This stochastic basis for interaction between island biotas, with likelihood of dispersal from one to another based on their distance and respective areas, offers one means whereby biogeographic pattern in dispersive groups of organisms can provide pointers to geological history in complex archipelagic systems. But this is realised through the methods of phenetic, rather than cladistic biogeography. There is the additional prospect, discussed by Roger Butlin at the meeting that led to this book, of modelling dispersal and speciation processes over different geological scenarios for the region, contrasting the resulting distribution patterns with actual ones. Again, phenetic methods of analysis may be as appropriate for this as cladistic ones. Such modelling has yet to be undertaken in a biogeographic context, but has been deployed to help understand Polynesian colonisation of the Pacific and options for coconut dispersal across it (Levison *et al.*, 1969; Ward and Brakefield, 1992).

Phenetic methods have been applied to data sets for mobile animal groups in the Indo-Australian tropics. Holloway and Jardine (1968) derived distance measures between major islands in terms of their faunal (species-level) similarities using data for butterflies, birds and bats. They used non-metric multidimensional scaling to find two-dimensional plots of points representing the islands that best summarised this array of 'faunal distances'. The plots for butterflies and birds showed high correlation with current geography, support for the predominance of a stochastic, dispersal mode of pattern generation within the group.

The authors went on to suggest that, where the plots departed from such correlation, this might reflect some aspects of past geography, particularly where plots for different groups of organisms showed the same trends. Thus, close association of peninsular Malaysia, Sumatra, Borneo and Java was interpreted in terms of faunal intermingling facilitated by union of these lands on the Sunda shelf during Pleistocene low sea-levels. Sulawesi was placed well separate from Borneo, with the Philippines in an intermediate position, perhaps indicative of greater isolation for the former in the past, interaction being primarily through the latter, perhaps through greater exposure of geological structures between the Minahasa peninsula of Sulawesi, Sangihe and Mindanao. Relationships between the northern and southern Moluccas, New Guinea, the Bismarck Islands and the Solomons also departed from current geography, with the scaling method indicating high stress values between the two-dimensional summary and the raw data, perhaps a measure of the complex geological history of what is now known to be a composite zone of several different structures, even within the island of New Guinea itself, as will be seen later.

Indeed, the history of the archipelago is as much one of fusion and convergence of terranes and other structures such as island arcs as of their fragmentation and divergence (Polhemus, 1996). Hence the focus of the cladistic approach to biogeography, attempting to portray area relationships in the form of a dichotomous tree structure from analysis of biogeographic pattern in plant and animal groups with taxa endemic to those areas needs to be examined carefully. It may thus not be possible to represent the hypotheses of geological evolution of the Indo-Australian archipelago in dichotomous tree form, though attempts have been made (Turner, 1995; Boer, 1995b). Yet this is seen as an important prerequisite for application of cladistic biogeographic methods (Morrone and Carpenter, 1994). Page and Lydeard (1994: their Fig.1) present what are stated to be geological area cladograms for the Caribbean from an earlier paper by Rosen. These lack a time dimension, each perhaps representing a snapshot of geographical juxtaposition of areas of endemism at five points in geological time from the mid-Cenozoic to the present.

Most recent references to the analysis of Holloway and Jardine (1968) have been in relation to its use of phenetic methodology (cluster analysis of the matrix of faunal dissimilarity coefficients that formed the basis for the non-metric multidimensional scaling just described) to generate just such a tree structure, and this has been compared with results deriving tree structures using the methods of cladistics (*e.g.*, Holloway (1991), using Lepidoptera data for the same area). A point of interest in this comparison relevant to the chapter by Jong (1998 this volume) is that the phenetic analysis of Lepidoptera distributions, reflecting overall faunal similarities, paired the northern and southern Moluccas, whereas the cladistic analyses, reflecting phylogenetic relationships, did not, grouping them independently with New Guinea (Holloway, 1991): this is entirely consistent with the conclusions of Jong.

But the primary purpose of Holloway and Jardine (1968) was to illustrate the complementarity of two approaches to analysis of distributional data, classifying areas in terms of relationships of their biotas (Q-mode), and taxa in terms of their distribution amongst areas (R-mode). This distinction perhaps also encapsulates the divergence between the cladistic and panbiogeographic schools of biogeography, a distinction rarely referred to in the literature, exceptions being Simberloff and Connor (1980) and Page (1989).

It is the purpose of this chapter to return to this complementarity and to suggest it offers an escape from the search for a unique dichotomous tree hypothesis of area relationships that is often the focus of cladistic biogeographic methods. It does not foreclose on the possibility discussed earlier that data to hand may contain several different sets of area relationships that may derive from different periods in geological time, offering a series of biological shapshots of past geography: the dispersive vicariance of Duffels (1983).

In addition, by comparing data from the Lepidoptera (the speciality of the author), offering a wide range of dispersal abilities across the taxa, with those from the cicadas, with a much higher degree of local endemism and very few extremely widespread taxa and therefore generally probably much less dispersive, it is possible to compare pattern in a group where stochastic pattern generation predominates with one where vicariant pattern generation is far more prevalent, as suggested by Boer (1995b).

Though groups of the second type may proffer pattern that is much more informative in terms of geology, their limited powers of dispersal mean that their diversity in archipelagos will be very much lower than in mainland areas and lower than in other, more mobile groups among the islands, and hence the sample of area cladograms available for analysis of congruence will be relatively small. An extreme example of this is shown by major freshwater fish groups, virtually restricted to continental shelf areas in the Indo-Australian tropics (Darlington, 1957). Thus, the cicadas are represented in the archipelagos east of the Sunda shelf and north of Australia mostly by two major lineages (Boer, 1995b). The Lepidoptera are represented by hundreds, though a number of families, such as the Lasiocampidae, Bombycidae, Eupterotidae and Limacodidae, also exhibit significantly reduced diversity in the more isolated archipelagos. Page and Lydeard (1994) advocated selecting groups with maximal endemism amongst their taxa but, as will be seen in the analyses, groups showing endemicity in the remoter archipelagos tend to have taxa that are uninformatively widespread in those adjacent to continental areas, and those with endemism in the latter rarely penetrate the remoter archipelagos.

An R-mode method of pattern analysis

The phylogenetic data for Indo-Australian insect groups, particularly Lepidoptera, that are accumulating do not lend themselves easily to Q-mode cladistic biogeographic analysis. Derivation of a single pattern of area relationships is virtually impossible using Component Analysis (too much sympatry, too many widespread species), and Parsimony Analysis methods yield results that appear to offer little advance over phenetic ones (Holloway, 1991). Morrone and Carpenter (1994) compared these and a further method in relation to a range of data sets and concluded that there still remained considerable problems in their application, and that these derived primarily from the effects of dispersal. There are also major computational problems in the application of such methods. Nelson and Ladiges (1996) have suggested that their protocol for eliminating geographical paralogy (repetition) in cladograms circumvents these problems. Turner (1995: p.103 *et seq.*) encountered similar problems in his analyses of Australasian plant data. The data suggest there may indeed by a multiplicity of contrasting patterns, possibly of different ages and certainly blurred by dispersal events. In addition, there is no concordance of pattern between montane-restricted and lowland restricted groups (Holloway, 1970, 1986a), though this would be predicted under a purely vicariant model of biological response to geological change.

An R-mode approach sets out to identify distinct groupings of patterns and will therefore offer a better prospect for at least a preliminary biogeographic analysis in the Indo-Australian tropics. Identification of such groupings could provide a prelude to a set of Q-mode area analyses incorporating taxa from each general pattern so identified: the complementarity of methodology advocated by Holloway and Jardine (1968). Trees for these taxa could be assessed for, and 'cleaned' of paralogy using the approach of Nelson and Ladiges (1996), prior to an analysis of congruence in area-relationships. However, this chapter will focus only on the preliminary R-mode analysis, as the sample of cladistic analyses of groups is still relatively small. When more analyses are available, it may prove that some of the R-mode derived groups are distinguished from each other merely by differential paralogy rather than by inherently different structure of area relationships, but this should not be an *a priori* assumption. The development of paralogy itself may be of interest for investigation of aspects of speciation and enrichment of biotas in biodiversity studies (*e.g.*, the widely different species-richness of Australian versus New Guinea genera in the cicada tribe Chlorocystini referred to later).

The initial R-mode method of Holloway and Jardine (applied to further Lepidoptera data by Holloway, 1973, 1974, 1979) classified taxa into faunal elements merely on similarity of presence or absence in the areas incorporated in the study (effectively areas of endemism). The results of such analyses are not dissimilar to the sort of track analyses that have been made by panbiogeographers (*e.g.*, Craw, 1989). An extension of this (Holloway, 1969) grouped higher taxa, mostly genera, in terms not just of presence or absence in each area but also of richness in species – a faunal centre analysis. This approach grouped genera together that shared centres of richness, or massing centres, as well as overall distribution.

A modified method is presented here that incorporates aspects of both these approaches. It classifies higher taxa as in the second method, but their component species are first assigned to distributional categories (faunal elements or generalised tracks) identified by the first. Representation of species in each higher taxon across these categories is tabulated and converted to a percentage. Pairwise comparisons of the higher taxa give a measure of similarity of representation across the categories for each pair. These similarity coefficients are then submitted to cluster analysis to identify significant groupings of the higher taxa, where similarity of representation of their component species amongst the distributional categories is high.

The single-link cluster analysis method is used. A linkage diagram is also constructed to assess further the structure and cohesion of the single-link clusters. This approach, although somewhat cumbersome, gives better retrieval of the information content of the array of similarity coefficients than do averaging and centroid clustering methods, and is an *ad hoc* and illustrative version of the non-hierarchic clustering method of Jardine and Sibson (1968). A more detailed discussion of the merits of this approach was presented by Holloway (1977, pp. 163-169).

At this stage no information on phylogenetic structure within the higher taxonomic groups is incorporated in the analysis. However, the application of this R-mode method is facilitated by advances in the systematics of the Lepidoptera since the earlier analyses referred to above, and also improved data on their distribution. The arrival of cladistic methodology and the sharpened concept of monophyly associated with it has meant that it is possible to identify and select more reliably groups that are likely to be monophyletic. The same applies to the cicada data.

The data

Lepidoptera

Genera, subgenera or clades (monophyletic groups of species within genera) with five or more species (up to about 30) were selected from recent taxonomic publications. Data for 85 monophyletic groups were located. Phylogenetic hypotheses (cladograms) were available for 38 of these: the rest are from recent revisions where the morphological definition of the groups is unambiguous. However, a number of these are rather large, and the method might most appropriately be applied to groups of a more uniform size range (*e.g.*, 5-15 species), equivalent to the subtrees of Nelson and Ladiges (1996). All these groups are restricted to the Indo-Australian tropics and subtropics. The data are drawn from the following:

Arctiidae: *Byrsia, Neoscaptia* (Holloway, 1984).
Drepanidae: *Macrauzata* (Inoue, 1993a); *Tridrepana* (Watson, 1957).
Geometridae: *Tanaorhinus rafflesii* group (Holloway, 1982); *Crasilogia, Nadagarodes, Polyacme* (Holloway, 1984); *Astygisa vexillaria* group, *Bracca* (3 clades), *Probithia* (Holloway, 1991); *Ectropidia, Zeheba* (Holloway, 1991, 1993); *Omiza, Peratophyga, Petelia medardaria* group (Holloway, 1993); *Dindica* (Inoue, 1990); *Dasyboarmia* (Sato, 1987; Holloway, 1993).
Hesperiidae: *Matapa* (Jong, 1983).
Lasiocampidae: *Arguda, Lajonquierea, Radhica, Syrastrena* (Holloway, 1987a).
Limacodidae: *Susica* (Holloway, 1982, 1986b); *Darna* (Holloway, 1986b; Holloway *et al.*, 1987); *Setora, Thosea* (Holloway *et al.*, 1987); *Narosa concinna* group (Holloway, 1991).
Lycaenidae: *Drupadia* (Cowan, 1974); *Curetis* (Eliot, 1990); *Allotinus, Logania, Miletus* (Eliot, 1986); *Caleta, Catochrysops, Catopyrops, Danis, Jamides* (*bochus* and *celeno* groups), *Nacaduba, Prosotas, Psychonotis, Udara* (subgenera *Udara, Perivaga, Selmanix*) (Hirowatari, 1992); *Callictita* (Parsons, 1986).
Noctuidae: *Lacera* (Holloway, 1979); *Achaea serva* group, *Avitta, Ophyx* (Holloway, 1984); *Aegilia, Paectes cristatrix* group (Holloway, 1985); *Chasmina* (Holloway, 1989); *Anomis* subgenus *Rusicada* (Holloway and Nielsen, 1998).
Notodontidae: *Besida, Cerura kandyia* group, *Phalera* subgenus *Erconholda, Phalera grotei* group, *Teleclita strigata* group (Holloway, 1987b).
Nymphalidae: *Euploea* (2 clades), *Ideopsis, Parantica* (4 clades) (Ackery and Vane-Wright, 1984); *Tellervo* (Ackery, 1987); *Ptychandra* (Banks *et al*, 1976); *Chersonesia, Cyrestis* (Holloway, 1973); *Idea* (Kitching *et al.*, 1986); *Polyura* (2 clades) (Smiles, 1982).
Papilionidae: *Graphium* (3 clades) (Saigusa *et al.*, 1977).
Pieridae: *Eurema* (2 clades) (Yata, 1990).
Pyralidae: *Vitessa* (Munroe and Shaffer, 1980).
Saturniidae: *Attacus + Coscinocera* (Peigler, 1989).
Thyrididae: *Herdonia* (Inoue, 1993b); *Misalina* (Whalley, 1976).

The distributional categories recognised include endemism to individual islands or tight island groups, together with wider categories (*e.g.*, Sundaland, Melanesian, Coral Sea, Indo-Australian) identified by the faunal element analyses for Lepidoptera by Holloway (1973, 1974, 1979).

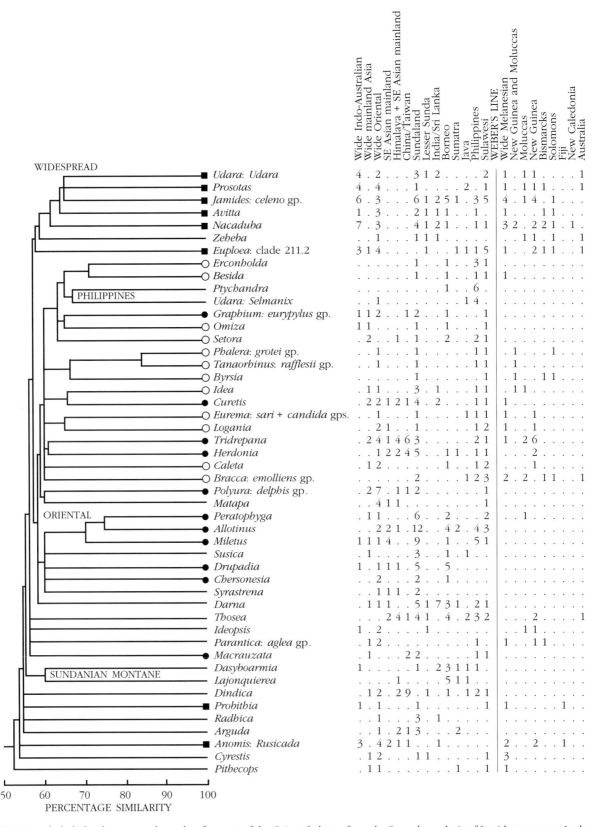

Fig.1. Single-link dendrogram and raw data for most of the Oriental cluster from the R-mode analysis of Lepidoptera taxa. In the table of raw data the numbers of species of each taxon falling into each distributional category (listed at top) are indicated. Symbols are used to facilitate cross-reference with the clusters recognised on the linkage diagram of Figs.3 and 4 and discussed in the text. The rest of the dendrogram and data are illustrated in Fig.2.

Pattern in Lepidoptera and cicadas

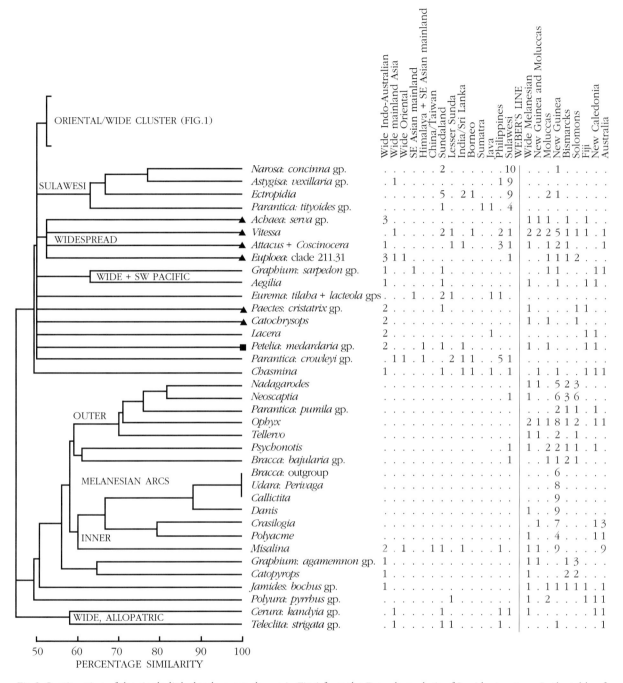

Fig.2. Continuation of the single-link dendrogram shown in Fig.1 from the R-mode analysis of Lepidoptera taxa. In the table of raw data the numbers of species of each taxon falling into each distributional category (listed at top) are indicated. Symbols are used to facilitate cross-reference with the clusters recognised on the linkage diagram of Figs.3 and 4 and discussed in the text.

Cicadas

The majority of the data for the cicadas is published by the group of researchers at the University of Amsterdam who have developed phylogenies for virtually all cicada groups in Sulawesi and from the Moluccas eastwards.

These workers have recently published major biogeographic syntheses of this work (Boer, 1995a, b; Boer and Duffels, 1996a, b), that also give comprehensive reference to the original taxonomic monographs and distributional data. Additional Oriental groups that extend into the archipelago in the genus *Cryptotympana*, revised

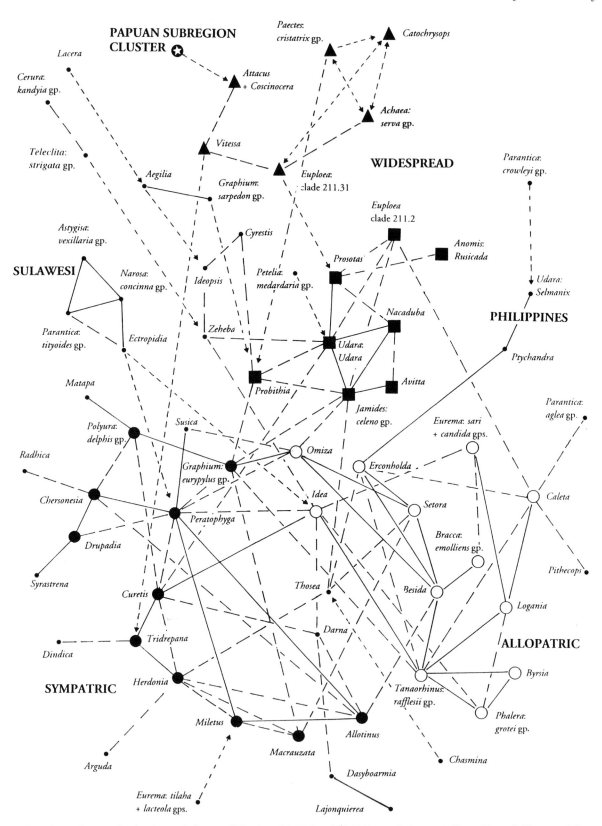

Fig.3. Linkage diagram for the Oriental cluster of Figs.1 and 2. Links of 60-100% similarity are indicated by solid lines and those of 51-59% similarity by long dashed lines. Clustering of taxa at a higher level than this is indicated by short dashed lines with arrows. Large symbols indicate members of the clusters or groupings of taxa mentioned in the text: the same symbols indicate members of these groups in Figs.1 and 2. Outlying taxa that do not fall into these groups are indicated by smaller solid circles. Linkage with the Papuan subregion cluster is indicated by the encircled star.

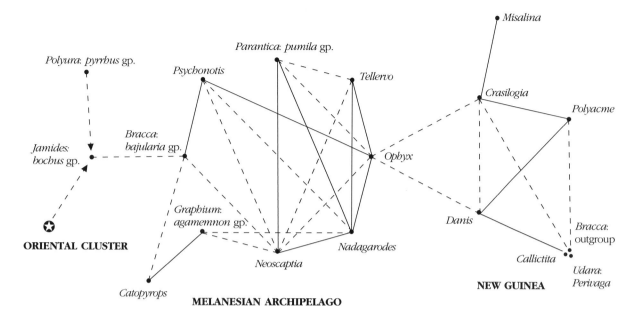

Fig.4. Linkage diagram for the Papuan cluster of Fig.2. Conventions are as for Fig.3.

by Hayashi (1987a, b), *Chremistica* (Bregman, 1985) and *Dundubia* (Bloem and Duffels, 1976) were also included.

These data were initially analysed using the same distributional categories as for the butterflies, but cicada species were virtually unrepresented in the more widespread faunal elements. Within the more easterly archipelagos cicada species are often restricted to one or two islands (including one or two endemic to very small islands, excluded from the analysis), and they are also localised within the island of New Guinea (Duffels and Boer, 1990). Therefore a second analysis was performed on Papuan subregion cicada genera using a much finer-grained set of distributional categories that are virtually those identified in Boer (1995b) and combinations of them. This finer-grained analysis provides an opportunity to test objectively the coincidence of cicada groups as a whole over areas of endemism recognised subjectively by Boer (1995b).

Results: Lepidoptera

The dendrogram resulting from the single-link cluster analysis, and the raw data from which it is derived, are shown in Figs.1 and 2. The dendrogram indicates a primary segregation of Papuan subregion (Melanesian archipelagos) groups from a combination of Oriental and widespread ones. Many of the Oriental groups extend into the Papuan subregion (bounded to the west by Weber's Line of Faunal Balance after the findings of Holloway and Jardine, 1968). The widespread groups are represented fairly evenly throughout the whole of the Indo-Australian tropics and make a major contribution to the faunas of Pacific archipelagos.

Linkage diagrams for these two main clusters are shown in Figs.3 and 4, with links down to 50% similarity being illustrated: clustering in of outlying taxa above that level is also shown. The widespread taxa (solid squares) form a moderately cohesive cluster distinct from a much larger grouping of taxa with much stronger representation (species richness) in the Oriental region. Within this larger grouping there is no definite substructure, but there is some polarization of groups (open and solid circles) across a continuum. There are various small groupings peripheral to these main clusters: a tight quartet of taxa with high endemism in Sulawesi (a comparable situation in the Philippines is much more weakly defined); a loose association of taxa (solid triangles) with some affinity to the widespread cluster, but of a more Melanesian character.

Oriental taxa: sympatric v. allopatric groups

The polarization of Oriental taxa is between taxa (solid circles) that are particularly rich in species

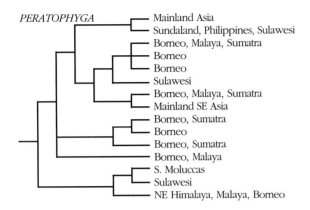

Fig.5. Cladogram with areas for the ennomine geometrid genus *Peratophyga*, with groupings as justified by Holloway (1993), exemplifying the sympatric Oriental type of genus.

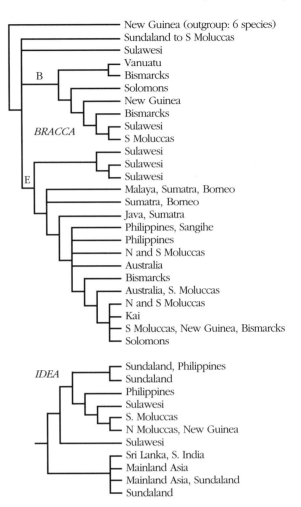

Fig 6. Cladograms with areas for the ennomine geometrid genus *Bracca* and the danaine butterfly genus *Idea*, from Holloway (1991). The latter and the *emolliens* clade (E) of *Bracca* exemplify the allopatric Oriental type of genus. The *bajularia* clade of *Bracca* (B) exemplifies the Melanesian archipelago type of genus in the Papuan subregion cluster, the *Bracca* outgroup being in the New Guinea subcluster.

in the mainland Asian and Sundanian distributional categories, and those (open circles) that are less well represented on the Asian mainland but have higher endemism in the Philippines and Sulawesi and are more likely than the other group to be represented in the Melanesian archipelagos. The first group of taxa has some endemism in Borneo, less in the Philippines and Sulawesi, and often shows a high degree of sympatry in mainland or Sundanian areas. Those in the second group have a more even distribution of their species amongst the distributional categories and exhibit greater allopatry, and may therefore offer a better chance of detecting compatibility of area relationships in their cladistic structure. For convenience in the text following, these groups will be referred to as sympatric versus allopatric Oriental groups.

The ennomine genus *Peratophyga* illustrates the sympatric group (Fig.5). Relationships within the genus are as described by Holloway (1993). The genus is particularly rich in various Sundanian categories, including Bornean endemics: ten out of the fifteen species occur in Borneo, and the clades recognised within the genus are also both sympatric and allopatric in character. In the *eurypylus* group of the swallowtail butterfly genus *Graphium* (Saigusa *et al.* 1977), the component species are generally much more widespread, but again the majority are represented in Borneo.

Within the allopatric group, different degrees of complexity are shown. In *Tanaorhinus*, the *Phalera grotei* group and *Byrsia* there is very little overlap amongst an array of species that fall into distributional categories of intermediate character, *e.g.*, mainland Oriental, Sundanian, Melanesian, as well as island endemics. Holloway (1982, 1987b) grouped these together subjectively with the *Teleclita strigata* and *Cerura kandyia* groups, but these are segregated in the analysis by virtue of representation in Australia and greater allopatry in the western part of their range. Holloway suggested that, as there were no obvious characters to indicate a strongly dichotomous cladistic structure within any of them, and the islands grouped by the more widespread elements largely reflected modern geography, these patterns probably resulted from relatively recent episodes of rapid ancestral dispersal, followed by vicariance over the islands in modern juxtaposition, perhaps in response to Pleistocene climatic changes.

More complex patterns are exhibited by the *Eurema sari* + *candida* group, *Besida*, *Erconholda*, the *Bracca emolliens* group and *Idea*, though in none of these is there any strong evidence of commonality of pattern, merely that of trends. In the *Eurema* example a Melanesian sister-pair is sister to an Oriental complex, with one allopatric trio overlapping the composite range of the unresolved remainder extensively. *Besida* is sister to a widespread Melanesian genus (added to the data), but has two Sundanian species and a Philippines-Sulawesi sister-pair. *Erconholda* also exhibits the Sulawesi-Philippines relationship but its interaction with Sundanian areas is different (Holloway, 1987b).

Phylogenetic hypotheses for the *Bracca* group and *Idea* are shown in Fig.6. In both, there can be seen a sort of progression from west to east within the cladogram, the earliest branches being the more easterly species. However, that for *Idea* 'commences' in India and mainland Asia and 'terminates' in New Guinea, and that for the *emolliens* group of *Bracca* 'commences' in Sundaland and 'terminates' in the Solomons, with earlier branches Melanesian rather than continental Asian. Even within the area of overlap of the two cladograms, that for the *Bracca* group is more species-rich, has a similar outlying taxon in Sulawesi, but lacks corresponding representation of Sulawesi further up the sequence.

Nevertheless, this general west to east progressive pattern of areas is seen also in a Brooks Parsimony Analysis for Lepidoptera where areas lacking a particular clade were coded '?' (uninformative) rather than zero (primitively absent) when data for that clade were tabulated (Holloway, 1991), and in general area cladograms for Hemiptera (Schuh and Stonedahl, 1986) and butterflies (Vane-Wright, 1990), also discussed and illustrated by Holloway (1991). This pattern will be discussed further below.

Groups of widespread taxa

The major widespread cluster contains groups (solid squares) with a high proportion of species in the most widespread category (through the Indo-Australian tropics) and in other widespread categories. Endemism in the Oriental region tends to be low except sometimes in the Philippines and Sulawesi, but the groups include a number of Melanesian archipelago endemics. Several are species-rich genera that, when subjected to cladistic analysis, may subdi-

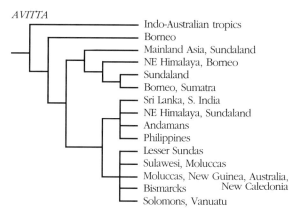

Fig. 7. Cladogram with areas for the noctuid genus *Avitta*, exemplifying a widespread Oriental cluster genus. From Holloway (1984).

vide into species groups that fall within some of the other clusters in the analysis. For example, in the noctuid genus *Avitta* (Fig.7), allopatric and sympatric Oriental clades occur as sister-groups.

The looser association of widespread taxa (solid triangles) consists of less species-rich groups where the proportion of Melanesian and Pacific endemics is equal to, or greater than that of widespread taxa, with more allopatry overall. This character is also seen in the *Graphium sarpedon* clade and *Aegilia* pair. Distributions of species of *Paectes* in the 'solid triangle' group and of *Aegilia* were mapped by Holloway (1985).

The widespread groups just discussed are generally considered to be biogeographically uninformative, an assertion that is generally true for much of their distribution through the Indo-Australian archipelago, but the mobility of these groups has also led to them being relatively well represented in remoter Pacific archipelagos where they often show a much higher degree of endemism. There will undoubtedly be a strong stochastic element in any pattern in this endemism but within this, as indicated earlier, some geological signal may be observed. For example, patterns involving New Caledonia, Vanuatu, Rotuma, Samoa and Fiji may reflect changes in the relative positions of these island groups in the past (Holloway 1979: pp. 211-213, 1983; Duffels, 1988).

Sulawesi taxa

The small cluster of taxa with high endemism in Sulawesi consists mostly of those with extensive radiations of species within Sulawesi, with a few species from these radiations occurring in some cases further east in the Moluccas or New Guinea, and with a sister-relationship to Oriental taxa to the west. These could be considered further examples of west-to-east progressive patterns. An exception within this cluster is the *tityoides* group of *Parantica* where affinities are with the Lesser Sundas and Banda island arcs generally.

Papuan subregion taxa: New Guinea and Melanesian archipelago groups

Groups in the Papuan subregion cluster segregate into New Guinea and Melanesian archipelago components, a segregation supported by the linkage diagram (Fig.4), where the two groups remain separate except for linkage through the noctuid genus *Ophyx*. Members of the New Guinea cluster are virtually restricted to that island. Species in these groups outside New Guinea tend to be in Australia and New Caledonia or in the Moluccas. Any representation further afield in the Melanesian archipelagos is by species in the widespread Melanesian category, presumably relatively dispersive. Examples from the Geometridae, the genera *Polyacme* and *Crasilogia* (including *Papuanticlea*), are shown in Fig.8.

The Melanesian archipelago groups have their species evenly distributed from the Moluccas through New Guinea into the archipelagos to the east. This distribution type is exemplified by the *bajularia* clade of *Bracca* (Fig.6). A phylogenetic hypothesis for *Ophyx*, the genus intermediate between the two groups, is shown in Fig.9. It segregates into two clades, one that would probably fall within the New Guinea cluster if coded separately, and the other an unusual combination of a small Melanesian archipelago group that is sister to an Australia/New Caledonia pair.

Associated with the Melanesian cluster is a pairing of the lycaenid butterfly genus *Catopyrops* with the *agamemnon* group of *Graphium*, where several species endemic to various Melanesian archipelagos are associated with species in more widespread categories. The case of *Graphium* will be revisited later.

These distinct New Guinea and Melanesian patterns are often cited as evidence for the convergence of two distinct geological systems, sometimes termed Inner and Outer Melanesian arcs (Holloway, 1984; Boer, 1995b; references

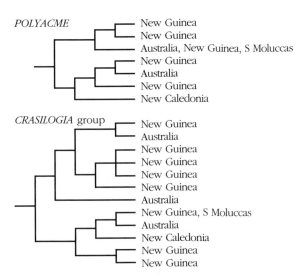

Fig.8. Cladograms with areas for two geometrid genera exemplifying the New Guinea subcluster of the Papuan subregion cluster: *Polyacme* (Ennominae); *Crasilogia* with *Papuanticlea* (upper and lower clades respectively; Larentiinae). From Holloway (1984).

therein), but, given current hypotheses of the geological complexity of the area, this is something of a misnomer (Polhemus, 1996; Polhemus and Polhemus, 1998 this volume), and Inner and Outer Melanesian archipelagos might be preferable terminology. However, the cicada data suggest, along with geological evidence, that the situation is far more complex (Boer, 1995b), and this primary segregation is merely a 'coarse focus' pattern.

There is also the possibility that the apparent segregation of the two groupings has arisen through operation of factors other than Earth history: some genera may have responded to growth of the land area of New Guinea by speciation; others, perhaps more adapted ecologically to archipelagic conditions, including having greater powers of dispersal, may have speciated extensively only in the Melanesian island groups. For example, the geometrid genus *Nadagarodes*, placed in the Melanesian group here, has a cladistic structure (Holloway, 1984) with weak segregation within it of New Guinea and Melanesian clades, but also with indication of much interchange between New Guinea, the Moluccas and the Solomons. The sister-genus was suggested to be *Probithia*, falling into one of the more widespread categories among the Oriental groupings, extending from India to Fiji (discussed further by Holloway, 1991).

Two factors may strengthen a geological, rather than ecological, interpretation of this di-

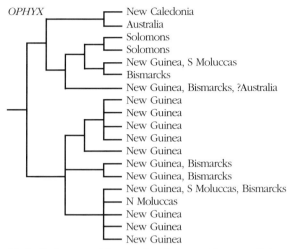

Fig.9. Cladogram with areas for the noctuid genus *Ophyx*, intermediate between the New Guinea and Melanesian archipelago subclusters of the Papuan subregion cluster. From Holloway (1984).

chotomy in pattern. The first is the extent to which putative Melanesian groups still show some restriction to the more recently accreted Melanesian terranes in New Guinea, such as the northern ranges of mountains. The second is the existence of sister-pairs of groups, one of each distribution type. There are few clear-cut examples of this in the Lepidoptera groups in this analysis, the best being the *Danis-Psychonotis* sister-pair in the Lycaenidae, though no phylogenetic analysis of this pairing has been undertaken. The situation in *Ophyx* is another possible example; Holloway (1984) also illustrated possible sister-groups within the arctiid genus *Cyana* where the Melanesian component does also show restriction to northern New Guinea.

Combinations of patterns

Both Melanesian and New Guinea patterns may also be represented amongst a number of closely related lineages, but not in a strict sister-relationship. There are a number of examples amongst the Lepidoptera groups included in the analysis, though sometimes the data were not good enough to permit all lineages to be included. They are drawn from a number of different families.

The swallowtail butterfly genus *Graphium* has already been mentioned, as has the geometrid genus *Bracca*. The analysis of *Graphium* by Saigusa *et al.* (1977) revealed three major lineages. Each of these includes one or more extremely widespread components. One is the largely Melanesian *agamemnon* group referred to above. It is sister to the other pair. One of these, the *eurypylus* group, has already been mentioned as an example of the sympatric Oriental type. The other, the *sarpedon* group, was also mentioned as one of the more widespread group that has a relatively high proportion of more localised species in the eastern part of its range: this is primarily a lineage distributed through the Moluccas, New Guinea, Australia and New Caledonia, a group, therefore, of weak New Guinea cluster character.

In *Bracca* (Fig.6), the Melanesian *bajularia* group is segregated from the *emolliens* group in an unresolved quadrichotomy with two other species, one widespread, one endemic to Sulawesi. The *emolliens* clade itself has already been discussed as illustrating a west to east progressive structure in a sister-relationship to a trio of species endemic to Sulawesi. This whole complex is probably sister to a clade of six species used as an outgroup in the analysis by Holloway (1991). This clade is restricted to New Guinea and falls within the New Guinea group in the analysis.

Holloway (1984) discussed four closely related genera of lithosiine arctiid. Two of these were included in the analysis. One falls into the Melanesian cluster (*Neoscaptia*) and the other, *Byrsia*, has already been noted as an allopatric widespread group of low diversity. The remaining two genera are *Scaptesyle*, probably a sympatric Oriental group, and *Damias*, with over thirty species restricted mostly to New Guinea, but with three species in the southern Moluccas (Seram, Buru). The genus has its greatest richness in the southeastern peninsula of New Guinea and is relatively weakly represented in the northern structures.

In the nymphalid butterfly genus *Parantica*, the *pumila* group is of Melanesian character, but the genus has no New Guinea components, the *pumila* group being sister to the *tityoides* and *crowleyi* groups, the former already mentioned for its high endemism in Sulawesi, and the latter having high endemism in the Philippines as well as a rather widespread Oriental character. This complex is sister to the *aglea* group which has a wide Indo-Australian distribution to as far east as the Solomons (see cladogram in Holloway (1984) from Ackery and Vane-Wright, 1984).

In another powerfully flighted nymphalid genus, *Polyura* (Smiles, 1982), the *pyrrhus* group, weakly associated with the Melanesian cluster in the analysis, is in an unresolved triplet with a

Solomons endemic and the sympatric Oriental *delphis* group. However, this triplet is sister to a single New Caledonian endemic.

Thus, though some gross pattern emerges from the distributions of Lepidoptera groups, that may represent geological signal, the phylogenetic structures within such groups, and between them when they come together in larger groupings, exhibit no strong pattern, indicative of process (*e.g.*, dispersal and speciation) that is probably of a highly stochastic nature. As stated in the introduction, it is rare to find Lepidoptera groups that range from India into the Pacific and yet show the maximal endemicity of their taxa throughout that range advocated by Page and Lydeard (1994) for biogeographical analysis.

Results: cicadas

The analysis of cicadas using the same distributional categories as for the Lepidoptera yielded similar clusters, but with virtually no overlap between Oriental and Papuan subregion groups, the link between the two being at almost zero percentage similarity (Fig.10). Indeed, the range of similarity values registered is considerably more extreme than that for the Lepidoptera.

The Oriental cluster shows more definite internal structure, with a tight trio of groups virtually restricted to Sulawesi and three pairings of groups centred on the Philippines, on Sundaland and mainland Asia, and on the Banda arcs. This is probably a reflection of the much smaller sample of Oriental groups than for the Lepidoptera, for amongst these pairings there is little commonality of area relationships. One of the Philippines groups is associated with Sulawesi and the other with Sundaland. One member of the Banda arc pair has its centre of richness at the western end of the geological structure, in Sumatra, and the other is more diverse at the eastern end, in the Lesser Sundas and around Timor: a further example of a Banda arc pattern is illustrated for a group of *Ptilomera* (waterstrider) species by Polhemus (1996).

The Papuan subregion groups are, like those for the Lepidoptera, clearly segregated into New Guinea and Melanesian archipelago clusters, the former relatively tight, the latter less so, with one outlier, *Aceropyga* + *Moana*, that has high endemism from the Bismarck Islands eastwards, but particularly in Fiji (*Aceropyga*).

As mentioned earlier, these cicada groups fall mainly within two major tribal groupings, the exception being the *Raiateana* group of genera (Solomons eastwards) that may be related to the Oriental *Cryptotympana* (Duffels, 1988, pp. 80-81) also included in the analysis. Most of the Chlorocystini groups (asterisks in Fig.10) fall within the New Guinea cluster, the exception being one group of the genus *Baeturia*. Two of the four Cosmopsaltriaria groups (daggers) fall within the Melanesian cluster, the others being the outlying *Aceropyga* + *Moana* group and the New Guinea cluster genus *Cosmopsaltria*, which bears a sister-relationship to the other Papuan subregion groups (Duffels, 1983). Both tribes have relatives in Sulawesi, the Chlorocystini being represented by their sister-tribe (Boer, 1995), the Prasiini (double asterisks). The Chlorocystini also have a number of small genera endemic to forests along the eastern seaboard of Australia from Cape York to southern New South Wales, a feature roughly paralleled by some New Guinea groups in the butterfly analysis. The significance of this will be assessed in the final, general discussion.

The areas of endemism recognised for the finer grained analysis of cicada groups are somewhat more numerous than those identified by Boer (1995b), but are generally similar, particularly within New Guinea (Fig.11). Species in each group were either endemic to these or represented in various pairs, triplets or more of them, yielding a total of 50 distributional categories into which all could be assigned. The results of the analysis are shown in Fig.12 both in the form of a dendrogram and a minimum spanning tree.

The dendrogram reveals no strong clustering structure, and similarity levels are generally low, not much exceeding 50%. The spanning tree confirms this, with the groups falling more or less into a chain from groups in New Guinea and the Moluccas through to the Melanesian archipelago groups. The spanning tree does bring together as neighbours the groups in the subclusters of the coarse-grained analysis. These groupings do not bear very close relationship to the subjective assignation of whole groups to ancestral areas of endemism by Boer (1995b). He referred *Cosmopsaltria* to central New Guinea, *Baeturia* as a whole, *Guineapsaltria* and *Mirabilopsaltria* to northern New Guinea, *Thaumastopsaltria*, *Papuapsaltria* and *Gymnotympana* to the Papuan peninsula, and *Aedeastria* to the Bird's Head. These assignations need not necessarily be incorrect, but the analysis here does suggest that stochastic processes such as dispersal have tended to blur any clear patterns that

Pattern in Lepidoptera and cicadas

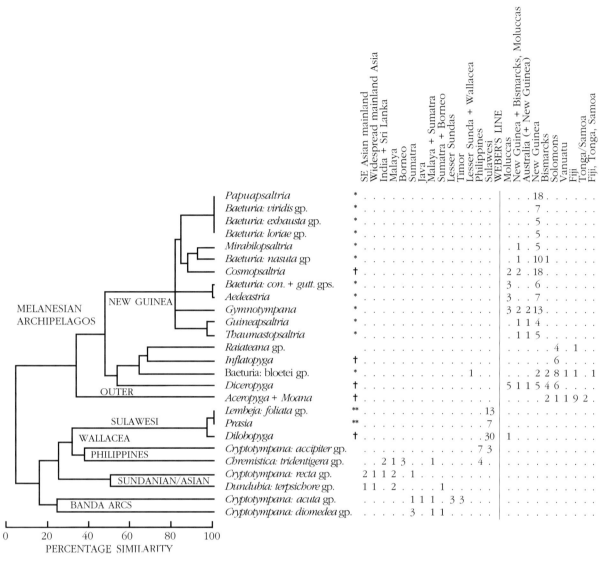

Fig. 10. Single-link dendrogram and data for the R-mode analysis of cicada taxa using the same distributional categories as for the Lepidoptera. Chlorocystini groups are indicated by *, Prasiini groups by ** and Cosmopsaltriaria groups by †.

may have existed, though on a much more localised scale than for the Lepidoptera. This is seen most strongly in *Baeturia* itself where, even if the overall 'weight' of the genus is in northern New Guinea, some groups (*nasuta*, *loriae*) approach *Cosmopsaltria* in their representation in central New Guinea, and another (*bloetei*) has a distribution resembling that of the Melanesian groups of the Cosmopsaltriaria.

The correlation of the analysis with ancestral areas of endemism categories of Boer can be assessed by examining the extent to which taxa in each category have high links primarily with other taxa in that category. Such links are indicated in bold for New Guinea groups in the list following. The list includes the genus or species group, area (or areas in ambiguous cases) of endemism suggested by Boer, followed by the highest few links (percentage similarity in brackets) with other taxa in order of similarity.

Gymnotympana (Papuan peninsula): **Papuapsaltria** (51); *Mirabilopsaltria* (56); *Cosmopsaltria* (44); **Guineapsaltria, Baeturia loriae group** (43).

Guineapsaltria (Papuan peninsula): *Mirabilopsaltria* (45); **Gymnotympana** (43); **Thaumastopsaltria** (40).

Thaumastopsaltria (Papuan peninsula):

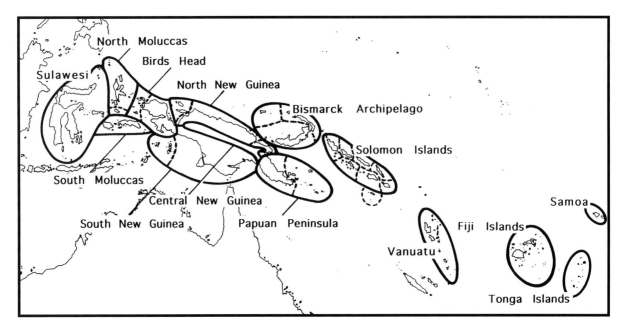

Fig.11. Areas of endemism recognised for the finer R-mode analysis of cicada patterns, after Boer (1995b). The solid lines delimit the areas of endemism of Boer; broken lines indicate further subdivision of these for this analysis.

Guineapsaltria (40); *Baeturia nasuta* group (38), **Papuapsaltria** (35).

Papuapsaltria (Papuan peninsula): ***Baeturia loriae* group** (56); *Mirabilopsaltria* (52); ***Gymnotympana*** (51); *Cosmopsaltria* (47).

Baeturia loriae group (Papuan peninsula, Central New Guinea): ***Baeturia nasuta* group** (59); **Papuapsaltria** (56); ***Gymnotympana*** (43); ***Cosmopsaltria*** (41).

Cosmopsaltria (Central New Guinea): *Mirabilopsaltria* (59); *Papuapsaltria* (47); ***Baeturia nasuta* group** (46); *Gymnotympana*, *Baeturia exhausta* group (44).

Baeturia nasuta group (Central New Guinea, Northern New Guinea): ***Baeturia loriae* group** (59); ***Cosmopsaltria*** (46); *Papuapsaltria* (42).

Mirabilopsaltria (Northern New Guinea): ***Baeturia exhausta* group** (73); *Cosmopsaltria* (59); *Papuapsaltria* (52); *Gymnotympana*, *Guineapsaltria* (45).

Baeturia exhausta group (Northern New Guinea): ***Mirabilopsaltria*** (73); *Cosmopsaltria* (44).

Baeturia viridis group (Northern New Guinea): ***Mirabilopsaltria***, *Papuapsaltria* (40); ***Baeturia nasuta* group** (37).

Baeturia conviva + *guttulinervis* group (Maluku, Northern New Guinea): *Aedeastria* (20).

Aedeastria (Bird's Head): *Baeturia exhausta* group, *Mirabilopsaltria* (34); *Cosmopsaltria* (31).

There is some general affinity amongst Papuan peninsula groups on the one hand and Northern New Guinea ones on the other, though *Mirabilopsaltria* of the latter also has links with the former, and *Cosmopsaltria* has its major links with members of both groups.

Discussion

Given this plethora of pattern variation in both insect groups, are there any basic themes within it that might have a geological basis, to which further detail may be added through considering fine-grained pattern in more localised and hopefully more informative groups such as cicadas?

Analyses for both groups do recognise a number of fairly gross areas of endemism within the Indo-Australian archipelago. In the Oriental region there is some segregation of Sundaland, Philippines and Sulawesi groups, with some interaction between Sulawesi and the Philippines, Sulawesi and the Moluccas and New Guinea, and Sulawesi and the Lesser Sundas. The cicada patterns, with finer detail, also include definite Banda arc groups. Some of the smaller subgroups of *Lembeja*, not included in the analysis, also feature Sulawesi and the Lesser Sundas (Jong, 1985, 1986, 1987). This variety of associations between Sulawesi and neighbouring areas

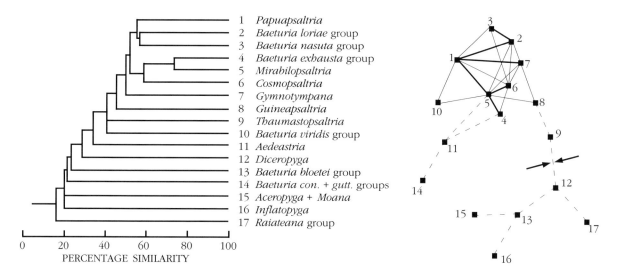

Fig.12. Single-link dendrogram (left) for the finer R-mode analysis of cicada patterns, with the linkage diagram to which it is related (right). In the linkage diagram heavy solid lines indicate links at 50% and above, light lines those at 40-49%. Broken lines indicate taxa clustering in (minimum spanning tree) at a lower level of similarity. Heavy arrows indicate the division between the New Guinea and Outer Melanesian Archipelago groups of Fig.10.

is of interest, and may suggest a composite origin for the current island, discussed briefly below. In the Papuan region there are two much clearer patterns: the New Guinea groups and Melanesian archipelago groups, the former often with Australian associations. Cicada groups sharing these patterns have very much more species richness than do Lepidoptera groups, but there are fewer of them. When this species richness occurs in larger islands, such as New Guinea and Sulawesi, the cicada taxa concerned are often localised within the island, and may reveal something of any geologically divergent histories among the components. In the Lepidoptera, such internal localisation, if it ever occurred, has usually been blurred by subsequent movements of the species, leading to a high degree of sympatry (Holloway, 1991).

Boer (1995b) focused mainly on the similarities between the biogeography of the Prasiini + Chlorocystini and the Cosmopsaltriaria cicada groups, but the differences, particularly in relation to Sulawesi and the composite nature of the Papuan subregion, may be as interesting. It is worth considering these groups in a little more detail (Fig.13), though the reader is recommended also to consult Boer (1995b), and Boer and Duffels (1996a, b).

As mentioned earlier, the Papuan subregion Cosmopsaltriaria groups, with the exception of *Cosmopsaltria* itself, are distributed mainly amongst the Melanesian archipelagos. Duffels (1990) proposed a possible phylogeny for the Cosmopsaltriaria group within Sulawesi, and reviewed the localisation of the species within the island. The group is represented by sister-genera, *Dilobopyga* and *Brachylobopyga*. *Dilobopyga* has three major species groups. Most of the species are localised within the island, and some subgroups within the major groups are concentrated in one or other of the geographical components. The genus is largely restricted to the north, centre and eastern peninsulas. Apart from an island-wide species, only two species are localised in southwest Sulawesi, a sister-pair, one on the island of Saleyer. There is one species in the southern Moluccas. *Brachylobopyga* has only a pair of species, one central and another in the southwest. So, as for the *Lembeja foliata* group of the Prasiini, discussed below, the greatest diversity is in the north and east of the island, though with richness also in the centre. Within both groups there are indications of vicariance between western and eastern groups of species.

The tribe Prasiini is strongly centred on Sulawesi. Its sister-group, the tribe Chlorocystini, occurs from the Sula Islands east to Samoa and northeastern Australia as discussed earlier. The sister relationship of the two together is thought

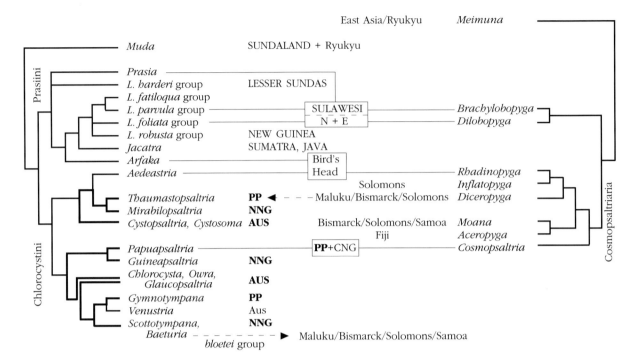

Fig.13. Diagrams of phylogenetic hypotheses for the Prasiini and Chlorocystini cicadas (left) and the Cosmopsaltriaria (right) from Boer (1995b), indicating major instances of overlap in their distributions. Areas listed to the left and right are possible Inner and Outer Melanesian archipelago ancestral ranges for each group. Central blocked areas indicate areas of overlap that are ambiguous, possibly indicating early dispersal between the two systems or colonisation of intermediately situated areas. Broken lines with arrows suggest instances where more recent dispersal may have occurred. The foci of endemism within the Chlorocystini follow those recognised by Boer (1995b). The repeated (paralogous) three-area pattern involving Australia is indicated by heavier lines and bold type: the double line to *Chlorocysta*, *Owra* and *Glaucopsaltria* represents a single branch on the cladogram, as the last genus is sister to the first two. From Holloway (1997).

to be the Sundanian genus *Muda* (Fig.13). Among the Prasiini, the genus *Prasia* itself is endemic to Sulawesi, with seven species (Jong, 1985). The genus *Arfaka* is found in the Bird's Head (Vogelkop) area of New Guinea, and a group of taxa currently in *Lembeja* is restricted to the Lesser Sunda Islands.

The Prasiini are completed by sister-genera, *Jacatra*, found in Java and Sumatra and thus the only exclusively Sundanian member of the group, and *Lembeja sensu stricto* that also contains groups endemic to Sulawesi, but ranges widely from northern Borneo and Mindanao to the Lesser Sundas, New Guinea and northern Queensland. This extensive range is exhibited by one group, the *fatiloqua* group (Jong, 1987; Boer, 1995b) that is found in northern Borneo, Mindanao, Sulawesi, Sumba, Sumbawa, south New Guinea and north Queensland. The connection with northern Borneo appears to have been via Mindanao rather than direct from Sulawesi – a distribution also shown by *Ayesha spathulata* (Duffels, 1990) – an extended Philippines-Sulawesi-Lesser Sunda pattern. Other groups of *Lembeja* include one in New Guinea, Obi and Queensland that is sister to a major group endemic to Sulawesi. Whilst the species of *Prasia* form an allopatric array distributed throughout the main island, those of the *Lembeja foliata* group are concentrated in the north and east of the island, also Sangihe Island (Jong, 1987), whereas those of the *fatiloqua* and *parvula* groups are found in the extreme north and the extreme southwest (Jong, 1986). There are two sister-pairs that span these extremes. The Lesser Sunda species appear to form an interrelated group, some still undescribed. Twenty species of *Lembeja* occur in Sulawesi (Duffels, 1990).

Phylogenetic and biogeographic analyses of the Chlorocystini by Boer (1995a, b) indicate a thrice repeated pattern involving Australia as sister area to a pair of areas of endemism in New Guinea: the southeast Papuan peninsula and the

Fig.14. Relationships within the Tmesisternini tribal complex of the cerambycid beetles suggested by Gressitt (1984), with distributions also included.

central northern part of New Guinea. This pattern is indicated by thick lines in Fig.13. In contrast to the Cosmopsaltriaria, only one of three of the northern New Guinea components shows a wide extension into neighbouring archipelagos, the genus *Baeturia* with a species group that ranges from the Moluccas to Samoa. Its Papuan peninsula sister-group also shows some extension, to the northern Moluccas and again to Australia. In one of the other trios there is unique representation of the New Guinea Bird's Head.

Thus, whilst the Chlorocystini have their main representation in Australia – at a generic level only, as species richness is an order of magnitude lower than in New Guinea: the relevance of this when considering area relationships is debatable (Nelson and Ladiges, 1996) – and southeastern and northern New Guinea, the Cosmopsaltriaria have their main representation in archipelagos from the Bismarck Islands to Fiji and Samoa in a sister-relationship to a grouping in the central and southeastern mountains of New Guinea (Fig.13). The archipelagic group overlaps principally with the Chlorocystini in the Bird's Head and Papuan peninsula, and is also represented in the Moluccas. The sister-group to this complex is in Sulawesi, discussed above, and the sister genus to the whole subtribe is suggested to be *Meimuna*, an east Asian genus extending no further south than the Ryukyu Islands. Thus the main feature in common between these two tribal groupings is rich representation in Sulawesi. Otherwise they appear to have developed roughly in parallel, the Chlorocystini + Prasiini along a more southerly axis than the Cosmopsaltriaria, alone of the two involving Sundaland and the Lesser Sundas.

Hence, despite considerable disparity in the geography of these two tribal groupings outside Sulawesi, they show at least partial convergence in their geography within the island. The higher taxonomic diversity of the Prasiini is as much in the Banda arcs (Sumatra to Lesser Sundas) as in Sulawesi, and also extends to the eastern and southern periphery of New Guinea, so the component of convergence with the Cosmopsaltriaria may postdate the initial development and radiation of ancestral lineages of the tribe. There is a further overlap of Cosmopsaltriaria, Prasiini and Chlorocystini in the New Guinea Bird's Head, with each contributing a genus endemic to, or centred on, that area: *Rhadinopyga, Arfaka* and *Aedeastria* respectively (Fig.13). There may also be an early north-south separation of the biota of Sulawesi, with the weak east-west one mentioned above developing later (see also Holloway, 1997).

Similar northern versus southern patterns can be seen in a number of weevil and longhorn beetle groups (Gressitt, 1956, 1982). The Pachyrhynchini weevils are a northern group. Both northern and southern groups are seen in the longhorn (Cerambycidae) Tmesisternini tribal complex. The northern groups are rarely represented on Sulawesi, if at all. But, unlike the Cosmopsaltriaria, all are represented in the Philippines, the weevil tribe being extremely diverse there. Another good example of a northern group ranging from the Philippines to Fiji excluding Sulawesi is the frog genus *Platymantis* (Allison, 1996). Morley (1998, this volume) suggests the flora of the eastern archipelagos could have been derived from floras in the Philippines and Sulawesi, an observation that is not inconsistent with such northern and southern patterns. The two lineages of *Aporosa* discussed by Schot (1998, this volume) also show parallels.

Gressitt (1984) had begun to investigate the Tmesisternini complex in detail just before his untimely death. In this work he gave an outline of a phylogenetic hypothesis for the complex. This is summarised in Fig.14. The northern groups referred to above are the Bumetopiini and Crinotarsini. The Tmesisternini are an essentially southern group, with a natural group of four genera in Queensland, a genus in Sulawesi and three distinct groups in New Guinea. One of these also includes a genus, *Sphingnotus*, that, like the *Baeturia bloetei* group in the Chlorocystini cicadas, is found in the outer Melanesian archipelagos, but consists of only three rather widespread species.

A novel feature of the beetle group relative to the Chlorocystini is the inclusion of New Caledonia in both lineages of the Tmesisternini. The sister-group of the Tmesisternini, the Homonoeini, ranges widely through the Indo-Australian tropics, but the Tmesisternini + Homonoeini pairing is itself sister to the Enicodini, a tribe endemic to New Caledonia. The outgroups include a tribe endemic to New Zealand as well as the two northern archipelagic distribution tribes mentioned above. A partial explanation of these differences might be that the ancient history of the beetle group is Australasian/Gondwanan, whereas that of the cicadas is Oriental, but this does not necessarily explain why New Caledonia features within the southern Queensland-New Guinea-Sulawesi (+ Melanesian archipelagos) patterns in the beetles but not in that for the Chlorocystini + Prasiini. The New Caledonian cicada fauna is entirely Australian in affinity (Holloway, 1979, pp. 234-235). A more exclusively Australian, New Caledonian and New Guinea pattern is seen in the plant genus *Arytera* (Sapindaceae; Turner, 1995).

The equation of gross pattern even between such biogeographically informative groups as the longhorn beetles and cicadas is thus not without problems: the development of these patterns may also involve stochastic processes, and this is not necessarily unexpected in complex archipelagic situations where emergence and convergence of geological structures, and change of relative positions, have been much more frequent than the sort of sundering and divergence that could lead to clear, repeated vicariant patterns. Objective comparison of biogeographic pattern with hypotheses of changes in geography due to plate tectonic processes thus is difficult: the paper by Boer (1995b) represents the most significant attempt to date. We still need a methodology for pattern analysis that does not introduce the sort of constraints that the past focus on a hypothetical unique set of area relationships has threatened to do. A complementary combination of R-mode and Q-mode approaches as first advocated by Holloway and Jardine (1968) may be one way forward.

Michaux (1991, 1994, 1998 this volume) has already made preliminary observations on the relationship of panbiogeographic, R-mode types of pattern to the geology of the region. The northern and southern types of distribution discussed above perhaps correlate with his (1994) Outer Melanesian track (the former) and a combination of his Inner Melanesian track with continental Australia and his Inner Banda track, excluding Borneo (the latter).

But there is a prime need for many more modern phylogenetic treatments of taxa from the region to add to our pattern sample and to enable us to explore further the complementarity of cladistic (Q-mode) and panbiogeographic (R-mode) approaches to analysis of biogeographic pattern in relation to Earth history. We also need a better understanding of biological process (dispersal, speciation, response to climatic factors) to facilitate interpretation of the patterns derived. Sneath (1982) and Page and Lydeard (1994) noted that the sort of geographically progressive cladistic pattern described earlier, shown by so many Indo-Australian biological groups, could equally well arise through dispersal processes as vicariant ones: gradual biological dispersal through the archipelago from west to east or progressive vicariance of land areas through introduction of sea barriers in the same direction.

Even if there is evidence of correlation between biological and geological data in the structure of area relationships, geologists frequently ask whether it is possible to date divergence in biological lineages so as to permit direct comparison with geological dates. This is a topic where there is still much disagreement. Holloway and Nielsen (1998), reviewing cladistic biogeographic analyses of the Mediterranean, noted that area relationships established to be common to a number of biological groups and related to the post-Oligocene history of the area by some authors were also manifest in a phylogenetic hypothesis based on molecular data for another group that suggested the events concerned were much more recent.

Dating of speciation events that relies on assumption of a molecular clock and uses distance data has been strongly criticized by cladistic biogeographers (*e.g.*, Morrone and Carpenter, 1994), but should not be rejected entirely. As Page and Lydeard (1994) indicated, the distance measures must be ultrametric for them to be clock-like (discussed also by Jardine *et al.*, 1969), and care must be taken with the analyses. If these conditions are satisfied, such analyses could well provide useful additional information to test concordance between biogeographic pattern and geological hypotheses (Page and Lydeard, 1994; Hedges *et al.*, 1994).

The complexity of the Indo-Australian tropics will undoubtedly require the sort of pragmatic approach adopted by Funk and Wagner (1995) in interpreting pattern in various plant and animal groups that have speciated in the Hawaiian archipelago. They developed a series of hypothetical patterns that might be expected under different process assumptions (rate of evolution relative to dispersal, mode and time of colonisation, direction of dispersals, etc.) in relation to current knowledge of the geological evolution of the Hawaiian island chain, and compared them with actual patterns. The most frequent pattern was of the progression type, with dispersal from older to younger islands, this progression being in the form of clades or grades in relation to the islands, dependent on the relative frequency of speciation relative to dispersal. This last distinction might equally be applicable to the differences observed within similar gross patterns in Lepidoptera and cicadas, leading to much greater species-richness within each pattern element of the latter, the paralogies of Nelson and Ladiges (1996).

Is it possible, therefore, to represent the patterns discussed in this chapter in some sort of sequence that might have some bearing on the geological history of the area? The basal groupings in many of the more complex patterns in the Lepidoptera or cicadas are: Oriental-Sundanian; tropical Australia with New Guinea or New Guinea alone, possibly with a focus on the Papuan peninsula and the Central and Sepik structures of the latter; outer Melanesian archipelagos, possibly also including some components now seen as northern structures within New Guinea; Sulawesi, possibly a component of the current island that was relatively isolated from both Oriental and Papuan land areas at that time. Establishment of these patterns is therefore likely to have been relatively early.

Intermediate patterns include development of an area of endemism in the Philippines that would appear to have had some interaction with a component of Sulawesi, such as its northern peninsula, development of patterns in the Banda arcs that also included interaction with a component of Sulawesi, probably the south-western peninsula, and progressive breakdown of the segregation of the New Guinea areas of endemism from those of the Melanesian archipelagos. The development of the *Cosmopsaltria* pattern endemic to the New Guinea areas could have preceded considerably the expansion of the Chlorocystini into the outer archipelagos in the form of the *Baeturia bloetei* group, and also the extension of *Diceropyga* of the Cosmopsaltriaria into the Papuan peninsula. This interpretation departs somewhat from that of Boer (1995b), though he also considered differential timing of development of the various patterns. The interaction of structures between Sulawesi and New Guinea with the major foci of endemism is still not clear. Cicada patterns involving the Bird's Head of New Guinea are complex, with interaction seen with Sulawesi, New Guinea and Melanesian archipelago (Solomons) groups (Fig.13). None of the islands comprising the current Moluccas appears to have acted as a strong centre of endemism, their faunas bearing a general relationship to neighbouring areas, though more so with areas to the east than to the west (see also Jong, 1998 this volume).

Patterns that are probably progressively more recent are the allopatric and widespread categories among Oriental taxa. The former, where progressive, do not appear to discriminate separate areas of endemism, or composites in Sulawesi or New Guinea. The species of the latter, more often than not, transcend the older areas of endemism and, through stochastic interchanges between islands, enable a strong signal from modern geography to be recaptured in scaling analyses such as in Holloway and Jardine (1968). Whilst tending to obscure signals from the past, these groups may well be establishing the foundations of signals that will be detected in the distant future.

Acknowledgements

This chapter is based in part on a presentation made in 1993 at the 13th meeting of the Willi Hennig Society in Copenhagen and in part on that made at the formative meeting for this book in 1996. It owes much to interaction through the years with Hans Duffels, Arnold de Boer, Dan Polhemus, Dick Vane-Wright, Rienk de Jong and others perplexed by the complexities encountered in the biogeography of the Indo-Australian tropics, and therefore embracing the 'heresy' of searching beyond the goal of a unique structure for area relationship within the archipelago. The coefficients for the analyses were calculated using software written by Gaden Robinson. I am grateful to my wife, Phillipa, for help with keyboarding the text and the lettering for the illustrations and to Robert Hall for reconstructing some of the latter in electronic format.

References

Ackery, P. R. 1987. The danaid genu *Tellervo* (Lepidoptera: Nymphalidae) – a cladistic approach. Zoological Journal of the Linnean Society 89: 203-274.

Ackery, P. R. and Vane-Wright, R. I. 1984. Milkweed Butterflies. British Museum (Natural History), London.

Allison, A. 1996. Zoogeography of amphibians and reptiles of New Guinea and the Pacific region. *In* The Origin and Evolution of Pacific Island Biotas, New Guinea to Eastern Polynesia: Patterns and Processes. pp. 407-436. Edited by A. Keast and S. E. Miller. SPB Academic Publishing, Amsterdam.

Banks, H. J., Holloway, J. D. and Barlow, H. S. 1976. A revision of the genus *Ptychandra* (Lepidoptera: Nymphalidae). Bulletin of the British Museum (Natural History) (Entomology) 32: 217-252.

Bloem, J. H. and Duffels, J. P. 1976. The *terpsichore*-group of the genus *Dundubia* Amyot and Serville, 1843 (Homoptera, Cicadidae). Bulletin Zoologisch Museum, Universiteit van Amsterdam 5: 141-154.

Boer, A. J. de 1995a. The phylogeny and taxonomic status of the Chlorocystini (sensu stricto) (Homoptera, Tibicinidae). Contributions to Zoology 65: 201-231.

Boer, A. J. de 1995b. Islands and cicadas adrift in the west-Pacific. Biogeographic patterns related to plate tectonics. Tijdschrift voor Entomologie 138: 169-241.

Boer, A. J. de and Duffels, J. P. 1996a. Historical biogeography of the cicadas of Wallacea, New Guinea and the West Pacific: a geotectonic exploration. Palaeogeography, Palaeoclimatology, Palaeoecology 124: 153-177.

Boer, A. J. de and Duffels, J. P. 1996b. Biogeography of Indo-Pacific cicadas east of Wallace's Line. *In* The Origin and Evolution of Pacific Island Biotas, New Guinea to Eastern Polynesia: Patterns and Processes. pp. 297-330. Edited by A. Keast and S. E. Miller. SPB Academic Publishing, Amsterdam.

Bregman, R. 1985. Taxonomy, phylogeny and biogeography of the *tridentigera* group of the genus *Chremistica* Stål 1870 (Homoptera, Cicadidae). Beaufortia 35: 37-60.

Cowan, C. F. 1974. The Indo-Oriental genus *Drupadia* Moore (Lepidoptera: Lycaenidae). Bulletin of the British Museum (Natural History), (Entomology) 29: 281-356.

Craw, R. C. 1989. New Zealand biogeography: a panbiogeographic approach. New Zealand Journal of Zoology 16: 527-547.

Crisci, J. V., Cigliano, M. M., Morrone, J. J. and Roig Juñent, S. 1991. A comparative review of cladistic approaches to historical biogeography of southern South America. Australian Systematic Botany 4: 117-128.

Darlington, P. J. Jr. 1957. Zoogeography: The Geographical Distribution of Animals. John Wiley and Sons, New York.

Duffels, J. P. 1983. Taxonomy, phylogeny and biogeography of the genus *Cosmopsaltria*, with remarks on the historic biogeography of the subtribe Cosmopsaltriaria (Homoptera: Cicadidae). Pacific Insects Monographs 39: 1-127.

Duffels, J. P. 1988. The cicadas of the Fiji, Samoa and Tonga islands, their taxonomy and biogeography (Homoptera, Cicadoidea). Entomonograph 10.

Duffels, J. P. 1990. Biogeography of Sulawesi cicadas (Homoptera: Cicadoidea). *In* Insects and the Rain Forests of South East Asia (Wallacea). pp. 63-72. Edited by W. J. Knight and J. D. Holloway. Royal Entomological Society, London.

Duffels, J. P. and Boer, A. J. de 1990. Areas of endemism and composite areas in East Malesia. *In* The Plant Diversity of Malesia. pp. 249-272. Edited by P. Baas, C. Kalkman and R. Geesink. Kluwer, Dordrecht.

Eliot, J. N. 1986. A review of the Miletini (Lepidoptera: Lycaenidae). Bulletin of the British Museum (Natural History), (Entomology) 53: 1-105.

Eliot, J. N. 1990. Notes on the genus *Curetis* Hübner (Lepidoptera, Lycaenidae). Tyô to Ga 41: 201-225.

Funk, V. A. and Wagner, W. L. 1995. Biogeographic patterns in the Hawaiian islands. *In* Hawaiian Biogeography, Evolution on a Hot Spot Archipelago. pp. 379-419. Edited by W. L. Wagner and V. A. Funk. Smithsonian Institution Press, Washington.

Gressitt, J. L. 1956. Some distribution patterns of Pacific island faunas. Systematic Zoology 5: 11-47.

Gressitt, J. L. 1982. Zoogeographical summary. Monographiae biologicae 42: 897-918.

Gressitt, J. L. 1984. Systematics and biogeography of the longicorn beetle tribe Tmesisternini. Pacific Insects Monograph 41.

Hayashi, M. 1987a. A revision of the genus *Cryptotympana* (Homoptera, Cicadidae). Part 1. Bulletin Kitakyoshi Museum of Natural History 6: 119-212.

Hayashi, M. 1987b. A revision of the genus *Cryptotympana* (Homoptera, Cicadidae). Part 2. Bulletin Kitakyoshi Museum of Natural History 7: 1-109.

Hedges, S. B., Hass, C. A. and Maxson, L. R. 1994. Reply: towards a biogeography of the Caribbean. Cladistics 10: 43-55.

Hirowatari, T. 1992. A generic classification of the tribe Polyommatini of the Oriental and Australian Regions (Lepidoptera, Lycaenidae, Polyommatinae). Bulletin of University of Osaka Prefecture. Series B 44 Supplement.

Holloway, J. D. 1969. A numerical investigation of the biogeography of the butterfly fauna of India, and its relation to continental drift. Biological Journal of the Linnean Society 1: 373-385.

Holloway, J. D. 1970. The biogeographical analysis of a transect sample of the moth fauna of Mt. Kinabalu, Sabah, using numerical methods. Biological Journal of the Linnean Society 2: 259-286.

Holloway, J. D. 1973. The taxonomy of four groups of butterflies (Lepidoptera) in relation to general butterfly distribution in the Indo-Australian area. Transactions of the Royal Entomological Society of London 125: 125-176.

Holloway, J. D. 1974. The biogeography of Indian butterflies. *In* Ecology and Biogeography in India. pp. 473-499. Edited by M. S. Mani. W. Junk, The Hague.

Holloway, J. D. 1977. The Lepidoptera of Norfolk Island, their biogeography and ecology. Series Entomologica 13. W. Junk, The Hague.

Holloway, J. D. 1979. A Survey of the Lepidoptera,

Biogeography and Ecology of New Caledonia. Series Entomologica 15. W. Junk, The Hague.

Holloway, J. D. 1982. Mobile organisms in a geologically complex area: Lepidoptera in the Indo-Australian tropics. Zoological Journal of the Linnean Society 76: 353-373.

Holloway, J. D. 1983. The biogeography of the macrolepidoptera of south-eastern Polynesia. GeoJournal 7: 517-525.

Holloway, J. D. 1984. Lepidoptera and the Melanesian Arcs. In Biogeography of the Tropical Pacific. Edited by F. J. Radovsky et al., Bishop Museum Special Publication 72: 129-169.

Holloway, J. D. 1985. The Moths of Borneo: family Noctuidae: Subfamilies Euteliinae, Stictopterinae, Plusiinae, Pantheinae. Malayan Nature Journal 38: 157-317.

Holloway, J. D. 1986a. Lepidoptera faunas of high mountains in the Indo-Australian tropics. In High altitude Tropical Biogeography. pp. 533-556. Edited by F. Vuilleumier and M. Monasterio. Oxford University Press, New York.

Holloway, J. D. 1986b. The Moths of Borneo: Key to families; Families Cossidae, Metarbelidae, Ratardidae, Dudgeoneidae, Epipyropidae and Limacodidae. Malayan Nature Journal 40: 1-165.

Holloway, J. D. 1987a. The Moths of Borneo: Superfamily Bombycoidea: families Lasiocampidae, Eupterotidae, Bombycidae, Brahmaeidae, Saturniidae, Sphingidae. Southdene, Kuala Lumpur, 199 pp.

Holloway, J. D. 1987b. Lepidoptera patterns involving Sulawesi: what do they indicate of past geography? In Biogeographical Evolution of the Malay Archipelago. pp. 103-118. Edited by T. C. Whitmore. Clarendon Press, Oxford.

Holloway, J. D. 1989. The Moths of Borneo: family Noctuidae, trifine subfamilies: Noctuinae, Heliothinae, Hadeninae, Acronictinae, Amphipyrinae, Agaristinae. Malayan Nature Journal 42: 57-226.

Holloway, J. D. 1991. Patterns of moth speciation in the Indo-Australian archipelago. In The Unity of Evolutionary Biology. Proceedings of IVth International Congress of Systematic and Evolutionary Biology. Pp. 340-372. Edited by E. C. Dudley. Dioscorides Press, Portland, Oregon.

Holloway, J. D. 1993. The moths of Borneo: family Geometridae, subfamily Ennominae. Malayan Nature Journal 47: 1–309.

Holloway, J. D., 1996. The Lepidoptera of Norfolk Island, actual and potential, their origins and dynamics. In The Origin and Evolution of Pacific Island Biotas, New Guinea to Eastern Polynesia: Patterns and Processes. pp. 123-151. Edited by A. Keast and S. E. Miller. SPB Academic Publishing, Amsterdam.

Holloway, J. D., 1997. Sundaland, Sulawesi and eastwards: a zoogeographic perspective. Malayan Nature Journal 50: 207-227.

Holloway, J. D., Cock, M. J. W. and Desmier de Chenon, R. 1987. Systematic account of south-east Asian pest Limacodidae. In Slug and Nettle Caterpillars, the Biology, Taxonomy and Control of the Limacodidae of Economic Importance on Palms in south-east Asia. pp. 15-117. Edited by M. J. W. Cock, H. C. J. Godfray and J. D. Holloway. CAB International, Wallingford.

Holloway, J. D. and Jardine, N. 1968. Two approaches to zoogeography: a study based on the distributions of butterflies, birds and bats in the Indo-Australian area. Proceedings of the Linnean Society of London 179: 153-188.

Holloway, J. D. and Nielsen, E. S. 1998. Biogeography of the Lepidoptera. In Lepidoptera, Volume 1. Handbuch der Zoologie 35. Edited by N. P. Kristensen. Walter de Gruyter, Berlin (in press).

Inoue, H. 1990. A revision of the genus *Dindica* Moore (Lepidoptera: Geometridae). Bulletin of the Faculty of domestic Science, Otsuma Women's University 26: 221–161.

Inoue, H. 1993a. A revision of the genus *Macrauzata* Butler (Lepidoptera, Drepanidae, Drepaninae). Bulletin of Otsuma Women's University, Home Economics, 29: 217-234.

Inoue, H. 1993b. A revision of the genus *Herdonia* Walker (Lepidoptera: Thyrididae, Siculodinae). Tyô to Ga, 44: 127-151.

Kitching, I. J., Vane-Wright, R. I. and Ackery, P. R. 1987. The cladistics of *Ideas*. Cladistics 3: 14-34.

Jardine, N. and Sibson, R. 1968. The construction of hierarchic and non-hierarchic classifications. Computer Journal 11: 177-184.

Jardine, N, van Rijsbergen, C. J. and Jardine, C. J. 1969. Evolutionary rates and the inference of evolutionary tree forms. Nature, London 224: 195.

Jong, M. R. de 1985. Taxonomy and biogeography of Oriental Prasiini 1: The genus *Prasia* Stål, 1863 (Homoptera, Tibicinidae). Tijdschrift voor Entomologie 128: 165-191.

Jong, M. R. de 1986. Taxonomy and biogeography of Oriental Prasiini 2: The *foliata* group of the genus *Lembeja* Distant, 1892 (Homoptera, Tibicinidae). Tijdschrift voor Entomologie 129: 141-180.

Jong, M. R. de 1987. Taxonomy and biogeography of Oriental Prasiini 3: *fatiloqua* and *parvula* groups of the genus *Lembeja* Distant, 1892 (Homoptera, Tibicinidae). Tijdschrift voor Entomologie 130: 177-209.

Jong, R. de 1983. Revision of the Oriental genus *Matapa* Moore (Lepidoptera, Hesperiidae) with discussion of its phylogeny and geographic history. Zoologische Mededelingen 57: 43-270.

Jong, R. de 1998. Halmahera and Seram: different histories, but similar butterfly faunas. In Biogeography and Geological Evolution of SE Asia. Edited by R. Hall and J. D. Holloway (this volume).

Levison, M. Fenner, T. I., Sentance, W. A., Ward, R. G. and Webb, J. W. 1969. A model of accidental drift voyaging in the Pacific Ocean with applications to the Polynesian colonization problem. Information Processing 68: 1521-1526.

MacArthur, R. H. and Wilson, E. O. 1967. The Theory of Island Biogeography. Princeton University Press, New Jersey.

Michaux, B. 1991. Distributional patterns and tectonic development in Indonesia: Wallace re-interpreted. Australian Systematic Botany 4: 25-36.

Michaux, B. 1994. Land movements and animal distributions in east Wallacea, eastern Indonesia, Papua New Guinea and Melanesia. Palaeogeography, Palaeoclimatology, Palaeoecology 112: 323-343.

Michaux, B. 1998. Terrestrial birds of the western Pacific. In Biogeography and Geological Evolution of SE Asia. Edited by R. Hall and J. D. Holloway (this volume).

Morley, R. J. 1998. Palynological evidence for Tertiary plant dispersals in the SE Asia region in relation to plate tectonics and climate. In Biogeography and Geological Evolution of SE Asia. Edited by R. Hall and J. D. Holloway (this volume).

Morrone, J. J. and Carpenter, J. M. 1994. In search of a method for cladistic biogeography: an empirical comparison of component analysis, Brooks parsimony analysis and three area statements. Cladistics 10: 99-153.

Munroe, E. G. and Shaffer, M. 1980. A revision of *Vitessidia* Rothschild and Jordan and *Vitessa* Moore (Lepidoptera:

Pyralidae), Bulletin of the British Museum (Natural History), (Entomology) 39: 241-360.
Nelson, G. and Ladiges, P. Y. 1996. Paralogy in cladistic biogeography and analysis of paralogy-free subtrees. American Museum Novitates 3167: 1-58.
Page, R. D. M. 1989. New Zealand and the new biogeography. New Zealand Journal of Zoology 16: 471-483.
Page, R. D. M. and Lydeard, C. 1994. Towards a cladistic biogeography of the Caribbean. Cladistics 10: 21-41.
Parsons, M. 1986. A revision of the genus *Callictita* Bethune-Baker (Lepidoptera: Lycaenidae) from the mountains of New Guinea. Bulletin of the Allyn Museum 103: 1-27.
Peigler, R. S. 1989. A Revision of the Indo-Australian genus *Attacus* (Lepidoptera: Saturniidae). Lepidoptera Research Foundation, Santa Barbara.
Polhemus, D. A. 1996. Island arcs, and their influence on Indo-Pacific biogeography. *In* The Origin and Evolution of Pacific Island Biotas, New Guinea to Eastern Polynesia: Patterns and Processes. pp. 51-66. Edited by A. Keast and S. E. Miller. SPB Academic Publishing, Amsterdam.
Polhemus, D. A. and Polhemus, J. T. 1998. Assembling New Guinea – 40 million years of arc accretion as indicated by distributions of aquatic Heteroptera (Insecta). *In* Biogeography and Geological Evolution of SE Asia. Edited by R. Hall and J. D. Holloway (this volume).
Saigusa, T., Nakanishi, A., Shima, H. and Yata, D. 1977. Phylogeny and biogeography of the subgenus *Graphium* Scopoli (Lepidoptera: Papilionidae, *Graphium*). Acta Rhopalocerologica 1: 1-32.
Sato, R. 1987. Taxonomic notes on *Menophra delineata* (Walker) (Geometridae: Ennominae) and its allies from Indo-Malayan region. Tinea 12 Supplement: 249-258.
Schot, A. 1998. Biogeography of *Aporosa* (Euphorbiaceae): testing a phylogenetic hypothesis using geology and distribution patterns. *In* Biogeography and Geological Evolution of SE Asia. Edited by R. Hall and J. D. Holloway (this volume).
Schuh, R. T. and Stonedahl, G. 1986. Historical biogeography in the Indo-Pacific: a cladistic approach. Cladistics 2: 337-355.
Simberloff, D. and Connor, E. F. 1980. Q-mode and R-mode analyses of island biogeographic distributions: null hypotheses based on random colonization. *In* Contemporary Quantitative Ecology and Related Econometrics. pp. 123-138. Edited by G. P. Dahl and M. L. Rosenzweig. International Cooperative Publishing House, Fairland, Maryland.
Smiles, R. L. 1982. The taxonomy and phylogeny of the genus *Polyura* Billberg (Lepidoptera: Nymphalidae). Bulletin of the British Museum (Natural History), (Entomology) 44: 115-237.
Sneath, P. H. A. 1982. Review of Nelson, G. and Platnick, N. 1981. Systematics and Biogeography: Cladistics and Vicariance. Systematic Zoology 31: 208-217.
Turner, H. 1995. Cladistic and biogeographic analyses of *Arytera* Blume and *Mischarytera* gen. nov. (Sapindaceae), with notes on methodology and a full taxonomic revision. Blumea, Supplement 9: 1-230.
Vane-Wright, R. I. (1990) The Philippines - key to the biogeography of Wallacea? *In* Insects and the Rain Forests of South East Asia (Wallacea). pp. 19-34. Edited by W. J. Knight and J. D. Holloway. Royal Entomological Society, London.
Vane-Wright, R. I. (1991) Transcending the Wallace Line: do the western edges of the Australian Region and Australian Plate coincide? Australian Systematic Botany 4: 183-197.
Wagner, W. L. and Funk, V. A. (Editors) 1995. Hawaiian Biogeography. Evolution on a Hot Spot Archipelago. Smithsonian Institution Press, Washington.
Ward, R. G. and Brockfield, M. 1992. The dispersal of the coconut: did it float or was it carried to Panama? Journal of Biogeography 19: 467-480.
Watson, A. 1957. A revision of the genus *Tridrepana* Swinhoe (Lepidoptera: Drepanidae). Bulletin of the British Museum (Natural History) (Entomology) 4: 409-500.
Whalley, P. E. S. 1976. Tropical Leaf Moths, a Monograph of the Subfamily Striglininae (Lepidoptera, Thyrididae). British Museum (Natural History), London.
Wilson, J. B. 1991. A comparison of biogeographic models: migration, vicariance and panbiogeography. Global Ecology and Biogeography Letters 1: 84-87.
Wilson, J. B., Rayson, G. l., Sykes, M. T., Watkins, A. J. and Williams, P. A. 1992. Distributions and climatic correlations of some exotic species along roadsides in the South Island, New Zealand. Journal of Biogeography 19: 183-193.
Yata, O. 1990. Cladistic biogeography of *Eurema* butterflies belonging to the subgenus *Terias* (Lepidoptera: Pieridae), with particular reference to Wallacea. *In* Insects and the Rain Forests of South East Asia (Wallacea). pp. 43-48. Edited by W. J. Knight and J. D. Holloway. Royal Entomological Society, London.

Halmahera and Seram: different histories, but similar butterfly faunas

Rienk de Jong
Nationaal Natuurhistorisch Museum, Department of Entomology, P O Box 9517, 2300 RA Leiden, The Netherlands. Email: jong@nnm.nl

Key words: historical biogeography, faunal similarity, Moluccas, Halmahera, Seram, butterflies

Abstract

The islands of Halmahera and Seram in the Moluccas (east Indonesia) have very different geological histories. Their present relative proximity is the result of a long westward journey of Halmahera. It is uncertain when the island emerged, and thus was first able to support a terrestrial fauna. Seram emerged at c.5 Ma at approximately its present location. On the basis of current geological scenarios, expected distribution patterns for terrestrial organisms are proposed. These patterns are tested using information on species distribution, endemism and phylogeny of butterflies.

The butterfly faunas of Halmahera and Seram are remarkably similar for faunas that have never been closer to each other than today. This is attributed to independent colonization by the same species from Sulawesi and particularly from New Guinea. The different geological histories of the two islands are not reflected in their butterfly faunas. Two possible reasons for this are suggested: either the older fauna of Halmahera was wiped out by later colonists, or Halmahera only emerged as recently or even later than Seram.

Endemism in the Moluccas is high (21.3% of the 385 species). Distribution patterns of the endemic species indicate that faunal exchange between the northern and southern islands of the Moluccas has been limited to less than 15% of the 82 endemic species. This is a further indication that the similarity between Halmahera and Seram is not due to a direct exchange of species, but to independent invasions from outside by the same species.

Introduction

Halmahera and Seram are the largest islands of the east Indonesian province of Maluku (Fig.1). This province does not completely coincide with the old notion of Moluccas. Following Vane-Wright and Peggie (1994) the Moluccas have been taken here to include the sub-provinces of N and C Maluku, but excluding Kepulauan Sula, which biogeographically forms part of the Sulawesi region, and the Banda and Gorong archipelagos. Halmahera is, at 17,400 km², the larger island, but at 16,720 km², Seram is only slightly smaller. The greater part of Halmahera lies north of the equator. Seram is situated well to the south, about 220 km from the southern tip of Halmahera. The island of Obi lies at about one third of this distance from Halmahera to Seram. Halmahera and Seram share the same humid and hot climate, with an average yearly rainfall of 1,500-2,000 mm and lowland temperatures varying between 25°C and 30°C. Monsoons and mountains affect the rainfall, but it is still rather generally distributed throughout the year. The natural vegetation is rain forest. There is still untouched forest left, but, particularly in Halmahera, deforestation is proceeding fast. There is a marked difference in height, the highest peak in Halmahera, Gunung Gamkunoro, reaching an altitude of 1635 m, while Gunung Binaiya, the highest mountain of Seram, ascends to 3027 m. As a consequence, Seram is more diverse in terms of habitat.

Halmahera and Seram are situated more or less halfway between Sulawesi and New Guinea, but smaller islands to the east and west of Halmahera and Seram reduce their degree of isolation from these far larger islands. Between the northeastern tip of Sulawesi and Halmahera, a slightly larger distance than between Halmahera and Seram, there are only two tiny islands, Mayu and Tifore, whereas between Halmahera and New Guinea, also a slightly

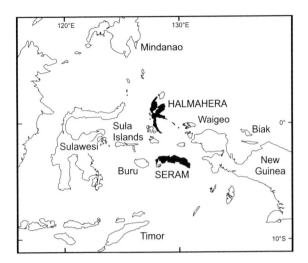

Fig.1. Location of Halmahera and Seram in east Indonesia.

larger stretch of water than between Halmahera and Seram, there are several islands, such as Gebe, Gag and Waigeo. Seram is separated from New Guinea by about 140 km, but the distance to Sulawesi is considerably greater, about 550 km. There are, however, several islands between Seram and Sulawesi, including Buru, and the Sula and Banggai archipelagos, which could serve as stepping stones.

In spite of their relative isolation, Halmahera and Seram have remarkably similar butterfly faunas. Of the 200 species known from Halmahera 58.5% are also found in Seram, while 52% of the 225 butterfly species of Seram is also found in Halmahera. Applying Jaccard's coefficient of similarity (Krebs, 1989), the butterfly faunas of Halmahera and Seram (total fauna 304 species) are exactly as similar ($Sj = 0.39$) as the butterfly faunas of Spain and Northwest Africa (total fauna 243 species; data from Tennent, 1996, and Vives Moreno, 1994), which are, however, only 15 km apart, and not 220 km like Halmahera and Seram. Knowing that the two islands have very different geological histories, it is of interest to investigate the possible causes of this similarity. The history of the areas is taken as a starting point rather than something to be derived from the distribution of organisms. Thus, the present contribution is not concerned with the relationship of areas, but with the history of existing faunas and how this may have been affected by the history of the areas. Since the areas concerned (Halmahera, Seram, Sulawesi and New Guinea) have never been connected by land, similarities between any two of the areas in terms of shared taxa must be the result of long distance dispersal, and not of range fragmentation. A fauna typically consists of species exhibiting a spectrum of range sizes, from narrow endemics to wide-ranging or even cosmopolitan species. A study of the history of a fauna should address all elements of this fauna, and not just the narrow endemics. This will be elaborated further in the section below on Methods.

Wallace (1869) claimed a predominantly New Guinea origin for the avifauna of the Moluccas, but later students (e.g., Bowler and Taylor, 1993) stressed the strong Asian representation in the bird faunas of the Moluccas and New Guinea, and Gressitt (1982) estimated that 60-95% of the Papuan fauna (depending on the group of animals under study) had Oriental origins. The present paper is not concerned with such larger scale distributional histories, but investigates only the biogeographic links in the butterfly faunas of Halmahera and Seram, and with Sulawesi and New Guinea. Even so, this may help to understand the biogeographic history of the whole of the Moluccas.

Methods

In biogeographic studies of the last 15 years, an approach has dominated in which the history of areas is considered the crucial factor in the distribution of life. This approach originated from the work of Croizat (e.g., 1962). The coupling of phylogenetic data to geological history, the basis of cladistic biogeography, has led to very fruitful research, partly thanks to the development of computer hardware and software able to evaluate large amounts of data (e.g., the COMPONENT program of Page, 1993). In this approach areas are considered related in terms of a shared biological history (as expressed in the phylogenies of the studied organisms). It leads to a hypothesis of area history, which can be corroborated or falsified by geological data. This approach, though robust, does not deal with faunas, but with informative taxa. In this context, informative means that taxa may reflect the history of the areas where they occur. The more widespread a taxon is, the less informative it is in this sense. As a consequence, cladistic biogeography may reveal and explain historical patterns of areas and distributions, but only part of the distributional history of the fauna as a whole, which indeed is a mixture of old and new species, with old and new distributional patterns.

There are some practical problems in the application of the method, particularly in the detection and meaning of areas of endemism, and in the synchrony of speciation patterns. By its procedure, cladistic biogeography is an inductive method. It leads to hypotheses which can only be tested by evidence from outside, e.g., geology. This can be a mutually beneficial enterprise. However, if the geological history is well-established or if there are competing geological hypotheses, it may be more effective to apply a hypothetico-deductive method, meaning that on the basis of geological knowledge, hypotheses of distributional patterns are formulated, which then can be tested against actual patterns. Such patterns can, but need not, be phylogenetic, and no part of the fauna is excluded beforehand. The author has applied this approach before (e.g., de Jong and Congdon, 1993; de Jong and Treadaway, 1993; de Jong, 1997a), and it will be followed here.

The overall question to be answered in this contribution is: are the different geological histories of Halmahera and Seram reflected in different distributional patterns in the butterfly faunas of these islands? If not, what possible causes could there be? Problems not addressed here include the origin of the faunas of Sulawesi and New Guinea. After a short explanation of the geological histories of Halmahera and Seram, the patterns to be tested are formulated.

Geological history of Halmahera and Seram

Halmahera consists of a younger western volcanic part and an eastern part. The two parts collided between 3 and 1 Ma (Hall and Nichols, 1990). The oldest rocks in the basement of eastern Halmahera are of Cretaceous or greater age (see Hall *et al.* 1995, for a summary of the geology of the Halmahera region). The island originated far east of its present position as part of an arc system bordering the Pacific plate. The Halmahera islands have a stratigraphic record which indicates that for most of the Cenozoic they were submerged although with a prolonged, albeit discontinuous, volcanic arc history and were at least locally emergent for short intervals (Hall, 1987; Hall *et al.*, 1995 and references therein). The east Philippine islands (the east Philippine arc of Hall, 1996) also belonged to this arc system. Halmahera and surrounding islands (*e.g.*, Waigeo, Biak, Yapen and parts of present-day northern New Guinea, such as the Gauttier Mountains) formed a broadly continuous arc during the Oligocene (Hall, 1998 this volume).

It is not certain from the literature when the islands of the Halmahera arc emerged. In a Late Eocene reconstruction of the New Guinea-Bismarck region, Struckmeyer *et al.* (1993) supposed that in the Halmahera arc parts of Halmahera, Waigeo and Gauttier were dry land. At the time, the northern half of New Guinea was little more than a number of allochthonous terranes carrying partly widely separate, submerged or emerged islands (see Boer, 1995, for a summary). The terranes successively accreted to the northern craton of the Australian plate, starting with the Sepik terrane, about 25 Ma, while the Arfak terrane, supposed to have originally been part of the Halmahera arc, only accreted to New Guinea at 2 Ma according to Pigram and Davies, 1987 (cf. Hall, 1998 this volume). Accretion and uplift finally shaped present-day New Guinea but, at the time of origin of Halmahera, New Guinea did not exist in its present form. What did exist were some emergent islands of the Sepik terrane (Central New Guinea) to the southwest of the Halmahera arc while the Kemum terrane, with a relatively large emergent part (according to Struckmeyer *et al.*, 1993) and the Misool terrane, with a small emergent part, were located far to the west. The Kemum terrane forms the greater part of the present-day Bird's Head of New Guinea, situated to the east of Misool.

According to Rangin *et al.* (1990) and others, at 10 Ma Halmahera lay approximately 800 km southeast of its present position. Between 8 and 4 Ma it (or rather the still separated western and eastern parts) passed by the Bird's Head. It reached its present position recently, and is still moving westward. During its entire life Halmahera has never been close to any part of Sulawesi; it has always been closer to, or associated with, terranes that finally accreted with New Guinea. Furthermore, the east Philippine islands have always been closer to Halmahera than to Sulawesi (see reconstructions in Hall, 1996).

The origin of Seram is still a matter of debate, although it is not doubted that its roots are Australian. Some authors suppose that it formed a microcontinent with Buru. This block probably rifted from the Australian continent, possibly from as far east as central Papua New Guinea (*e.g.*, Pigram and Panggabean, 1984), or from northwestern Australia (*e.g.*, Daly *et al.*, 1991). According to Hall (1996), Seram moved eastward relative to the Bird's Head, from 12-4 Ma, and northward to its present position relative to the Bird's Head, from 4-0 Ma. Biogeographically the uncertainties about the early evolution of Seram are not important. According to Audley-Charles (1986, 1993), there was no land between the coast of Australia-New Guinea and Sulawesi before the Miocene emergence. As shown by sedimentary rocks, Seram was beneath the sea until about 6 Ma. Fortuin and de Smet (1991) estimate the time of emergence of Seram at 5 Ma. It emerged by uplift, and it has never been connected to other land areas, a deep marine trough separating it from New Guinea and Misool. Nevertheless, New Guinea was the closest land area. According to the reconstructions by Hall (1996), at 5 Ma Buru and the Sula archipelago, which at present form the connecting link with Sulawesi, were still far to the north of Seram.

Patterns to be expected

Based on different geological histories and different or similar ages for the terrestrial faunas of Halmahera and Seram, the following biogeographic patterns can be expected.

General

Different histories are supposed to lead to different faunas. Conversely, do similar faunas have similar histories? The similarity between faunas can be expressed in several ways. It is usually measured at the species level, but in addition there are other ways of measuring similarity. One way is similarity above the species level, as will be explained below. The other way is by comparing what is here called distribution profiles, a breakdown of a regional fauna according to distribution patterns. It is an overall comparison from which some conclusions can readily be drawn. Faunas of historically different areas are supposed to have different distribution profiles. In other words, Halmahera and Seram would be predicted to have different profiles, unless the early history of Halmahera has not left any traces, because of extinction or because Halmahera was submerged (pattern A).

Age

In the absence of fossils or other clues to (relative or absolute) age, we must resort to more indirect clues. Supposing that some part of (proto-)Halmahera has always remained subaerial, a greater age and longer period of isolation could have had the following effects on the fauna of Halmahera relative to Seram:

B1: Presence of older species. This can only be checked in monophyletic groups with endemic representatives in Halmahera, Seram and at least one other area.

B2: Higher endemism. Apart from speciation rate, degree of endemism depends on diversity of habitats and degree of isolation, either in time or in space. Since Seram has more diverse habitats due to higher altitudes, it could be expected to have more endemics than Halmahera, if other things, including age and degree of isolation, were equal. If Halmahera has an equal or higher endemism than Seram, either it is older or it has been more strongly isolated in space, or both.

B3: Reduced similarity to New Guinea. Due to speciation and extinction, not only in Halmahera but in New Guinea as well. This would also apply to Sulawesi if Halmahera had originated in its present position, which we know it did not. Clearly this pattern only makes sense if proto-Halmahera was close enough to proto-New Guinea to exchange organisms.

Topographic origin and subsequent history

The differences between Halmahera and Seram in history could have led to the following different patterns:

C1: There has not been much difference between Halmahera and Seram in the degree of isolation from Sulawesi. When Seram emerged from the sea it was probably closer to Sulawesi than Halmahera, then to the north of the Bird's Head, but Buru and the Sula islands were still so far to the north that they could not function as stepping stones between Sulawesi and Seram. When they reached that position, Halmahera also reached its present position. Consequently, the expected pattern is a comparable degree of similarity between the faunas of Halmahera and Sulawesi on one hand, and between those of Seram and Sulawesi on the other.

C2: If Halmahera was subaerial from Eocene times and it received butterflies at such an early date, the butterfly fauna would not have been very different from the fauna found on other islands of the Halmahera arc, like Waigeo and Gauttier, and after emergence also Biak, Yapen and Arfak. An indication for the existence of such a fauna would be distribution patterns involving some of these areas and excluding other areas.

C3: Whatever the time of emergence, Halmahera has always been closer to New Guinea than to Sulawesi; consequently, it can be expected that the similarity of the Halmahera fauna with the New Guinea fauna should be much greater than with the Sulawesi fauna.

C4: Since Seram emerged much closer to New Guinea than to other areas of dry land, it can be expected to carry a fauna that is much more similar to the fauna of New Guinea than to the fauna of Sulawesi.

C5: If the fauna of Seram actually came from New Guinea, the ties with New Guinea are older than with Sulawesi, since Buru and the Sula archipelago could only act as stepping stones at a much later date. Consequently, among the New Guinea links would be older species than among the Sulawesi links. This can only be tested in cases of endemics in the three areas forming a monophyletic or paraphyletic group. Biogeographically it seems an unlikely pattern.

C6: Halmahera and Seram have never been closer to each other than nowadays, and have never been connected directly or indirectly. Similarity between the two islands can be expected to involve mainly widespread species (at least covering New Guinea, in addition to Halmahera and Seram), and species only occurring in Halmahera and Seram are not to be expected, unless by chance dispersal.

The expected patterns are based on the supposition that butterflies have existed in the area at least since the origin of Halmahera. The supposition may be self-evident, but the presence of the butterflies is not. The oldest fossil butterfly remains are about 50 Ma old (de Jong, 1997b) but the butterflies as a taxon must be much older. Undoubtedly, when (proto-) Halmahera and surrounding islands emerged in the Tertiary, the butterflies as a group already existed, but the young islands may have been too isolated for the first butterflies to reach them at a very early date.

Patterns actually found

Distribution profile

Table 1 gives the number of species per butterfly family in Halmahera and Seram, and the number of species also occurring in Sulawesi and New Guinea. In Tables 2 and 3 these numbers have been broken down to distribution pat-

Table 1. Number of butterfly species in Halmahera and Seram, with the numbers of these species also occurring in Sulawesi and/or New Guinea.

	Sulawesi	**Halmahera**	**Seram**	New Guinea
Hesperiidae	19	**25**	**32**	30
Papilionidae	10	**20**	**17**	13
Pieridae	12	**23**	**30**	12
Lycaenidae	34	**54**	**72**	60
Nymphalidae	38	**78**	**74**	62
Total	113	**200**	**225**	177

terns, and for each distribution pattern the percentages of the species relative to the total number of species in Halmahera and Seram, respectively, are given. In Fig.2 the percentages of Tables 2 and 3 are represented graphically in a distribution profile, which allows a quick comparison of the faunas of Halmahera and Seram. Although there are slight differences in the percentages, the profiles for the two islands are remarkably similar, suggesting a largely similar biogeographic history for the butterflies. It is also clear that a greater part of the faunas (55.6% in Seram, 59.8% in Halmahera) is widespread, *i.e.,* involving at least three areas. This is to be expected in islands that were colonized by long distance dispersal over about the same length of time. The profiles further show that the Sulawesi connection excluding New Guinea is much weaker (Halmahera 26 species or 13.3%, Seram 27 species or 12.2%) than the New Guinea connection excluding Sulawesi (Halmahera 70 species or 35.3%, Seram 76 species or 33.7%). It not only agrees with the supposition that Halmahera and Seram have always been closer to New Guinea, but also shows that the degree of isolation with respect to Sulawesi and New Guinea has been about the same for the two islands (see above, expected general pattern).

Similarity at the species level

Many coefficients are available to express the similarity of two regional faunas, or two samples in general (Krebs, 1989). Applying Jaccard's coefficient (proportion of shared species relative to total number of species in the two areas) we find that the butterfly faunas of Halmahera and Sulawesi, and those of Seram and Sulawesi are equally similar, in both cases Sj = 0.04. The same

Table 2. Numbers and percentages of non-endemic butterfly species of Halmahera, divided over the seven possible distribution patterns. For coding of distribution patterns, see Fig.2.

Halmahera	1	2	3	4	5	6	7
Hesperiidae	4	0	1	2	1	8	9
Papilionidae	2	2	1	2	1	4	3
Pieridae	1	2	1	2	0	3	5
Lycaenidae	1	1	12	3	6	11	10
Nymphalidae	3	3	11	6	2	18	22
total	11	8	26	15	10	44	49
percentage	5.6	4.1	13.3	7.7	5.1	22	25

Table 3. Numbers and percentages of non-endemic butterfly species of Seram, divided over the seven possible distribution patterns. For coding of distribution patterns, see Fig.2.

Seram	1	2	3	4	5	6	7
Hesperiidae	1	0	9	2	2	8	9
Papilionidae	0	2	2	2	2	4	3
Pieridae	2	2	1	2	2	3	5
Lycaenidae	6	1	13	3	8	11	10
Nymphalidae	3	3	7	6	2	18	22
total	12	8	32	15	16	44	49
percentage	5.4	3.6	14.2	6.8	7.2	19.5	22.1

applies to the similarity of the butterfly faunas of Halmahera and New Guinea on one hand, and that of Seram and New Guinea on the other, in both cases the coefficient being Sj = 0.08. Thus we find that the butterfly faunas of Halmahera and Seram are (a) equally similar to the faunas of Sulawesi and New Guinea respectively (contra expected pattern B3, but in agreement with expected pattern C1), and (b) have a much greater similarity to New Guinea than to Sulawesi. The latter finding is not surprising since the New Guinea fauna is much richer (estimated to be at least 800 species) than the Sulawesi fauna (470 species). However, New Guinea is a very large island and many of its species occur further away from Halmahera than the average Sulawesi species. Thus, the greater similarity of the Moluccan islands to New Guinea than to Sulawesi appears to be the result not only of the richer New Guinea fauna but of the shorter distance. This is in agreement with expected patterns C3 and C4.

Similarity above the species level

Similarity at the species level may be reduced by extinction and speciation after the dispersal event. Since Halmahera and Seram have endemic species, it is clear that speciation has occurred in the islands. Finding their sister species either in New Guinea or Sulawesi would strengthen the ties with these areas. For this we need phylogenetic studies. Phylogenetic information on the endemics is available for very few species only. These cases will be discussed. In some other cases an approximation can be made by checking the overall distribution of the genus; if it occurs only either east or west of the Moluccas, we can be reasonably sure that the sister species of the relevant endemic also occurs only east or west of the islands. These will be discussed here as well.

Ornithoptera croesus Wallace, 1859. The species occurs on Halmahera, Bacan, Ternate, Tidore and Morotai. The occurrence on Sanana in the Sula archipelago, from where only a single female is known, needs checking (Collins and Morris, 1985), but is almost certainly erroneous (Vane-Wright, pers. comm., 1997). The taxon is on the verge of specific distinctness from *O. priamus* (Linnaeus, 1758) (S Moluccas and New Guinea), and the two, together with *O. aesacus* (Ney, 1903) (Obi) and *O. urvillianus* (Guérin-Méneville, 1830) (Bismarck archipelago, Bougainville, Solomon Islands), can be considered to form a superspecies (Parsons,

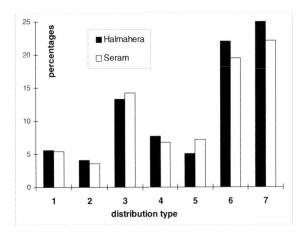

Fig.2. Distribution profiles for Halmahera and Seram. The distribution types are numbered 1-7, meaning Halmahera or Seram with: 1, Sulawesi; 2, Seram, respectively Halmahera; 3, New Guinea; 4, Sulawesi and Seram, respectively Halmahera; 5, Sulawesi and New Guinea; 6, New Guinea and Seram, respectively Halmahera; 7, Halmahera, Seram, Sulawesi and New Guinea. The frequency of the distribution types is given as percentages of the fauna of Halmahera and Seram, respectively, without endemics.

1996). Thus, biogeographically, *O. croesus* is linked with New Guinea (directly or through S Moluccas), and not with Sulawesi.

Troides criton (C. and R. Felder, 1860). According to Tsukada and Nishiyama (1980), *T. criton* (Halmahera, Bacan, Obi) forms an unresolved trichotomy with *T. plato* Wallace, 1865 (Timor) and *T. haliphron* (Boisduval, 1836) (Sumba to Tanimbar, S Sulawesi). Together they form an unresolved trichotomy with *T. riedeli* (Kirsch, 1885) (Tanimbar) and *T. vandepolli* (Snellen, 1890) (Java, Sumatra). Although with the present distribution area of this group of species we must assume extinction in the greater part of Sulawesi to understand how *T. criton* could ever reach the northern Moluccas, the absence of any related species in New Guinea indicates that the link is to the west and not to the east.

Papilio deiphobus Linnaeus, 1758. According to Tsukada and Nishiyama (1980) this widespread Moluccan endemic forms part of an unresolved monophyletic group of seven species, distributed throughout the Oriental region, including Sulawesi and the Lesser Sunda Islands as far east as Timor. No relatives occur in New Guinea.

Papilio heringi Niepelt, 1924. Hancock (1983) believes this endemic of Halmahera to be a natural hybrid between *P. tydeus* (C. and R. Felder, 1860) and *P. fuscus* Goeze, 1779, whereas Racheli and Haugum (1983) reassessed its specific status in accordance with Tsukada and Nishiyama (1980). If considered a proper species, it belongs to a supposedly monophyletic group of eight species with a distribution from the Moluccas eastward. In this case it is clearly one of the New Guinea links of Halmahera.

Papilio gambrisius Cramer, 1777 (S Moluccas), and *P. tydeus* C. and R. Felder, 1860 (N Moluccas). Hancock (1983) showed the two species (Fig.3) to be sister species. Together they have as a sister species *P. aegeus* Donovan, 1805, which is distributed from New Guinea and satellite islands to Australia and the Pacific. The three species together have as sister group two species from the Solomon Islands and Admiralty Island, respectively. They all belong to the *aegeus* group of nine species, which together have an eastern distribution; the Moluccas and islands around Tanimbar being the western outposts.

Graphium stresemanni Rothschild, 1916. This species is either an endemic of Seram, with *G. batjanensis* Okano, 1984 (Bacan and, probably, Halmahera) as sister, or the two taxa are conspecific. Its or their sister is *G. weiskei* Ribbe, 1900 from southeastern New Guinea (or the whole of New Guinea; Collins and Morris 1985). These taxa combined have as sister *G. macleayanum* Leach, 1814 from E Australia, Tasmania, Lord Howe Island and Norfolk Island (extinct now, Holloway, pers. comm. 1997; recently also known from New Guinea; Collins and Morris, 1985), and sister to all these taxa is *G. gelon* Boisduval, 1859 from New Caledonia.

Hypochrysops. The genus includes 57 species in 20 species groups (Sands, 1986). Except one species, *H. coelisparsus* Butler, 1883, in S Thailand, Malaysia, Sumatra and some smaller islands, and a yet undescribed species from Sulawesi, the genus is restricted to the Australian region (including the Moluccas). Six species are endemic to the Moluccas. *H. chrysanthis* C. Felder, 1865, endemic to Seram and Ambon, forms a species group together with *H. coelisparsus* and an Australian species, and thus has links to the west and to the east. All other Moluccan species have only links to the east, except that further knowledge of the Sulawesi species may disturb this pattern.

Luthrodes. This genus contains, in addition to the Moluccan endemic, *L. buruana* Holland, 1900 (Obi, Seram, Buru), only a single species in New Guinea and surrounding islands, and in Timor.

Philiris. A large genus of 63 species, mainly endemics of New Guinea, but one species is restricted to Seram and Ambon, and another to Obi and S Moluccas; there are no species west of the Moluccas.

Ariadne. The Moluccas form the easternmost locality for this genus, which contains 10 species. The single Moluccan endemic, *A. obscura* Felder, 1865, is found on Halmahera, Bacan and Buru.

Mynes. Another genus of 10 species, but this one does not occur west of the Moluccas (with the exception of one species in Flores and Timor). One Moluccan endemic, *M. plateni* Staudinger, 1877, occurs on Bacan and Halmahera.

Sumalia. An Oriental genus of five species. Although the genus does not occur in Sulawesi, the Moluccan endemic, *S. staudingeri* Ribbe, 1898, from Seram and Ambon can have only western links.

Taenaris. A typical genus (25 species) of the Papuan sub-region, with only a single species (*T. horsfieldii* Swainson, 1820) in the Oriental region (Malaysia, Java, Borneo, Palawan). The

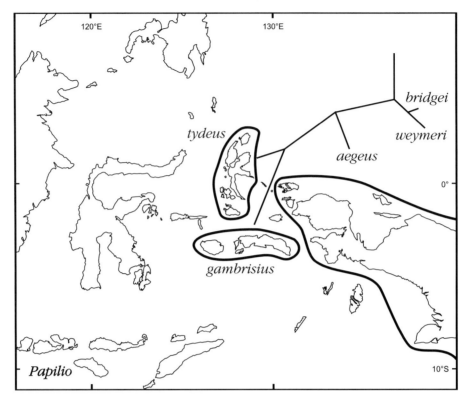

Fig.3. Phylogeny and distribution of the *Papilio aegeus* group (after Hancock, 1983).

five Moluccan species belong to three species groups (Stichel, 1912), which have only representatives to the east of the Moluccas.

Suniana. The three species in this genus are restricted to the Australian region, with one species extending to Timor, and another to Sumba. One species, *S. subfasciata* Rothschild 1915, is endemic to Seram and Ambon. The second Moluccan species, *S. sunias* Felder, 1860, is found in the N Moluccas, and as far east and south as Fiji and New South Wales in Australia. The genus is sister to another genus, *Ocybadistes*, with similar distribution (de Jong, 1990).

In addition to the genera dealt with here, there are other widespread genera which do not occur east of the Moluccas, but lack endemic species in the Moluccas, *e.g., Erionota, Iraota, Discolampa, Lethe.* Similarly, there are a number of eastern genera which do not occur west of the Moluccas, but lack endemic species in the Moluccas, *e.g., Netrocoryne, Hypochlorosis, Praetaxilia, Eutraeropis, Apaturina, Tellervo.* The relevant species have been included in the similarity assessments of the regional faunas.

In summary, of the endemic species of the Moluccas (at least for those which occur in Halmahera or Seram), four show exclusively western links, while 21 indicate a relationship with New Guinea. Halmahera and Seram differ slightly in the proportions: west-east scores 3-10 for Halmahera, and 3-12 for Seram, but the numbers are too low to attach much meaning to. In some cases, like *Papilio gambrisius/tydeus*, direct links between Halmahera and Seram are apparent. These results reinforce the conclusions based on similarity at the species level. For all other endemics there is currently no clue as to the geographic link.

Patterns of endemism

Most of the 82 endemic Moluccan species occur in two or more islands. The distribution is not random. Dividing the Moluccas into northern and southern groups of islands along a line between Seram and Obi (as done by Vane-Wright and Peggie, 1994), we find that 34 of the endemics occur only in the northern islands, 47 only on the southern islands, and 12 species (14.6%) on both sides of the dividing line. Moreover, four of the 12 widespread endemics occur north of the line only in Obi, the southernmost island of the northern group. The pattern of endemism, thus, supports the division of the Moluccas into a northern and a southern group. This agrees well with the finding that the distribution pattern of the Moluccan endemics is non-random. Apparently, faunal interchange has been easier within the northern islands and within the southern islands, than it has been between these two island groups.

Only five (13.9%) of the 36 endemic Moluccan species in Halmahera are restricted to this island. For Seram these figures are 6 (13.6%) of the 44 endemic Moluccan species. Since there are only eight endemics occurring in both Halmahera and Seram, Halmahera has 28 endemics not occurring in Seram, and Seram has 36 endemics not occurring in Halmahera. This strengthens the impression that the line between Obi and Seram dividing the Moluccas into northern and southern islands, is not imaginary. It is in agreement with expected pattern C6.

Even if we suppose that all endemics occurring in Halmahera, except those also occurring in Seram, actually originated in Halmahera, and do the same for Seram, then Seram is still a richer centre of origin for endemics than Halmahera. This agrees with the expected pattern for faunas of comparable age, but differing in diversity of habitats (see expected pattern B2), and not with an older age for the Halmahera fauna.

Phylogenetic patterns

The little information available on the phylogeny of endemic species has already been discussed above. The only case known in which a monophyletic group of three endemic species can be distinguished, one in Halmahera, one in Seram, and the third in New Guinea (with possible extensions further south and east), is *Papilio gambrisius-tydeus-aegeus*. Since the two Moluccan species are sister species, they have the same age. A comparable case could be *Graphium stresemanni-batjanensis-weiskei*, depending on taxonomic decisions and exact distributions. Therefore, there is no phylogenetic evidence of an older age for the Halmahera endemics relative to the Seram endemics (see expected pattern B1). Similarly, there is no evidence for older connections between Seram and New Guinea than between Seram and Sulawesi (see expected pattern C5). The genera *Delias, Arhopala, Hypochrysops, Jamides* and *Taenaris* are potential sources for this kind of information, since they all have endemics in Halmahera, Seram and New Guinea. Up to now, however, the great morphological similarity in these genera has frustrated attempts to analyse these taxa phylogenetically.

Halmahera arc pattern

This pattern could either involve species or monophyletic groups of species restricted to the former islands of the Halmahera arc (see expected pattern C2). As discussed in the previous paragraph, phylogenetic information is still too meagre to be of much help. The author knows of only one species with such a distribution, *Euploea morosa* Butler, 1866, which is found in the northern Moluccas, Misool, Biak and Numfoor. With eleven other *Euploea* species, this species forms an almost completely unresolved monophyletic group (Ackery and Vane-Wright, 1984). The existence of only one species (among the 196 species of Halmahera) showing this pattern is considered insufficient evidence for support. The nominate subspecies of *Erysichton palmyra* Felder, 1860 is found in the N and S Moluccas, Numfoor and Biak, but the species as a whole is widespread in the Papuan and Australian regions, and its sub-specific differentiation needs re-examination.

Conclusions

Despite distance and different histories, Halmahera and Seram have remarkably similar butterfly faunas. This can be attributed to similar positions relative to New Guinea and Sulawesi, which acted as source areas. New Guinea, with a much larger fauna than Sulawesi, contributed many more species. As shown by the pattern of endemism within the Moluccan islands, direct exchange between the northern and southern Moluccas apparently played a minor role. In the butterfly fauna there are no traces evident of the older history of Halmahera, but a number of possibly informative genera have not yet been analysed phylogenetically. As far as available evidence goes, we can only conclude that Halmahera remained submerged for most of its life and did not emerge before Seram, or traces of an older butterfly fauna have been wiped out by later immigrants. In both cases, the butterfly fauna of Halmahera, like that of Seram, is younger than that of New Guinea or Sulawesi.

Consequently, exchange will have been largely unidirectional, *i.e.*, into the Moluccas, although at a later stage exchange in the opposite direction may have taken place. With regard to the regional links of the faunas of the northern and southern Moluccas the conclusions generally agree with those of Vane-Wright and Peggie (1994) who studied the butterflies of the whole of the Moluccan islands with a different objective, and applied a slightly different approach.

Acknowledgements

The author is grateful to the organisers of the meeting on the Biogeography and Geological Evolution of SE Asia in London, March 6-7, 1996, for the opportunity to read a paper there and to have it published in the present volume. He wants to acknowledge the long-standing co-operation with R. I. Vane-Wright in the study of SE Asian butterflies, and he thanks R. I. Vane-Wright and N. Møller Andersen for fruitful comments on the manuscript.

References

Ackery, P. R. and Vane-Wright, R. I. 1984. Milkweed butterflies, their cladistics and biology. British Museum (Natural History), London.

Audley-Charles, M. G. 1986. Timor-Tanimbar Trough: the foreland basin of the evolving Banda orogen. Special Publications: International Association of Sedimentologists 8: 91-102.

Audley-Charles, M. G. 1993. Geological evidence bearing upon the Pliocene emergence of Seram, an island colonizable by land plants and animals. *In* Natural History of Seram. Edited by I. D. Edwards, A. A. Macdonald and J. Proctor. Intercept Ltd, Andover.

Boer, A. J. de. 1995. Islands and cicadas adrift in the West-Pacific. Biogeographic patterns related to plate tectonics. Tijdschrift voor Entomologie 138: 169-244.

Bowler, J. and Taylor, J. 1993. The avifauna of Seram. *In* Natural History of Seram. Edited by I. D. Edwards, A. A. Macdonald and J. Proctor. Intercept Ltd, Andover.

Collins, N. M. and Morris, M. G. 1985. Threatened Swallowtail Butterflies of the World. IUCN, Gland and Cambridge.

Croizat, L. 1962. Space, Time and Form, the biological synthesis. Published by the author, Caracas.

Daly, M. C., Cooper, M. A., Wilson, I., Smith, D. G. and Hooper, B. G. D. 1991. Cenozoic plate tectonics and basin evolution in Indonesia. Marine and Petroleum Geology 8: 1-21.

de Jong, R. 1990. Some aspects of the biogeography of the Hesperiidae (Lepidoptera, Rhopalocera) of Sulawesi. *In* Insects and the Rain Forests of South East Asia (Wallacea). Edited by W. J. Knight, and J. D. Holloway. RESL, London.

de Jong, R. 1997a. The continental Asian element in the fauna of the Philippines as exemplified by *Coladenia* Moore, 1881 (Lepidoptera, Hesperiidae). Cladistics 12: 323-348.

de Jong, R. 1997b. Fossil butterflies in phylogenetic perspective. Entomologica Scandinavica: in press.

de Jong, R. and Congdon, T. C. E. 1993. The montane butterflies of the eastern Afrotropics. *In* Biogeography and ecology of the rain forests of eastern Africa. Edited by J. C. Lovett and S. K. Wasser. Cambridge University Press, Cambridge.

de Jong, R. and Treadaway, C. G. 1993. The Hesperiidae (Lepidoptera) of the Philippines. Zoologische Verhandelingen 288: 1-115.

Fortuin, A. R. and de Smet, M. E. M. 1991. Rates and magnitudes of late Cenozoic vertical movements in the Indonesian Banda Arc and the distinction of eustatic effects. Special Publications of the International Association of Sedimentology 12: 79-89.

Gressitt, J. L. 1982. Zoogeographic summary. *In* Biogeography and ecology of New Guinea. Edited by J. L. Gressitt. Junk, The Hague.

Hall, R. 1987. Plate boundary evolution in the Halmahera region, Indonesia. Tectonophysics 144: 337-352.

Hall, R. 1996. Reconstructing Cenozoic SE Asia. *In* Tectonic Evolution of Southeast Asia. Edited by R. Hall and D. Blundell. Geological Society Special Publication, 106: 153-184.

Hall, R. 1998. The plate tectonics of Cenozoic SE Asia and the distribution of land and sea. *In* Biogeography and Geological Evolution of SE Asia. Edited by R. Hall and J. D. Holloway (this volume).

Hall, R. and Nichols, G. J. 1990. Terrane amalgamation in the Philippine Sea margin. Tectonophysics 181: 207-222.

Hall, R., Ali, J. R. Anderson, C. D. and Baker, S. J. 1995. Origin and motion history of the Philippine Sea Plate. Tectonophysics 251: 229-250.

Hancock, D. L. 1983. *Princeps aegeus* (Donovan) and its allies (Lepidoptera: Papilionidae): systematics, phylogeny and biogeography. Australian Journal of Zoology 31: 771-197.

Krebs, C. J. 1989. Ecological Methodology. Harper and Row, New York.

Page, R. D. M. 1993. COMPONENT, version 2.0. Computer program and manual. Natural History Museum, London.

Parsons, M. J. 1996. Gondwanan evolution of the troidine swallowtails (Lepidoptera: Papilionidae): cladistic reappraisals using mainly immature stage characters, with focus on the birdwings *Ornithoptera* Boisduval. Bulletin Kitakyushu Museum of Natural History 15: 43-118.

Pigram, C. J. and P. J. Davies 1987. Terranes and the accretion history of the New Guinea orogen. B.M.R. Journal of Australian Geology and Geophysics 10: 193-212.

Pigram, C. J. and H. Panggabean, 1984. Rifting of the northern margin of the Australian continent and the origin of some microcontinents in eastern Indonesia. Tectonophysics 107: 331-353.

Racheli, T. and Haugum, J. 1983. On the status of *Papilio heringi* Niepelt, 1924. Papilio International 1: 37-45.

Rangin, G., Jolivet, L. and Pubellier, M. 1990. A simple model for the tectonic evolution of the southeast Asia and Indonesia region for the past 43 m.y. Bulletin de la Société géologique de France 8 (6): 889-905.

Sands, D. P. A. 1986. A revision of the genus *Hypochrysops* C. and R. Felder. Entomonograph 7. E. J. Brill/Scandinavian Science Press, Leiden, Copenhagen.

Stichel, H. 1912. Amathusiidae. *In* Das Tierreich. Vol. 34. Edited by F. E. Schulze. R. Friedlènder und Sohn, Berlin.

Struckmeyer, H. I. M., Young, M. and Pigram, C. J. 1993. Mesozoic to Cainozoic plate tectonic and palaeographic evolution of the New Guinea region. *In* Petroleum Exploration and Development in Papua New Guinea. Proceedings of the Second PNG Petroleum Convention, Port Moresby 1993. pp. 261-290. Edited by G. J. and Z. Carman.

Tennent, J. 1996. The Butterflies of Morocco, Algeria and Tunisia. Gem Publishing Company, Brightwell cum Sotwell, Wallingford.

Tsukada, E. and Nishiyama, Y. 1980. Papilionidae. *In* Butterflies of the South East Asian Islands. Vol. I. Edited by E. Tsukada. Plapac Co., Ltd., Tokyo. (Japanese edition; English edition 1982).

Vane-Wright, R. I. and Peggie, D. 1994. The butterflies of Northern and Central Maluku: diversity, endemism, biogeography, and conservation priorities. Tropical Biodiversity 2: 212-230.

Vives Moreno, A. 1994. Catalogo sistematico y sinonimico de los lepidopteros de la Peninsula Iberica y Baleares (Insecta: Lepidoptera) (Segunda Parte). Ministerio di Agricultura, Madrid.

Wallace, A. R. 1869. The Malay Archipelago. MacMillan, London.

Assembling New Guinea: 40 million years of island arc accretion as indicated by the distributions of aquatic Heteroptera (Insecta)

Dan A. Polhemus[1] and John T. Polhemus[2]
[1]Department of Entomology, MRC 105, National Museum of Natural History, Smithsonian Institution, Washington, D. C., USA 20560
[2]Colorado Entomological Museum, 3115 S. York St., Englewood, CO, USA 80110

Key words: New Guinea, island arcs, geology, biogeography, aquatic Heteroptera

Abstract

The island of New Guinea consists of the northern margin of the Australian continental craton that has collided over the past 40 million years with a series of migrating island arcs. Each of these arcs has had a separate tectonic history and carried a correspondingly different biota. Recent faunal surveys and phylogenetic analyses, coupled with evolving tectonic knowledge, are now allowing us to identify elements within the New Guinea aquatic Heteroptera biota that may be correlated with particular accreted arc systems. It appears that many Asian-derived groups arrived via a Papuan arc that collided obliquely with the northern margin of Australia between the Late Eocene and Early Oligocene, while other distinctively Melanesian groups evolved on an isolated Solomons arc that was initiated in the Oligocene and accreted terranes to northeastern New Guinea from the Miocene into the Pliocene. This gradually improving understanding of the island's interrelated tectonic and biotic history is permitting a better comprehension of the complex patterns of faunal fusion and disjunction currently present within the Melanesian region.

Introduction

A long standing goal of zoogeographers and historical geologists has been to integrate biological information and Earth history models. Nowhere is this better pursued than in the young and tectonically complex region surrounding New Guinea. The island represents the northern margin of the Australian continental craton, which has been uplifted through a series of collisions with southward-migrating island arcs that have formed along the boundary between the Australian and Pacific tectonic plates (Hamilton, 1979, 1988; Kroenke, 1984). Recent advances in our understanding of the island's geological history have revealed that at least two major episodes of collision and terrane accretion have occurred in the northern half of the island (Pigram and Davies, 1987; Davies, 1990; Smith, 1990; Pigram and Symonds, 1991), and each of these has in turn left a biological signature. The current challenge for regional biogeographers lies in deciphering these signatures, and determining, if possible, which plant and animal groups aggregated to the present faunal mixture via which arcs.

In a previous paper, Polhemus (1996) reviewed the history of biogeographic concepts relating to New Guinea, in particular the development of the 'Inner arc' versus 'Outer arc' paradigm that has dominated most modern analyses of the island's biogeography. The underlying geological concepts upon which this hypothesis was based were shown to be correct in the broad sense, but flawed in many other aspects. In particular, the concept of a single 'Outer arc' extending from New Guinea to Tonga was shown to be simplistic, since the arcs grouped under this concept represented at least four separate systems with dissimilar ages, origins, and subduction polarities (those readers unfamiliar with the nomenclature, geological structure, and mobilistic nature of island arcs are also referred to the paper cited above, which contains a review of these subjects). In this paper we seek to correlate particular groups of aquatic and semi-aquatic true bugs (referred to hereafter as aquatic Heteroptera) with particular arc systems and terranes, assessing their probable time of arrival in New Guinea, and their subsequent speciation or dispersal within the island.

In the analysis below, brief synopses of current geological hypotheses for the assembly of New Guinea are presented, noting also areas of current uncertainty. One difficulty in providing such a discussion lies in the confusing number of different names that have been given by various authors to the Tertiary arc systems that formed north of New Guinea. In the present paper we follow the terminology of Kroenke (1984), recognizing both his Papuan and Solomons arcs. The former is equivalent to the Salumei System of Davies (1990), while the latter corresponds to the Melanesian arc of Smith (1990) and Weiland and Cloos (1996).

The known distributions of aquatic Heteroptera are compared to these geological models, and the degree of congruence discussed. The analysis of such patterns is far from complete. Several of the most potentially informative groups, such as *Rhagovelia* (Veliidae) and *Ptilomera* (Gerridae) have not yet been taxonomically revised, much less cladistically analyzed, and as a result this paper relies heavily on currently unpublished data (*i.e.*, undescribed species in the collections of the authors). The analysis should thus be seen as a preliminary evaluation of biogeographic hypotheses based on work in progress, rather than a validation of such hypotheses; the latter can only be based on phylogenetic analyses of the insect groups involved. Many such analyses are in the process of being completed, however, and in the near future it should be possible to provide more rigorous tests of the hypotheses presented below.

Geological setting

The northern half of New Guinea is a composite of island arcs that have accreted to the northern margin of Australia at various times over the last 40 million years. Kroenke (1984), in a synthesis covering the eastern half of the island, recognized three such arc systems, the Papuan, Trobriand, and Solomons arcs, that had already collided with New Guinea, and a fourth, the Bismarck arc, that was in the process of doing so. In the west, Hamilton (1979, 1988) discussed the incipient collision of a fifth system, the Banda arc, that is approaching New Guinea from the southwest.

The Banda arc system in the west is formed along the boundary of the Australian and Eurasian plates above the well defined Sunda subduction zone. By contrast, the arcs lying to the east of New Guinea appear to have been formed during alternating episodes of volcanism along two distinct zones of crustal weakness in the Solomons and northern New Guinea. According to Kroenke's (1984) model, each time that subduction along one of these zones became inactivated, it would reactivate along the other. The sequence below has been postulated (terminology for trenches and arc systems follows Kroenke, 1984).

Northern New Guinea subduction episodes and corresponding trench systems:
1. Early Eocene to Early Oligocene: Aure-Moresby-Pocklington trenches.
2. Early Miocene to Late Miocene: Wewak-Trobriand trenches (= the New Guinea trench of Davies, 1990).

Solomons subduction episodes and corresponding trench systems:
1. Early Oligocene to Early Miocene: North Solomon trench.
2. Late Miocene to Holocene: South Solomon trench.

These alternating episodes of subduction activity in eastern New Guinea are correlated with the following tectonic events (for a CD ROM video of the hypothesized plate and arc motions consult Yan and Kroenke, 1993):

Cretaceous – Rifting and basin formation occurred along the passive northern Australian margin. This rifting, which took place in several phases, isolated small slivers of continental crust outboard of marginal basins. These slivers have been referred to as the 'Inner Melanesian arc' by many previous biogeographers. While there may have been some arc related activity and back-arc spreading in the marginal basins during the final phases of this process, it was for the most part a rifting event, similar to the process seen in current day East Africa.

Middle Eocene – Eastward subduction below the Pacific plate along the Aure-Moresby-Pocklington trenches, lying well north and east of the rifted Australian margin, produced a southward migrating Papuan arc in an oceanic setting.

Early Oligocene – The eastward subduction below the Papuan arc ceased as the arc collided with the Australian margin, causing overthrusting of arc terranes onto the Australian craton. New westward subduction was initiated below the Australian plate in the Solomons zone along the North Solomon trench, forming the eastward migrating Solomons arc in an isolated oceanic setting. Western extensions of this arc, linked by transforms, appear to have extended to the area north of Irian Jaya.

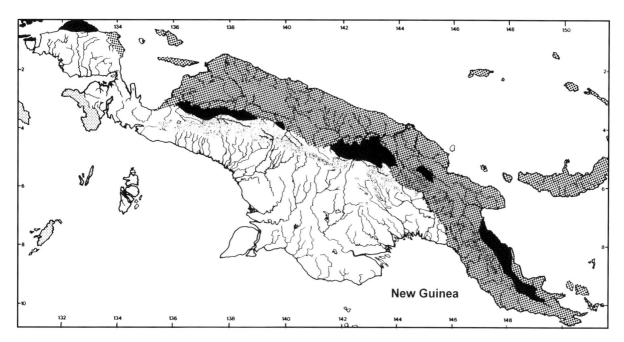

Fig.1. Map of New Guinea showing geological provinces discussed in the text. Unstippled areas = Australian craton and Vogelkop craton fragment; black areas = Papuan arc terranes, accreted in the Eocene and Oligocene; dark stippled areas = Solomons arc terranes, accreted in the Miocene; light stippled areas = Banda arc terranes and associated uplifts, currently in the process of accretion; very finely shaded areas = mountainous regions at elevations above 2000 metres. Areas in the Bird's Neck are of uncertain geological affinity; although assigned to the Australian craton in this figure, they are also likely to contain accreted terranes from other sources.

Early Miocene – The Ontong Java plateau jammed the North Solomons trench, ending this episode of subduction in the Solomons zone. South or west-dipping subduction was reactivated in the New Guinea zone along the Wewak and Trobriand trenches (Hall, 1997), producing onshore volcanism in New Guinea, forming the Trobriand arc to the east, and gradually consuming the sea floor separating the Solomons arc islands from northern New Guinea.

Middle Miocene – Terranes associated with the Solomons arc system collided obliquely with northern New Guinea from west to east (Davies et al., 1996; Hall, 1997), ending subduction in the New Guinea zone for a second time. This collision continued into the Pliocene, with the last unit to be sutured consisting of the Adelbert-Finisterre terrane. Subduction then reactivated in the Solomons zone along the South Solomons trench, but in an eastward direction opposite to the previous polarity of the North Solomons trench. In the west, the Vogelkop peninsula was sutured to the main body of New Guinea.

Holocene – Complex fracturing of plates northeast of New Guinea created many small arcs and subduction zones, including the New Britain arc, which began to advance southeastward over the Solomon Sea. In the west, the accreted terranes of northern Vogelkop were dismembered by left lateral faulting, while the Banda arc began the first stages of collision from the southwest.

All the arc systems associated with these episodes of subduction discussed above, with the exceptions of the recent New Britain arc and the eastward migrating Banda arc, appear to have contributed accreted terranes, and thus accreted biota, to the modern island of New Guinea (Fig.1), although the sequence of accretion is still poorly understood in the western half of the island. New Guinea as we see it today thus consists of a complex fusion of geological fragments derived from the sources summarised below.

The Australian craton and Vogelkop

The southern half of New Guinea consists of the Australian continental craton, which was a passive margin until the Cretaceous, when renewed rifting began along its eastern side, followed in the Eocene by the series of arc collisions discussed above. Although the margin itself lay in a roughly east-west orientation, a basement high extended north in the area of the current Papua New Guinea-Irian Jaya border as far as the Border Mountains west of the Sepik basin, and remained subaerial during the Mesozoic when most of the remaining platform was submerged (Davies, 1990). Evidence for this high can still be seen in current drainage patterns, which flow away from it to both the east and west. This structure has had an important and previously underestimated influence on aquatic zoogeography in the region, since it represented a subaerial salient of Australia that projected northward into the tropics and subtropics when the rest of the continent was still in cooler climatic zones. It also seems to have acted like a wedge, splitting the accreted terranes of the Oligocene Papuan arc.

The Vogelkop peninsula for most of its history was a detached piece of the Australian continental craton (the Kemum terrane) that separated in the Mesozoic (Hamilton, 1979; Pigram and Davies, 1987). It was probably never far away from the northwestern margin of the main Australian land mass, however, since its post-Cretaceous tectonic history has been substantially similar to that of the main body of New Guinea; it has received several waves of arc terrane laminations onto the northern and eastern margins of its continental core, and in generally the same historical sequence. The Vogelkop, after a considerable period of isolation, was finally fused to the remainder of New Guinea in the Miocene. The suture zone is marked by the Wandamen peninsula and adjoining Mangguar Terrane of Pigram and Davies (1987).

The 'Inner Melanesian arc'

The 'Inner Melanesian arc' (a term that is misleading and probably ought to be dropped, except for its long history of usage; see Polhemus, 1996) was not an arc at all, but a set of thin strips of continental crust rifted from the eastern margin of the Australian continent during the development of the Tasman and Coral Sea basins in the Late Cretaceous and Early Tertiary. The New Guinea sector of this 'arc' is represented by Pigram and Davies (1987) as a long, narrow salient of Australian continental craton lying north of the Coral Sea basin (the Eastern and Papuan plateaus). These crustal slivers north of the Coral Sea were apparently never emergent, but did constitute a topographic barricade against which the southward migrating arc systems that formed the Papuan peninsula eventually collided. In addition, another cratonic fragment, now present in the Jimi Terrane of Papua New Guinea (*i.e.*, the Bismarck-Kubor block), rifted from the Australian margin and functioned as a separate island during the Late Cretaceous before being reintegrated into New Guinea at or near the time of the collision with the Papuan arc (Pigram and Davies, 1987; Davies, 1990); this Bismarck-Kubor block appears to have remained subaerial from the Cretaceous onward (Davies, 1990). Other subaerial blocks of continental crust may also have been rifted from northern Australia and followed similar histories as small islands north of the main craton during the Late Cretaceous, but if so, then all of them were trapped and crushed between the Papuan arc and the Australian margin in the Eocene (Rogerson and Hilyard, 1990).

The Papuan arc

The first arc collision event in the assembly of New Guinea occurred when the Papuan arc collided with the northern margins of Vogelkop and Australia. This collision was oblique from west to east, thus the initial contact appears to have been possibly latest Cretaceous or Early Eocene in Irian Jaya (Davies *et al.*, 1996), extending through the Late Eocene and Early Oligocene in Papua New Guinea (Kroenke, 1984; Pigram and Davies, 1987; Davies, 1990, Pigram and Symonds, 1991). The collision emplaced the overthrust central Irian Jaya ophiolite and metamorphic belts (the Rouffaer terrane of Pigram and Davies, 1987), along with correlative April Ultramafics, Marum Complex, and Papuan Ophiolite Belt in Papua New Guinea (the Sepik, Marum, and Bowutu terranes of Pigram and Davies, 1987), and apparently the Tamrau Mountains of Vogelkop. This collision also resulted in the depression of the Australian continental margin south of the overthrust terranes, which were thus separated from the subaerial craton by a water gap through most of the Miocene (Pigram and Symonds, 1991). The limestones that formed in this shallow sea would later end up on the summit of the New Guinea central ranges after a second episode of orogeny driven by the collision of the Solomons arc in the Late Miocene (Nash *et al.*, 1993).

Waigeo is of the correct age and composition to be part of the Oligocene episode of terrane collision (Charlton *et al.*, 1991), and has had clear historical associations with greater New Guinea based on its biota, but its high degree of species level endemism also implies an extended period of isolation by water gaps. The presence of thick and extensively exposed limestone strata also indicates that much of the island was submerged during the Miocene (Charlton et al., 1991). It seems possible that Waigeo was a part of the Papuan arc that was accreted to northern Vogelkop in the area west of the Tamrau Mountains during the Oligocene, then sheared off and carried further westward along the Sorong Fault zone to its present position during the Pliocene.

The Solomons arc

A large number of terranes unrelated to those of the Papuan arc accreted to northern New Guinea during the Miocene and into the Pliocene. These were formed along a separate arc system that lay north of the Papuan arc, possibly a transform connected western extension of Kroenke's (1984) Solomons arc subduction system. They are hypothesized to include, from west to east, the Arfak Mountains, Biak, Yapen, the Van Rees, Foja, Cyclops, Torricelli, Prince Alexander, Adelbert, Finisterre and Saruwaged Mountains, and possibly the Papuan peninsula. As with the terranes of the Papuan arc that preceded them, these terranes converged obliquely from west to east, with collision beginning in the Late Miocene in Irian Jaya, but continuing until the Pliocene in Papua New Guinea (Cooper and Taylor, 1987).

Kroenke (1984) considered the Papuan peninsula to have been accreted to New Guinea in the Late Eocene or Early Oligocene, at the same time that the ophiolite belts of the central ranges were emplaced, and this interpretation is also implied in the maps presented by Weiland and Cloos (1996). Pigram and Davies (1987), by contrast, considered the Papuan peninsula to consist of a complex terrane composed of multiple arc fragments (much like the present day Philippines) that was assembled offshore in the Oligocene, then accreted to the remnant submarine continental sliver of the 'Inner Melanesian arc' lying north of the Coral Sea basin in the early to middle Miocene, and finally sutured to the remainder of New Guinea.

Nearly all tectonic reconstructions agree that the Adelbert-Finisterre terrane accreted to New Guinea in the Late Miocene or Early Pliocene, and was the last of the Miocene terranes to arrive. Kroenke (1984) provides a structural and tectonic analysis linking this terrane to the Solomons arc system via a transform, and Pigram and Davies (1987) similarly suggest that it was previously part of a system that included the Bismarcks and Solomons, and formed in an isolated oceanic setting to the northeast of New Guinea. Given its late arrival in comparison to other terranes linked to the Solomons arc system, it is possible that this terrane occupied an isolated position on a transform connected sector of the arc lying between the Solomons to the east and the other Miocene terranes to the west.

In the far west, all of the Miocene accreted terranes of the Vogelkop have been highly fragmented by left-lateral shearing. For instance, Pigram and Davies (1987) hypothesized that the Arfak Mountains, Yapen, and parts of Biak are all portions of the same original terrane that was ripped apart by faulting. In central Irian Jaya, the Miocene arc collision also caused a massive deformation of the Australian continental craton south of the overthrust ophiolitic and metamorphic terranes of the Irian Jaya Mobile Belt that had been emplaced in the Late Eocene or Early Oligocene (see previous discussion), pushing the Miocene limestones that had formed south of this belt to elevations of over 5000 metres, and creating a large, overthrust anticlinal structure, the Mapenduma anticline (Nash et al., 1993). This anticline, which appears to be structurally related to the Muller anticline of west central Papua New Guinea, was then subsequently eroded in an asymmetrical fashion from the south, due to orographically induced rainfall, producing the huge escarpments that form the southern flank of the present Irian Jaya central ranges (Weiland and Cloos, 1996).

The tectonic history of western New Guinea in the Early to Middle Miocene is still poorly understood, particularly in regard to the accretion of terranes in the northern coastal ranges of Irian Jaya and the formation of the Weyland Mountains in the Bird's Neck region (Milsom, 1991). Our assignment of Irian Jaya's Miocene terranes to a westerly extension of the Solomons arc system is thus tentative, pending more detailed geological and biogeographic analyses.

Other arc systems

In addition to the systems discussed above, two other arcs, the New Britain arc and the Banda arc, are advancing toward New Guinea, and will collide with the island over the next ten million years if current plate motions continue. Although these incipient collisions have not yet

Table 1. Currently known distributions of aquatic Heteroptera genera occurring on New Guinea in relation to arc system terranes and other geographic areas discussed in the text (see Fig.1). Codes used in table: AC = Australian craton; VK = Vogelkop craton fragment; IA = 'Inner Melanesian arc' terranes (ie., Jimi terrane); PA = Papuan arc terranes; SA = Solomons arc terranes; NB = New Britain arc and Bismarck Islands; SL = Solomon Islands; BA = Banda arc.

GENUS	AREA							
	AC	VK	IA	PA	SA	NB	SL	BA
BELOSTOMATIDAE								
Appasus	X	X						X
Lethocerus	X	X						
CORIXIDAE								
Cnethocymatia	X							
Micronecta	X	X						
Sigara				X				
GELASTOCORIDAE								
Nerthra	X	X	X	X	X	X	X	X
GERRIDAE								
Andersenella					X			
Calyptobates	X							
Ciliometra	X				X			
Halobates	X	X			X	X	X	X
Iobates	X	X			X			
Limnogonus	X	X	X	X	X	X	X	X
Limnometra	X	X	X	X	X	X	X	X
Metrobatoides					X			
Metrobatopsis					X	X	X	
Neogerris	X	X						
Pseudohalobates		X			X			
Ptilomera	X	X	X	X	X	X	X	X
Rhagdotarsus	X	X			X	X		X
Rheumatometroides	X	X			X	X	X	X
Stenobates	X	X			X			
Stygiobates		X			X			
Tenagogonus	X	X			X	X		X
Thetibates					X	X	X	
HEBRIDAE								
Hebrus	X	X	X	X	X	X	X	X
HERMATOBATIDAE								
Hermatobates	X	X			X	X	X	X
HYDROMETRIDAE								
Hydrometra	X	X	X	X	X	X	X	X
LEPTOPODIDAE								
Valleriola	X	X			X			
MESOVELIIDAE								
Mesovelia	X	X			X	X	X	X
Phrynovelia					X			
NAUCORIDAE								
Aphelocheirus	X			X	X			
Aptinocoris		X			X			
Cavocoris				X	X			
Idiocarus	X	X	X	X	X			
Nesocricos			X	X				
Quadricoris					X			
Sagocoris	X		X	X	X			
Tanycricos			X	X				
Warisia					X			
NEPIDAE								
Cercotmetus	X	X			X			X
Laccotrephes	X							
Ranatra	X	X			X			X
NOTONECTIDAE								
Anisops	X	X	X	X	X	X	X	X
Enithares	X	X	X	X	X	X	X	X
OCHTERIDAE								
Ochterus	X	X	X	X	X	X	X	X
PLEIDAE								
Paraplea	X				X			

Table 1. Continued.

GENUS	AREA							
	AC	VK	IA	PA	SA	NB	SL	BA
SALDIDAE								
Pentacora	X				X		X	
Saldula	X			X	X			
VELIIDAE								
Aegilipsicola				X				
Halovelia	X	X			X	X	X	X
Halovelioides					X			X
Microvelia	X	X	X	X	X	X	X	X
Neusterensifer	X	X			X			
Rhagovelia	X	X	X	X	X	X	X	X
Strongylovelia	X	X			X	X	X	X
Tanyvelia	X				X			
Tarsovelia	X	X	X	X	X			
Veliohebra					X			
Xenobates					X			X

emplaced terranes into the main body of the island, they have provided dispersal corridors for biota, and produced buckling of the continental shelf limestones in the west (Polhemus, 1996), creating young uplifts such as the Fakfak peninsula. These unaccreted arc systems are not treated in further detail in the current paper; readers wishing additional background are instead referred to Hamilton (1979, 1988) and Polhemus (1996), who provide discussions of the incipient collision of the Banda arc with western New Guinea, and the resulting biogeographic consequences.

Biological responses

The general distribution patterns displayed by plants and animals on island arc systems were reviewed by Polhemus (1996). In the section below, we attempt to identify aquatic Heteroptera distributions within New Guinea that appear to have been influenced by particular episodes of terrane accretion discussed in the section above. These distributions are further summarized in Table 1.

Australian craton and Vogelkop distributions

Even though the geological evidence indicates that Vogelkop was a separate island lying west of the main Australian craton for a considerable period between the Mesozoic and the Miocene, the biological evidence from aquatic Heteroptera clearly indicates that it could not have been very far away during this period. Although the Vogelkop fauna shows strong local endemism at the species level, it shares most of the same regionally endemic genera found on the main body of New Guinea (Table 1), while having very few endemic genera of its own. One exception occurs in the gerrid genus *Stygiobates*, which is currently known only from Vogelkop and Morotai (Polhemus and Polhemus, 1993).

Inner Melanesian arc distributions

The evidence for distributional patterns in aquatic Heteroptera potentially influenced by terranes of an 'Inner Melanesian arc' has been reviewed in detail by Polhemus (1995, 1996), and is not repeated at length here. The currently accepted geological hypothesis of marginal basin formation isolating continental slivers along the Australian continental margin in the late Cretaceous to early Paleocene is consistent with the biological evidence. Polhemus (1995, 1996) argued on the basis of aquatic Heteroptera distributions that such slivers must have been to some extent emergent above sea level, carrying a distinctive and in some cases relict (*i.e.*, New Caledonia gymnosperms) biotic assemblage eastward with them into the Pacific at large. Groups of aquatic Heteroptera conforming to

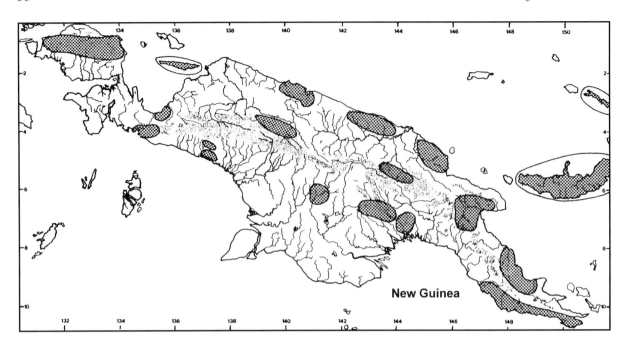

Fig. 2. Currently known distributions of *Ptilomera* water strider species occuring on New Guinea and associated arcs. Most of the nineteen species recognized from this region on the basis of recent character analysis are still undescribed. This fast water genus is absent in Australia and the North Moluccas, and current sampling also indicates that it does not occur on Biak, Salawati, or the Fakfak peninsula. The highly localized distributions of species in the northern and central ranges indicate that speciation in this group may have been influenced by the accretion of terranes associated with the Papuan and Solomons arc systems (see text).

such patterns typically show di- or tricentric distributions involving widely separated geographical areas along the western margin of the Pacific plate, including the southern Philippines. Three groups are notable in this context:

Rhagovelia (Veliidae), *novacaledonica* species group – members of this group occur in the uplands of New Caledonia, New Guinea, and Mindanao (Polhemus, 1995).

Rhagovelia (Veliidae), *caesius* species group – members of this group occur in New Guinea and the southern Philippines (Polhemus, 1995).

Naucoridae, subfamily Cheirochelinae, tribe Sagocorini – creeping water bugs in this tribe are found in New Guinea and the Philippines (Polhemus and Polhemus, 1987; Polhemus, 1996).

In none of the above instances are members of these groups present in Australia, the Moluccas, or Sulawesi. Such patterns probably date from the Late Cretaceous, and are visible now only at the tribal or intrageneric species group level. Since the terranes these species occupied formed the initial cores of what would eventually become larger islands, such as New Guinea and Mindanao, their biological signatures have been somewhat obscured by subsequent geological and biological events.

Papuan arc distributions

The accretion of the Papuan arc in the late Eocene and early Oligocene was a complex event whose biological signature as reflected by aquatic Heteroptera distributions is still being worked out. Groups influenced by terranes from this arc might be expected to show localized species level endemism in the central New Guinea ophiolite and metamorphic belts and in the mountains of the northern Vogelkop peninsula, and to have outgroup relationships with taxa occurring to the west, in the Malay archipelago and mainland SE Asia. Several groups appear to fall in this category:

Microveliinae (Veliidae) – the genus *Aegilipsicola* is currently known only from the northern slopes of the central New Guinea highlands. In addition, another currently undescribed microveliine genus, referred to in manuscript as 'Papuavelia', is known only from the central Irian Jaya ophiolite and metamorphic belts.

Ptilomera (Gerridae) – several undescribed species in this genus appear on the basis of current collections to be localized endemics with distributions centred on the Rouffaer and Sepik terranes of Pigram and Davies (1987). The situation is not clear-cut, however, since other endemic species have ranges centred around terranes associated with the Solomons arc, which accreted in the Miocene, indicating that members of this genus may have been distributed across multiple arc systems (as in fact they are today; see Polhemus, 1996 and Fig.2).

Nesocricos (Naucoridae) – members of this genus are entirely confined to the central mountains of New Guinea, and certain species appear to have ranges correlating with terranes defined by Pigram and Davies (1987). For instance, a currently undescribed species is apparently confined to the central Irian Jaya metamorphic and ophiolite belts that comprise the Rouffaer terrane of the above authors.

Tanycricos (Naucoridae) – as with *Nesocricos*, members of this genus are endemic to New Guinea, confined to the central mountains of the island, and unknown from any of the north coastal terranes that accreted in the Miocene. This suggests that they or their ancestors also arrived via the Papuan arc.

The metamorphic and ophiolite belts of central New Guinea are large, relatively cohesive geological units, and this is reflected in their aquatic Heteroptera biotas. In Irian Jaya, for instance, the mid-montane aquatic Heteroptera assemblage typical of these exposures is distinctive but broadly distributed, occurring throughout the upper basin of the Mamberamo River. The local dispersal of this biota has no doubt been facilitated by the long river valleys that trend east to west along strike through the Derewo Fault Zone.

Interestingly, the Heteroptera biota of the ophiolite and metamorphic belts emplaced in the Oligocene seems to be considerably richer than that of the highland limestone belt lying immediately to the south, which was elevated later, in the Late Miocene and Pliocene. This is partly due to elevation; the limestones frequently lie at elevations ranging from 3000 to 5000 metres, and the jagged crest they form represents an insurmountable cold water barrier to aquatic Heteroptera, breached only by the Baliem Valley to the east and the Paniai basin to the west. Both of these gaps in the high crest drain to the south, and, as shown by the distribution of crayfishes (Holthuis, 1982), have clearly acquired their aquatic biotas from that direction; as such, they have offered little opportunity for north to south faunal interchange.

In addition, the limestones produce stream basin characteristics that many aquatic species find unsuitable. Streams in these limestone regions usually have marly beds, basic water chemistries, and occupy catchments characterized by high gradients, frequent waterfalls, and poor integration of drainage; in numerous instances these streams disappear into caves in the karst terrain, then reappear great distances away as resurgences in the form of springs, often bursting as high waterfalls from sheer cliffs. Such habitats, though scenically spectacular, are not particularly easy for aquatic Heteroptera to colonize, since they present many significant topographic barriers. In addition, during dry periods much of the water retreats to underground conduits, and those permanent surface watercourses that do exist are often discontinuous.

Stream basins in the ophiolite and metamorphic terrains, by contrast, show very different characteristics. They tend to have relatively pH neutral waters flowing in well sorted beds with high substrate heterogeneity, and occupying well integrated catchments with moderate gradients. This allows the development of a far richer aquatic Heteroptera biota than streams in limestone at similar elevations. The dichotomy is particularly well illustrated by a transect over the central ranges just west of Puncak Jaya. At Tembagapura, on the south slope at an elevation of 2000 metres, the mountain streams are high gradient, originate in limestone catchments, and support no aquatic Heteroptera. At Bilogai, on the north slope at 2000 metres, the mountain streams are moderate gradient, originate in metamorphic catchments, and contain a diverse aquatic Heteroptera biota. This leads to an interesting possibility – that, based on samples of aquatic insects, inferences about the petrology of a catchment can be made *a priori*, even when the geological information is not precisely known.

Not only are the aquatic Heteroptera biotas of ophiolitic and metamorphic terrains richer than those of other areas, they are also regionally distinctive. This is hypothesized to result from the fact that, in addition to providing suitable physical habitats, the ophiolites are in many instances markers for accreted terranes that have travelled considerable distances from their original point of formation, bringing isolated suites of aquatic insect species with them (Polhemus, 1996). Even after an arc terrane has been incorporated into a larger island such as New Guinea, its associated aquatic insects often appear to retain a high fidelity to the accreted block, probably because such accretions are frequently surrounded by topographic basins containing lakes and slow water streams that are barriers to the dispersal of upland species, or because the

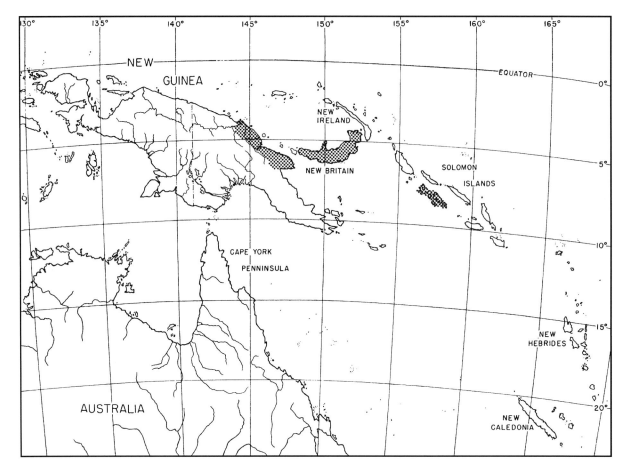

Fig.3. Distribution of the near-shore marine water strider genus *Thetibates* on New Guinea and surrounding arcs.

accreted ophiolitic terranes are often closely juxtaposed with adjacent limestone terranes also uplifted in the course of the accretion (as in the central mountain chain discussed above), which once again serve as effective barriers to dispersal of many aquatic species.

Solomons arc distributions

There is abundant evidence to indicate that the accreted terranes of Miocene age in the northern coastal ranges of New Guinea and the Papuan peninsula harbour distinctive suites of endemic aquatic Heteroptera species that may be linked to their former history as isolated land masses. The situation is made complex, however, by the large number of terranes, and the resulting large number of endemic species. Several groups have the potential to be particularly informative in this regard, but since two of the most critical are in the process of taxonomic revision, only a limited analysis of patterns within the Miocene terranes can be made at this time:

Ptilomera (Gerridae) – this genus occurs from India through Indochina and westward through the Malay archipelago to New Guinea, New Britain, New Ireland, and Bougainville; it is absent in Australia, the Moluccas and Philippines (except Palawan). Such a distribution implies that the group is of Asian affinities, and arrived in the New Guinea area from the west. Patterns from the central New Guinea ophiolites suggest that this genus was distributed on the Papuan arc prior to its accretion, and the degree of localized speciation in the northern coastal ranges strongly suggests that species occupied the Solomons arc as well. A distribution map of species endemism within New Guinea, including undescribed species, was presented by Polhemus (1996); although accurate in its broad details, this figure was overly simplistic. Ongoing character analysis has shown that many additional species are present, and that terrane fidelity among them is even more pronounced than was suggested in that paper (Fig.2).

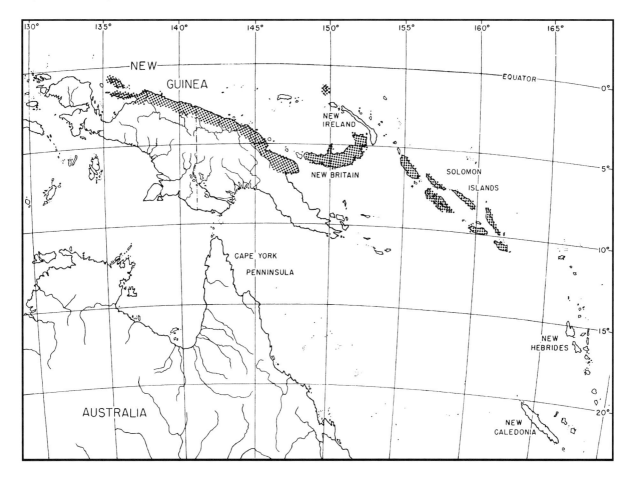

Fig.4. Distribution of the water strider genus *Metrobatopsis* on New Guinea and surrounding arcs.

Rhagovelia (Veliidae) – a large number of species in this group show patterns of localized speciation centred around accreted terranes of Miocene age within New Guinea, suggesting that their distributions have been strongly arc influenced, but only a handful of species have been described (Lansbury, 1993). The genus promises to provide excellent resolution of terrane linked distributions, but this must await completion of a taxonomic and phylogenetic revision currently underway by the authors.

Metrobatini (Gerridae) – two genera in this tribe, *Andersenella* and *Metrobatoides*, are confined to Miocene terranes in northern and eastern New Guinea. A phylogenetic analysis currently in progress indicates that both genera are most closely related to groups occurring in New Guinea, Australia, and the Solomons, supporting the hypothesis of a Melanesian derivation for some of the Miocene arc biota.

Although the biological patterns present among aquatic Heteroptera occurring on the Miocene terranes are still for the most part in the process of being deciphered, the biological evidence from water bugs does strongly indicate that the northern and southern sections of the Papuan peninsula have functioned as separate units, although it does not provide direct support for whether the final amalgamated unit was assembled on or offshore. The following genera have endemic species pairs distributed on the northern and southern sides of the peninsula, and thus conform to the predictions of the terrane assembly model: *Ptilomera* (Gerridae; see Hungerford and Matsuda, 1965); *Ciliometra* (Gerridae; see Polhemus and Polhemus, 1993; Lansbury, 1996); *Rhagovelia* (Veliidae); *Aptinocoris* (Naucoridae); *Sagocoris* (Naucoridae).

In all cases these species pairs have their closest putative relatives in the main body of New Guinea. Although these patterns indicate that the northern and southern sections of the peninsula functioned as separate biological units, they do not strongly support the idea that this complex was assembled at any great distance from the main land mass of New Guinea. In addition,

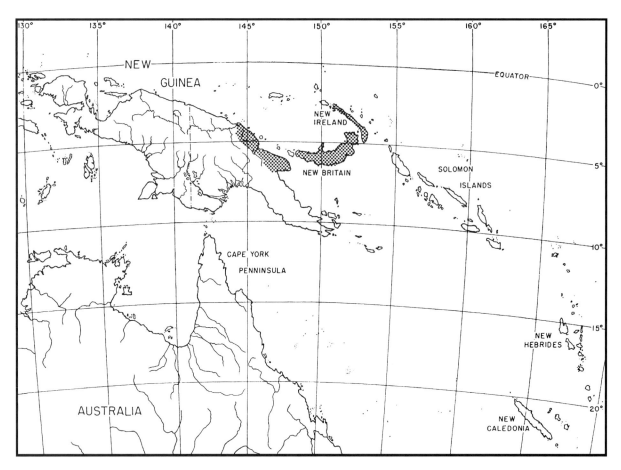

Fig.5. Distribution of the riffle bug *Rhagovelia biroi* on New Guinea and surrounding arcs.

the evidence does not permit rejection of an alternative hypothesis to the terrane assembly model, namely, that the current north to south disjunction may be the result of vicariance due to the formation of the Owen Stanley Range *in situ*.

Another interesting set of patterns is present in regard to the Adelbert-Finisterre terrane. If one accepts the geological hypothesis of Kroenke (1984) regarding the formation of the Solomon Islands and the eventual accretion of a western extension of the Solomons arc into northeast New Guinea, and assumes that the Adelbert-Finisterre terrane was an isolated sector of this arc, then taxa associated with this system should have distributions encompassing the Solomons, the Bismarcks, and the Adelbert, Finisterre, and Saruwaged ranges of mainland New Guinea, and should show outgroup relationships including taxa in Melanesia and Australia. Three separate lineages of aquatic Heteroptera are known to display such patterns:

Thetibates (Gerridae) – this marine genus (Fig.3) is known only from the Solomons, New Britain, and the coastal zones of the Adelbert-Finisterre terrane on New Guinea (Polhemus and Polhemus, 1996).

Metrobatopsis (Gerridae) – members of this freshwater genus (Fig.4) occur on the Solomons, New Britain, Mussau, the Adelbert-Finisterre terrane; future collecting will undoubtedly reveal them on New Ireland as well. Several species are also distributed along the north coast of New Guinea westward to Yapen and Biak (Polhemus and Polhemus, 1993). This could indicate that some of the Miocene terranes and islands in this latter area represent westward extensions of the Miocene Solomons arc, but the pattern could also be the result of dispersal following terrane accretion.

Rhagovelia (Veliidae) – *R. biroi* (Fig.5) occurs on New Ireland and New Britain, plus the Adelbert-Finisterre terrane of New Guinea (Lansbury, 1993); extensive surveys have not found it elsewhere on the latter island.

These patterns suggest that the distributions of these groups were linked to the tectonic history of the Solomons arc, and support Kroenke's (1984) hypothesis that the Adelbert-Finisterre terrane is an accreted portion of this arc.

Conclusions

Based on current geological hypotheses and known species distributions, it has been possible to hypothesize linkages between the distributions of certain groups of aquatic Heteroptera present on New Guinea and particular island arc terranes that accreted to the northern margin of the island during the Tertiary.

The Vogelkop peninsula possesses a biota typical of that seen on New Guinea as a whole, indicating that it and the main body of New Guinea have been in close proximity since at least the beginning of the Tertiary.

The central ophiolite belt of New Guinea possesses a number of distinctive genera not seen in the north coast ranges; these may have been derived from the Papuan arc which accreted in the Oligocene.

The terranes of the north coastal ranges, derived from a Miocene arc, show pronounced local endemism, with many species restricted to single terranes. This indicates that speciation may have occurred while these terranes were separate islands in an arc system north of New Guinea.

The northern and southern portions of the Papuan peninsula also show marked regional endemism, and have basically behaved as two separate islands from a biogeographic standpoint. This is compatible with the hypothesis of offshore assembly followed by subsequent accretion to the main body of New Guinea.

The Adelbert-Finisterre terrane has a distinct biota which frequently shows sister group relationships with the Solomons, rather than with the remainder of New Guinea. This accords well with the hypothesis that it formed as part of the Solomons arc system.

Such hypotheses, although interesting, carry little weight until they are subsequently tested by cladistic analyses of the groups involved. Analyses of this type are now underway, and promise to provide a more detailed picture of the relationships between the species assemblages occurring on the various terranes. In the near future, we may thus have a much better comprehension of the biogeographic interactions that took place during the assembly of modern New Guinea.

Acknowledgements

The authors wish to thank Dr Warren Hamilton, of the U. S. Geological Survey, Denver; Dr Loren W. Kroenke of the University of Hawaii, Manoa; and Dr Kevin T. M. Johnson of the Bishop Museum, Honolulu, all of whom generously took time out of their busy schedules to educate us in the complexities of island arc geology. In addition, Kevin Hill of La Trobe University, provided a useful review of the geological discussion and suggested many areas for improvement. This research was sponsored in part by a series of grants from the National Geographic Society, Washington, D. C. (1806-77, 2698-83, 3053-85, 4537-91), and by grant BSR-9020442 from the National Science Foundation, Washington, D. C.

References

Charlton, T. R., Hall, R. and Partoyo, E. 1991. The geology and tectonic evolution of Waigeo Island, NE Indonesia. Journal of Southeast Asian Sciences 6: 289–297.

Cooper, P. and Taylor, B. 1987. Seismotectonics of New Guinea: a model for arc reversal following arc-continent collision. Tectonics 6: 53–67.

Davies, H. L. 1990. Structure and evolution of the border region of Papua New Guinea. In Petroleum Exploration in Papua New Guinea: Proceedings of the First PNG Petroleum Convention, Port Moresby, 12–14th February 1990. pp. 245-270. Edited by G. J. Carman and Z. Carman. PNG Chamber of Mines and Petroleum, Papua New Guinea.

Davies, H. L., Winn, R. D., and KenGemar, P. 1996. Evolution of the Papuan Basin – a view from the orogen. In Petroleum Exploration, Development and Production in Papua New Guinea: Proceedings of the 3rd PNG Petroleum Convention, Port Moresby, 9th–11th September 1996. pp. 53-62. Edited by P. G. Buchanan. PNG Chamber of Mines and Petroleum, Papua New Guinea.

Hall, R. 1997. Cenozoic tectonics of SE Asia and Australasia. In Petroleum Systems of SE Asia and Australasia. pp. 47-62. Edited by J. V. C. Howes and R. A. Noble. Indonesian Petroleum Association, Jakarta.

Hamilton, W. B. 1979. Tectonics of the Indonesian region. U. S. Geol. Survey Professional Paper 1078. U. S. Government Printing Office, Washington, D. C. 345 pp.

Hamilton, W. B. 1988. Plate tectonics and island arcs. Geological Society of America Bulletin 100: 1503–1527.

Holthuis, L. B. 1982. Freshwater Crustacea Decapoda of New Guinea. In Biogeography and Ecology of New Guinea. Edited by J. L. Gressitt. Dr. W. Junk, The Hague.

Hungerford, H. B. and Matsuda, R. 1965. The genus Ptilomera Amyot and Serville (Gerridae: Hemiptera). University of Kansas Science Bulletin 45: 397–515.

Kroenke, L. W. 1984. Cenozoic development of the Southwest Pacific. United Nations Economic and Social Commission for Asia and the Pacific, Committee for Co-ordination of Joint Prospecting for Mineral Resources in South Pacific Offshore Areas, Technical Bulletin 6: 122 pp.

Lansbury, I. 1993. *Rhagovelia* of Papua New Guinea, Solomon Islands, and Australia (Hemiptera-Veliidae). Tijdschrift voor Entomologie 136: 23–54.

Lansbury, I. 1996. Two new species of *Ciliometra* Polhemus (Hem., Gerridae) from Papua New Guinea. Entomologist's Monthly Magazine 132: 55–60.

Milsom, J. 1991. Gravity measurements and terrane tectonics in the New Guinea region. Journal of Southeast Asian Earth Sciences 6: 319–328.

Nash, C. R., Artmont, G., Gillan, M. L., Lennie, D., O'Connor, G., and Parris, K. R. 1993. Structure of the Irian Jaya mobile belt, Irian Jaya, Indonesia. Tectonics 12: 519–535.

Pigram, C. J. and Davies, H. L. 1987. Terranes and the accretion history of the New Guinea orogen. Bureau of Mineral Resources, Journal of Australian Geology and Geophysics 10: 193–211.

Pigram, C. J. and Symonds, P. A. 1991. A review of the timing of the major tectonic events in the New Guinea orogen. Journal of Southeast Asian Earth Sciences 6: 307–318.

Polhemus, D. A. 1995. Two new species of *Rhagovelia* from the Philippines, with a discussion of zoogeographic relationships between the Philippines and New Guinea. Journal of the New York Entomological Society 103: 55-58.

Polhemus, D. A. 1996. Island arcs, and their influence on Indo-Pacific biogeography. In The Origin and Evolution of Pacific Island Biotas, New Guinea to Eastern Polynesia: Patterns and Processes. pp. 51-66. Edited by A. Keast and S. E. Miller. SPB Academic Publishing bv, Amsterdam.

Polhemus, D. A. and Polhemus, J. T. 1987. A new genus of Naucoridae (Hemiptera) from the Philippines, with comments on zoogeography. Pan-Pacific Entomologist 63: 265-269.

Polhemus, J. T. and Polhemus, D. A. 1993. The Trepobatinae (Heteroptera: Gerridae) of New Guinea and surrounding regions, with a review of the World Fauna. Part 1. Tribe Metrobatini. Entomologica Scandinavica 24: 241–284.

Polhemus, J. T. and Polhemus, D. A. 1996. The Trepobatinae (Heteroptera: Gerridae) of New Guinea and surrounding regions, with a review of the World Fauna. Part 4. The marine tribe Stenobatini. Entomologica Scandinavica 27: 279–346.

Rogerson, R. J. and Hilyard, D. B. 1990. Scrapland: a suspect composite terrane in Papua New Guinea. In Petroleum Exploration in Papua New Guinea: Proceedings of the First PNG Petroleum Convention, Port Moresby, 12–14th February 1990. pp. 271-282. Edited by G. J. Carman and Z. Carman. PNG Chamber of Mines and Petroleum, Papua New Guinea.

Smith, R. I. 1990. Tertiary plate tectonic setting and evolution of Papua New Guinea. In Petroleum Exploration in Papua New Guinea: Proceedings of the First PNG Petroleum Convention, Port Moresby, 12–14th February 1990. pp. 229-244. Edited by G. J. Carman and Z. Carman. PNG Chamber of Mines and Petroleum, Papua New Guinea.

Weiland, R. J. and Cloos, M. 1996. Pliocene-Pleistocene asymmetric unroofing of the Irian fold belt, Irian Jaya, Indonesia: Apatite fission-track thermochronology. Geological Society of America Bulletin 108: 1438–1449.

Yan, C. Y. and Kroenke, L. W. 1993. A plate tectonic reconstruction of the SW Pacific 0-100 Ma. In Proceedings of the Ocean Drilling Program, Scientific Results 130: 697-709. Edited by T. Berger, L. W. Kroenke, L. Mayer. *et al.*

Marine water striders (Heteroptera, Gerromorpha) of the Indo-Pacific: cladistic biogeography and Cenozoic palaeogeography

N. Møller Andersen
Zoological Museum, University of Copenhagen, Universitetsparken 15, DK-2100 Copenhagen Ø, Denmark
Email: nmandersen@zmuc.ku.dk

Key words: cladistic biogeography, Cenozoic palaeogeography, Indo-Pacific, marine water striders, Heteroptera, Gerromorpha

Abstract

More than 140 species of marine water striders (Heteroptera, Gerromorpha), representing three families and 11 genera, are distributed throughout the Indo-Pacific region. The largest genera are: *Hermatobates* (Hermatobatidae), *Halovelia*, *Haloveloides*, *Xenobates* (Veliidae), *Asclepios*, *Halobates*, *Rheumatometroides*, and *Stenobates* (Gerridae). Marine water striders live in estuaries, mangroves, intertidal coral reef flats, and on the sea surface near coral reefs and rocky coasts. Adult marine water striders are always wingless but may disperse along coasts and chains of islands. Five species of sea skaters, *Halobates*, have colonized the surface of the open ocean. The present work updates and extends previous work on the cladistic biogeography of Indo-Pacific marine water striders. The method of paralogy-free subtree analysis developed by Nelson and Ladiges is applied to taxon-area cladograms for four monophyletic groups, and the results are combined to yield general area cladograms depicting the relationships between eight areas of endemism of the Indo-Pacific region. Finally, hypotheses of area relationships are discussed in the light of available knowledge about the distribution of fossil marine water striders, the Cenozoic palaeogeography of the Indo-Pacific region, and the palaeoecology of mangroves and reef-building corals.

Introduction

The Indo-Pacific comprises land areas bordering the Indian and west Pacific Oceans (Fig.1) and has traditionally been divided into the Ethiopian, Oriental, and Australian regions. Studies using the methods of cladistic biogeography (Nelson and Platnick, 1981; Humphries and Parenti, 1986; Humphries *et al.*, 1988), however, do not support this division (*e.g.*, Schuh and Stonedahl, 1986; Andersen, 1991a; Muona, 1991; Parenti, 1991; Vane-Wright, 1991; Boer, 1995; Boer and Duffels, 1996). In general, patterns of distribution seem to be compatible with a set of hierarchical relationships between more restricted areas of endemism. These are quite similar to those delimited by Gressitt (1956) and Gressitt *et al.* (1961), although the areas of the Indian Ocean were not recognized as part of their 'Pacific' region. Malesia (as defined by Whitmore, 1981, 1987) is seemingly not a genuine, monophyletic area of endemism (Andersen, 1991a).

Most studies of Indo-Pacific biogeography have been based upon terrestrial animals and plants. In this chapter, the results of biogeographical studies of a group of marine insects belonging to the heteropterous infraorder Gerromorpha are presented. Although the majority of the about 1,600 gerromorphan species are limnic, about 180 species belonging to five families and seven subfamilies occur in the marine environment and reach their greatest diversity in the Indo-Pacific region (Andersen and Polhemus, 1976; Andersen, 1982; Cheng, 1985). The present work updates and extends previous works on the cladistic biogeography of Indo-Pacific marine water striders (Andersen, 1991a, 1991b; Andersen and Weir, 1994a, 1994b). The method of paralogy-free subtree analyses (Nelson and Ladiges, 1996) is applied to taxon-area cladograms for four monophyletic groups of marine water striders and the results are combined to yield general area cladograms depicting the relationships between areas of endemism of the Indo-Pacific region. Finally, the hypotheses of area interrelationships are discussed

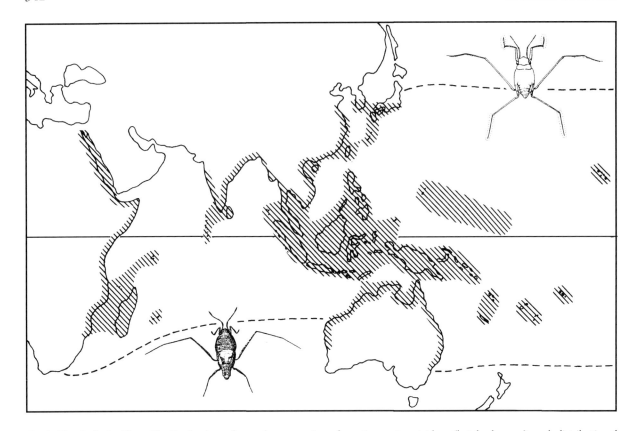

Fig.1. The Indo-Pacific with distribution of nearshore species of marine water striders (hatched areas) and distributional boundaries of oceanic *Halobates* (dashed line). Insertions show *Halovelia malaya* Esaki, length 2.5 mm (bottom left) and *Halobates sericeus* Eschscholtz, length 3.4 mm (top right).

in the light of available knowledge about the Cenozoic palaeogeography of the region (Boer, 1995; Boer and Duffels, 1996; Packham, 1996; Hall, 1996, and references therein).

Marine water striders and their distribution

General

More than 140 species of marine water striders, representing three families and 11 genera, are distributed throughout the Indo-Pacific region. They live in estuaries, mangroves, intertidal coral reef flats, and on the sea surface near coral reefs and rocky coasts (Fig.1). Five species of sea skaters, *Halobates*, have colonized the surface of the open ocean (Cheng, 1985, 1989). Adult marine water striders are always wingless but may disperse along coasts, chains of islands and, for a few species of *Halobates*, possibly across wider stretches of open sea. Although some species belonging to the genera *Halovelia* and *Halobates* are widespread, most species of marine water striders have rather restricted areas of distribution.

Hermatobates

The genus *Hermatobates* contains eight described species, but a taxonomic revision is required. The biology of these odd marine insects was studied by Foster (1989), who named them coral treaders because they live on the tidal flats of coral reefs. Both adults and nymphs retreat to holes in porous blocks of dead coral during high tides. *Hermatobates* is unique (Andersen, 1982), belonging to its own family, one of few exclusively marine insect families. It has no close freshwater relatives and it is difficult to trace its geographical origin. *Hermatobates* species are found along continental coasts and islands throughout the Indo-Pacific. Widespread species are *H. djiboutensis* Coutière and Martin (Red Sea, East Africa, Seychelles, Maldives), *H. marchei* Coutière and Martin (Ryukyu Islands, Philippines, Indonesia, Australia, islands of the

West Pacific), and *H. hawaiiensis* China (Hawaiian Islands and islands of Central Pacific). *H. breddini* Herring (Caribbean) is the only species found outside the Indo-Pacific region.

Halovelia

The subfamily Haloveliinae belongs to the Veliidae, one of the most speciose families of water striders, with more than 600 species. Most haloveliines are marine. Freshwater relatives belong to two genera: *Entomovelia* with one described and some undescribed species in Burma, Malaya and Borneo, and *Strongylovelia* with several, mostly undescribed, species in the Indo-Australian region. The marine Haloveliinae have been classified into three genera, *Halovelia*, *Xenobates*, and *Haloveloides* (Andersen, 1989a, 1989b, 1992; Lansbury, 1989, 1996). So far, 45 species have been described, but during the past 10 years, marine haloveliines have been collected in many new localities, increasing the number of known species to more than 60.

Species of *Halovelia* or coral bugs inhabit the intertidal zone of rocky coasts or coral reefs where they can be found on the surface of tidal pools. At high tide they retreat to holes in blocks of coral or other porous rocks where they rest, surrounded by an air bubble, until the next low tide (Andersen, 1989b). The 32 described species of the genus *Halovelia* range from the Red Sea and East African coast to the West Pacific islands as far as Samoa (Andersen, 1989a, 1989b). Species belonging to the *H. esakii* group occur along the coasts of the Philippines, Sulawesi, New Guinea, the Solomons, northern Australia, and the Fiji and Tonga Islands. Species of the *H. bergrothi* group have a similar distribution but two species are widespread in Australia and the islands of the West Pacific, respectively. Two closely related species, the *H. lannae* group, are found on both sides of Wallace's line. Species belonging to the *H. malaya* group are found along the coasts of East Africa, southern Asia, and island groups of the Indian Ocean.

Xenobates and Haloveloides

Species of *Xenobates* typically live in mangroves where they can be found on the surface of tidal canals. Until recently this genus was thought to be monotypic, but now more than 25 species have been collected throughout the Indo-Australian region. Other species have been placed in *Halovelia* (China, 1957; Polhemus, 1982; Lansbury, 1989). A preliminary classification of *Xenobates* species has revealed a number of monophyletic species groups (Andersen, unpublished). The *X. seminulum* (Esaki) group (with four species) occurs in Maluku, northern New Guinea, the Bismarck archipelago and Solomon Islands. A related species group is found in Maluku, Sulawesi, southern Philippines, Java, Singapore, and Sri Lanka. Another group occurs in Maluku, Sulawesi, southern Philippines, North Borneo, and Malaya. Eight species of *Xenobates*, most of them undescribed, are found along the coast of tropical Australia (Andersen and Weir, unpublished). Finally, *X. loyaltiensis* (China) is endemic to New Caledonia and the Loyalty Islands.

A separate genus, *Haloveloides*, with seven species, was erected for the '*Halovelia*' *papuensis* Esaki group (Andersen, 1992). It is closely related to *Xenobates* but some species have left the protecting mangroves and live on the sea surface at some distance from the coast (Lansbury, 1996). One species is found in areas of the Sunda shelf, another one in Sulawesi, Maluku, and southern Philippines, three species in Palawan and Luzon, and another two species in northern New Guinea, the Bismarck archipelago and the Solomons.

Stenobates and Rheumatometroides

Another large family of water striders, the Gerridae, with more than 500 species, contains several groups of marine species. The tribe Stenobatini has 21 mangrove-inhabiting species in Malesia (Polhemus and Polhemus, 1996). They belong to the subfamily Trepobatinae and are related to the freshwater genus *Naboandelus* found in Africa, India, and SE Asia (Andersen, 1982; Polhemus and Polhemus, 1993). *Stenobates* (10 species) is distributed from Singapore to Australia, while the distribution of *Rheumatometroides* (7 species) ranges from Singapore to the Solomons. *Thetibates* (2 species) and the monotypic genera *Pseudohalobates* and *Stenobatoides* have more restricted distributions within Malesia.

Halobates and Asclepios

The subfamily Halobatinae, tribe Halobatini, contains the well-known sea skaters, genus *Halobates*, with 42 described species (Herring,

1961; Polhemus and Polhemus, 1991; Andersen and Foster, 1992; Andersen and Weir, 1994b). Most species are confined to coastal, marine habitats such as estuaries, mangroves and coral reefs. Freshwater relatives belong to the tribe Metrocorini (Andersen, 1982) and live in running freshwater throughout Africa, South and SE Asia, and the Malay archipelago, including Sulawesi (Polhemus, 1990). The closest relatives of *Halobates* are *Austrobates*, a monotypic, freshwater-inhabiting genus endemic to the Cape York peninsula, Australia (Andersen and Weir, 1994a), and *Asclepios* with three species distributed along the coasts of SE and East Asia. These genera are more primitive than *Halobates* in some characters but otherwise difficult to separate from the most primitive species, *H. mjobergi* Hale and *H. lannae* Andersen and Weir, found in tropical Australia (Andersen and Weir, 1994b).

A cladistic analysis of *Halobates* (Andersen, 1991b), allows the definition of a number of monophyletic species-groups, each showing a characteristic distribution pattern. Most species of the *H. regalis* Carpenter group occur in tropical Australia but two species (*H. zephyrus* Herring and *H. whiteleggei* Skuse) live along the coast of New South Wales as far as 100 km south of Sydney. There is considerable distributional overlap between species. *H. peronis* Herring, ranging from the Philippines to the Solomons, and *H. sexualis* Distant, from Sri Lanka and Malaya, are both closely related to species from Australia (*H. darwini* Herring and *H. herringi* Polhemus and Cheng).

Species belonging to the *H. princeps* White group occur throughout the Indo-West Pacific region. Most species are endemic to single islands or island groups. There is one subgroup of closely related species in the Indian Ocean (*H. alluaudi* Bergroth group) including the west coast of India, and another subgroup (*H. mariannarum* Esaki group) on islands of the western Pacific. *H. princeps* White, the sister taxon of these subgroups, is widely distributed in Malesia. Many *Halobates* species are endemic to single islands or island groups like *H. robustus* Barber (Galapagos), *H. kelleni* Herring (Samoa), *H. salotae* Herring (Tonga), *H. bryani* Herring and *fijiensis* (Fiji), and *H. tethys* Herring (Mauritius), but a few coastal species are widespread like *H. hayanus* White (Red Sea to New Guinea) and *H. flaviventris* Eschscholtz (East Africa to Vanuatu).

Five species of *Halobates* spend their entire life on the surface of the open ocean, sometimes hundreds of kilometres from the nearest coast (Cheng, 1985, 1989): *H. germanus* White (Indian Ocean, West and Central Pacific Ocean), *H. sericeus* Eschscholtz (Pacific Ocean), *H. sobrinus* White (East Pacific Ocean, off the coasts of Middle America), *H. splendens* Witlaczil (East Pacific Ocean, off the coast of South America), and *H. micans* (all tropical oceans). Herring (1961) considered this group monophyletic but the cladistic analysis of *Halobates* (Andersen, 1991b) shows that two oceanic species are closer to some coastal, but widespread species (*H. flaviventris* and *hawaiiensis* Usinger), than to other oceanic species.

Methods

Areas of endemism

For the present study, the Indo-Pacific region is divided into eight areas of endemism following Andersen (1991a: Fig.9). These areas are delimited as follows (with abbreviations used throughout the chapter and in the figures):
Australia (Aust): The coasts of the Australian continent, but also including southern New Guinea, New Caledonia, and Vanuatu (New Hebrides).
East Asia (EAsi): The coasts of continental Asia including Vietnam, China, and Korea, Taiwan, Japan, and Ryukyu Islands.
Indian Ocean (IndO): Red Sea, East African coast, Madagascar, Seychelles, Mascarenes, Maldives, and west coast of India.
Malayan (Mala): East coast of India, Burma, Thailand, Cambodia, Malaysia, Singapore, and Indonesia as far east as Wallace's line.
Papuasia (Papu): Northern New Guinea, Belau (Palau Islands), the Bismarck archipelago, and Solomon Islands.
Philippines (Phil): The Philippine Islands.
Sulawesi (Sula): Sulawesi (Celebes), the Lesser Sunda Islands (except Bali), and Maluku (the Moluccas).
West Pacific (WPac): Islands of the west Pacific Ocean as far east as Samoa and the Hawaiian Islands.

Analytical methods

For each monophyletic group of marine water striders, the distributions of species or species groups were recorded as taxon-area cladograms

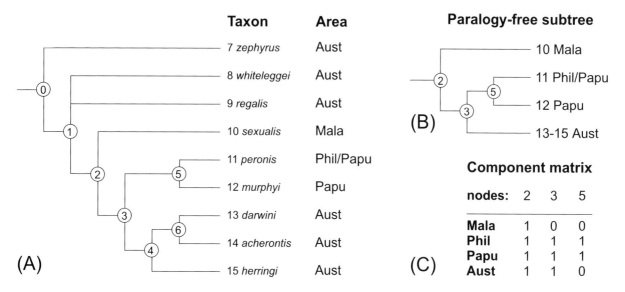

Fig.2. Paralogy-free subtree analysis. (A) Taxon-area cladogram for *Halobates zephyrus* and the *H. regalis* group (Gerridae, Halobatinae). (B) Paralogy-free subtree derived from (A). (C) Component matrix derived from the nodes of (B). Further explanation in text.

of the eight areas of endemism defined above. Two methods of biogeographic analysis were applied to these taxon-area cladograms. First, they were subjected to component analysis (Nelson and Platnick, 1981; Humphries and Parenti, 1986) using the computer program COMPONENT, version 1.5 (Page, 1989, 1990). The cladograms were analyzed by the options *Build* and Assumption 0, 1, and 2 (which stipulate different ways of treating widespread taxa). The results are cladograms showing the relationships between the eight areas of endemism. The options *Compare, Consensus,* and *Strict* were used in cases which yielded more than a single area cladogram, obtaining a strict consensus cladogram for all possible area cladograms.

Second, the taxon-area cladograms were analyzed by the method of paralogy-free subtree analysis (Nelson and Ladiges, 1996). The subtree algorithm builds subtrees from a taxon-area cladogram, starting at each terminal node and progressing to the base of the cladogram. A node (taxon) that relates organisms that, as different taxa, do not overlap in geographic distribution is associated with the non-overlapping geographic data. A node (taxon) that relates organisms that overlap in distribution is deemed paralogous (analogous to the molecular phenomenon) and is not generally associated with geographic data.

Fig.2A shows an example of a taxon-area cladogram for the monophyletic *Halobates regalis* (Gerridae, Halobatinae) group with its outgroup species *H. zephyrus*. The terminal taxa 13-15 are all endemic to Aust which therefore is the only geographic data associated with nodes 4 and 6. The distribution of taxon 11 (Phil/Papu) and 12 (Papu) overlaps. Node 5 is therefore paralogous in the strict sense, yet node 5 evidently relates Phil more close to Papu than to the areas Mala and Aust. Following Assumption 2 (which treats areas for widespread taxa as redundant and not necessarily indicating close area relationships), the widespread distribution of taxon 11, *H. peronis*, can be reduced to Phil and node 5 thereby made paralogy-free. Likewise, the nodes 2 and 3 (Fig.2A) are paralogy-free, associated with the geographic information Mala/Phil/Papu/Aust and Phil/Papu/Aust, respectively. In contrast, nodes 0, 1, and 2 are paralogous because the area Aust already is associated with the more terminal nodes 4 and 6. Thus, the original taxon-area cladogram (Fig.2A) yields only one, fully resolved paralogy-free subtree (Fig.2B) showing unique relationships between the areas Aust, Mala, Phil, and Papu.

A component matrix can be derived from each subtree by scoring geographical information associated with the subtree nodes (Fig.2C). Matrices derived from two or more subtrees can be combined and subjected to a parsimony analysis using programs like Hennig86 (Farris, 1988) or PAUP (Swofford, 1993). For each analysis, the result is given as the number of most

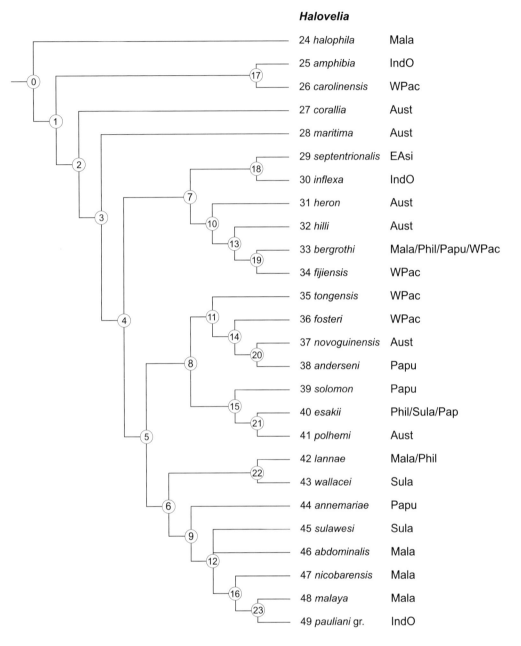

Fig.3. Taxon-area cladogram for species of *Halovelia* (Veliidae, Haloveliinae). Further explanation in text.

parsimonious trees, length, consistency index (ci), and retention index (ri) of these trees. The geographical information may also be scored as three item statements (Nelson and Ladiges, 1991), but parsimony analyses of matrices of component data and three item statements usually yield the same trees (Nelson and Ladiges, 1996). This is not the case, however, for the taxon-area cladogram used as an example above (Fig.2A). A parsimony analysis for three item statements yields two area cladograms, viz., (Mala (Aust (Phil, Papu))) and (Aust (Mala (Phil, Papu))). This suggests that the area Aust may have a dual history.

Results

Hermatobates

Classified as a family of its own (Hermatobatidae), *Hermatobates* is probably the oldest group

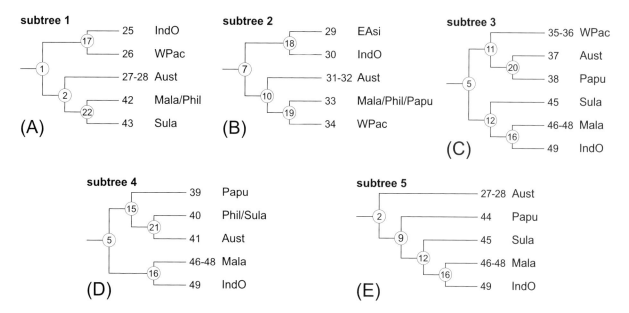

Fig.4. Paralogy-free subtrees derived from the taxon-area cladogram for *Halovelia* (Fig.3). Further explanation in text.

of marine water striders. Knowledge about the phylogenetic relationships between the species of *Hermatobates* is not available at present, precluding a cladistic biogeographical analysis. The occurrence of one species in the Caribbean indicates that the genus was already widely distributed before the final closure of the Isthmus of Panama (since 5 Ma), and probably much earlier than this event.

Halovelia

The taxon-area cladogram for *Halovelia* (Fig.3) has 26 terminal taxa. Five species occurring in East Africa and various Indian Ocean islands are joined in the *H. pauliani* group. For subtree analysis, informative nodes of the taxon-area cladogram for *Halovelia* (Fig.3) are 7, 9, and 10-23. Five paralogy-free subtrees (Fig.4) can be derived from this cladogram using assumption 2 when distributions overlap. The component matrix for the nodes of individual subtrees is given in Table 1. Parsimony analysis of this matrix give 9 trees (length = 22, ci = 59, and ri = 60). All of these trees have IndO and EAsi as sister-areas. The relationships between other areas vary and are unresolved in a strict consensus tree (Fig.7A). The COMPONENT analysis for *Halovelia* yields 4 trees (assumption 0), 30 trees (assumption 1), and 11 trees (assumption 2), all resulting in highly unresolved consensus trees for the 8 areas involved.

Xenobates and *Haloveloides*

Species of *Xenobates* are only recorded from Mala, Phil, Sula, Papu, and Aust. Three paralogy-free subtrees (nos. 6-8) can be derived from a preliminary taxon-area cladogram for *Xenobates* (Andersen, unpublished) using assumption 2. The taxon-area cladogram for *Haloveloides* (Andersen, 1992) has 7 terminal taxa with the same distributional areas as *Xenobates* (except Aust). Only one subtree can be derived from this cladogram. The component matrix for the nodes of individual subtrees for *Xenobates* and *Haloveloides* is listed in Table 1. Parsimony analysis of this matrix gives one, completely resolved tree (length = 20, ci = 70, and ri = 62). In contrast, the COMPONENT analysis for *Xenobates* + *Haloveloides* yields 2 trees (assumption 0), 3 trees (assumption 1), and 6 trees (assumption 2), all resulting in highly unresolved consensus trees for the 5 areas involved.

Stenobates and *Rheumatometroides*

Species of the five genera of the tribe Stenobatini are only recorded from Mala, Phil, Sula, Papu, and Aust. A preliminary taxon-area cladogram for species belonging to the tribe Stenobatini (Andersen, unpublished) has 22 terminal taxa. Three paralogy-free subtrees (nos. 9-

Table 1. Component matrices for subtrees derived from taxon-area cladograms for marine water striders. Nodes of subtrees nos. 1 - 19. OG = "outgroup" area. Further explanation in text.

	Halovelia					*Xenobates & Haloveloides*				Stenobatini			Halobatini						
Tree No.	1	2	3	4	5	6	7	8	9	10	11	12	13	14	15	16	17	18	19
Node	012 272	111 089	1112 1260	112 561	011 926	001 296	001 496	011 518	00 24	000 489	001 494	001 375	1112 3673	1112 3673	0112 9673	012 182	11 49	0012 7963	1122 1734
OG	000	000	0000	000	000	000	000	000	00	000	000	000	0000	0000	0000	000	00	0000	0000
IndO	010	010	0110	010	111	???	???	???	??	???	???	???	1010	1010	1010	???	??	0000	1101
Mala	101	101	0110	010	111	111	111	100	11	110	100	111	1111	1111	1111	010	00	1111	1010
EAsi	???	010	????	???	???	???	???	???	??	???	???	???	1111	1111	1111	010	??	1111	1010
Phil	101	101	????	101	???	111	111	111	00	000	000	111	????	????	????	101	11	????	????
Sula	101	???	0100	101	110	110	110	111	10	101	110	100	1110	1110	1110	101	??	1110	1101
Papu	???	101	1001	100	100	000	000	101	11	101	111	000	1110	1110	1110	101	11	1110	1101
Aust	100	000	1001	101	000	100	100	000	??	110	111	110	0000	0000	1000	100	10	1100	0000
WPac	010	101	1000	???	???	???	???	???	??	???	???	???	0010	0010	0000	101	??	1000	1100

11) can be derived from this cladogram using assumption 2. The component matrix for the nodes of individual subtrees for the Stenobatini is listed in Table 1. Parsimony analysis of this matrix gives two trees (length = 14, ci = 64, and ri = 54). The strict consensus tree is only partly resolved, with Phil as the basal area and Sula and Papu as sister areas. The COMPONENT analysis yields 1 tree (assumption 0 and 1) and 14 trees (assumption 2), the latter resulting in a completely unresolved consensus tree for the 5 areas involved. The trees resulting from applying assumptions 0 and 1 are different, placing Aust close to either Papu/Sula or Mala/Phil, respectively.

Halobates and Asclepios

The taxon-area cladogram for the three genera of the tribe Halobatini (including *Austrobates*; Fig.5) has 29 terminal taxa. Only *Halobates germanus* and *flaviventris* were included from the chiefly oceanic *H. micans* group, either because they are widespread (*H. micans* and *H. sericeus*) or because their distribution falls outside the Indo-Pacific region (*H. sobrinus* and *splendens*) *H. robustus* is endemic to the Galapagos Islands which also is outside the boundary of the Indo-Pacific region (but see discussion below). Informative nodes of the taxon-area cladogram (Fig.5) are 12, 14, 16, 17, 18, 19, 22, 23, 24. Seven paralogy-free subtrees (Fig.6, A-G) can be derived from this cladogram using assumption 2 when distributions overlap. For example, geographic data associated with node 18 overlaps with data associated with node 22. Since Mala is part of a widespread distribution of taxon 40 (*H. proavus*), the geographic data at node 22 are reduced to Phil/Sula/Papu/WPac.

The component matrix for the nodes of individual subtrees (Fig.6, A-G) for the Halobatini is given in Table 1. A parsimony analysis of this matrix give 5 trees (length = 33, ci = 75, ri = 81). All trees have Aust, WPac, and IndO (in that order) in a basal position and Mala and EAsi as sister areas. The relationships between Mala/EAsi, Phil, Sula, and Papu vary and are unresolved in a strict consensus tree (Fig.7B).

The COMPONENT analysis for the Halobatini yields 2 trees under assumption 0 with a consensus tree showing EAsi, WPac, Sula, and Aust (in that order) in basal positions and the relationships between Mala, Phil, and Papu unresolved. No less than 819 trees are found under assumption 1 which gives a completely unresolved consensus tree. Finally, 17 trees are found under assumption 2 with a unresolved strict consensus tree except for a close relationship between EAsi and Mala.

All groups combined

Halovelia and the two genera of Halobatini occur in all of the eight areas of endemism considered here. When the informative subtree nodes for these taxa are analyzed together, one fully resolved tree is found (Fig.7, C; length = 61, ci = 62, and ri = 65).

Xenobates, Haloveloides, and the five genera

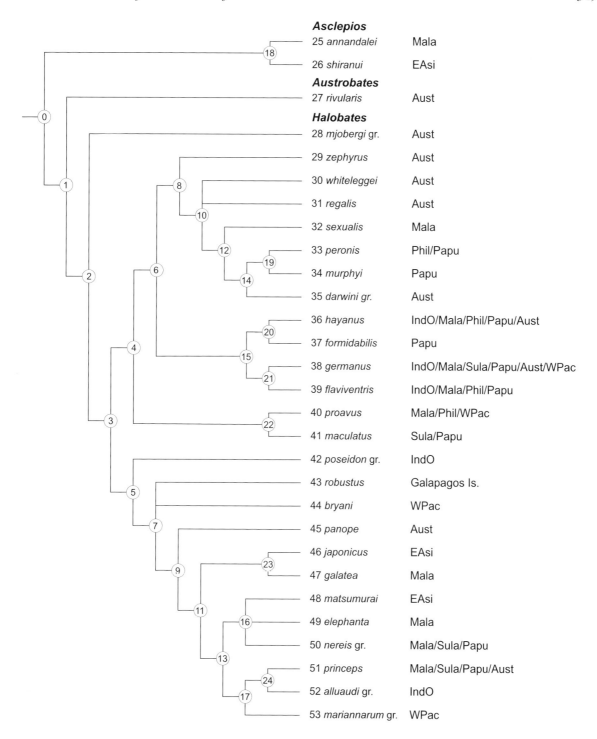

Fig. 5. Taxon-area cladogram for species of *Asclepios*, *Austrobates*, and *Halobates* (Gerridae, Halobatinae). Further explanation in text.

of Stenobatini only occur in five out of the eight areas of endemism (EAsi, IndO, and WPac excluded). A parsimony analysis of informative subtree nodes for these genera yields two trees (length = 37, ci = 62, and ri = 48). The consensus tree (Fig. 7D) is partly resolved.

A parsimony analysis of the complete component matrix (Table 1) yields only two trees (length = 103, ci = 59, ri = 55). One tree (Fig. 7E) has Aust, WPac, and IndO (in that order) in ba-

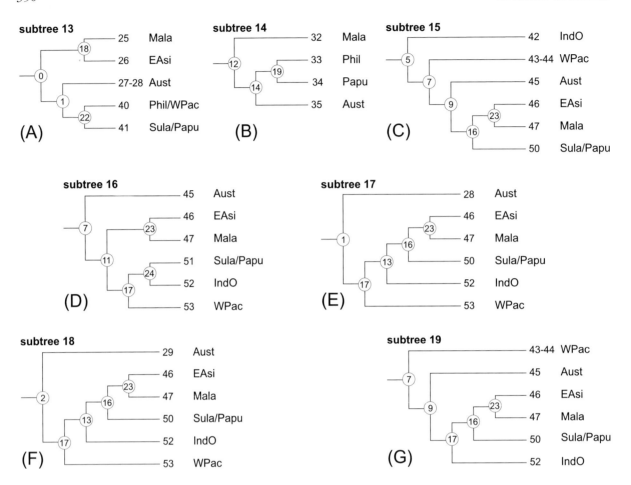

Fig.6. Paralogy-free subtrees derived from the taxon-area cladogram for *Asclepios, Austrobates,* and *Halobates* (Fig.5). Further explanation in text.

sal positions. The other tree (Fig.7F) has Aust, WPac, and Papu (in that order) in basal positions and IndO in a position close to the sister areas Mala and EAsi.

Discussion

Parsimony analyses of matrices derived from individual (not combined), paralogy-free subtrees generally yield cladograms with better resolution among areas than COMPONENT analyses of taxon-area cladograms for the same groups. While a strict consensus of the most parsimonious area cladograms for *Halovelia* (Fig.7A) is largely unresolved, a hierarchic structure is more apparent in the cladograms for the other groups of marine water striders, although the area relationships are not completely congruent. When data for *Halovelia* and the Halobatini are combined, the area relationships among all eight areas are completely resolved (Fig.7C). When data for *Xenobates, Haloveloides,* and the Stenobatini are combined, the tree shows partly resolved area relationships among the areas Aust (Australia), Mala (the Malayan subregion), Papu (Papuasia), Phil (Philippines), and Sula (Sulawesi) (Fig.7D).Combining data for all four groups of marine water striders yields two, completely resolved area cladograms (Fig.7, E-F) which only differ in the position of IndO (Indian Ocean) in relation to the areas Papu, Sula, Phil, and Mala/EAsi (East Asia).

A strict consensus of the two, completely resolved area cladograms results in completely unresolved relationships between these areas which is not a fair representation of the information conveyed by the original area cladograms. In general, consensus trees are not well suited to summarize information about relationships between taxa or areas and should therefore be avoided. Instead, one should seek of possible

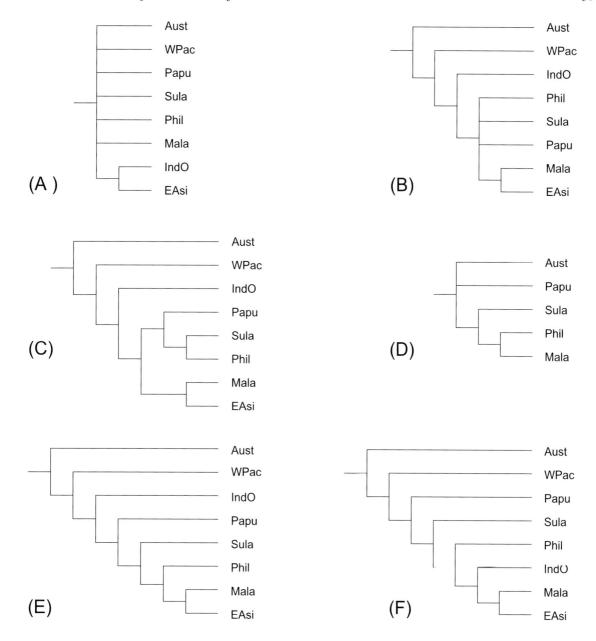

Fig. 7. Area cladograms for marine water striders. (A) Consensus tree of 9 most parsimonious trees for *Halovelia* (Fig.4 and Table 1). (B) Consensus tree of 5 most parsimonious trees for *Asclepios*, *Austrobates*, and *Halobates* (Fig.6 and Table 1). (C) Most parsimonious area cladogram derived from subtrees for *Halovelia* and the Halobatini (Figs.4, 6, and Table 1). (D) Most parsimonious area cladogram derived from subtrees for *Haloveloides*, *Xenobates*, and the Stenobatini (Table 1). (E) One of two most parsimonious area cladograms derived from subtrees for all groups (Table 1). (F) Another most parsimonious area cladogram. Further explanation in text.

causes for the ambiguous area relationships. Unlike organisms, geographical areas may have more than one history. For example, most *Halovelia* species found along the coasts of East Africa, Madagascar, and the islands of the Indian Ocean (Mascarenes, Seychelles, and the Maldives) belong to the SE Asian *H. malaya* group (Andersen, 1989b) and may represent relatively late dispersal and subsequent speciation. On the other hand, the disjunct distributions of the most basal clades of *Halovelia* may be relicts of an ancient, much wider distribution of the genus as supported by the finding of a fossil *Halovelia* in the Oligo-Miocene Dominican amber (20-30 Ma; Andersen and Poinar, 1997).

Phylogenetic relationships provide clues to the evolution of present-day marine water striders, suggesting the number of independent origins of modern taxa, possible pathways of habitat changes and adaptive evolution, and the approximate geographical location of this evolution. The sister group of the marine Haloveliinae is a clade composed of two limnic genera, *Entomovelia* and *Strongylovelia* found in India, Indo-China, and Malesia (Andersen, 1982). The distribution of *Xenobates* and *Haloveloides* covers the same geographical areas, while the present distribution of *Halovelia* includes the whole Indo-Pacific region. Under the premise that marine species evolved from freshwater species occupying the same or an adjacent geographical area, marine haloveliines probably evolved somewhere in the Indo-Australian region.

The sister group of the Halobatini (Gerridae-Halobatinae) is the tribe Metrocorini with about 80 species distributed in tropical Africa, continental Asia, and the Indo-Malayan archipelago (including Sulawesi). The genus *Asclepios* (with three marine species) is confined to Asia. The sister group of *Halobates* is the monotypic genus *Austrobates*, endemic to tropical Australia (Andersen and Weir, 1994a) where also the most basal clade of *Halobates*, the subgenus *Hilliella* (with two species), is found (Andersen and Weir, 1994b). Under the assumption that ancestral taxa were no more widespread than their descendants, one may hypothesize that *Halobates* evolved along the coasts of Australia-New Guinea and/or in the Indo-Malayan archipelago and subsequently dispersed to other parts of the Indo-Pacific. The record of a Middle Eocene *Halobates* from modern northern Italy (45 Ma; Andersen *et al.*, 1994) suggests that these events took place in the early Tertiary, before the closure of Tethys and the formation of the present-day Middle East.

The most significant palaeogeographic event in the Indo-Pacific region was the northward movement of the Australian continent during the Cenozoic to its collision with an intra-oceanic arc some 25 Ma (Boer, 1995; Boer and Duffels, 1996; Hall, 1996, 1998 this volume; Packham, 1996). Before this event, the subduction of the oceanic lithosphere under the Pacific plate was accompanied with volcanic activity forming an island arc system comprised by present-day central Philippines and northern and southeastern New Guinea. The northward movement of Australia was preceded by continental fragments which form the present-day Birds Head peninsula of New Guinea and parts of the Moluccas. Throughout most of the Cenozoic, the area between Asia and Australia was not an open ocean, but probably an archipelago of volcanic islands and microcontinents with rapidly changing areas of land and coastline contours. This may have been the perfect conditions for allopatric speciation and diversification among organisms of shallow seas including marine water striders.

Marine water striders are chiefly confined to the tropics, and their distributions seem to be limited by the same temperature regimes as reef-building corals and mangrove trees, i.e., in water bounded by the 20°C isotherms. Fossil evidence suggests major changes throughout the Tertiary in the climatic conditions favouring the presence of organisms of tropical shallow seas, including marine water striders. The mangrove palm, *Nypa*, had an extensive distribution in the Atlantic/Caribbean/East Pacific region in early Tertiary and once occurred at high latitudes in Europe (Ricklefs and Latham, 1993). Throughout much of the Tertiary, the reef coral belt appears to have been wider in latitude than it is today, though it was noticeably narrowed to something like its present limits by the late Neogene, and was reduced still further during the Pleistocene when it was narrower than today. The diversity of corals declined during the Neogene in the Atlantic/Caribbean and rose in the Indo-Pacific region, while in the Mediterranean, reef corals disappeared completely by the end of the Miocene (Rosen, 1988). Information derived from the few fossil marine water striders (Andersen *et al.*, 1994; Andersen and Poinar, 1997), suggests that both *Halovelia* and *Halobates* occupied wider geographical ranges in the past and that extinction may have played a significant role in shaping the present distributional patterns of marine water striders.

Acknowledgements

I thank Hans Duffels and Jeremy Holloway for inviting me to present a talk at the symposium: 'Biodiversity and Biogeography of Southeast Asia and Southwest Pacific' organized during the XX International Congress of Entomology, Florence, August 1996. I also thank the editors of this volume for the opportunity to publish the results of my work. Robert Hall, Jeremy Holloway, Toby Schuh, and an anonymous referee gave useful comments on an earlier draft of the manuscript. This chapter is part of a project supported by grants from the Danish Natural Science Research Council (Grant No. 9502155).

References

Andersen, N. M. 1982. The Semiaquatic Bugs (Hemiptera, Gerromorpha). Phylogeny, adaptations, biogeography, and classification. Entomonograph 3: 1-455.

Andersen, N. M. 1989a. The coral bugs, genus *Halovelia* Bergroth (Hemiptera, Veliidae). I. History, classification, and taxonomy of species except the *H. malaya* group. Entomologica Scandinavica 20: 75-120.

Andersen, N. M. 1989b. The coral bugs, genus *Halovelia* Bergroth (Hemiptera, Veliidae). II. Taxonomy of the *H. malaya*-group, cladistics, ecology, biology, and biogeography. Entomologica Scandinavica 20: 179-227.

Andersen, N. M. 1991a. Cladistic biogeography of marine water striders (Hemiptera, Gerromorpha) in the Indo-Pacific. Australian Systematic Botany 4: 151-163.

Andersen, N. M. 1991b. Marine insects: genital morphology, phylogeny and evolution of sea skaters, genus *Halobates* (Hemiptera, Gerridae). Zoological Journal of the Linnean Society 103: 21-60.

Andersen, N. M. 1992. A new genus of marine water striders (Hemiptera, Veliidae) with five new species from Malesia. Entomologica Scandinavica 22: 389-404.

Andersen, N. M. and Foster, W. A. 1992. Sea skaters of India, Sri Lanka, and the Maldives, with a new species and a revised key to Indian Ocean species of *Halobates* and *Asclepios* (Hemiptera, Gerridae). Journal of Natural History 26: 533-553.

Andersen, N. M. and Poinar, G. O. 1997. A marine water strider (Hemiptera, Veliidae) from Dominican amber. Entomologica Scandinavica (in press).

Andersen, N. M. and Polhemus, J. T. 1976. Water-striders (Hemiptera: Gerridae, Veliidae, etc.). *In* Marine Insects. pp. 187-224. Edited by L. Cheng. North-Holland Publishing Company, Amsterdam.

Andersen, N. M. and Weir, T. A. 1994a. *Austrobates rivularis* gen. et sp. nov., a freshwater relative of *Halobates* (Hemiptera, Gerridae) with a new perspective on the evolution of sea skaters. Invertebrate Taxonomy 8: 1-15.

Andersen, N. M. and Weir, T. A. 1994b. The sea skaters, genus *Halobates* Eschscholtz (Hemiptera, Gerridae), of Australia: taxonomy, phylogeny, and zoogeography. Invertebrate Taxonomy 8: 861-909.

Andersen, N. M., Farma, A., Minelli, A. and Piccoli, G. 1994. A fossil *Halobates* from the Mediterranean and the origin of sea skaters (Hemiptera, Gerridae). Zoological Journal of the Linnean Society 112: 479-489.

Boer, A. J. de, 1995. Islands and cicadas adrift in the West-Pacific. Biogeographic patterns related to plate tectonics. Tijdschrift voor Entomologie 138: 169-244.

Boer, A. J. de, and Duffels, J. P. 1996. Historical biogeography of the cicadas of Wallacea, New Guinea and the West Pacific: a geotectonic explanation. Palaeogeography, Palaeoclimatology, Palaeoecology 124: 153-177.

Cheng, L. 1985. Biology of *Halobates* (Heteroptera: Gerridae). Annual Reviews of Entomology 30: 111-135.

Cheng, L. 1989. Factors limiting the distribution of *Halobates* species. *In* Reproduction, genetics and distributions of marine organisms. pp. 357-362. Edited by J. S. Ryland and P. A. Tyler. Fredensborg, Denmark.

China, W. E. 1957. The marine Hemiptera of the Monte Bello Islands, with descriptions of some allied species. Zoological Journal of the Linnean Society 40: 342-357.

Farris, J. S. 1988. Hennig86 Reference, Version 1.5. 41 Admiral Street, Port Jefferson Station, New York.

Foster, W. A. 1989. Zonation, behaviour and morphology of the intertidal coral-treader *Hermatobates* (Hemiptera: Hermatobatidae) in the south-west Pacific. Zoological Journal of the Linnean Society 96: 87-105.

Gressitt, J. L. 1956. Some distribution patterns of Pacific island faunae. Systematic Zoology 6: 12-32.

Gressitt, J. L., Maa, T. C., Mackerras, I. M., Nakata, S. and Quate, L. W. 1961. Problems in the zoogeography of Pacific and Antarctic insects. Pacific Insects Monograph 2: 1-127.

Hall, R. 1996. Reconstructing Cenozoic SE Asia. *In* Tectonic Evolution of Southeast Asia. Edited by R. Hall and D. J. Blundell. Geological Society Special Publication 106: 153-184.

Hall, R. 1998. The plate tectonics of Cenozoic SE Asia and the distribution of land and sea. *In* Biogeography and Geological Evolution of SE Asia. Edited by R. Hall and J. D. Holloway (this volume).

Herring, J. L. 1961. The genus *Halobates* (Hemiptera: Gerridae). Pacific Insects 3: 223-305.

Humphries, C. J. and Parenti, L. E. 1986. Cladistic Biogeography. Oxford Monographs in Biogeography. Clarendon Press, Oxford.

Humphries, C. J., Ladiges, P. Y., Roos, M. and Zandee, M. 1988. Cladistic biogeography. *In* Analytical Biogeography. An integrated approach to the study of animal and plant distributions. pp. 371-404. Edited by A. A. Myers and P. S. Giller. Chapman and Hall, London, New York.

Lansbury, I. 1989. Notes on the Haloveliinae of Australia and the Solomon Islands (Insecta, Hemiptera, Heteroptera: Veliidae). Reichenbachia 26: 93-109.

Lansbury, I. 1996. Notes on the marine veliid genera *Haloveloides*, *Halovelia* and *Xenobates* (Hemiptera-Heteroptera) of Papua New Guinea. Tijdschrift voor Entomologie 139: 17-28.

Muona, J. 1991. The Eucnemidae of South-east Asia and the western Pacific — a biogeographic study. Australian Systematic Botany 4: 165-182.

Nelson, G. and Ladiges, P. Y. 1991. Three-area statements: standard assumptions for biogeographic analysis. Systematic Zoology 40: 470-485.

Nelson, G. and Ladiges, P. Y. 1996. Paralogy in cladistic biogeography and analysis of paralogy-free subtrees. American Museum Novitates 3167: 1-58.

Nelson, G. and Platnick, N. 1981. Systematics and Biogeography. Cladistics and Vicariance. Columbia University Press, New York.

Packham, G. 1996. Cenozoic SE Asia: reconstructing its aggregation and reorganization. *In* Tectonic Evolution of Southeast Asia. Edited by R. Hall and D. J. Blundell. Geological Society Special Publication 106: 123-152.

Page, R. D. M. 1989. Component User's Manual. Release 1.5. Auckland: R. D. M. Page.

Page, R. D. M. 1990. Component analysis: a valiant failure? Cladistics 6: 119-136.

Parenti, L. R. 1991. Ocean basins and the biogeography of freshwater fishes. Australian Systematic Botany 4: 137-149.

Polhemus, D. A. 1990. A revision of the genus *Metrocoris* Mayr (Heteroptera: Gerridae) in the Malay Archipelago and the Philippines. Entomologica Scandinavica 21: 1-28.

Polhemus, J. T. 1982. Marine Hemiptera of the Northern Territory, including the first fresh-water species of *Halobates* Eschscholtz (Gerridae, Veliidae, Hermatobatidae and Corixidae). Journal of the Australian Entomological Society 21: 5-11.

Polhemus, J. T. and Polhemus, D. A. 1991. Three new species of marine water-striders from the Australasian region, with notes on other species (Gerridae: Halobatinae, Trepobatinae). Raffles Bulletin of Zoology 39: 1-13.

Polhemus, J. T. and Polhemus, D. A. 1993. The Trepobatinae (Heteroptera: Gerridae) of New Guinea and surrounding regions, with a review of the world fauna. Part 1. Tribe Metrobatini. Entomologica Scandinavica 24: 241-284.

Polhemus, J. T. and Polhemus, D. A. 1996. The Trepobatinae (Heteroptera: Gerridae) of New Guinea and surrounding regions, with a review of the world fauna. Part 4. Tribe Stenobatini. Entomologica Scandinavica 27: 279-346.

Ricklefs, R. E. and Latham, R. E. 1993. Global patterns of diversity in mangrove floras. In Species Diversity in Ecological Communities. Historical and Geographical Perspectives. pp. 215-229. Edited by R. E. Ricklefs and D. Schluter. University of Chicago Press, Chicago.

Rosen, B. R. 1988. Progress, problems and patterns in the biogeography of reef corals and other tropical marine organisms. Helgoländer Meeresuntersuchungen 42: 269-301.

Schuh, R. T. and Stonedahl, G. 1986. Historical biogeography in the Indo-Pacific: A cladistic approach. Cladistics 2: 337-355.

Swofford, D. L. 1993. PAUP: Phylogenetic analysis using Parsimony, version 3.1. Illinois Natural History Survey, Chanpaign, Illinois.

Vane-Wright, R. I. 1991. Transcending the Wallace line: do the western edges of the Australian region and the Australian plate coincide? Australian Systematic Botany 4: 183-197.

Whitmore, T. C. (Editor) 1981. Wallace's line and plate tectonics. Oxford Monographs on Biogeography. 1. Clarendon Press, Oxford.

Whitmore, T. C. (Editor) 1987. Biogeographical Evolution of the Malay Archipelago. Oxford Monographs on Biogeography. 4. Clarendon Press, Oxford.

Biogeography of Sulawesi grasshoppers, genus *Chitaura*, using DNA sequence data

R. K. Butlin[1], C. Walton[1], K. A. Monk[2] and J. R. Bridle[1]
[1]*Department of Biology, The University of Leeds, Leeds LS2 9JT, UK*
[2]*Department of Zoology, The University of Reading, Whiteknights, Reading RG6 2AJ, UK*
Email: r.k.butlin@leeds.ac.uk

Key words: molecular phylogeny, Orthoptera, speciation, mitochondrial DNA

Abstract

A molecular phylogeny is presented for part of the genus *Chitaura* (Orthoptera: Acrididae) based on DNA sequence data from the mitochondrial cytochrome oxidase 1 locus. *Chitaura* has at least 20 parapatrically distributed colour forms in Sulawesi and the Moluccas. While the status of these forms is uncertain, they behave as genetically isolated species where they are in contact. Molecular data are consistent with speciation *in situ* with isolation mainly due to wet rainforest on mountain ridges. *Chitaura* has apparently been diversifying within Sulawesi for 7-14 Ma and reached the Moluccas from North Sulawesi some 2.5-5 Ma ago.

Introduction

Grasshoppers in the genus *Chitaura* I. Bolivar, 1918 (Orthoptera: Acrididae) are flightless, brightly-coloured inhabitants of forest edge, stream-side and light gap habitats in primary and secondary forest. The genus has 10 described species from Sulawesi, 9 of which are endemic and one occurs also in the Moluccas. Outside Sulawesi, the genus is recorded from southern India, Java, Bali, Palawan and several Moluccan islands (Ambon, Haruku, Saparua, Seram, and the Kai Islands) but has only one or two species in each locality. The closely related, fully winged genus *Oxytauchira* Ramme, 1941 has a single representative in Sulawesi (Monk and Butlin, 1990).

Within Sulawesi, the *Chitaura* species are apparently parapatric, suggesting speciation *in situ* as a result of island or habitat fragmentation in the past. This pattern is found in several other taxa. It has been documented most clearly in the Sulawesi macaques but has also been described in pond-skaters, cicadas, carpenter bees, butterflies, limacodid moths and tiger beetles (Whitten *et al.*, 1987; Ciani *et al.*, 1988; Knight and Holloway, 1990; Cassola, 1996). If these current distributions do reflect past land or habitat islands, they should be broadly coincident across taxa and related to topographic features unless the patterns have been obscured by subsequent dispersal and/or hybridisation. The phylogenetic relationships of the species or populations may reflect the history of vicariance within taxa.

Thus *Chitaura* provides a model for the within-island diversification which is a major contributor to the high proportion of endemic species in the Sulawesi fauna in general. However, it is difficult to relate patterns of present-day distribution, or phylogenetic relationships, to past geological or climatic events without an independent timescale. The 'molecular clock' (Avise, 1994) provides an opportunity to determine *relative* timings of branching events with confidence, and *absolute* timings within the limits set by the available calibrations. Here we describe a phylogeny for *Chitaura* based on mitochondrial DNA sequence data from 27 individuals representing seven of the described species and up to nine additional species (plus one individual of *Oxytauchira gracilia* and an outgroup). This is the first molecular phylogeny for a SE Asian insect group. Unfortunately, there is currently no morphologically-based phylogenetic hypothesis for the genus, and species definitions are in need of revision, as discussed below.

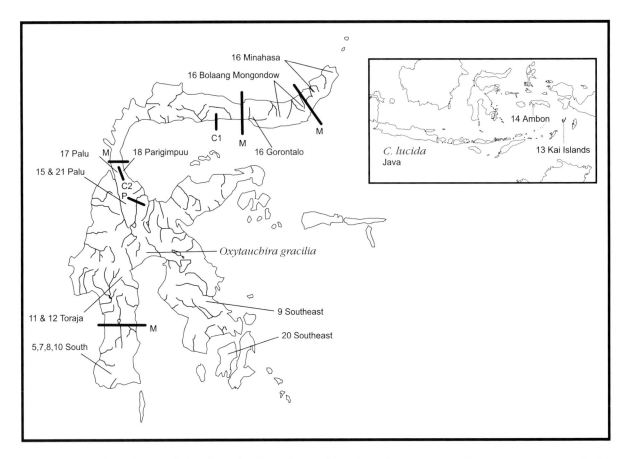

Fig.1. Locations of samples included in the molecular analysis and positions of contact zones. Contact zones are marked with heavy lines for *Chitaura* (C1 between *Chitaura 16* and *18* and C2 between *Chitaura 18* and *17*) in relation to those in macaques (M) and one in the pond-skater *Ptilomera* (P) (see text for references). Although parapatric distributions have been described for other taxa, contact zones have yet to be identified. Inset: Part of SE Asia showing the locations of *Chitaura* samples from outside Sulawesi.

Distribution and contact zones

The currently described species are defined, often from only a few specimens, on the basis of the distinctive adult coloration (Willemse, 1956; Hollis, 1975). Our extensive collections from Sulawesi (Fig.1) suggest that the actual number of endemic colour forms is at least 20. For present purposes, these forms are identified by numbers, some of which are tentatively equated with described species (Fig.2). In more than 50 collecting sites, we have always found only one form, confirming that distributions are parapatric. However, the geographical extent of individual forms apparently varies widely. At least four, and possibly seven, forms exist in the Gunung Lompobatang area of South Sulawesi, while a single colour form is found over the entire eastern half of North Sulawesi. Given this taxonomic uncertainty, the molecular phylogeny should be interpreted as a phylogeny of individuals rather than species.

Defining the extent and status of these forms depends on the identification of contact zones. To date, we have identified two such zones (Fig.1). In both cases there is an abrupt transition over a distance of about 200 m, coincident for several colour pattern elements and with no intermediates detected. In north Sulawesi, *Chitaura 16* (= *C. brachyptera* I. Bolivar) has hind femora that are red proximally fading to greenish-yellow distally and without spots. The thoracic sternites are red. These and other colour pattern characteristics distinguish it from *Chitaura 18* whose hind femora are orange-yellow with three distinctive black markings and whose thoracic sternites are white with narrow dark margins. The two forms have been found

DNA sequence data and Sulawesi grasshoppers

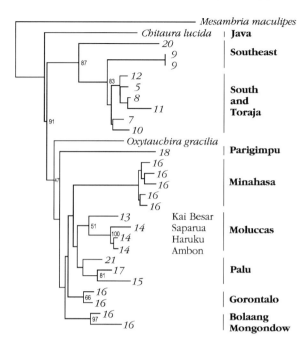

Fig.2. Neighbour-joining tree based on Jukes-Cantor distances between sequences from the mitochondrial CO1 locus, rooted by outgroup. The tree topology includes all of the well supported nodes in a Wagner parsimony tree. Bootstrap values from the parsimony tree are included on these nodes. Tentative associations of numbered colour forms with described species are as follows: 5 - *C. elegans*; 7 - *C. vidua*; 10 - *C. poecila*; 13 - *C. striata*; 14 - *C. moluccensis*; 16 - *C. brachyptera*. The remaining colour forms do not correspond to any of the descriptions in Willemse (1956).

on either side of a small river near Tilamuta, west of Gorontalo. No site has been found with both forms present. The differences between forms *Chitaura 18* and *17* are less striking but, nevertheless, include several apparently independent pattern elements on the legs, head, thoracic sternites and abdomen. The forms occur together in a small, recently cut clearing where the main Palu-Gorontalo road crosses the mountains of the isthmus north of Palu. Despite occurring together, no intermediates have been found. These observations strongly suggest genetic independence between the taxa and are most readily explained by secondary contact following allopatric divergence.

Although the two contact zones are in the same general areas as contacts in macaques, there are substantial differences. The *Chitaura 18/16* contact in North Sulawesi is displaced some 50 km west of a macaque contact zone which is located at the lowland area around Gorontalo (Ciani *et al.*, 1988). The *Chitaura 18/17* contact near Palu (Fig.1) is orientated east-west across the mountain ridge rather than north-south along the isthmus. In this respect it is more similar to the contact zone in *Ptilomera* pond-skaters (Polhemus and Polhemus, 1990).

Molecular analysis

A molecular phylogenetic analysis has been conducted on 29 individuals using 290 base pairs of DNA sequence from the mitochondrial cytochrome oxidase 1 (CO1) locus. DNA sequence was obtained from either dried or ethanol-preserved specimens. DNA was purified from a single hind femur by a silica-based extraction method and then the sequence of interest was amplified by the Polymerase Chain Reaction (PCR). The amplified DNA was manually sequenced directly from the PCR product following separation of the two strands. Details of these methods are given in Walton *et al.* (1997) and general descriptions can be found in Avise (1994). Mitochondrial DNA was chosen because of its maternal inheritance and high copy number, and because many 'universal' PCR primers are available. The CO1 locus has been amplified previously from a range of insect taxa which aids interpretation of sequence data (Lunt *et al.*, 1996).

The resulting sequence data can be analysed in two ways: cladistically, treating the 290 base-pairs as characters with four possible states corresponding to the four nucleotides of DNA, or phenetically, by calculating distances based on the proportion of bases that differ between each pair of sequences. Here we consider the results of the distance-based methods, which have the advantage of providing divergence time estimates, but the major features of the phylogenetic relationships inferred from cladistic approaches are concordant with these results (Walton *et al.*, 1997). Divergence time estimates assume a 'molecular clock', that is they assume that, for a given locus, sequence divergence accumulates at a steady rate and that this rate is similar in related taxa (Avise, 1994).

Twenty-three individuals from 13 colour forms from within Sulawesi, 4 Moluccan individuals, *Chitaura lucida* from Java, and an outgroup were included in this analysis (Figs.1 and 2). The analysis (Fig.2) demonstrates that the Sulawesi species are monophyletic and that the Moluccan species fall within the Sulawesi clade. This includes the geographically isolated Kai Island species which has the most distinctive colour pattern (lateral white or yellow stripes on the head and thorax, present in all other Sulawesi and Moluccan forms, are replaced by a broad central stripe). Within Sulawesi, relationships are strongly patterned geographically, with a deep division between South + SE Sulawesi and North Sulawesi, in particular. This geographic pattern reinforces the implication of speciation *in situ*. The Moluccan species fall within the North Sulawesi clade suggesting colonisation from this part of the island, perhaps most probably from the Banggai peninsula which we have been unable to sample so far. Surprisingly, *Oxytauchira gracilia* falls within the Sulawesi *Chitaura* clade suggesting that a revision of generic boundaries is necessary.

Within colour forms, it is only for *Chitaura 16* (= *C. brachyptera*) that we have sequenced sufficient individuals to make inferences about their status. Genetically similar individuals are clearly grouped geographically but distances between these groups are substantial, reaching as much as 10%. This is greater than the distances between some other colour forms and greater than the distances between well-differentiated species in other taxa (Avise, 1994), including other grasshoppers (Hewitt, 1996). It seems likely that cryptic species, separated by contact zones, exist within this colour form which is one of the most widespread.

Maximum sequence divergence within the Sulawesi clade is 14%. This figure can be used to provide a minimum estimate for the timescale of diversification within the island of Sulawesi by comparison with calibrations derived from other insects. Using a range of 1-2% divergence per million years for mitochondrial DNA coding regions (Avise, 1994) the timescale is 7 to 14 Ma. Maximum divergence within the Moluccan clade is 4.5% and this is only slightly less than the minimum divergence between Moluccan and North Sulawesi specimens. Thus, colonisation of the Moluccas appear to have taken place 2.5 to 5 Ma ago.

Within Sulawesi, the major branches in the phylogenetic tree fit better with an hypothesis of vicariance due to mountain barriers than an alternative hypothesis based on inundation of lowland areas, or break-up of forest. For example, taxa from the Gunung Lompobatang and Toraja areas are closely related despite separation by the lowland area near Parepare where a macaque contact zone is located (Fig.1). This is probably the only area to have been inundated by the sea during the Pleistocene. Breakup of rainforest during drier climatic intervals might explain the positions of other macaque contact zones. However, the large-scale distribution of *Chitaura* (present in Java, Sulawesi, the Moluccas and Palawan, absent in Sumatra, Borneo and peninsular Malaysia) suggests that it is excluded from perennially wet forests and occurs only in drier monsoon forest habitats. Thus dry intervals may not have been as important an influence on distribution of *Chitaura* as they were for macaques. Forms *17* and *18* are separated only by a narrow mountain chain, yet are quite distinct in DNA sequence. Form *16* includes several divergent clades in the molecular tree including the Gorontalo and Bolaang Mongondow samples from opposite ends of the same mountain block. These patterns are consistent with persistent separation by wet rainforest on mountains. Contact *17/18* north of Palu may be a recent consequence of deforestation along the road.

Conclusions

Our molecular data for *Chitaura* are consistent with the presence of the genus on the island since its formation by rifting of South Sulawesi away from Borneo some time more than 10 Ma ago (Hall, 1998 this volume). They do not place any upper limit on this figure since populations on the island may have contracted at any time post-colonisation. However, this may be viewed as a lower limit, albeit with quite large potential error because of the reliance on external calibration of the molecular clock. It is interesting that the divergence between Javan and Sulawesi individuals is only slightly greater than the divergence within Sulawesi. This suggests a relatively homogeneous population across the Sunda shelf from Java, through Borneo to Sulawesi (and Palawan?) around 10-15 Ma ago. Presumably *Chitaura* has since been excluded from Borneo by climatic change.

Within Sulawesi, the distribution patterns and relationships are not concordant either with the macaque distributions or with expected positions of separate islands in the past. They are more consistent with the positions of current extrinsic barriers to gene exchange. The large number of forms on the single isolated Gunung Lompobatang, and the molecular divergence within form *16* in North Sulawesi, are suggestive of divergence across local mountain barriers. Although *Chitaura* can be found at high elevations (at least 2000 m) they are most common by larger water courses or at forest edges. Thus forested mountain ridges may represent substantial barriers to dispersal. The deep separation of some of the forms within the phylogenetic tree suggests that these barriers have been stable for at least 5 Ma. The Moluccan islands appear to have been colonised from North Sulawesi or the Banggai peninsula relatively recently (2.5-5 Ma ago). This is consistent with recent geological reconstructions which suggest that these islands have only been above water and close to Sulawesi on approximately this timescale (Hall, 1998 this volume).

Acknowledgements

We are grateful to all those who helped with collection of insects, on Project Wallace 1985, on a TrekForce expedition to Sulawesi in 1991, and while RKB and JRB visited Indonesia in 1994 and 1995, especially Professor Jatna Supriatna and Dr Dantje Sembel, and also to the Leverhulme Trust, The Royal Society and the British Ecological Society for funding.

References

Avise J. C. 1994. Molecular markers, natural history and evolution. Chapman and Hall, New York.

Cassola, F. 1996. Studies on tiger beetles: LXXXIV: Additions to the tiger beetle fauna of Sulawesi, Indonesia (Coleoptera: Cicindelidae). Zoologische Mededelingen (Leiden) 70: 145-153.

Ciani, A. C., Stanyon, R., Scheffrahn, W. and Sampurno, B. 1988. Evidence of gene flow between Sulawesi macaques. American Journal of Primatology 17: 257-270.

Hall, R. 1998. The plate tectonics of Cenozoic SE Asia and the distribution of land and sea. *In* Biogeography and Geological Evolution of SE Asia. Edited by R. Hall and J. D. Holloway (this volume).

Hewitt, G. M. 1996. Some genetic consequences of ice ages, and their role in divergence and speciation. Biological Journal of the Linnean Society 58: 247-276.

Hollis, D. 1975. A review of the subfamily Oxyinae (Orthoptera: Acridoidea). Bulletin of the British Museum (Natural History) Entomology 31: 189-234.

Knight, W. J. and Holloway, J. D. (Editors) 1990. Insects and the Rain Forests of South East Asia (Wallacea). Royal Entomological Society of London.

Lunt, D. H., Zhang, D-X., Szymura, J. M. and Hewitt, G. M. 1996. The insect cytochrome oxidase I gene: evolutionary patterns and conserved primers for phylogenetic studies. Insect Molecular Biology 5: 153-166.

Monk, K. A. and Butlin, R. K. 1990. A biogeographic account of the grasshoppers (Orthoptera: Acridoidea) of Sulawesi, Indonesia. Tijdschrift voor Entomologie 133: 31-38.

Polhemus, J. T. and Polhemus, D. A. 1990. Zoogeography of the aquatic Heteroptera of Celebes: regional relationships versus insular endemism. *In* Insects and the Rain Forests of South East Asia (Wallacea). pp.73-86. Edited by W. J. Knight and J. D. Holloway. Royal Entomological Society of London.

Walton, C., Butlin, R. K. and Monk, K. A. 1997. A phylogeny for grasshoppers of the genus *Chitaura* (Orthoptera: Acrididae) from Sulawesi, Indonesia, based on mitochondrial DNA sequence data. Biological Journal of the Linnean Society 62: 356-382.

Whitten, A. J., Mustafa, M. and Henderson, G. S. 1987. The Ecology of Sulawesi. Gadjah Mada University Press, Indonesia.

Willemse, C. 1956. Synopsis of the Acridoidea of the Indo-Malayan and adjacent regions. Part II. Fam. Acrididae, subfam. Catantopinae, part I. Publicaties van het natuurhistorisch Genootschap in Limburg 8: 3-225.

Terrestrial birds of the Indo-Pacific

B. Michaux
Private Bag, Kaukapakapa, New Zealand

Key words: Indo-Pacific biogeography, fragmentation of east Gondwana, nightjars, *Gallirallus philippensis*

Abstract

The avifaunas of New Zealand, New Caledonia, central Polynesia, New Guinea, Maluku and Sulawesi are discussed. Distributional patterns within and between avifaunas are described and related to Mesozoic tectonic activity along the east Gondwana margin. Spreading, rift formation, subduction, obduction and mobile arc systems were all important components of this activity. Avian distributional patterns at family, genus and species levels are discussed in terms of rift-arc interactions within modern zones of active plate convergence.

Introduction

I'm sure that Leon Croizat would have derived some satisfaction at this coming together of biologists and geologists to discuss matters of mutual interest regarding the Indo-Pacific region. For biologists the Indo-Pacific is an evolutionary laboratory in which taxonomic diversification has occurred on a dynamic and complex stage. For geologists it is a region where a modern orogeny can be studied and its development through time reconstructed. The key question for me is to what extent are distributional patterns consistent with the reconstructions provided by geologists?

Even though most birds are highly mobile, a component of an island's avifauna is sedentary. What proportion of an avifauna is sedentary depends on several factors, including distance to areas that could be colonised. Mayr (1941) discussed the efficiency of sea barriers to bird colonisation from New Guinea to New Britain and said:

".... of the 265 species of land birds which are known from that part of New Guinea which is opposite New Britain, only about 80 species have a representative on New Britain. In other words, the 45 mile [70 km] stretch of water which separates the two islands has prevented the crossing over of 70 percent of the New Guinean species."

Clearly there is more to colonisation than crossing barriers because most birds should be able to cross 70 km of sea. Yet according to Mayr only 30% of the Huon peninsula avifauna has shown evidence of colonising New Britain. Mayr (1941) thought that 70% 'sedentary' species was general for avifaunas within the Indo-Australian archipelago. Poor colonising ability and a developed taxonomic and systematic base make birds excellent biogeographical tools.

The avifaunas of New Zealand, New Guinea, Fiji, Tonga, Samoa, Maluku and Sulawesi are described and broad patterns in the distributional data discussed. These patterns link Indo-Pacific islands into a number of groupings. These groupings do not always follow convention, for example Timor is linked to south Maluku. Island groupings based on avifaunal distributions are related to three broad geological systems. These are the Melanesian rift-arc system, the Banda rift-arc system and the Sumba terrane. The geological histories of these structures are described and differences in geological interpretations discussed.

Finally, distributional patterns are interpreted

in the light of geologists' reconstructions. These interpretations are necessarily general for two reasons. Firstly, cladistic bird phylogenies of appropriate scope are not available. In most cases the taxa used in describing patterns are either species or genera for which monophyly is likely, but as yet undemonstrated cladistically. However, Sibley and Ahlquist (1990) have produced a systematic treatment which is used to examine the relationship between nightjar evolution and the Mesozoic and Tertiary history of the Indo-Pacific in some detail. The second reason is that the geology of the Indo-Pacific is complicated and many geological questions are yet to be resolved. Definitive answers to problems of Indo-Pacific biogeography are not available at present, but this is precisely what makes the Indo-Pacific such an interesting area for research.

Avifaunas

New Zealand

Extant and extinct New Zealand (Fig.1: 1) terrestrial birds are listed in Table 1. The data came from Falla *et al.* (1979), Fordyce (1982), Fuller (1987) and Turbott (1990). Species unknown to pre-European Maori (*i.e.*, lack a Maori name) and self-introduced since European arrival are not included in Table 1, because many of these species appear to be associated with man-made habitats. The presence of these species may be due to the availability of grassland habitats, introduced food sources or nesting sites, and they have been excluded. Table 1 shows two characteristics of New Zealand's terrestrial avifauna, a high degree of endemism and low taxonomic diversity.

Five endemic terrestrial avian families are listed in Table 1. The Apterygidae (*kiwi*), Acanthisittidae (New Zealand wrens) and Callaeatidae (New Zealand wattle birds, including the extinct *huia*) are extant, while two families of *moa*, the Emeidae and Dinornithidae, are extinct. New Zealand thrushes were represented by a single extinct species — the *piopio* — regarded by Turbott (1990) as a member of the Paradisaeidae, but better placed either in the Turnagridae or Ptilonorhynchidae (Sibley, pers. comm., 1996). Thirty five genera and 79 species listed in Table 1 are endemic. Only seven species are shared with Australia and three of these are also found elsewhere. A further twenty species are shared with Australia, southwest Pacific islands such as New Caledonia, Vanuatu, the Solomon Islands, Fiji, Samoa and Tonga, or Indonesia. The genera *Nestor* and *Cyanoramphus* (Psittacidae) and the species *Eudynamis taitensis* (Cuculidae) are endemic to the southwest Pacific.

Low diversity at family, generic and (with few exceptions) specific levels is also a characteristic of the New Zealand terrestrial bird fauna. While it is true that extinction has reduced taxonomic diversity (the families Dinornithidae, Emeidae, Pelagornithidae, Pelecanidae, Phasianidae, Aptornithidae, Aegothelidae, Turnagridae and Corvidae are no longer part of New Zealand's endemic fauna), it would seem that low diversity has always been a characteristic of New Zealand's terrestrial avifauna. Bird families that one might have expected to be part of New Zealand's pre-European avifauna are missing. For example, some self- or deliberately introduced members of the families Cracticidae (Australian Magpie), Hirundinidae (Welcome Swallow) and Turdidae (Blackbird and Song Thrush) have established themselves in New Zealand.

Within families there is a general paucity of genera and species. This point is illustrated by the family Columbidae which is represented in New Zealand by a single species *Hemiphaga novaeseelandiae*. The genus *Hemiphaga* is now confined to New Zealand, but an extinct subspecies, *H. novaeseelandiae spadicae*, was found on Norfolk Island (Schodde *et al.*, 1983). This can be contrasted with the Columbidae in New Caledonia, where there are six species in five genera, or in Australia where there are 23 species in eleven genera. The exceptions to this pattern are the Anatidae, Rallidae and Phalacrocoracidae. This latter family reaches its greatest specific diversity in New Zealand waters. Cormorants have been included in this study because four species (*Phalacrocorax carbo, P. sulcirostris, P. melanoleucos,* and *P. varius*) are predominantly estuarine or freshwater, but other members of the family are marine or oceanic species.

New Caledonia

The terrestrial avifauna of New Caledonia (Fig.1: 2) and the Loyalty Islands is listed in Table 2. The data for Table 2 are from Mayr (1945). Table 2 shows that the number of avian taxa endemic in New Caledonia is less than in New Zealand. There are five endemic genera (Keast (1996) records eight), 21 endemic species and a single

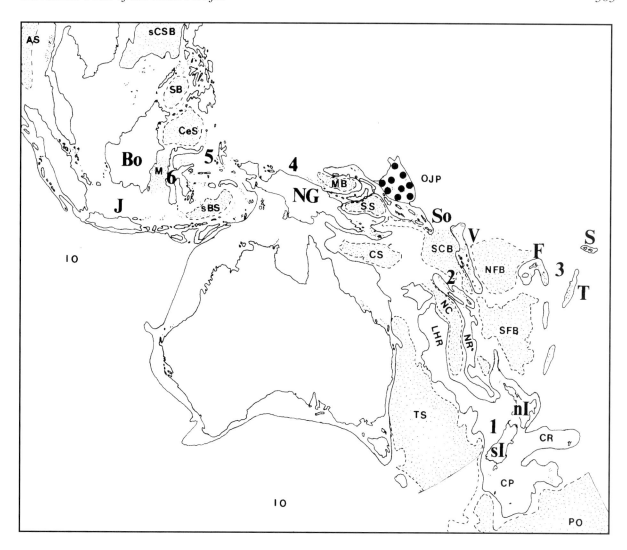

Fig. 1. The Indo-Pacific region showing extent of continental crust and marginal oceanic basins (stippled). Avifaunas discussed in text: 1 = New Zealand, 2 = New Caledonia, 3 = Fiji, Tonga and Samoa, 4 = New Guinea, 5 = Maluku, 6 = Sulawesi. Marginal basins: AS = Andaman Sea, sBS = south Banda Sea, CS = Coral Sea, CeS = Celebes Sea, sCSB = South China Sea, M = Makassar Strait, MB = Manus Basin, NC = New Caledonian Basin, NFB = North Fiji Basin, PO = SW Pacific, SB = Sulu Basin, SCB = Santa Cruz and Loyalty Basins, SFB = south Fiji Basin, SS = Solomon Sea, TS = Tasman Sea. CP = Campbell Plateau, CR = Chatham Rise, LHR = Lord Howe Rise, NR = Norfolk Ridge, OJP = Ontong Java Plateau, IO = Indian Ocean. Bo = Borneo, F = Fiji, nI = North Island, New Zealand, sI = South Island, New Zealand, J = Java, NG = New Guinea, S = Samoa, So = Solomons, T = Tonga, V = Vanuatu.

endemic family — the Rhynochetidae — in New Caledonia. The number of endemic taxa has been greater in the past. Balouet and Olson (1989) describe a large, flightless species *Sylviornis neocaledonica* (Megapodiidae?) from Holocene deposits, together with eleven extinct non-passerine species that have relatives on smaller islands off New Zealand, northern Melanesia and Asia. Two extinct birds of prey have been recorded by Thiollay (1993). New Zealand and New Caledonia have few avian families, which indicates a long isolation from the Australian avifauna.

Within the non-endemic element of New Caledonia's avifauna, distributional patterns are similar to those discussed for New Zealand. Only three species are shared exclusively with Australia despite New Caledonia's proximity to the Australian continent. Thirty three species (44% of the total avifauna) have widespread distributions, listed in Table 2 as Australia-southwest Pacific-Indonesia. Five of these species are absent from Australia and ten are present in New Zealand. There is also a pronounced south-

Table 1. Terrestrial birds of New Zealand. Bold entries are extinct taxa. Note on flycatchers and warblers: in Table 1 and following, *Rhipidura* is included with other monarch flycatchers in the Monarchidae and *Gerygone* with other Australian wren-warblers in the Acanthizidae. Sibley and Ahlquist (1990) include the former in the expanded family Corvidae and the latter in the family Pardalotidae.

Family	Endemic	Shared with Australia	Shared with Australia +SW Pacific/ New Guinea/Indonesia	SW Pacific
Emeidae	***Anomalopteryx***			
	Megalopteryx			
	Pachyornis			
	Emeus			
	Euryapteryx			
Dinornithidae	***Dinornis***			
Apterygidae	*Apteryx*			
Acanthisittidae	*Acanthisitta chloris*			
	Xenicus longipes			
	Xenicus gilviventris			
	Traversia lyalli			
	Pachyplichas yaldwyni			
	Pachyplichas jagmi			
Callaeatidae	*Callaeas cinerea*			
	Philesturnus carunculatus			
	Heteralocha acutirostris			
Podicipedidae	*Poliocephalus rufopectus*	*Podiceps cristatus*		
Pelagornithidae	***Pelagornis miocaenus***			
	Pseudodontornis stirtoni			
Pelecanidae	***Pelecanus novaezealandiae***			
Phalacrocoracidae	*Leucocarbo carunculatus*	*Phalacrocorax varius*	*Phalacrocorax carbo*	
	Leucocarbo chalconotus		*Phalacrocorax sulcirostris*	
	Leucocarbo onslowi		*Phalacrocorax melanoleucos*	
	Leucocarbo ranfurlyi			
	Leucocarbo colensoi			
	Leucocarbo campbelli			
	Leucocarbo atriceps			
	Stictocarbo punctatus			
	Stictocarbo punctatus steadi			
	Stictocarbo featherstoni			
Ardeidae	*Ixobrychus novaezelandiae*	*Egretta alba*	*Egretta sacra*	
			Botaurus poiciloptilus	
Threskiornithidae			*Platalea regia*	
Anatidae	***Pachyanas chathamica***	***Oxyura australis***	*Anas gibberifrons*	
	Euryanas finschi	*Anas rhynchotis*	*Anas superciliosa*	
	Cnemiornis gracilis	*Anas gracilis*		
	Hymenolaimus malacorhynchos			
	Tadorna variegata			
	Anas chlorotis			
	Anas aucklandica			
	Aythya novaeseelandiae			
	Biziura delautouri			
	Malacorhynchus scarletti			
	Cygnus sumnerensis			
	Mergus australis			
Accipitridae	***Circus eylesi***		*Circus approximans*	
	Harpagornis moorei			
	Haliaeetus australis			
Falconidae	*Falco novaeseelandiae*			
Phasianidae	***Coturnix novaezealandiae***			
Rallidae	***Fulica chathamensis***		*Porzana tabuensis*	
	Fulica prisca		*Porzana pusilla*	
	Gallinula hodgeni		*Porphyrio porphyrio*	
	Capellirallus karamu		*Gallirallus philippensis*	
	Diaphorapteryx hawkingi		*Rallus pectoralis*	
	Gallirallus australis			
	Rallus modestus			
	Porphyrio mantelli			

Table 1. Continued.

Family	Endemic	Shared with Australia	Shared with Australia +SW Pacific/ New Guinea/Indonesia	SW Pacific
Aptornithidae	**Aptornis otidiformis**			
	Aptornis defossor			
Charadriidae	Charadrius obscurus			
	Charadrius bicinctus			
	Anarhynchus frontalis			
	Thinornis novaeseelandiae			
Scolopacidae	Coenocorypha aucklandica			
	Coenocorypha pusilla			
	Coenocorypha chathamica			
Haematopodidae	Haematopus unicolor	Haematopus ostralegus		
	Haematopus chathamensis			
Columbidae	Hemiphaga novaeseelandiae			
Psittacidae	Strigops habroptilus			Nestor
	Nestor meridionalis			Cyanoramphus
	Nestor notabilis			novaezelandiae
	Cyanoramphus unicolor			
	Cyanoramphus auriceps			
Cuculidae			Chrysococcyx lucidus	Eudynamis taitensis
Strigidae	**Sceloglaux albifacies**		Ninox novaeseelandiae	
Aegothelidae	**Megaegotheles novaezealandiae**			
Alcedinidae			Halcyon sancta	
Eopsaltriidae	Petroica macrocephala			
	Petroica australis			
	Petroica traversi			
Monarchidae			Rhipidura fuliginosa	
Sylviidae	Bowdleria punctata			
	Bowdleria rufescens			
Pachycephalidae	Mohoua albicilla			
	Mohoua ochrocephala			
	Mohoua novaeseelandiae			
Acanthizidae	Gerygone igata			
	Gerygone albofronta			
Motacillidae			Anthus novaeseelandiae	
Meliphagidae	Notiomystis cincta			
	Anthornis melanura			
	Prosthemadera novaeseelandiae			
Zosteropidae			Zosterops lateralis	
Turnagridae	**Turnagra capensis**			
Corvidae	**Palaeocorax moriorum**			

west Pacific influence in the New Caledonian avifauna, with four genera, eighteen species and two subspecies linking New Caledonia exclusively to other southwest Pacific islands.

The avifauna of Norfolk Island, which lies between New Zealand and New Caledonia, has a number of southwest Pacific genera in common with New Caledonia. It has links to New Zealand with the occurrence of the genera *Hemiphaga* (Columbidae) and *Nestor* (Psittacidae). The majority of species found on Norfolk Island are widespread. As one would expect for such a small island the endemic element is small, Schodde *et al.* (1983) listing only five endemic species (including two *Zosterops* spp.) and nine endemic subspecies.

Fiji, Tonga and Samoa

The avifaunas of Fiji, Tonga and Samoa (Fig.1: 3) are listed in Table 3. The data are from Mayr (1945) and Watling (1982). The terrestrial avifauna of these central Polynesian islands, like that of New Zealand and New Caledonia, is depauperate in avian families. While species endemism is high, there are only six endemic genera. These are the Collared Lory (*Phigys solitarius*), Red-breasted Musk Parrot (*Prosopeia tabuensis*) and Kadavu Honeyeater (*Xanthotis provocator*) of Fiji and the Tooth-billed Pigeon (*Didunculus strigirostris*) of Samoa. An extinct species, *Didunculus* sp. nov. has been reported from 'Eua, Tonga by Steadman (1995). The Silktail (*Lam-*

Table 2. Terrestrial birds of New Caledonia and the Loyalty Islands.

Family	Endemic	Shared with Australia	Shared with Australia +SW Pacific/ New Guinea/Indonesia	SW Pacific
Podicipedidae			*Podiceps novaehollandiae*	
Phalacrocoracidae			*Phalacrocorax melanoleucus*	
Ardeidae		*Botaurus poiciloptilus*	*Ardea novaehollandiae*	
			Butorides striatus	
			Egretta sacra	
			Nycticorax caledonicus	
Anatidae		*Anas rhynchotis*	*Dendrocygna arcuata*	
			Anas superciliosa	
			Anas gibberifrons	
			Aythya australis	
Accipitridae	*Accipiter haplochrous*		*Haliaster sphenurus*	
			Accipiter fasciatus	
			Circus approximans	
			Pandion haliaetus	
Falconidae			*Falco peregrinus*	
Turnicidae		*Turnix varia*		
Rallidae	*Tricholimnas lafresnayanus*		*Gallirallus philippensis*	
			Porzana tabuensis	
			Poliolimnas cinereus	
			Porphyrio porphyrio	
Rhynochetidae	*Rhynochetos jubatus*			
Columbidae	*Drepanoptila holosericea*		*Chalcophaps indica*	*Columba vitiensis*
	Ducula goliath			*Ducula pacifica*
				Ptilinopus greyii
Psittacidae	*Vini diadema*			*Trichoglossus haematodus*
	Eunymphicus cornutus			*Cyanorhamphus novaezelandiae*
Cuculidae			*Cacomantis pyrrhophanus*	*Eudynamis taitensis*
			Chrysococcyx lucidus	
Tytonidae			*Tyto alba*	
			Tyto longimembris	
Aegothelidae	*Aegotheles savesi*			
Caprimulgidae			*Eurostopodus mystacalis*	
Apodidae			*Collocalia spodiopyga*	*C. s. leucopygia*
			Collocalia esculenta	*C. e. uropygalis*
Alcedinidae			*Halcyon sancta*	
Hirundinidae				*Hirundo tahitica*
Campephagidae	*Coracina analis*			*Coracina caledonica*
				Lalage leucopyga
Turdidae				*Turdus poliocephalus*
Sylviidae	*Megalurulus mariei*			
Acanthizidae				*Gerygone flavolateralis*
Eopsaltriidae	*Eopsaltria flaviventris*			
Monarchidae				*Rhipidura spilodera*
				Myiagra caledonica
				Clytorhynchus pachycephalodes
Pachycephalidae	*Pachycephala caledonica*		*Pachycephala pectoralis*	*Pachycephala rufiventris*
Artamidae			*Artamus leucorhynchus*	
Sturnidae	*Aplonis striatus*			
Corvidae	*Corvus moneduloides*			
Meliphagidae	*Phylidonyris undulata*		*Myzomela sanguinolenta*	*Myzomela cardinalis*
	Gymnomyza aubryana			*Lichmera incana*
	Philemon diemenensis			*Gymnomyza*
Zosteropidae	*Zosterops minuta*		*Zosterops lateralis*	
	Zosterops xanthochroa			
	Zosterops inorta			
Estrildidae	*Erythrura psittacea*		*Erythrura trichroa*	

Table 3. Terrestrial birds of Fiji, Tonga and Samoa. (S) = Samoan endemics, (T) = Tongan endemics.

Family	Endemic	Shared with Australia +SW Pacific/ New Guinea/Indonesia	Fiji/Tonga/Samoa	SW Pacific
Ardeidae		*Egretta sacra*		
		Butorides striatus		
Anatidae		*Dendrocygna arcuata*		
		Anas superciliosa		
Accipitridae	*Accipiter rufitorques*	*Circus approximans*		
Falconidae		*Falco peregrinus*		
Megapodiidae	*Megapodius pritchardii* (T)			
Rallidae	*Nesoclopeus poecilopterus*	*Gallirallus philippensis*		*Nesoclopeus*
	Pareudiastes pacificus (S)	*Porzana tabuensis*		*Pareudiastes*
		Poliolimnas cinereus		
		Porphyrio porphyrio		
Columbidae	*Ptilinopus luteovirens*		*Ptilinopus perousii*	*Ptilinopus porphyraceus*
	Ptilinopus layardi		*Gallicolumba stairii*	*Columba vitiensis*
	Ptilinopus victor			*Ducula pacifica*
	Ducula latrans			
	Didunculus strigirostris (S)			
Psittacidae	*Charmosyna amabilis*			*Vini australis*
	Phigys solitarius			*Prosopeia tabuensis*
	Prosopeia personata			
Cuculidae		*Cacomantis pyrrhophanus*		*Eudynamis taitensis*
Tytonidae		*Tyto alba*		
		Tyto longimembris		
Apodidae		*Collocalia spodiopygia*		
Alcedinidae	*Halcyon recurvirostris* (S)	*Halcyon chloris*		
Hirundinidae		*Hirundo tahitica*		
Campephagidae	*Lalage sharpei* (S)			*Lalage maculosa*
Turdidae				*Turdus poliocephalus*
Sylviidae	*Vitia ruficapilla*			*Vitia*
	Trichocichla rufa			
Eopsaltriidae		*Petroica multicolor*		
Monarchidae	*Rhipidura personata*		*Clytorhynchus vitiensis*	*Rhipidura spilodera*
	Rhipidura nebulosa (S)			*Mayrornis*
	Mayrornis lessoni			*Clytorhynchus*
	Mayrornis versicolor			
	Clytorhynchus nigrogularis			
	Myiagra albiventris (S)			
	Myiagra azureocapilla			
	Lamprolia victoriae			
Pachycephalidae	*Pachycephala flavifrons* (S)	*Pachycephala pectoralis*		
Artamidae		*Artamus leucorhynchus*		
Sturnidae	*Aplonis atrifuscus* (S)			*Aplonis tabuensis*
Meliphagidae	*Gymnomyza samoensis* (S)		*Foulehaio carunculata*	*Myzomela cardinalis*
	Gymnomyza viridis			
	Xanthotis provocator			
	Myzomela jugularis			
Zosteropidae	*Zosterops samoensis* (S)	*Zosterops lateralis*		
	Zosterops explorator			
Estrildidae	*Erythrura kleinschmidti*		*Erythrura cyanovirens*	

prolia victoriae) of Fiji was placed by Mayr (1945) in 'uncertain family', although Sibley and Ahlquist (1990) regarded it as a monarch flycatcher on DNA evidence. The genus *Foulehaio* (Meliphagidae) is endemic to Fiji and Samoa and was formerly found in Tonga (Steadman, 1995). Both the pigeon (Columbidae) and flycatcher (Monarchidae) families are diverse in species, a characteristic of the New Guinean avifauna discussed below. The occurrence of megapodes (Megapodiidae: one endemic species in Tonga) is also a characteristic of tropical Indo-Pacific islands.

No species is shared exclusively with Australia, but a number of widespread species (Australia-southwest Pacific-New Guinea-Indonesia)

Table 4. Terrestrial birds of New Guinea that are either not found or are poorly represented in Australia. X = presence; MP = Malay peninsula; W.Is. = western islands of New Guinea; CY = one species restricted to Cape York; numbers refer to the number of species found in Australia; ? = unknown.

Family	Species	Africa/Eurasia	India/S China	SEA	Andamans/Nicobars	Greater Sundas	Sulawesi	Philippines	Lesser Sundas	S Maluku	N Maluku	N, C & SE New Guinea	S New Guinea	Bismarck Archipelago	Solomons	Vanuatu	Pacific	Micronesia	New Caledonia	New Zealand	Australia
Casuariidae	*Casuarius bennetti*											X	X								
Podicipedidae	*Tachybaptus ruficollis*	X	X	X		X	X	X	X	X	X		X	X							
Anatidae	*Dendrocygna guttata*						X	X		X	X	X	X	X							
Accipitridae	*Macheiramphus alcinus*	X		X		X	X					X	X								
	Circus spilonotus	X	X	X		X		X				X	X								
	Accipiter meyerianus											X	X	X	X	X					
Falconidae	*Falco severus*		X	X		X	X	X		X	X	X	X	X	?						
Megapodiidae	*Megapodius*				X	X	X	X	X	X	X	X		X	X	X	X				1
Rallidae	*Gymnocrex plumbeiventris*											X	X	X							
	Gallirallus torquatus					X	X		Sula			X									
Columbidae	*Columba vitiensis*					X	X	X	X	X	X	X	X	X	X	X			X		
	Macropygia		X	X	X	X	X		X	X	X	X	X	X	X						1
	M. nigrirostris											X	X	X							
	M. mackinlayi											X		X	X	X					
	Reinwardtoena										X	X	X	X	X	X					
	R. reinwardtii										X	X	X								
	Chalcophaps stephani						X			X		X		X	X						
	Caloenas nicobarica			X	X	X	X	X	X	X	X			X		X					
	Gallicolumba						X	X			X	X		X		X	X				
	G. beccarii											X	X	X	X						
	G. jobiensis											X	X	X	X						
	Ptilinopus			MP		X	X	X	X	X	X	X	X	X	X	X	X	X	X		4
	P. wallacii									X		X	X								
	P. rivoli									X	X	X		X							
	P. solomonensis											X		X	X						
	P. viridis									X	X	X		X							
	Ducula		X	X	X	X	X	X	X	X	X	X	X	X	X	X	X				1
	D. concinna						X			X		X									
	D. pacifica											X				X					
	D. myristicivora										X	X									
	D. rufigaster											X									
	D. pistrinaria											X		X	X						
	Gymnophaps											X	X	X	X	X					
Psittacidae	*Chalcopsitta*											X	X	X							
	Eos											X	X	X							
	Trichoglossus					Bali	X	X	X	X	X	X	X	X	X	X		X			3
	Lorius											X	X	X	X	X					
	L. hypoinochrous											X		X							
	Charmosyna									X		X	X	X	X	X	X		X		
	C. rubrigularis											X		X							
	C. placentis									X	X	X	X	X							
	Micropsitta											X		X	X	X					
	Tanygnathus					X	X	?		X	X	W.Is									
	T. megalorynchos						X			X	X	W.Is									
	Alisterus amboinensis						X			X	X	X									
	Loriculus		X	X	X	X	X	X		X	X	X									
	L. aurantiifrons											X	X	X							
	Eclectus roratus								X	X	X	X		X	X						CY
	Geoffroyus								X	X	X	X		X	X						CY
Cuculidae	*Centropus*	X	X	X	X	X	X	X	?	X	X	X	X	X							1
Strigidae	*Otus*	X	X	X	X	X	?	X	X	X	X										
	Otus magicus						X			X	X	X	X								

Table 4. Continued.

		Africa/Eurasia	India/S China	SEA	Andamans/Nicobars	Greater Sundas	Sulawesi	Philippines	Lesser Sundas	S Maluku	N Maluku	N, C & SE New Guinea	S New Guinea	Bismarck Archipelago	Solomons	Vanuatu	Pacific	Micronesia	New Caledonia	New Zealand	Australia
Hemiprocnidae	*Hemiprocne*			X	X	X	X	X	X	?	X	X	X	X	X	X					
	Hemiprocne mystacea										X	X	X	X	X	X					
Apodidae	*Collocalia*			X	X	X	X	X	X	?	X	X	X	X	X	X	X	X	X	X	1
	C. esculenta				X	X	X			X	X	X	X	X	X	X	X		X		
	Aerodramus vanikorensis						X				X	X	X	X	?	X	X		X	X	CY
	A. whiteheadi							X				X		X		X					
Alcedinidae	*Tanysiptera*											X	X	X	X	X					CY
	T. galatea											X	X	X	X						
	Halcyon saurophaga											X	X		X	X					
	Ceyx lepidus							X				X	X	X	X	X					
	Alcedo	X	X	X	X	X	X	X	X	X	X	X	X	X	X	X					
	Alcedo atthis	X	X	X	X			X				X	X	X		X	X				
	M. philippinus			X	X	X	X	X	X	X	X		X	X	X						
Coraciidae	*Eurystomus orientalis*			X	X	X	X	?	X	?		X	X	X	X	X					X
Bucerotidae	*Rhyticeros*			X	X	X	X	X	X	X	X	X	X	X	X	X					
	R. plicatus										X	X	X	X	X	X					
Pittidae	*Pitta erythrogaster*							X	X		X	X	X								?
	P. sordida			X	X	X	X	X				X	X								
Hirundinidae	*Hirundo tahitica*			X	X	X	X		X	X	X	X	X	X	X	X	X	X		X	
Campephagidae	*Coracina morio*						X	X				X									
	Lalage atrovirens								X			X									
Laniidae	*Lanius*	X	X	X		X		X				X									
	L. schach		X	X		X		X				X									
Turdidae	*Saxicola*	X	X	X		X	X	X	X			X		X							
	S. caprata	X	X	X		X	X	X	X			X		X							
	Turdus	X	X	X		X	X	X	X			X		X	X	X	X		X		
	T. poliocephalus					X	X	X				X		X	X	X	X		X		
Sylviidae	*Phylloscopus*	X	X	X	X		X					X	X	X		X	X				
	P. trivirgatus			MP		X		X	X	X	X		X		X	X					
Monarchidae	*Monarcha cinerascens*						X		X	X	X	X		X	X						
	Rhipidura			X	X		X	X	X		X	X	X	X	X	X	X	X	X	X	3
Zosteropidae		X	X	X	X	X	X	X	X	X	X	X	X	X	X	X	X	X	X	X	4
Nectariniidae	*Nectarinia sericea*						X			X	X	X	X	X							
Sturnidae	*Aplonis*			X	X	X	X	X	X	X	X	X	X	X	X	X	X	X	X	X	CY
	A. cantoroides											X	X	X	X						
	A. mysolensis								X		X	X	X								
	Mino dumontii											X	X	X	X						
Oriolidae	*Oriolus*	X	X	X	X	X			X	X	X	X	X	X							2
Dicruridae	*Chaetorhynchus*											X									
	Dicrurus	X	X	X	X	X	X	X	?		X	X	X	X	X						1

form an important element of the central Polynesian avifauna. A southwest Pacific element forms a second important grouping. Seven species and one genus are endemic to Fiji + Tonga + Samoa, while nine species and five genera are endemic to the southwest Pacific. Niue, which lies to the east of Tonga, shares six of these southwest Pacific endemics (Wodzicki, 1971).

New Guinea

Bird life in New Guinea (Fig.1: 4) is varied with 708 species officially recorded (Beehler *et al.*, 1986). Besides birds of paradise (Paradisaeidae: 38 species) and bower birds (Ptilonorhynchidae: 11 species), hawks and eagles (Accipitridae: 25 species), parrots (Psittacidae: 46 species), pigeons (Columbidae: 45 species), king-

Table 5. Terrestrial birds of Maluku. S Maluku = Sula + Buru + Seram + Ambon + Kai + Timor; N Maluku = Morotai + Halmahera + Ternate + Bacan + Obi; * = S Maluku only; # = N Maluku only; (T) = Timor only;

Family	Endemic S Maluku	Endemic N Maluku	Endemic Maluku	S Maluku-S NG/Aust
Casuariidae				*Casuarius casuarius*
Podicipedidae				
Phalacrocoracidae				
Ardeidae			*Butorides striata moluccarum*	*Ardea novaehollandiae*
				Egretta intermedia
Threskiornithidae				*Threskiornis moluccus*
Anatidae				*Anas gibberifrons* (T)
				Nettapus pulchellus
Accipitridae		*Accipiter henicogrammus*	*Accipiter erythrauchen*	*Accipiter fasciatus*
Falconidae				
Megapodiidae			*Megapodius wallacei*	
Phasianidae				
Turnicidae				*Turnix maculosa*
Rallidae		*Habroptila wallacii*		*Gallinula tenebrosa*
Jacanidae				
Columbidae	*Ducula cineracea* (T)	*Ducula basilica*	*Ducula melanura*	*Ducula concinna*
	Gymnophaps mada	*Ptilinopus granulifrons*	*Ducula perspicillata*	*Ptilinopus regina*
	Treron psittacea	*Ptilinopus hyogaster*		*Ptilinopus wallacii*
	Turacoena modesta (T)	*Ptilinopus monarcha*		*Ptilinopus viridis*
	Gallicolumba hoedtii	*Ptilinopus bernsteinii*		
Psittacidae	*Charmosyna toxopei*	*Loriculus a. amabilis*		*Trichoglossus haematodus*
	Eos bornea	*Lorius garrulus*		*Micropsitta keiensis*
	Eos reticulata	*Cacatua alba*		
	Eos semilarvata			
	Lorius domicella			
	Cacatua moluccensis			
	Cacatua goffini			
	Prioniturus mada			
	Tanygnathus gramineus			
	Psitteuteles iris (T)			
	Aprosmictus jonquillaceus (T)			
Cuculidae	*Centropus spilopterus*	*Centropus goliath*		*Centropus phasianinus*
	Chrysococcyx crassirostris	*Cuculus heinrichi*		
Tytonidae	*Tyto sororcula*			*Tyto novaehollandiae*
Strigidae				*Ninox novaeseelandiae*
Aegothelidae		*Aegotheles crinifrons*		
Caprimulgidae				
Apodidae				
Hemiprocnidae				
Alcedinidae	*Halcyon lazuli*	*Halcyon funebris*		*Halcyon macleayii*
		Halcyon diops		
Meropidae				
Coraciidae		*Eurystomus azureus*		
Bucerotidae				
Pittidae	*Pitta versicolor elegans*	*Pitta maxima*		
Campephagidae	*Coracina ceramensis*	*Coracina parvula*	*Coracina atriceps*	
	Coracina fortis	*Lalage aurea*		
	Coracina dispar			

Terrestrial birds of the Indo-Pacific 371

West = Sundaland/Asia; East = New Guinea/Australia/Pacific; Widespread = found either side of Wallace's Line.

N Maluku-NG	West	East	Widespread
			Tachybaptus novaehollandiae#
			Tachybaptus ruficollis
			Phalacrocorax sulcirostris
			Phalacrocorax melanoleucos
	Ardea purpurea*	*Nycticorax caledonicus*	*Ardea sumatrana*
		Ixobrychus flavicollis australis	*Ardea alba modesta**
			Bubulcus ibis
			Egretta sacra
	Dendrocygna a. arcuata (T)	*Anas superciliosa*	*Dendrocygna guttata**
		Tadorna radjah	
Aquila gurneyi	*Ictinaetus malayensis*	*Accipiter meyerianus*	*Haliaster indus*
		Accipiter novaehollandiae	*Alanus caeruleus*
		Aviceda subcristata	
	Falco moluccensis		*Falco severus*
Megapodius freycinet		*Megapodus reinwardt*	
		Synoicus ypsilophorus (T)	
Gymnocrex plumbeiventris	*Amaurornis phoenicurus**	*Amaurornis olivacus*	*Gallirallus philippensis*
	Rallina fasciata	*Rallina tricolor**	*Porphyrio porphyrio**
			Irediparra gallinacea*
Ducula myristicivora	*Ducula rosacea*	*Ptilinopus superbus*	*Ducula bicolor*
Gymnophaps albertisii	*Ptilinopus cinctus*	*Ptilinopus rivoli*	*Caloenas nicobarica*
	Ptinopus melanospila*	*Reinwardtoena reinwardtsii*	*Chalcophaps indica*
	Streptopelia bitorquata (T)		*Macropygia amboinensis*
	Geopelia striata*		*Columba vitiensis*
	Macropygia ruficeps (T)		
	Macropygia magna*		
	Treron pompadora*		
Eos squamata		*Eclectus roratus*	*Tanygnathus megalorhynchos*
		Alisterus amboinensis	
		Micropsitta bruynii*	
		Charmosyna placentis	
		Cacatua galerita*	
		Geoffroyus geoffroyi	
	Centropus bengalensis	*Cuculus variolosus*	
	Eudynamis scolopacea	*Chrysococcyx crassirostris*	
		Scythrops novaehollandiae	
Ninox connivens	*Otus magicus*		*Tyto alba**
	Ninox squamipila		
	Caprimulgus affinis*		*Caprimulgus macrurus*
	Aerodramus fulciphagus (T)		*Collocalia esculenta*
	Aerodramus infuscatus		*Aerodramus vanikorensis*
		Hemiprocne mystacea	
Halcyon saurophaga	*Halcyon australasia**	*Ceyx lepidus*	*Halcyon chloris*
Alcedo azurea		*Tanysiptera galatea*	*Halcyon sancta*
		Alcedo pusilla	*Alcedo atthis*
			Merops superciliosus
			Eurystomas orientalis
		Rhyticeros plicatus	
			Pitta erythrogaster
		Coracina novaehollandiae*	*Coracina tenuirostris*
		Coracina papuensis	
		Lalage atrovirens*	
		Lalage leucomela	

Table 5. Continued.

Family	S Maluku	Endemic N Maluku	Maluku	S Maluku-S NG/Aust
Pycnonotidae			Ixos affinis	
Turdidae	Zoothera schistacea			
	Zoothera dumasi			
	Zoothera machiki			
	Saxicola gutturalis (T)			
Timalidae				
Sylviidae	Buettikoferela bivittata (T)			
	Urosphena subulata			
Acanthizidae	Gerygone inornata (T)			
Muscicapidae	Rhinomyias addita			
	Ficedula buruensis			
	Ficedula timorensis (T)			
	Cyornis hyacinthina (T)			
	Microecia hemixantha			
Monarchidae	Rhipidura superflua		Myiagra galeata	
	Rhipidura dedemi		Monarcha pileatus	
	Rhipidura opistherythra			
	Rhipidura fuscorufa			
	Monarcha leucurus			
	Monarcha mundus			
	Monarcha loricatus			
Pachycephalidae	Pacycephala orpheus (T)		Pachycephala griseonata	Pachycephala simplex
Dicaeidae	Dicaeum vulneratum		Dicaeum erythrothorax	Dicaeum hirundinaceum
Nectariniidae				
Zosteropidae	Madanga ruficollis	Zosterops atriceps		
	Lophozosterops pinaiae			
	Tephrozosterops stalkeri			
	Zosterops buruensis			
	Zosterops grayi			
	Zosterops uropygialis			
	Zosterops kuehni			
	Heleia muelleri (T)			
Meliphagidae	Lichmera monticola	Philemon fuscicapillus		
	Lichmera squamata	Melitograis gilolensis		
	Lichmera deningeri			
	Lichmera flavicans (T)			
	Lichmera notablis (T)			
	Myzomela blassii			
	Myzomela vulnerata (T)			
	Philemon moluccensis			
	Philemon subcorniculatus			
	Philemon inornatus (T)			
	Philemon citreogularis			
	Meliphaga reticulata (T)			
Estrildidae	Padda fuscata (T)			
	Erythrura tricolor			
Sturnidae	Basilornis corythaix			
	Aplonis crassa			
Oriolidae	Oriolus bouroensis	Oriolus phaeochromus		Oriolus flavocinctus
	Oriolus forsteni			
	Sphecotheres viridis			
Dicruridae				
Artamidae				
Paradisaeidae		Lycocorax pyrrhopterus		
		Semioptera wallacei		
Corvidae		Corvus validus		

N Maluku-NG	West	East	Widespread
	Brachypteryx leucophyrs (T)		*Turdus poliocephalus**
	Zoothera peronii (T)		
	Zoothera doherty (T)		
	Zoothera andromedae (T)		
	Pneopyga pusilla (T)		
	Cettia vulcania (T)	*Megalurus timoriensis**	*Acrocephalus stentoreus**
	Orthotomus cuculatus	*Phylloscopus poliocephala*	*Cisticola exilis**
	Bradypterus castaneus*		*Cisticola juncidis**
	Bradypterus seebohmi (T)		
	Phylloscopus presbytes (T)		
	Seicercus montis*		
	Eumyias panayensis		
	Ficedula dumetoria*		
	Ficedula hyperythra		
	Ficedula westermanni*		
		Rhipidura rufiventris*	
		Rhipidura leucophrys	
		Rhipidura rufifrons*	
		Myiagra ruficollis*	
		Monarcha cinerascens*	
		Monarcha trivirgatus	
		Piezorhynchus alecto	
		Pachycephala phaionotus	
		Pachycephala rufiventris	*Pachycephala pectoralis*
		Pacycephala leucogastra	
	Dicaeum agile (T)		
	Dicaeum maugi*		
	Dicaeum sanguinolentum (T)		
	Nectarinia solaris (T)		*Nectarinia jugularis*
	Zosterops montanus	*Zosterops atrifrons**	*Zosterops chloris*
		Lichmera argentauris	
		Lichmera indistincta (T)	
		Myzomela sanguinolenta	
		Myzomela obscura#	
		Philemon buceroides (T)	
	Lonchura quinticolor (T)		*Erythrura trichroa**
	Lonchura molucca		*Poephila guttata*
	Lonchura punctulata*		
	Amandava amandava		
		Aplonis metallica	
		Aplonis mysolensis	
	Dicrurus densus	*Dicrurus bracteatus*	
	Dicrurus hottentottus		
		Artamus cinereus*	*Artamus leucorhynchus*
	Corvus macrorhynchos (T)	*Corvus orru*	
	Corvus enca*		

fishers (Alcedinidae: 22 species), cuckoos (Cuculidae: 21 species), Australian warblers (Acanthizidae: 20 species), flowerpeckers (Dicaeidae: 10 species), whistlers (Pachycephalidae: 26 species), flycatchers (Monarchidae: 30 species) and honeyeaters (Meliphagidae: 65 species) are all highly diversified. In addition to the Ptilorhynchidae there are six other endemic New Guinea/Australian bird families (Casuariidae, Maluridae, Climacteridae, Grallinidae, Cracticidae and Orthonychidae) Notable absences from the New Guinean avifauna are Old World flycatchers (Muscicapidae), trogons (Trogonidae), barbets (Capitonidae), woodpeckers (Picidae), bulbuls (Pycnonotidae) and broadbills (Eurylaimidae).

Although New Guinea's avifauna shows strong similarity to that of Australia, particularly to northern and northeastern regions, there are New Guinean taxa that are not found in Australia or are poorly represented and often restricted to Cape York. Table 4 lists species from many bird families that are present in New Guinea and areas to the east and west, but absent from Australia. Many of these species are widespread. Examples of these taxa include a hornbill (Bucerotidae) the genera *Caloenas, Reinwardtoena, Gallicolumba* (Columbidae), *Chalcopsitta, Eos, Lorius, Charmosyna, Tanygnathus, Micropsitta, Loriculus* (Psittacidae), *Otus* (Strigidae), *Hemiprocne* (Hemiprocnidae), *Alcedo* (Alcedinidae), *Lanius* (Laniidae), *Saxicola, Turdus* (Turdidae) and *Phylloscopus* (Sylviidae). Other examples in which there is a single Australian species, often restricted to Cape York or the Queensland coast, include *Megapodius* (Megapodiidae), *Macropygia, Ducula* (Columbidae), *Geoffroyus* (Psittacidae), *Centropus* (Cuculidae), *Collocalia* (Apodidae), *Tanysiptera* (Alcedinidae), *Oriolus* (Oriolidae), *Merops* (Meropidae), *Aplonis* (Sturnidae), *Nectarina* (Nectariniidae) and *Dicrurus* (Dicruridae).

Maluku

The avifauna of Maluku (Fig.1: 5), listed in Table 5, is based on the distributional data in Bemmel (1948), Bemmel and Voous (1953), Peters (1934-86) and White and Bruce (1986). Distributional data were cross-referenced with MacKinnon and Phillipps (1993) and Beehler *et al.* (1986). Wallace (1880) recognised that the avifauna of the northern Maluku islands was distinct from that of the southern islands. In this paper Ternate, Halmahera, Morotai, and Bacan are classed as north Maluku. The island of Obi has an uncertain biogeographical status (Duffels and Boer, 1990), but for the present is regarded as part of north Maluku. Southern Maluku is composed of the islands of Buru, Seram, Ambon and Kai. White and Bruce (1986) argued for the inclusion of Tanimbar in southern Maluku and Michaux (1994) for the inclusion of Timor.

The data in Table 5 are arranged to demonstrate several distributional patterns. These patterns involve an endemic element (endemic), taxa distributed between south Maluku and continental Australia (S Maluku-S NG/Aust) and taxa that link north Maluku to New Guinea (N Maluku-NG). Many birds found in Maluku are widely distributed. Some of these are Asian or Sundaic (West in Table 5), some Australasian (East), while still others are found both east and west of Maluku (Widespread).

White and Bruce (1986) discussed endemism in Maluku and noted that while there are few endemic genera, species endemism is high. Table 5 lists 72 endemic species in south Maluku with three endemic, monotypic genera — *Tephrozosterops, Madanga* (Zosteropidae) and *Buettikoferela* (Sylviidae). In north Maluku there are 26 endemic species or subspecies and two endemic genera — *Lycocorax* and *Semioptera* (Paradisaeidae). Table 5 also lists eleven species endemic to Maluku as a whole.

North and south Maluku have different relationships with surrounding areas. Wallace (1880) noted that the fauna of north Maluku shows many similarities with New Guinea, while the fauna of south Maluku is closer to that of continental Australia (including southern New Guinea). White and Bruce (1986) confirmed that the avifauna of north Maluku has a New Guinean element, and listed 23 species they regarded as New Guinean (White and Bruce, 1986: 44). Apart from the nine species listed as N Maluku-NG in Table 5, a number of north Maluku endemics can also be classed as New Guinean. These include *Habroptila wallacii* (Rallidae), *Centropus goliath* (Cuculidae), *Halcyon diops* (Alcedinidae), as well as various taxa from the families Columbidae, Psittacidae, Meliphagidae and Paradisaeidae.

The south Maluku avifauna shows a relationship to cratonic areas of southern New Guinea and Australia (Table 5: S Maluku-S NG/Aust and entries marked * under East). The families Acanthizidae, Meliphagidae, Monarchidae and Aegothelidae and the genera *Gymnophaps, Ducula, Gallicolumba* (Columbidae), *Charmosyna, Eos* and *Lorius* (Psittacidae) illustrate this pattern. A second pattern in the south Maluku distributions links south Maluku to Sundaland or Asia (Table 5: West). Examples illustrating this pattern include the families Timaliidae, Sylviidae, Pycnonotidae and Muscicapidae and the genera *Treron* (Columbidae), *Brachypteryx, Saxicola, Zoothera* (Turdidae) and *Oriolus* (Oriolidae).

Sulawesi

A list of non-endemic birds of Sulawesi (Fig.1: 6) is provided in Table 6. The distributional data are from Walters (1980), White and Bruce (1986) and MacKinnon and Phillipps (1993). Thirty six of 140 species can be classed as Asian. Twenty five of these species extend from mainland Asia to Sulawesi, which is at the eastern extremity of their ranges. A further eleven species extend to the Sula Islands and/or the Lesser Sundas. Twenty five species are Australasian, with Sulawesi at the western extremity of their ranges. A further 26 species are endemic to Wallacea and the remaining 53 species cross Wallace's line. Some 60% of the species listed in Table 6 are also found in the Philippines. Other areas that share species with Sulawesi include the Greater Sunda islands (particularly Java, western Sumatra and north Borneo), the Lesser Sundas and south Maluku.

White and Bruce (1986) discussed the high degree of generic and specific endemism shown by the Sulawesian avifauna. Twelve of the endemic genera are monotypic: *Macrocephalon* (Megapodiidae), *Aramidopsis* (Rallidae), *Cryptophaps* (Columbidae), *Cittura* (Alcedinidae), *Meropogon* (Meropidae), *Cataponera, Heinrichia* (Turdidae), *Malia, Geomalia* (Timaliidae), *Coracornis, Hylocitrea* (Pachycephalidae), *Enodes* and *Scissirostrum* (Sturnidae). There are two species belonging to the endemic genus *Myza* (Meliphagidae). In addition, the genera *Prioniturus, Tanygnathus* (Psittacidae), *Streptocitta* and *Basilornis* (Sturnidae) are restricted to Sulawesi + Philippines + S Maluku, the genus *Turacoena* (Columbidae) to Sulawesi and Timor and the genus *Penelopides* (Bucerotidae) to Sulawesi and the Philippines.

The number of endemic species in Sulawesi is not known for certain. Stresemann (reported in White and Bruce, 1986) concluded that 84 out of 220 species were endemic (31%), but White and Bruce (1986) suggested that this figure was too high. Walters (1980) listed 80 Sulawesian endemics. The Sulawesian fauna is not only phylogenetically isolated, but also geographically isolated. The island of Borneo, separated from Sulawesi by the Makassar Strait, is only 105 km distant (reduced to 40 km during the Pleistocene), yet many Bornean species are absent. White and Bruce (1986) discussed the differences in avifauna between Borneo and Sulawesi. They provided the following list showing the reduction between Borneo and Sulawesi (species numbers in brackets):Phasianidae (12), Strigidae (8), Podargidae (6), Trogonidae (6), Bucerotidae (8), Capitonidae (9), Picidae (17), Eurylaimidae (8), Aegithinidae (6), Pycnonotidae (23), Timaliidae (36), and *Arachnothera* (7). The families Podargidae, Trogonidae, Capitonidae, Eurylaimidae, Aegithinidae and Pycnonotidae are absent from Sulawesi.

Geological histories and local avifaunas

The New Zealand subcontinent

Fig.1 is a simplified version of the Plate-Tectonic Map of the Circum-Pacific Region (Southwest Quadrant) by AAPG (1981). Fig.1 shows an extensive area of predominantly submerged continental crust off the east coast of Australia (Fig.1: LHR + NR + nI + sI + CR + CP). Kamp (1986) referred to this area as the New Zealand subcontinent. The New Zealand subcontinent is separated from Australia by the Tasman and Coral Seas (Fig.1: TS and CS), which are underlain by oceanic crust. Fig.1 shows that the New Zealand subcontinent has been disrupted in the north by the New Caledonian basin (Fig.1: NC). According to Wood and Uruski (1990) this basin represents a failed rift underlain by thinned continental crust.

A series of small basins, composed of oceanic crust, are shown in Fig.1 (SFB, NFB, SCB, SS and MB). A summary of basin ages is given in Fig.2. The ages given in Fig.2 are from Yan and Kroenke (1993). There are discrepancies in dates quoted in Yan and Kroenke (1993) and AAPG (1981). When dates differ those from AAPG (1981) are indicated by horizontal bars in Fig.2. A series of archipelagos and island groups are found on the Pacific side of these basins (Fig.1: T, S, F, V and So). Vanuatu (Fig.1: V), the Solomons (Fig.1: So) and Tonga (Fig.1: T) are island arcs situated above subduction zones. These archipelagos and island groups straddle the present boundary between the Pacific and Indo-Australian plates. The relationship between oceanic basins, the New Zealand subcon-

Table 6. Non-endemic terrestrial birds of Sulawesi. X = present; X* = present in S Maluku; [X] = not present in Malay peninsula.

Family	Species	Africa	Eurasia	India	S China	SEA	Greater Sundas	Philippines	Lesser Sundas	Maluku	New Guinea	Australia	Pacific	New Zealand
Phalacrocoracidae	Phalacrocorax melanoleucos						X				?	X	X	X
	P. sulcirostris						X		X	X	X	X		X
	Anhinga melanogaster			X		[X]	X	X						
Ardeidae	Hydranassa picata										?	X		
	Ardea purpurea	X	X	X	X	X	X	X		X*				
	A. novaehollandia								X	X	X	X	X	X
	A. speciosa					[X]	X	X						
	Egretta intermedia	X	X	X	X	X	X	X	X	X	X	X	X	
	Ixobrychus cinnamomeus		X	X	X		X	X	X	X*				
	Nycticorax caledonicus						X	X	X	X	X	X	X	X
Ciconidae	Ciconia episcopus	X		X		[X]	X	X						
	Mycteria cinerea					[X]	X							
Threskionithidae	Plegadis perigrinus						X	X						
	Platalea regia										X	X	?	X
Anatidae	Anas gibberifrons						X		X	X*	X	X	?	X
	A. superciliosa						X		X	X	X	X	X	X
	Dendrocygna guttata							X		X				
	D. arcuata						X	X		X*	X	X	X	
	Netapus pulchellus							X		X*	X	X		
Accipitridae	Aviceda jerdoni			X	X	[X]	X	X		Sula				
	Elanus caeruleus	X	X	X		X		X	X	X*	X			
	Macheiramphus alcinus	X					X	X			X			
	Ichthyophaga humilis			X			X	X		X*				
	Haliaeetus leucogaster			X	X		X	X	X	X				
	Butaster liventer				X	[X]	X			X*				
	Ictineatus malayensis			X	X	X	X			X				
	Hieraatus kienerii			X		[X]		X						
	Circus assimilis											X		
	Pernis celebensis							X						
Falconidae	Falco moluccensis						X		X	X				
	F. severus			X			X	X		X	X			
Megapodiidae	Megapodius cumingii						X	X						
Turnicidae	Turnix suscitator			X	X	X	X	X	X					
Rallidae	Gallinula chloropus			X		X	X	X						
	G. tenebrosa						X			X*	X	X		
	Gallirallus striatus			X	X	X	X	X						
	G. torquatus								X	Sula	X			
	Rallina eurizonoides			X	X	[X]	X	X		Sula				
	Porzana fusca			X	X	X	X	X	X					
	P. cinerea						X	X	X	X	X	X	X	
	Gallicrex cinerea			X	X	X	X	X						
	Amaurornis phoenicurus			X	X	X	X	X						
	Fulica atra		X	X	X	[X]	X							
	Porphyrio porphyrio	X		X			X		X	X	X	X	X	X
Jacanidae	Irediparra gallinacea						X	X	X	X*	X	X		
Charadriidae	Charadrius peronii					X	X	X	X					
Burhinidae	Burhinus giganteus					[X]	X	X		X	X	X		
Columbidae	Treron griseicauda						X			Sula				
	T. vernans					X	X	X	X					
	Ptilinopus superbus									X	X	X		
	P. melanospila						X	X	X	X*				
	Ducula aenea			X		X	X	X	X	Sula				
	D. pickeringi						X	X						
	D. luctuosa									Sula				
	D. bicolor			X			X			X	X	X		
	Chalcophaps stephani									X*	X			
	C. indica			X		X	X	X		X	X	X	X	
	Macropygia amboinensis									X	X	X		
	M. magna									X*	X			
	Columba vitiensis						X	X		X	X		X	
Psittacidae	Cacatua sulphurea								X					
	Tanygnathus megalorynchos							X		X*	X			
	T. sumatranus						X	X						
Cuculidae	Scythrops novaehollandiae									X*	X	X		
	Clamator coromandus			X		[X]	X	X						
	C. russatus						X	X	X		X	X		
	Surniculus lugubris			X	X		X	X		X#				
	Centropus bengalensis			X	X	X	X	X	X	X				
	Cuculus sepulcralis					[X]	X	X	X	X*				
	C. merulinus			X	X		X	X						

Table 6. Continued.

Family	Species	Africa	Eurasia	India	S China	SEA	Greater Sundas	Philippines	Lesser Sundas	Maluku	New Guinea	Australia	Pacific	New Zealand
Strigidae	*Ninox scutulata*			X		X	X	X	X	X*				
	Tyto capensis	X	X	X	X	X	X		X		X	X	X	
Caprimulgidae	*Eurostopodus macrotis*			X	X	X	X	X						
	Caprimulgus affinis			X		[X]	X	X	X	X*				
	C. macrurus			X		X	X	X	X	X	X	X		
Apodidae	*Collocalia salangana*						X							
	C. esculenta			X			X	X		X	X		X	
	Aerodramus vanikorensis									X	X		X	
	Hirundapus giganteus			X		X	X	X						
	H. celebensis							X						
Hemiprocnidae	*Hemiprocne longipennis*					X	X	X		Sula				
Alcedinidae	*Alcedo meninting*			X	X	X	X	X		Sula				
	A. atthis	X	X	X	X	X	X	X	X	X	X	X		
	H. coromanda			X	X	X	X	X		Sula				
Meropidae	*Merops philippinus*				?	[X]		X			X			
Coraciidae	*Eurystomus orientalis*				X	X	X	X	X	X	X		X	
Pittidae	*Pitta erythrogaster*						X	X		X	X	X		
	P. sordida			X	X		X	X			X			
Campephagidae	*Lalage sueurii*							X	X		X	X		
	L. nigra					X	X	X						
	Coracina tenuirostris								X	X	X	X	X	
	C. morio							X		X*	X			
Dicruridae	*Dicrurus bracteatus*							X	X	X	X			
Oriolidae	*Oriolus chinensis*			X	X	X	X	X	X	Sula				
Corvidae	*Corvus enca*					X	X	X		X*				
	C. macrorhynchos		X	X	X	X	X	X	X	X				
Turdidae	*Saxicola caprata*		X	X	X	[X]	X	X	X	X	X			
	Turdus poliocephalus						X	X		X*	X		X	
	T. obscurus								X					
Sylviidae	*Cisticola juncidis*	X	X	X	X	X	X	X	X	X	X	X		
	C. exilis			X	X	[X]	X	X	X	X	X	X		
	Culicicapa helianthea							X						
	Bradypterus castaneus									X*				
	Orthotomus sepium						X		X					
	O. cuculatus			X	X	X	X		X	X*				
	Megalurus timorensis							X		X*	X	X		
Acanthizidae	*Gerygone sulphurea*					[X]	X	X	X					
Monarchidae	*Culicicapa helianthea*							X						
	Hypothymis azurea			X	X	X	X	X	X					
Muscicapidae	*Ficedula hyperythra*			X	X	X	X	X	X	X				
	F. westermanni			X	X	X	X	X	X	X*				
	Cyornis rufigastra					X	X	X						
	Eumyias panayensis							X		X*				
	Muscicapa basilanica							X						
	M. griseisticta				X	X	X	X		X	X			
Motacillidae	*Anthus novaeseelandiae*	X		X	X	X	X	X			X	X		X
Pachycephalidae	*Colluricincla megarhyncha*										X	X		
Sturnidae	*Aplonis minor*						X	X	X					
	A. mysolensis									X	X			
	A. panayensis			X		X	X	X						
	Acridotheres javanicus			?	X	[X]	X							
	A. fuscus			X		X	X							
Nectariniidae	*Anthreptes malacensis*					X	X	X		Sula				
	Nectarinia sperata			X		X	X	X						
	N. jugularis					X			X	X	X		X	
	N. aspasia									X*	X			
	Aethopyga siparaja			X	X	X	X	X						
Meliphagidae	*Myzomela sanguinolenta*									X*	X		X	X
Zosteropidae	*Zosterops montanus*						X	X	X	X				
	Z. atrifrons									X*	X			
	Z. chloris							X		X	X*	X		
Ploceidae	*Erythrura trichroa*									X	X	X	X	
	E. hyperythra					X	X	X	X					
	Lonchura molucca								X	X				
	L. punctulata	X		X	X	X	X	X	X					
	L. malacca			X	X	X	X	X		X#				
	L. pallida								X					
Fringillidae	*Serinus estherae*						X	X						

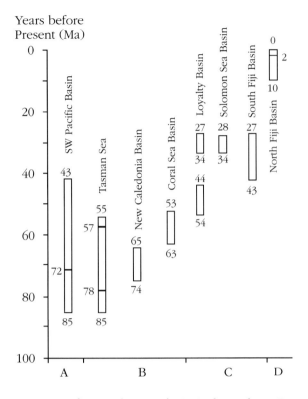

Fig. 2. Age of marginal oceanic basins in the southwest Pacific. A = formation of South Island, New Zealand-Campbell Plateau-Chatham Rise block. B = formation of the Melanesian rift system. C = fragmentation of the Melanesian rift into a rift and arc system. D = fragmentation of the Melanesian arc, Fiji moves away from Vanuatu.

tinent and these outer islands is critical in the interpretation of Pacific biogeography.

The oldest basins are the Tasman Sea and southwest Pacific basin (Fig.1: TS and PO and Fig.2). There are many published reconstructions of the New Zealand subcontinent prior to their opening (Kamp, 1986). While all models show the New Zealand subcontinent adjacent to the Australian/Antarctic section of east Gondwana at 100 Ma, there are differences in the details of various reconstructions. Some reconstructions place the Chatham Rise and Campbell Plateau (Fig.1: CR and CP) adjacent to either the North Island, New Zealand (Fig 1: nI), or to both North and South Islands (Fig.1: nI and sI) (Borg and Depado, 1991; Kamp, 1986). In these reconstructions the Chatham Rise and Campbell blocks have always been adjacent to the rest of the New Zealand subcontinent.

In other reconstructions (Yan and Kroenke, 1993) the Chatham Rise-Campbell Plateau-South Island block (Fig.1: CR + CP + sI) was derived from Marie Byrd Land (Antarctica). In the reconstruction of Yan and Kroenke (1993) this block started separating from Antarctica at 85 Ma (oldest magnetic anomaly 72 Ma) and moved north paralleling Pacific plate movement. It united with the rest of the New Zealand subcontinent (Fig.1: LHR + NR + nI) in the Oligocene (40 Ma). Where the intervening ocean between this block and the rest of the New Zealand subcontinent was subducted, what (if any) movement occurred along the Campbell fault, and how the collision between the two blocks relates to the on-shore geology of New Zealand are not clear from Yan and Kroenke's (1993) account.

According to Yan and Kroenke (1993) the Lord Howe Rise-Challenger Plateau-Norfolk Ridge-North Island, New Zealand block separated from the Australian margin about 85 Ma (oldest magnetic anomaly 78 Ma) and moved away in a northeasterly direction. This implies that the boundary between the Indo-Australian and Pacific plates at the time was a west-dipping subduction zone east of the New Caledonian Ridge. The Lord Howe Rise and New Caledonian basin quickly submerged to abyssal depths following thermal relaxation (Wood and Uruski, 1990). New Caledonia and parts of New Zealand are emergent segments of the New Caledonian Rise today. It is probable, given the tectonic activity described above, that parts of this ridge have been emergent since the beginning of the Tertiary. The New Caledonian Ridge is identified as the Melanesian rift in this paper. The Melanesian rift is equivalent to the inner Melanesian arc of others, but this name is inappropriate (Polhemus, 1996) because the Norfolk Ridge may not be the remnant of an island arc, but a rifted continental block.

The Solomons, Vanuatu, Fiji, Tonga and Samoa form the Melanesian arc. The history of this arc system is complicated and poorly resolved. The arc may represent a single structure that has been disrupted (Burrett *et al.*, 1991), a multiple arc system presently juxtaposed by tectonic dynamics (Polhemus, 1996), or an arc system with embedded microcontinents in Fiji. With our present state of knowledge such detail is secondary to knowing where the arc or arcs were formed. There are two schools of thought on the origin of the Melanesian arc, both of which have different implications for Pacific biogeography.

One hypothesis is that the Melanesian arc(s) was formed within the Pacific plate far from any land. This implies that the Melanesian arc biota was derived by over-water colonisation (*e.g.*, Boer and Duffels, 1995). A second hypothesis is that the Melanesian arc (or parts thereof) was formed along the eastern edge of the Melanesian rift between 50 and 40 Ma (Hall, 1998 this volume). The present position of this arc system is seen as the result of opening the marginal basins discussed above, which transported the arc to the east. The biogeographical implication of this is that the Melanesian arc biota is original and derived from that of the Melanesian rift. The composite rift-arc structure is referred to here as the Melanesian rift-arc.

Implications for southwest Pacific biogeography

Craw (1988) interpreted New Zealand biogeography in terms of a "parallel arcs model". In this model the western 'arc' (and its associated biota) was related to cratonic Gondwana fragments and the eastern 'arc' to what were, prior to the Rangitata orogeny in the early Cretaceous, geosynclinal sediments. This model generates a number of systematic hypotheses. The biota associated with cratonic Gondwana fragments should show relationships to other cratonic areas of Gondwana such as Australia and South America. The relationship of the biota associated with what I have termed the Melanesian rift-arc system (equivalent to Craw's (1988) eastern arc) should show closer relationships to other rift and arc fragments to the north and northeast. The lack of appropriate systematic data makes an unequivocal test of such hypotheses impossible at present, but the data presented below are suggestive.

Although the New Zealand avifauna can be generally characterised as depauperate, a number of families such as the Psittacidae and Phalacrocoracidae are diverse. The genus *Cyanoramphus* (Psittacidae) is a southwest Pacific endemic (Table 1) and well represented in the New Zealand region. The red-crowned parakeet (*C. novaezelandiae*) is found from the Three Kings Islands in the north to the Auckland Islands in the south. It is now rare on the New Zealand mainland, but was once common throughout the country. A number of subspecies are recognised which are found in the Kermadecs (*cyanurus*), Chathams (*chathamensis*), Antipodes (*hochstetteri*), Macquarie (*erythrotis* extinct), Lord Howe (*subflavescens* extinct), Norfolk Island (*cooki*) and New Caledonia (*saisetti*). The distribution of this species is remarkable, because it is found further south than any other parrot and on Southern Ocean islands where environmental conditions could hardly be less favourable for a family generally found in more equitable climes.

Five other species within the genus are recognised. The Yellow-crowned parakeet (*C. auriceps*) is distributed throughout mainland New Zealand and a number of off-shore islands, although its numbers on the mainland had already sharply declined by the turn of the century (Hutton and Drummond, 1904). A subspecies, *forbesi*, is recognised on Chatham, Pitt and Mangere islands. The Orange-fronted parakeet (*C. malherbi*) is found in a few sites in the South Island, while *C. unicolor* is restricted to the Antipodes Islands. Two species, now extinct, were known from the Society Islands. *C. ulietanus* was found on Raiatea and *C. zealandicus* on Tahiti (Fuller, 1987).

Three other parrot species are found in New Zealand, the *kea* (*Nestor notabilis*), *kaka* (*Nestor meridionalis*), and the *kakapo* (*Strigops habroptilus*). The *kea* is restricted to the mountains of the South Island, while *kaka* subspecies were once widespread. An extinct subspecies of the *kaka* was found on Norfolk Island and nearby Phillip Island (*productus*). The *kakapo* is endemic to New Zealand where its continued long-term survival depends on populations established on predator-free islands. Although its relationship to other parrots is obscure, some workers suggest that the closest relative of the *kakapo* may be found among the Australian endemics *Pezoporus wallicus* (ground parrot), *P. occidentalis* (night parrot) and *Melopsittacus undulatus* (budgerigar) (MacDonald and Slater, 1992). The New Zealand parrot fauna does not appear to be particularly Australian with respect to phylogenetic affinity. *Cyanoramphus novaezelandiae* is endemic to the New Zealand subcontinent. The genus *Cyanoramphus* links the New Zealand subcontinent with eastern Polynesia. The genus *Nestor* is endemic to areas forming the southern parts of the Melanesian rift and, with *Cyanoramphus,* forms part of what is termed here as New Zealand's northern biota. The *kakapo's* relationship to other parrots needs clarification.

One bird family that does show clear austral relationships is the Phalacrocoracidae. New Zealand's cormorant fauna is very diverse with a greater number of species (14) being found in New Zealand waters than anywhere else in the world (Falla *et al.*, 1979). Three groups of cormorants are recognised on the basis of leg colour and crest characteristics. The first group of 4 species (*Phalacrocorax carbo, P. varius, P. sulcirostris,* and *P. melanoleucos*) are characterised by black legs and feet. A crest may or may not be present. They are fresh water, estuarine or coastal birds that nest in trees. The four *Phalacrocorax* species are also found in Sumatra, southern Borneo, Java, eastern Indonesia, New Guinea, Australia, and New Caledonia. The second group have pink legs and feet and a single crest. They are marine or oceanic species that nest on ledges. This group are placed in the genus *Leucocarbo* by Turbott (1990) and are found around the coasts of New Zealand and its subantarctic islands. The genus is also found in South America and Kerguelen Island. The third group have yellow legs and feet and are double crested. These species are found around New Zealand, Stewart Island and the Chathams. They are placed in the genus *Stictocarbo*. *S. gaimardi* is a South American representative of the double-crested New Zealand species.

A cladistic analysis of some external and ecological characters was carried out using PAUP (Swofford, 1991). The character matrix is shown in Table 7. The branch and bound option was used and multistate characters were treated as polymorphisms. Five minimal length trees were found (length 25, CI = 0.88, RI = 0.89). Fig.3 shows the strict consensus tree rooted using *Anhinga* as an outgroup (Sibley and Ahlquist, 1990). In Fig.3 *Phalacrocorax* is monophyletic

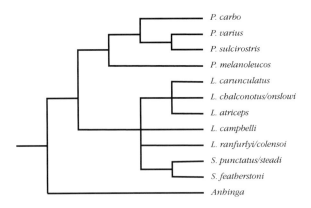

Fig. 3. Strict consensus cladogram of five minimal length trees (length = 25, CI = 0.88, RI = 0.89) for New Zealand cormorants. See text for details.

and a sister group to *Leucocarbo* + *Stictocarbo*. The New Zealand cormorant fauna is a mixture of 'northern' and 'southern' groups. The former are associated with the Melanesian rift-arc system, while the latter are associated with the Campbell plateau-Chatham Rise-South Island, New Zealand block. Although the above analysis was intended only as an interest exercise, the phylogenetic pattern within New Zealand cormorants is congruent with geological patterns.

The high degree of endemism of the New Zealand avifauna suggests that post-rifting isolation has been an important factor in its evolution. The fossil record shows influxes of northern marine invertebrates and plants into New Zealand during the upper Eocene and in the earliest Miocene (Fleming, 1980). Michaux (1989) interpreted the earlier event as a response to subsidence along the Norfolk Ridge. The Miocene event coincided with the final stages of the opening of the South Fiji basin and emplacement of the Northland allochthon. Sibley and Ahlquist (1990) suggested separation of the New Zealand Wrens (Acanthisittidae) from their ancestor at 42 Ma and *kiwi* from an Emu-Cassowary clade at 45 Ma. Both these avian families have Papuan phylogenetic affinities and their origin coincides with the Eocene event described above. Endemic elements of the New Zealand avifauna may date from 65 Ma (initial isolation after rifting), 40 Ma (Eocene influx) and 22 Ma (Miocene event)

The avifaunas of New Caledonia, Norfolk Island, Fiji/Tonga/Samoa and Niue (Tables 2 and 3) show a strong Melanesian arc influence. Ballance *et al.* (1982) suggested that the Melanesian arc separated from the Melanesian rift in the New Caledonia-Tongan region prior to the

Table 7. Characters and character-state matrix of New Zealand cormorants.

	A	B	C	D	E	F	G	H	I
Phalacrocorax carbo	1	1	1	1	0	0	2	1	1
P. varius	1	0	2	1	0	0	2	1	1
P. sulcirostris	1	0	3	1	0	0	2	1	1
P. melanoleucos	1	0,1	1	1	0	0	1	2	1
Leucocarbo carunculatus	2	0	4	2	1	1	1	1	2
L. chalconotus/onslowi	2	1	4	2	1	1	1	1	2
L. campbelli	2	1	4	2	0	0	1	3	2
L. ranfurlyi/colensoi	2	1	4	2	0	1	1	3	2
L. atriceps	2	1	5	3	1	1	1	1	2
Stictocarbo punctatus/steadi	3	2	6	4	0	1	1	3	2
S. featherstoni	3	2	7	?	0	0	1	3	2
Anhinga	2	0	1	0	0	0	1	0	1

A leg colour: 1 = black 2 = pink 3 = yellow/orange
B crest: 0 = absent 1 = one crest 2 = double crest
C facial skin: 1 = yellow 2 = blue 3 = olive 4 = red
    5 = brown 6 = green 7 = purple
D gular pouch: 1 = yellow 2 = red 3 = brown 4 = blue
E caruncles: 0 = absent 1 = present
F alar bar: 0 = absent 1 = present
G iris colour: 1 = brown 2 = green
H eye ring: 1 = blue 2 = brown 3 = purple
I habitat: 1 = fresh water 2 = marine

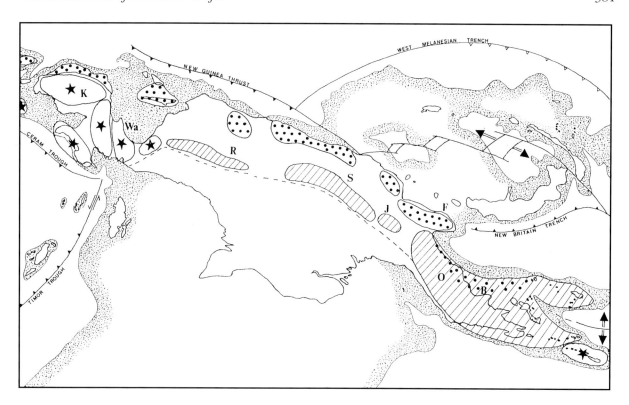

Fig. 4. New Guinea terranes. Dots = Melanesian arc terranes, stars = terranes from the Australian craton, slanted line = Melanesian rift terranes. B = Bowutu terrane, F = Finisterre terrane, J = Jimi, Schrader, Benabena and Marum terranes, K = Kemum terrane, O = Owen Stanley terrane, R = Rouffaer terrane, S = Sepik terrane, Wa = Wandamen terrane. Stipple = continental crust, arrows = opening basins, toothed lines = subduction zones. Closed teeth represent subduction zones marking plate boundaries and open teeth intra-plate subduction zones.

opening of the South Fiji basin at 35 Ma. According to Burrett *et al.* (1991) the Melanesian arc was still a relatively coherent entity until 7 Ma, when the southern section (Vanuatu, the Fiji Platform and the Tonga Ridge) became isolated from the Solomons. The opening of the North Fiji basin subsequently separated the Fiji Platform-Tongan Ridge-Samoa block from Vanuatu (Fig.2D). Tentative ages from which New Caledonian endemism date are 65 Ma (initial separation from Gondwana) and 35 Ma (separation of the Melanesian arc from the Melanesian ridge) and for Fiji/Samoa/Tonga from 35 Ma and 7 Ma.

New Guinea and the Melanesian rift-arc system

An interpretive summary of Pigram and Davies' (1987) account of the collision and amalgamation of a variety of terranes with the northern margin of the Australian craton is shown in Fig.4. In this paper these terranes are interpreted as lateral extensions of the Melanesian rift-arc system. The interaction at this margin was dominated by shear (the 'bacon slicer' effect) with fragments entering the shear zone transported westwards (Fig.4: filled circles). To the west of New Guinea, Melanesian arc fragments (or its lateral equivalents) appear to have been incorporated onto the Philippine plate and rotated clockwise to amalgamate with pre-existing terranes in north Maluku and the southern Philippines (Hall, 1987; 1996).

Hall *et al.*'s (1995: Fig.9) reconstruction of the Philippine Sea plate showed it to the north of Australia at 45 Ma and rotating clockwise. By 25 Ma this rotation had brought the southern boundary of this plate in contact with northern New Guinea, at a time when Pigram and Davies (1987) and Lee and Lawver (1995) suggested that exotic terranes had started to collide with the Australian craton. According to Hall *et al.* (1995) clockwise rotation continued until 5 Ma by which time the eastern boundary was to the northwest of New Guinea. Rangin *et al.* (1990) have described kinematic motions that are consistent with this model. Lee and Lawver (1995) placed the Sepik arc terrane (= Melanesian rift) to the northeast of Australia at 50 Ma. General westward translation of this terrane resulted in collision and amalgamation with the Australian craton at 30 Ma. Northern New Guinea terranes, which Lee and Lawver (1995) link to the Bismarcks and Solomon Islands (= Melanesian arc), also entered the collision zone from the east to amalgamate between 10 Ma and 5 Ma. Clockwise rotation of the Philippine Sea plate has been an important factor in the development of the New Guinea margin.

In Fig.4 the terranes that Pigram and Davies (1987) identified along the north coast have been interpreted as Melanesian arc islands, now incorporated as coastal ranges along the New Guinea margin. While each may have had a unique geological history, they are viewed here as members of a tectonic unit that docked after 10 Ma. New Britain and New Ireland are adjacent parts of the Melanesian arc that are currently sandwiched between the Manus basin, which is opening in a west-east direction, and a subduction zone along the northern boundary of the Solomon Sea (Fig.4).

The amalgamation of the northern section of the Melanesian rift with the Australian craton is dated by Pigram and Davies (1987) from 25 Ma in the west to 15 Ma in the east. By 15 Ma the east Papuan composite terrane had docked forming southeast New Guinea. The east Papuan composite terrane formed somewhere to the northeast of the Australian craton prior to its docking. In Fig.4 this terrane is shown composed of both rift and arc fragments (Michaux, 1994). The Louisiade plateau (Fig.4: star) and a section of the southern coast around Port Moresby are parts, according to Pigram and Davies (1987), of the Australian craton moved into their present position by the opening of the Coral Sea (63-53 Ma).

In the Bird's Head region, terranes have been identified by Pigram and Davies (1987) as fragments of the Australian craton (Fig.4: star). The largest of these terranes is the Kemum terrane which makes up most of the Bird's Head south of the Sorong fault. It is composed of Palaeozoic basement with only a thin, incomplete Mesozoic sequence. Pigram and Davies (1987) reported that polar wandering curves for the Kemum terrane are identical to those of Australia until their separation in the Cretaceous. Stratigraphic and palaeontological evidence also indicate an Australian origin for the Kemum terrane. During the Oligocene to Miocene the Kemum terrane amalgamated with other Bird's Head terranes, before docking with the Australian craton at 10 Ma. The relationship between these cratonic terranes of the Bird's Head, their lateral equivalents in the central ranges interpreted here as parts of the Melanesian rift (Fig.4: diagonal lines), and outer Banda rift-arc fragments (discussed later) is unclear.

A number of general points emerge from a study of the New Guinean avifauna. Firstly, many of New Guinean taxa are endemic. Secondly, the similarity to Australia is not as marked as one might expect, with many taxa in central and northern New Guinea showing phylogenetic links to the east and west, rather than to Australia in the south. The majority of New Guinean taxa with phylogenetic links to Australia are found in southern New Guinea. Within central and northern New Guinea taxonomic differentiation is strongly developed along an east-west axis.

Table 4, which lists New Guinean taxa absent from Australia, shows that northern and central New Guinea are linked to the southwest Pacific via the Melanesian arc islands of the Bismarck archipelago and the Solomons. Northern New Guinean taxa also have phylogenetic links to north Maluku and the Philippines. Asian influences are strong in central New Guinea. New Guinean montane flora shows a pronounced mixture of Asian and Austral taxa, being the only place in the world where southern beeches and northern oaks exist together. Asian Dipterocarpaceae dominate hill forest while *Araucaria* is found in lower montane zones with oaks at higher levels and *Nothofagus* above 2000 m. The higher moss forests are dominated by the Myrtaceae and at higher levels still by Eurasian genera such as *Vaccinium* and *Rhododendron* (Ericaceae) (Gressitt, 1982a). Similar patterns occur in other groups. Four anuran families are found in New Guinea (Menzies, 1975). The Leptodactylidae and Hylidae are shared with Australia and South America (Cracraft, 1980). The Ranidae and Microhylidae are characteristic of Indonesia and Asia. Gressitt has discussed the 'Oriental' character of New Guinea's rich and diverse beetle fauna in which similar patterns are also evident (Gressitt, 1982a, b).

An examination of birds shared by New Guinea and Australia shows that southern New Guinea and the Port Moresby area in the southeast peninsula feature in 42 of 121 of such distributions (MacDonald and Slater, 1992). The majority of these species are endemic to Australia and New Guinea, although five species are more widely found in New Caledonia, New Zealand, Timor and Aru. New Guinean species shared with Australia are often, but not exclusively, restricted to Cape York. Southern New Guinea is geologically part of the Australian craton. What is more surprising is the evidence for southern southeast peninsular Cape York to be an area of endemism, sharing the same or closely related species including the Magpie Goose (*Anseranus semipalmata*), which Sibley and Ahlquist (1990) place with the South American screamers in a sister group to all other ducks, swans and geese. Pigram and Davies' (1987) suggestion, that the southern coast of the southeast peninsula represented a part of the Australian craton detached and moved north by spreading in the Coral Sea at 63 Ma, seems to be borne out by bird distributional data.

Pigram and Davies' (1987) discussion of the composite geological nature of New Guinea also makes intelligible the pronounced east-west taxonomic differentiation reported for anurans (Menzies, 1975), murid rodents (Taylor *et al.*, 1982) and birds of paradise (Gilliard, 1969). For example, various species and subspecies of the Six-wired Bird of Paradise (*Parotia* spp.) appear to be associated with the following terranes:

P. wahnesi with Finisterre (Fig.4: F).
P. sefilata with Kemum and Wandamen (Fig.4: K and Wa).
P. carolae with Rouffaer and Sepik (Fig.4: R and S).
P. lawesi exhibita with Jimi, Schrader, Marum, and Benabena (Fig.4: J)
P. l. fuscior and *P. l. lawesi* with Owen Stanley (Fig.4: O).
P. l. helena with Bowutu (Fig.4: B)

These taxa link the Bird's Head terrane to Melanesian rift fragments. The occurrence of *P. l. helena* on the Bowutu terrane and *P. wahnesi* on the Finisterre terrane, parts of the Melanesian arc system, may be the result of dispersal of *P. lawesi* from the Owen Stanley ranges during collision. On the other hand an ancestral taxon may have been widely distributed and present on the Bird's Head terranes and Melanesian rift-arc system. Further investigation into the phylogenetic relationship between taxa and terrane analysis would be of interest to biogeographers and geologists.

Arc and rift systems in Maluku

North Maluku (Morotai, Halmahera, Ternate and Bacan) is situated near the southern end of the Philippine trench at the boundary of a complex collision zone. Hall (1987, 1996) has outlined the geological history of Halmahera, which he linked to the clockwise rotation of the Philippine Sea plate discussed previously. According to Hall (1987), Halmahera has a basement of oceanic crust imbricated with Cretaceous to Eocene sediments. This basement complex, which can be traced north into eastern Mindanao, is interpreted by Hall (1987) as the remnant of a late Cretaceous to early Tertiary fore-arc ridge. During the late Eocene and early Oligocene the east Halmahera basement underwent strong deformation and uplift, resulting in the deposition of river conglomerates in the Oligocene and Miocene. Synchronous deformation is recognised over an area stretching from western New Guinea to the western Pacific, leading Hall (1987) to view the east Halmahera-east Mindanao terrane as a westward continuation of a Papuan arc complex (= Melanesian arc).

Hall's reconstructions (1998 this volume) show east Halmahera continued to be translated west on the Philippine Sea plate. Eastward subduction of the Molucca Sea microplate beneath east Halmahera was initiated at c.10 Ma in response to collision of continental fragments with Sulawesi that stopped westward movement along splays of the Sorong fault. By 3 Ma subduction under east Halmahera led to the building of an active arc on east Halmahera. At 1 Ma arc volcanism ceased temporarily before shifting westwards, building a new arc on the eroded remnants of this Pliocene arc to form west Halmahera. Halmahera's continued western migration eventually closed the Molucca Sea and brought the island to its present position.

Metamorphosed continental rocks in central Bacan have been linked by Pb-Nd isotope ratios to northern Australia (Vroon *et al.*, 1996). Continental basement on Obi may also be a detached fragment of the Australian craton. Ali and Hall (1995) have discussed the geology of Obi and the islands of Bisa and Tapas which lie to the northwest. South Obi has a Palaeozoic continental metamorphic basement, while that in north Obi is composed of Jurassic ophiolite. The islands of Bisa and Tapas include high-grade Palaeozoic metamorphics which are linked by Ali and Hall (1995) to similar metamorphics on Bacan, 50 km to the north. According to Ali and Hall (1995) these terranes are derived from the Australian craton. On the basis of palaeomagnetic studies the Obi-Misool-Sula platform block was further south in Cretaceous times. Obi underwent northward translation during the Cretaceous, westward translation during the late Eocene-Oligocene (30 Ma) and northwest translation (north Obi) and northeast translation (south Obi) in the Neogene. These results suggest that Obi is a detached fragment of the Australian craton and may explain its anomalous biogeographical position noted before.

The islands of south Maluku are continental fragments derived from the Australian craton, overthrust by oceanic basement material (Hartono and Tjokosapoetro, 1986; Hamilton, 1988; Audley-Charles *et al.*, 1988; Lee and Lawver, 1995). The term Banda rift-arc is used here to describe this structure. In Seram, ophiolite material has been thrust onto Palaeozoic continental basement (Linthout *et al.*, 1996). Linthout *et al.* (1996) suggested this occurred as Seram was 'squeezed' between Australia and the inner Banda arc. Seram was translated northwards where it entered the Terera Aiduna fault zone (south of and parallel to the Sorong fault) where counter clockwise rotation and westward translation brought it to its present position. Linthout *et al.* (1996) dated these events as 8 Ma or younger.

In the Timor region a similar collision between continental fragment(s) and the Banda arc also commenced in the Miocene. In Timor, the interaction between rift and arc has led to a chaotic jumble of continental slivers, ophiolites and subduction related sediments as Australian continental crust is thrust under the Banda arc. Snyder *et al.* (1996) have interpreted seismic and gravity profiles across the cratonic margin behind Timor. These profiles suggest that this margin is rifting and providing further continental fragments for eventual inclusion into the collision zone. Their reconstructions show rifting in the Joseph Bonaparte Gulf, which is situated between the Kimberley and Sturt blocks of northwest Australia. This rifting appears to be detaching fresh continental crust (Sahal and Ashmore platforms) from the craton. The age of the detachment of the original Banda rift fragments is not clear. Jurassic to early Cretaceous rifting has been suggested by various authors, but this seems rather early on avian evidence.

The avifauna of north Maluku and the Melanesian/west Papuan arc

The northern island of Halmahera is part of an arc system and related to fragments now incorporated into northern New Guinea. Taxa shared by north Maluku and New Guinea have already been discussed. These phylogenetic patterns are intelligible in the light of Hall's (1987, 1996, 1998 this volume) description of Halmahera's geological history. The east Philippine-Halmahera-New Guinea connection is an important result of Hall's (1987) study as it explains similar distributional patterns in micro-lepidoptera (Diakonoff, 1967), murid rodents (Groves, 1984), weevils of the tribe Pachyrrhynchini (Gressitt, 1956) and other groups discussed in Holloway (1990). The endemic element of north Maluku's avifauna indicates a degree of isolation from other parts of the Melanesian arc and from New Guinea. According to Hall (1987) the Halmahera section of the Melanesian arc may have collided with New Guinea in the late Eocene or early Oligocene. It is possible that the origin of the genera *Lycocorax* and *Semioptera* (Paradisaeidae) dates from this time (c.40 Ma).

The avifauna of south Maluku and the Banda rift-arc system

There seems little doubt, given the preceding discussion of avian and geological links, that south Maluku was once part of the Australian craton and was derived from northwest Australia sometime in the Mesozoic. The high number of endemic species (Table 5) indicates that south Maluku and Australia have been isolated from each other, but the paucity of endemic genera is problematical.

Polhemus (1995) has shown that the waterstrider *Aquarius lili* from Timor is not closely related to Australian species, but is a member of an Asian/Sundaic clade. The non-endemic bird data in Table 5 also link south Maluku to Asia and Sundaland. One factor of potential importance in understanding this link between south Maluku and Asia is the collision of Timor with the Banda arc in the Pliocene (since 5 Ma) and the resulting dispersal between Flores and Timor discussed by Michaux (1994). A second factor is the possible relationship between south Maluku and the Sumba terrane which is discussed in the following sections.

Sulawesi as a geological composite

The geological development of Sulawesi has been discussed by a number of authors. Hutchison (1989) recognised that the southwest arm and central Sulawesi was a microcontinental fragment. Hutchison (1989) envisaged that an island arc was amalgamated to this terrane, together with the continental Sula platform terrane, in the late-Miocene. The northern arm of Sulawesi was also interpreted by Hutchison (1989) as an island arc which became sutured to the rest of Sulawesi towards the end of the Tertiary.

Hall (1996) illustrated a more complex history for Sulawesi. In Hall's (1996) model the central/ southwest Sulawesi terrane is sutured to Kalimantan as in Hutchison (1989). At 45 Ma the north arm of Sulawesi is shown as an arc close to its present position. From 45 Ma to 25 Ma the east Philippine-Halmahera arc (= Melanesian arc) was translated westwards in response to clockwise rotation of the Philippine Sea plate. The approach of the Philippine-Halmahera arc to the Northern Arm arc lead to its suture to southwest and central Sulawesi at 25 Ma. This time also marks the collision of the eastern arms of Sulawesi, which Hall (1996) derived from the south. At 20 Ma Australian continental crust was thrust under the eastern arms. Two further continental terranes, Buton and the Sula platform, became amalgamated at 10 Ma and 5 Ma respectively. Hall (1996) illustrated the derivation of the Buton terrane and Sula platform from the Bird's Head terrane by westward translation along splays of the Sorong fault system.

Implications for Sulawesi biogeography

Table 6 shows twenty five bird species that have their most westerly occurrence in Sulawesi. These species, together with thirteen bats listed in Cranbrook (1991) and Musser (1987) which have a Sulawesi/Maluku/New Guinea distribution, three marsupials of the genus *Phalanger* (Musser, 1987) and microhylid frogs of the genus *Oreophryne* (Kampen, 1923) can be classed as Australasian elements in the Sulawesian fauna. An explanation for the presence of these taxa in Sulawesi is that they (or their ancestors) were part of the Sula platform or Buton terrane fauna. There are also eleven Asian/Sundaic bird species which only extend to the Sula Islands (Table 6). These species appear to have colonised the Sula Islands from Sulawesi.

The southwest microcontinental terrane was originally attached to southeast Kalimantan, but became separated when the Makassar Strait opened in the mid-Tertiary (McCabe and Cole, 1989). Hutchison (1989) suggested that this fragment was derived from the north, but there is no support for this hypothesis from faunal analysis (Michaux, 1991, 1994). The distributional data in Table 6 confirm that the Sulawesian avifauna has close links with the Australian and Indian cratons. Musser (1987) regarded Sulawesian mammals, with their high degree of endemism and phylogenetic isolation from Sundaic taxa, as more typical of an oceanic island than one surrounded by lands rich in mammal families. The avifauna also shows a high degree of endemism (c.25%) as do other groups such as the butterfly families Papilionidae (47%) and Pieridae (56%) (Jong, 1990) and cicadas (95%) (Duffels, 1990). This degree of endemicity shown by Sulawesi's biota points to Sulawesi's isolation from other landmasses.

A key task is to understand the relationship of the southwest Sulawesian microcontinental terrane to other terrane/arcs in the region. Rangin *et al.* (1990) united the east Borneo terranes with southwest Sumatra, east Java, southwest Sulawesi, the "west Philippine islands" and Sumba as a Gondwana continental block, which they termed the Sumba terrane. This terrane became sutured to the margin of the Sunda shelf during the early Tertiary, as what remained of the Ceno-Tethys ocean (Metcalfe, 1996) was subducted. Rangin *et al.* (1990) regarded the ophiolite-bearing melanges found in Borneo, central Java, and Sumatra as evidence of this Tethys suture zone.

Michaux (1996) discussed a distributional pattern linking Sulawesi with Mindanao, north Borneo, east and southeast Kalimantan, Java, west Sumatra (and islands), the Andamans/Nicobars and Burma. Distributions of this type avoid the Malay peninsula and reach Asia (and sometimes beyond) via Burma. To the east of Java, this distributional pattern extends to the Lesser Sunda islands. The hypothesis of a Sumba terrane clearly has relevance for understanding the connections shown by the Sulawesian avifauna to surrounding avifaunas from Mindanao, north Borneo, the Lesser Sundas, west Sumatra, Andamans and Nicobars.

One might speculate that the fragments described above represented a rift-arc complex(es) of east Gondwana affinity, which was sutured to the Sundaland margin at the end of the Cretaceous or early Tertiary. Subsequent tectonic events have dispersed these fragments. Some of them may have been incorporated into the Banda arc which has transported them back towards Australia, a possible source area from which they were derived. Any relationship between the Sumba terrane of Rangin *et al.* (1990) and the Banda rift-arc system alluded to earlier would stem from the similarity of their respec-

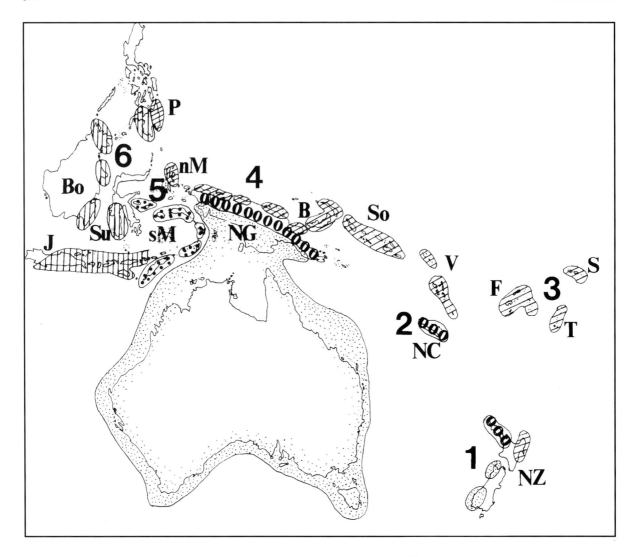

Fig. 5. Geological structures discussed in the text. Stippled = cratonic areas; diagonal lines = Melanesian arc; circles = Melanesian rift; dots = Banda rift-arc structure; vertical lines = Sumba terrane. B = Bismarck archipelago, Bo = Borneo, F = Fiji, J = Java, nM = north Maluku, sM = south Maluku, NC = New Caledonia, NG = New Guinea, NZ = New Zealand, P = Philippines, S = Samoa, So = Solomons, Su = Sulawesi, T = Tonga, V = Vanuatu. Numbers refer to areas for which geological histories are provided.

tive source areas.

Discussion

Clarifying the relationship between terranes, rift-arc complexes, and mobile arc systems is an important task in unravelling Indonesian biogeography (Polhemus, 1996). In turn, how are these structures related to the Indian and Australian cratonic regions derived from east Gondwana? What is the explanation for the origin of 'Oriental' elements in the New Guinean and Pacific avifaunas but their absence from Australia? A proper analysis of widespread Indo-Pacific groups has the potential to answer these and other questions.

As far as I understand the geological literature describing the breakup of Gondwanaland in the Mesozoic and the reassembly of drifted fragments in Asia, Southeast Asia and Indonesia during the Tertiary, an analysis using ridged plates models is not always appropriate. The breakup of east Gondwana appears to have involved two qualitatively different fragmentation regimes. Internal rifting produced large cratonic areas which act as rigid structures. Marginal rifting produced smaller, elongate rifts and/or microcontinents which are often (always?) associated with island arcs. As these products of breakup interact in collision zones, cratonic areas can either grow at the margins through accretion (*e.g.*, northern Australia) or their margins can fracture producing new rifts/microcontinents (*e.g.*, west and east Australia). Where major plates interact (*e.g.*, along the Banda arc) the movement of continental and arc structures caught up in collision zones between rigid plates appear fluid over a geological time scale. An example of this is way in which the northern section of the Banda rift-arc structure has been translated north and west to its present position. Other examples would include the movement of arcs across plate boundaries in the southwest Pacific and the lateral movement of rift-arc fragments along transform faults or in response to oblique subduction in Indonesia.

Marginal rifting occurred along the entire east Gondwana margin from the Indian to the southwest Pacific sector. According to Metcalfe (1996) a series of rifts were detached from the Indian margin from late Devonian (east Asia and Indochina) to early Cretaceous (Burma-west Sumatra-east Kalimantan-west Sulawesi-Lesser Sundas). The rifted fragments incorporated into the outer Banda arc islands of Timor, Tanimbar, Kai, Seram and Buru were detached from northwest Australia probably in the early Cretaceous. This pattern of marginal rifting is repeated to the east with the formation of continental fragments now in the Bird's Head, continental terranes in central New Guinea and the New Zealand subcontinent. Larger cratonic fragments (India and Australia) were detached and conveyed north into the Himalayan and Indonesian-Papuan collision zones. Significant lateral movement, development of marginal oceanic basins and complex plate margin responses have subsequently occurred in the Indonesian-Papuan collision zone.

In my view the birds of the Indo-Pacific region show clear evidence that their distributions have been determined by the geological events described above. Fig.5 shows areas in the modern Indo-Pacific region which share a degree of geological similarity and whose avifauna, or portion thereof, reflect this in their phylogenetic links. Such a hypothesis is likely to be more acceptable to botanists than to zoologists, who tend to view mammalian and bird families as relatively recent arrivals. However, Hedges *et al.* (1996) have recently contested this view and suggested not only that bird orders diversified much earlier at c.100 Ma, but that the fragmentation of continents has been an important factor in avian diversification. The evolution of the Nightjar family is discussed as an example below.

Nightjar evolution and the fragmentation of east Gondwana

Fig.6 shows the distribution of four avian families (nightjars) that Sibley and Ahlquist (1990) place in the Strigiformes. The Eurostopodidae (Fig.6: E) and the cosmopolitan Caprimulgidae are 'sister' taxa. The range of the Eurostopodidae defines the present day extent of former parts of east Gondwana. The restriction of this family to former east Gondwana fragments indicates that the origin and diversification of the Eurostopodidae are related to an ancestral taxon widely distributed in east Gondwana, and subsequent fragmentation of its range followed by taxonomic diversification.

A further 'sister' grouping identified by Sibley and Ahlquist (1990) is between the families Podargidae (Fig.6: P) and Batrachostomidae (Fig.6: B). The Batrachostomidae are distributed on the Indian craton and Gondwana terranes of Southeast Asia and Sundaland, while the Podargidae are found on the Australian craton and Melanesian rift-arc fragments. This implies that these two families date from the Cretaceous at c.90 Ma. According to Sibley and Ahlquist (1990), the Aegothelidae is the most primitive nightjar family. The range of this family is cratonic Australia plus the Melanesian rift-arc system (Fig.6: A). If Sibley and Ahlquist (1990) are correct in their placement of the Aegothelidae then its present distribution represents a relict. Two other central and south American families included in the Strigiformes are 'sister' taxa and together form the sister taxon to the Eurostopodidae plus Caprimulgidae. These two families, the Nyctibiidae and Steatornithidae, must date their origin to a time after south America had split from west Gondwana between 118-96 Ma (Veevers, 1988).

Widespread species — Gallirallus philippensis

The interpretation of widespread species is a key element in any biogeographical analysis. These species may be efficient dispersers, they may be unidentified species complexes that already had widespread ancestors on east Gondwana in the Cretaceous, they may represent a single species that has not responded to geological fragmentation of their original range, or they are not monophyletic and their wide distribution is an artefact.

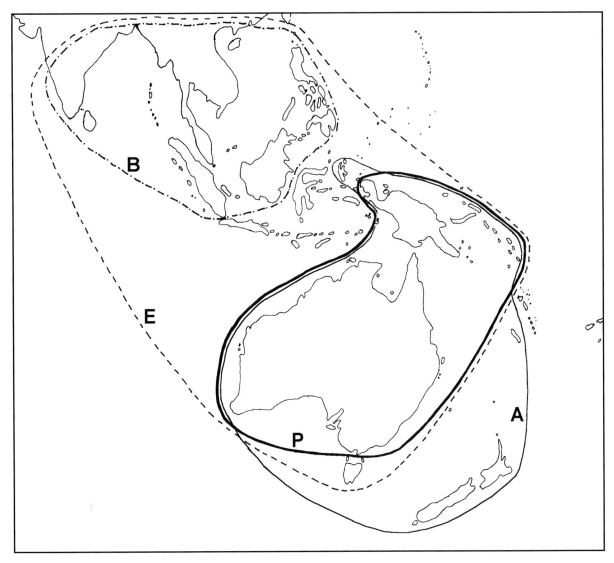

Fig. 6. Distributions of four night-jar families. A = Aegothelidae, B = Batrachostomidae, E = Eurostopodidae, P = Podargidae. See text for explanation.

Extreme evolutionary conservatism seems rather improbable given the time scales involved. As for widespread distributions being an artefact, Musser (1981) has speculated that the murid genus *Rattus* may not be monophyletic and Beehler *et al.* (1986) have separated the Australian false babblers (Pomatostomatidae) from African and Asian babblers (Timaliidae). Sibley and Ahlquist (1990) discussed the relationships between various Alcedinidae and showed the genus *Halcyon* to be paraphyletic. *H. sancta*, the Australian craton-Melanesian rift species, is most closely related to the monotypic *Melidora macrorrhina* (New Guinea) and these taxa are in turn the 'sister' group to Paradise Kingfishers (*Tanysiptera*: Maluku-New Guinea-north Queensland). An African species, *H. senegalensis*, is the outgroup to these Austropapuan species.

The occurrence of widespread, polymorphic species composed of allopatric populations provide conditions for investigating some of the taxonomic possibilities outlined above and as a test for geological models. Ripley (1977) discussed the taxonomy of the rail *Gallirallus philippensis* which is widespread across eastern Indonesia, Australia and the Pacific. Twenty four subspecies of *G. philippensis* are described by Ripley (1977) and their distributions shown in Fig.7. In Fig.7 the subspecies have been labelled

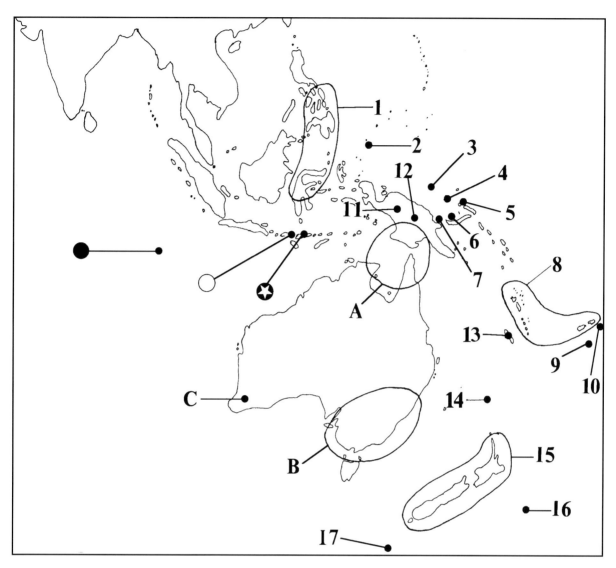

Fig. 7. Distributions of the subspecies of the rail *Gallirallus philippensis* 1 = *philippensis* (Philippines and Sulawesi), 2 = *pelewensis* (Palau), 3 = *anachoretae* (Anchorite Island), 4 = *admiralitatis* and *praedo* (Admiralty Island), 5 = *lesouefi* (New Hanover), 6 = *meyeri* (Witu Island), 7 = *reductus* (Long Island), 8 = *sethsmithi* (Vanuatu and Fiji), 9 = *ecaudatus* (Tonga), 10 = *goodsoni* (Samoa), 11 = *randi* (central mountains, Irian Jaya), 12 = *wahgiensis* (central mountains, PNG), 13 = *swindellsi* (New Caledonia), 14 = *norfolkensis* (Norfolk Island), 15 = *assimilis* (New Zealand), 16 = *dieffenbachii* (Chathams), 17 = *macquariensis* (Macquarie Island), A = *yorki* (northern Australia, Cape York and southern New Guinea), B = *australis* (NSW and South Australia), C = *mellori* (southwest Australia), filled circle = *andrewsi* (Cocos Island), open circle = *xerophilus* (Gunung Api), star in circle = *wilkinsoni* (Flores).

to indicate the degree of relationship that might be expected as a consequence of geological models discussed earlier. The Melanesian rift and arc taxa are numbered. The nominate species, *G. p. philippensis* (Fig.7: 1) has been included with other Melanesian rift and arc taxa, but its position is equivocal because Sulawesi and Mindanao can also be related to Sumba terrane fragments in the Lesser Sundas. Melanesian arc taxa (Fig 7: 2-10) might be expected to show the closest relationship to Melanesian rift taxa (Fig.7: 11-17). These groups should in turn show a sister group relationship to the Australian craton taxa (Fig.7: A-C). The position of the Sumba terrane taxa (in which the Cocos subspecies *andrewsi* is included (Fig.7: closed circle) depends on the geological relationship of this terrane to south Maluku, Australia and the Melanesian rift-arc system.

Acknowledgments

It is a great pleasure to acknowledge John Palmer of Arnold Books, Christchurch for the loan of some of the sources used in this study. Robert Hall, Hans Duffels, Arnold de Boer, Hubert Turner and Rod Hay provided valuable assistance in improving earlier drafts. The helpful comments and suggestions of the referees, Charles Sibley and Edward Dickinson, are gratefully acknowledged.

References

AAPG. 1981. Plate-tectonic map of the circum-Pacific region, southwest quadrant. American Association of Petroleum Geologists, Tulsa, Oklahoma.

Ali, J. R. and Hall, R. 1995. Evolution of the boundary between the Philippine Sea plate and Australia: palaeomagnetic evidence from eastern Indonesia. Tectonophysics 251: 251-275.

Audley-Charles, M. G., Ballantyne P. D., and Hall, R. 1988. Mesozoic-Cenozoic rift-drift sequence of Asian fragments from Gondwanaland. Tectonophysics 155: 317-330.

Ballance, P. F., Pettiga, J. R. and Webb, C. 1982. A model of the Cenozoic evolution of northern New Zealand and adjacent areas of the southwest Pacific. Tectonophysics 87: 37-48.

Balouet, J. C. and Olson, S. L. 1989. Fossil birds from the late Quaternary deposits of New Caledonia. Smithsonian Contributions to Zoology 469: 38 pp.

Beehler, B. M., Pratt, T. K. and Zimmerman, D. A. 1986. Birds of New Guinea. Wau Ecology Institute, Handbook Number 9. Princeton University Press, New Jersey, 293 pp.

Bemmel, A. C. V. van. 1948. A faunal list of the birds of the Moluccan Islands. Treubia 19: 323-402.

Bemmel, A. C. V. van. and Voous, K. H. 1953. Supplement to the faunal list of the birds of the Moluccan Islands. Beaufortia (miscellaneous publications) 4: 1-7.

Boer, A. J. de, and Duffels, A. J. 1995. Historical biogeography of the cicadas of Wallacea, New Guinea and the west Pacific: a geotectonic explanation. Palaeogeography, Palaeoclimatology, Palaeoecology 124: 153-177.

Borg, S. G. and Depado, D. J. 1991. A tectonic model of the Antarctic Gondwana margin with implications for southeastern Australia: isotopic and geochemical evidence. Tectonophysics 196: 339-358.

Burrett, C., Duhig, N., Berry, R. and Varne, R. 1991. Asian and south-western Pacific continental terranes derived from Gondwana, and their biogeographic significance. Australian Systematic Botany 4: 13-24.

Cracraft, J. 1980. Biogeographic patterns of terrestrial vertebrates of the southwest Pacific. Palaeogeography, Palaeoclimatology, Palaeoecology 31: 353-369.

Cranbrook, Earl of 1991. Mammals of South-East Asia. Oxford University Press, Oxford, 96 pp.

Craw, R. C. 1988. Continuing the synthesis between panbiogeography, phylogenetic systematics and geology as illustrated by empirical studies on the biogeography of New Zealand and the Chatham Islands. Systematic Zoology 37: 291-310.

Diakonoff, A. 1967. Microlepidoptera of the Philippine Islands. Smithsonian Institution Press, Washington, 484 pp.

Duffels, J. P. 1990. Biogeography of Sulawesi cicadas (Homoptera: Cicadoidea). In Insects and the rainforests of Southeast Asia. pp. 63-72. Edited by W. J. Knight and J. D. Holloway. Royal Entomological Society, London.

Duffels, J. P. and Boer, A. J. de, 1990. Areas of endemism and composite areas in east Malesia. In The plant diversity of Malesia, pp. 249-272. Edited by P. Baas et al. Kluwer, Dordrecht.

Falla, R. A., Sibson, R. B. and Turbott, E. G. 1979. Birds of New Zealand. Collins, Auckland, 247 pp.

Fleming, C. A. 1980. The geological history of New Zealand and its life. Auckland University Press, Auckland, 141 pp.

Fordyce, R. E. 1982. The fossil vertebrate record of New Zealand. In The fossil vertebrate record of Australasia, pp. 629-698. Edited by P. V. Richie and E. M. Thompson. Monash University Press, Victoria.

Fuller, E. 1987. Extinct Birds. Viking, London, 256 pp.

Gilliard, E. T. 1969. Birds of Paradise and Bower Birds. Weidenfeld and Nicholson, London, 485 pp.

Gressitt, J. L. 1956. Some distribution patterns of Pacific island faunae. Systematic Zoology 5: 11-32.

Gressitt, J. L. 1982a. Pacific-Asian biogeography with examples from the Coleoptera. Entomologia Generalis 8: 1-11.

Gressitt, J. L. 1982b, Zoogeographic Summary. Monographiae Biologicae 42: 897-917.

Groves, C. P. 1984. Mammal faunas and the palaeogeography of the Indo-Australian region. Courier Forschungsinstitut Senckenberg 69: 267-273.

Hall, R. 1987. Plate boundary evolution in the Halmahera region, Indonesia. Tectonophysics 144: 337-352.

Hall, R. 1996. Reconstructing Cenozoic SE Asia. In Tectonic evolution of Southeast Asia, pp. 153-184. Edited by R. Hall and D. J. Blundell. Geological Society Special Publication 106, London.

Hall, R. 1998. The plate tectonics of Cenozoic SE Asia and the distribution of land and sea. In Biogeography and Geological Evolution of SE Asia. Edited by R. Hall and J. D. Holloway (this volume).

Hall, R., Ali, J. R., Anderson, C. D. and Baker, S. J. 1995. Origin and motion history of the Philippine Sea plate. Tectonophysics 251: 229-250.

Hamilton, W. B. 1988. Plate tectonics and island arcs. Geological Society of America Bulletin 100: 1503-1527.

Hartono, H. M. S. and Tjokrosapoetro, S. 1986. Geological evolution of the Indonesian archipelago. Geological Society of Malaysia Bulletin 20: 97-136.

Hedges, S. B., Parker, P. H., Sibley, C. G. and Kumar, S. 1996. Continental breakup and the diversification of birds and mammals. Nature 381: 226-229.

Holloway, J. D. 1990. Sulawesi biogeography — discussion and summing up. In Insects and the rainforests of Southeast Asia pp. 95-102. Edited by W. J. Knight and J. D. Holloway. Royal Entomological Society, London.

Hutchison, C. S. 1989. Displaced terranes of the southwest Pacific. In The evolution of the Pacific Ocean margins pp. 161-75. Edited by Z. Ben-Avraham. Oxford Monographs in Geology and Geophysics vol. 8, Oxford University Press, New York.

Hutton, Captain F. W. and Drummond, J. 1904. Animals of New Zealand. Whitcombe and Tombs Ltd., Christchurch, 381 pp.

Jong, R., de 1990. Some aspects of the biogeography of the Hesperiidae (Lepidoptera, Rhopalocera) of Sulawesi. In Insects and the rainforests of Southeast Asia. pp. 35-42. Edited by W. J. Knight and J. D. Holloway. Royal Entomological Society, London.

Kamp, P. J. J. 1986. Late Cretaceous-Cenozoic tectonic development of the southwest Pacific region. Tectonophysics 121: 225-251.

Kampen, P. N. van. 1923. The amphibians of the Indo-Australian Archipelago. E. J. Brill, Leiden, 304 pp.

Keast, A. 1996. Avian geography: New Guinea to the eastern Pacific. In The origin and evolution of Pacific Island biotas, New Guinea to eastern Polynesia: patterns and processes pp. 373-398. Edited by A. Keast and S. E. Miller. SPB Academic Publishing, Netherlands.

Lee, T.-Y. and Lawver, L. A. 1995. Cenozoic plate reconstruction of Southeast Asia. Tectonophysics 251: 85-138.

Linthout, K., Helmers, H., Wijbrans, J. R. and Wees, J. D. van. 1996. ^{40}Ar/^{39}Ar constraints on obduction of the Seram ultramafic complex: consequences for the evolution of the southern Banda Sea. In Tectonic evolution of Southeast Asia, pp. 455-464. Edited by R. Hall and D. J. Blundell. Geological Society Special Publication 106, London.

McCabe, R. and Cole J. 1989. Speculations on the late Mesozoic and Cenozoic evolution of the Southeast Asian margin. In The evolution of the Pacific Ocean margins pp. 143-60. Edited by Z. Ben-Avraham. Oxford Monographs in Geology and Geophysics 8. Oxford University Press, New York.

MacDonald, J. D. and Slater, P. 1992. Birds of Australia. Reed, NSW, 552 pp.

MacKinnon, J. and Phillipps, K. 1993. The Birds of Borneo, Sumatra, Java, and Bali. Oxford University Press, Oxford, 491 pp.

Mayr, E. 1941. The origin and the history of the bird fauna of Polynesia. Proceedings of the 6th Pacific Science Congress 4: 197-216.

Mayr, E. 1945. Birds of the southwest Pacific. MacMillan, New York, 316 pp.

Menzies, J. I. 1975. Handbook of Common New Guinea Frogs. Wau Ecology Institute, Handbook Number 1, 75 pp.

Metcalfe, I. 1996. Pre-Cretaceous evolution of SE Asian terranes. In Tectonic evolution of Southeast Asia pp. 97-122. Edited by R. Hall and D. J. Blundell. Geological Society Special Publication 106, London.

Michaux, B. 1989. Generalized tracks and geology. Systematic Zoology 38: 390-398.

Michaux, B. 1991. Distributional patterns and tectonic development in Indonesia: Wallace reinterpreted. Australian Systematic Botany 4: 25-36.

Michaux, B. 1994. Land movements and animal distributions in east Malesia (eastern Indonesia, Papua New Guinea and Melanesia). Palaeogeography, Palaeoclimate, Palaeoecology 112: 323-343.

Michaux, B. 1996. The origin of southwest Sulawesi and other Indonesian terranes: a biological view. Palaeogeography, Palaeoclimatology, Palaeoecology 167: 167-183.

Musser, G. G. 1981. The giant rat of Flores and its relatives east of Borneo and Bali. Bulletin of the American Museum of Natural History 169: 69-176.

Musser, G. G. 1987. The mammals of Sulawesi. In Biogeographical Evolution of the Malay Archipelago pp. 73-94. Edited by T. C. Whitmore. Clarendon Press, Oxford.

Peters, J. L. 1934-86. Checklist of birds of the world, vols I to XV. Museum of Comparative Zoology, Cambridge, Massachusetts.

Pigram, C. J. and Davies, H. L. 1987. Terranes and the accretion history of the New Guinea orogen. BMR Journal of Australian Geology and Geophysics 10: 193-211.

Polhemus, D. A. 1995. A new species of *Aquarius* (Heteroptera: Gerridae) from Timor, with notes on Timorese Zoogeography. Proceedings of the Entomological Society of Washington 96: 54-62.

Polhemus, D. A. 1996. Island arcs, and their influence on Indo-Pacific biogeography. In The origin and evolution of Pacific Island biotas, New Guinea to eastern Polynesia: patterns and processes pp. 51-66. Edited by A. Keast and S. E. Miller. SPB Academic Publishing, Netherlands.

Rangin, C. and others 1990. The quest for Tethys in the western Pacific. 8 paleogeodynamic maps for Cenozoic time. Bullétin de la Société géologique Française 6: 907-913.

Ripley, S. D. 1977. Rails of the World. Godine, Boston.

Schodde, R., Fullagar, P. and Hermes, N. 1983. A Review of Norfolk Island Birds: Past and Present. Australian National Parks and Wildlife Service Special Publication number 8, Canberra, 119 pp.

Sibley, C. G. and Ahlquist, J. E. 1990. Phylogeny and Classification of Birds. Yale University Press, Connecticut, 976 pp.

Snyder, D. B., Milsom, J. and Prasetyo, H. 1996. Geophysical evidence for local indentor tectonics in the Banda arc east of Timor. In Tectonic evolution of Southeast Asia, pp. 61-73. Edited by R. Hall and D. J. Blundell. Geological Society Special Publication 106, London.

Steadman, D. V. 1995. Prehistoric extinctions of Pacific island birds: biodiversity meets zooarchaeology. Science 267: 1123-1131.

Swofford, D. L. 1991. PAUP: Phylogenetic Analysis Using Parsimony, v 3.1. Computer programme distributed by the Illinois Natural History Survey, Champaign, Illinois.

Taylor, J. M, Calaby, J. H. and Deusen, H. M. van. 1982. A revision of the genus *Rattus* (Rodentia, Muridae) in the New Guinea region. Bulletin of the American Museum of Natural History 173: 177-336.

Thiollay, J. M. H. 1993. Habitat segregation and the insular syndrome in two congeric raptors in New Caledonia, the White-belled Goshawk *Accipiter haplochrous* and the Brown Goshawk *A. fasciatus*. Ibis 135: 237-246

Turbott, E. G. 1990. Checklist of the Birds of New Zealand. Ornithological Society of New Zealand, Wellington, 247 pp.

Veevers, J. J. 1988. Morphotectonics of Australia's northwestern margin — a review. In The Northwest Shelf, Australia pp. 19-27. Edited by P. G. Purcell and R. R. Purcel. Proceedings of the Petroleum Exploration Society, Perth.

Vroon, P. Z., Bergen, M. J. van. and Forde, E. J. 1996. Pb and Nd isotope constraints on the provenance of tectonically dispersed continental fragments in east Indonesia. In Tectonic evolution of Southeast Asia. pp. 445-453. Edited by R. Hall and D. J. Blundell. Geological Society Special Publication 106, London.

Wallace, A. R. 1880. The Malay Archipelago. MacMillan, London, 653 pp.

Walters, M. 1980. The Complete Birds of the World. Reed, Sydney, 340 pp.

Watling, R. 1982. Birds of Fiji, Tonga, Samoa. Millwood Press, Wellington, 176 pp.

White, C. M. N. and Bruce, M. D. 1986. The Birds of Wallacea. British Ornithologists' Union Check-list 7, London, 524 pp.

Wodzicki, K. 1971. The birds of Niue Island, south Pacific: an annotated checklist. Notornis 18: 291-304.

Wood, R. and Uruski, C. 1990. The New Caledonian basin (abstract). In Recent developments in New Zealand Basin studies pp. 27-29. DSIR Geology and Geophysics, Wellington.

Yan, C. Y. and Kroenke, L. W. 1993. A plate tectonic reconstruction of the southwest Pacific 0-100 Ma. Proceedings of the Ocean Drilling Program, Scientific Results 30: 697-707.

Pre-glacial Bornean primate impoverishment and Wallace's line

Douglas Brandon-Jones
Department of Palaeontology, Natural History Museum, Cromwell Road, London SW7 5BD, UK

Key words: Asia, Australasia, Borneo, climate, dispersal barriers, island hopping, Java, Mentawai archipelago, Oriental biogeography, rafting, rainforest refugia, Sumatra, Wallacea

Abstract

Leaf monkeys (*Semnopithecus*, subgenus *Trachypithecus*) and lorises (*Loris* and *Nycticebus*) are both geographically disjunct between southern India and SE Asia, with endemic representatives in eastern Indochina. These parallels appear to result from restriction to, and re-expansion from, rainforest glacial refugia in southern India, northeast Indochina and west Java. Sureli (*Presbytis*) and gibbon (*Hylobates*) distributions reveal further refugia in north Borneo, north Sumatra and the Mentawai Islands. Modern Sumatran primate distribution was moulded by at least two cold dry glacial periods. The earlier one 190,000 years ago eliminated all Sumatran primate habitats whereas, after recolonization, the later one 80,000 years ago left a north Sumatran rainforest refugium. Not only did the Mentawai Islands provide a reservoir for the recolonization of Sumatra, but indirectly for an interglacial invasion of Borneo which, like Sulawesi, had previously been outside the range of *Presbytis* and gibbons. Bornean primate zoogeography indicates that before the first arid period there may have been fewer than four primate species on Borneo. Most of the present twelve or thirteen Bornean primate species rafted there interglacially or post-glacially from Sumatra. Pre-glacial Bornean primate impoverishment is primarily attributed to a suspected south coastal dry zone which would have inhibited or precluded colonization. Colonization of islands further east must generally have bypassed Borneo via Java or the Philippines. The Bornean climatic barrier presented a more severe impediment to faunal exchange across Wallace's line than did the sea depth along its course. Such climatic barriers, whose influences waxed and waned with the glacial cycles, would have affected most SE Asian islands and were the prime inhibitor of faunal and floral exchange between the Oriental and the Australasian regions.

Introduction

SE Asia has a rich primate fauna, comprising orang-utans (*Pongo*), gibbons (*Hylobates*), colobine monkeys (*Nasalis*, *Pygathrix*, *Presbytis* and *Semnopithecus*, subgenus *Trachypithecus*), macaques (*Macaca*), loris (*Nycticebus*) and tarsiers (*Tarsius*). Twelve (or thirteen if the presence of *Hylobates agilis* is accepted) primate species occur on the island of Borneo. And yet, despite the presence of suitable habitats, only the macaques on Sulawesi and the Lesser Sunda Islands and, to a much lesser extent, the leaf monkeys on Lombok (purportedly by human introduction), have crossed Wallace's line.

The effectiveness as a faunal barrier of this most widely-adopted division of the Oriental from the Australasian zoogeographic region, is generally attributed to the depth of the sea channel extending from the Bali-Lombok Strait, between Borneo and Sulawesi, to the east of the Philippines. The deep Makassar Strait remained a sea barrier when the Sunda and Sahul shelves were exposed during glacial sea-level depressions. Huxley's line coincides approximately with the eastern edge of the Sunda shelf, and Lydekker's line with the western edge of the Sahul shelf. These later variations on the division have been regarded as clear-cut faunal boundaries enclosing a transitional zone.

However, major recolonization of the volcanic Krakatau archipelago, 12 km away from the next nearest island, has occurred in only a matter of decades (Smith, 1943; Thornton, 1996). No convincing explanation has been offered as to how a sea barrier such as the Bali-Lombok Strait, little more than three times as wide, could have inhibited colonization for millennia. Floating islands capable of transporting a viable sample of flora and fauna have been reliably reported. Even if these crossed a strait only once a century, their impact on floral and faunal exchange between the two land masses would have been significant. Conditions favouring such rafting would have been enhanced during periods of climatic change (see the discussion below).

The inference (Brandon-Jones, 1996a) that during the glaciations Asian rainforest was reduced by drought to a few scattered pockets of distribution provides a more plausible explanation for these biogeographic barriers. Not only could rainforest communities on such islands as the Philippines and Sulawesi, have been entirely eliminated, leaving only severely degraded forest or no forest at all, but the recession of rainforest from some coastal areas, would have seriously impaired the ability of rainforest to disperse by rafting. Contraction and expansion of rainforest distribution has been the prime mediator of present primate species diversity on Borneo (Brandon-Jones, 1996b), in sharp contrast to the probable presence there, demonstrated in this paper, of only two primate species before the penultimate glaciation. If such a faunal turnover can be established for Borneo, seemingly in the heart of the Asian moist rainforest, extending such analysis to other Indo-Pacific islands should produce further insights into the effects of climate change on floral and faunal migration and diversity.

Glacial effects on Asian primate distribution

Both the pied leaf monkeys (*Semnopithecus auratus, S. francoisi, S. hatinhensis, S. laotum, S. delacouri* and *S. johnii*) of Java (Indonesia), northeast Indochina and southern India (Brandon-Jones, 1995), and the grizzled surelis (*Presbytis comata*) of Java, north Sumatra and north Borneo (Brandon-Jones, 1993, 1996a, b) display a tripartite disjunction. Brandon-Jones (1996a) inferred that the once continuous subcontinental Indian, Chinese and SE Asian rainforest was fragmented by a glacial drought 190,000 years ago. It subsequently re-expanded, although probably not to its former extent, only to contract again during a second, less severe drought 80,000 years ago. Asian colobine monkey zoogeography suggests that these droughts eliminated all but a few small pockets of rainforest. Such rainforest refugia survived in north Sumatra, the Mentawai Islands (off west Sumatra), north Borneo, west Java, northeast Indochina and southern India (Brandon-Jones, 1993, 1995, 1996a, b, 1997). Some of these refugia are located at one thousand metres or more in altitude, but most are areas which would have remained sea-bound, either as promontories or islands, or are coastal areas which retained a maritime climate during desiccative glacial sea-level depressions.

The dual contraction and re-expansion of the rainforest led to the fracturing of the distribution of the pied leaf monkeys from continuity between Java, northeast Indochina and southern India, to survival in those areas alone (Fig.1; Brandon-Jones, 1995). The prosimian loris has a parallel distribution, with an endemic representative, *Nycticebus pygmaeus*, in eastern Indochina (Fig.2). This suggests a similar history of disjunction and partial recolonization. In both cases recolonization has been northward, undoubtedly from Java in the case of the leaf monkeys, and probably so in the case of the loris. Gibbon (*Hylobates*) distribution is similar, but without an outlying population in southern India. Recolonization has been extensive but incomplete. At least four primate genera (*Hylobates, Presbytis, Semnopithecus* and *Macaca*) have added Borneo to their pre-glacial distribution, but others (*Pongo, Nasalis* and *Pygathrix*) have undergone little or no post-glacial dispersal.

The Mentawai archipelago is the key to interpreting the biogeography of Sumatra and Borneo. Ancestors of its endemic primates (*Hylobates klossii, Nasalis concolor, Presbytis potenziani* and *Macaca pagensis*) must formerly have existed on Sumatra, but no longer occur there. The ebony leaf monkey (*Semnopithecus auratus*) can only have reached Java by way of Sumatra, from which it is also now absent. *Presbytis potenziani* is sister-taxon to *P. comata* of north Sumatra, north Borneo and west Java (Brandon-Jones, 1993). An ancestral taxon similar to *P. potenziani* is presumed to have been the initial coloniser of SE Asia. *Hylobates klossii* and the gibbons of Java and (with some variation) north Borneo, are chromatically monomorphic, unlike those of the Malay peninsula and Sumatra. *H. 'lar' vestitus* of north Sumatra is replaced in southern Sumatra by the polymorphic *H. agilis*. The call of the south Bornean gibbon (whose specific allocation, like that of *H. 'lar' vestitus,* remains debatable) is virtually identical to that of *H. agilis* (Geissmann, 1995). This indicates a geographic relationship between the Mentawai Islands gibbon and other gibbons analogous to that between *P. potenziani* and *P. comata. H. klossii* today is suggested to be mor-

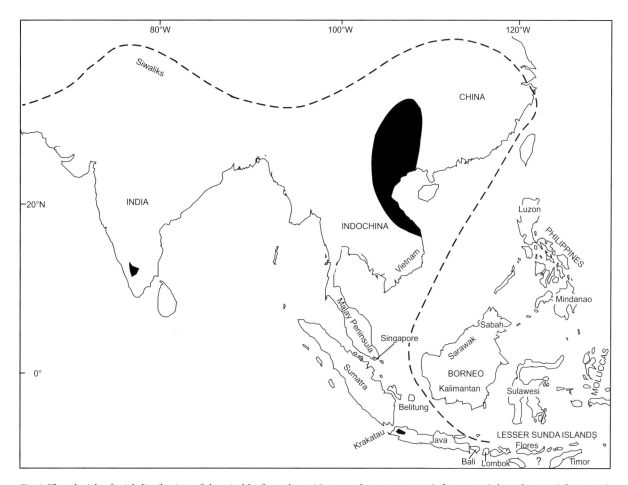

Fig.1. The glacial refugial distribution of the pied leaf monkeys (*Semnopithecus auratus*, *S. francoisi*, *S. hatinhensis*, *S. laotum*, *S. delacouri* and *S. johnii*) after deforestation 190,000 years ago, with an estimate (dashed line) of their distribution before that date.

phologically conservative, descended directly from the Mentawai progenitor of most, if not all other extant gibbons, except for the concolor gibbons and the siamang. Brandon-Jones (1993, 1996a) inferred that all Sumatran primates (and Sumatran primate habitats) disappeared during the earlier glacial drought, persisting only on the Mentawai Islands, whose maritime climate protected it from the desiccating effect of the glacial emergence of the Sunda shelf.

During the interglacial, *Presbytis comata* diverged from *P. potenziani*, and *Hylobates 'lar' vestitus*, *H. muelleri* and *H. moloch* diverged from *H. klossii*, as moist rainforest recolonisation of Sumatra, Java and southern Borneo facilitated dispersal from the Mentawai Islands. *Pongo* reinvaded Sumatra, probably from Borneo, but possibly from Indochina. During the later, and lesser drought, rainforest area contracted less than during the previous dry period, enabling these recolonisers to survive in north Sumatra,

north Borneo and west Java. The absence of *Nasalis*, and of endemic subspecies of *Semnopithecus* and *Macaca* in the Sumatran refugium, which sustained the more moist rainforest-associated *Pongo*, *Hylobates* and *Presbytis*, indicates that two deforestations occurred. After the second dry period, independent evolution of brownish species occurred on Sumatra (*P. femoralis*), Borneo (*P. frontata*) and Java (*P. fredericae*) from the populations of *Presbytis comata* within those islands. *P. femoralis* dispersed to Borneo. Two reddish species, the Sumatran *P. melalophos*, and the Bornean *P. rubicunda* are the end-products of chromatic successions from black, through grey, then brown to red, and in some cases to albinistic, which characterise colobine post-glacial dispersal (Brandon-Jones, 1996b). Each stage in the succession was probably correlated with a phase of rainforest regeneration. This suggests that climatic remission was punctuated, rather

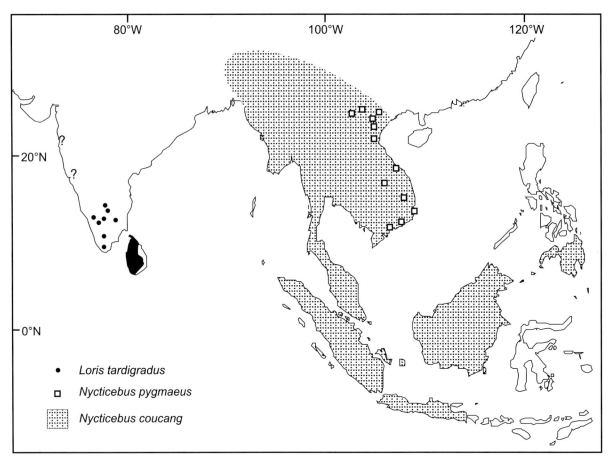

Fig.2. The geographic distribution of the lorises (*Loris* and *Nycticebus*).

than continuous.

The endemic Mentawai Islands macaque, *Macaca pagensis*, is closely related to the pigtailed macaque, *M. nemestrina*. Externally however, the population on Siberut (the northernmost Mentawai island) is more similar to other members of its species group, such as the lion-tailed macaque, *M. silenus*, isolated in southern India, and the Sulawesi macaque, *M. tonkeana*. These Siberut, Sulawesi and south Indian macaques probably also have a glacially-fragmented distribution (Fig.3; Brandon-Jones, 1998). The distribution of *Macaca nemestrina* dwindles to a narrow corridor as it enters Meghalaya, India (Fooden, 1975; Biswas and Diengdoh, 1978), indicating it is a northward-dispersing species, yet to colonize Java and north Vietnam. The northern subspecies is paler in pelage colour which, by Hershkovitz's (1968) principles of metachromism, supports this interpretation. Although it had an earlier common ancestor with the Sulawesi species, its most recent common ancestor was probably with the macaque of the two southernmost Mentawai islands.

Dating the deforestations

The silvered leaf monkey, *Semnopithecus cristatus* has reached islands such as Belitung and Serasan, and the long-tailed macaque, *Macaca fascicularis* many more. Their failure to colonize the Mentawai Islands seems to reflect a shortage of time rather than ability, and indicates that their geographic radiations occurred recently. *S. auratus* (a close relative of *S. cristatus*) evidently did not spread from Java between the two arid periods, suggesting this interval was short. *S. cristatus*, *Macaca fascicularis* and *M. nemestrina* are unrepresented by endemic taxa in the north Bornean refugium and thus appear to have been absent from Borneo until after the second arid period. Subfossil evidence from Niah Cave in northwest Borneo indicates the presence there of *Macaca*

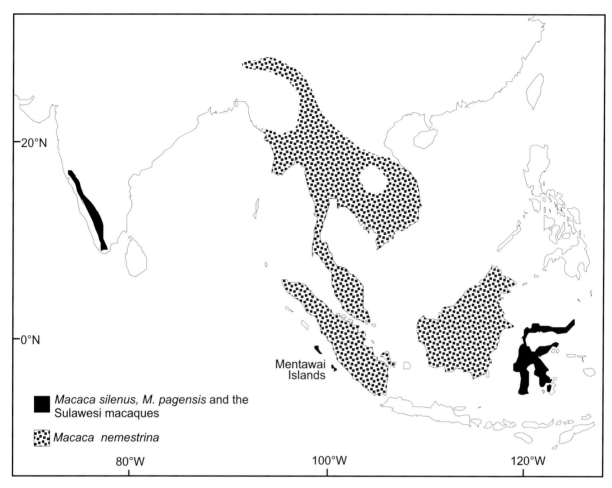

Fig.3. The geographic distribution of the *Macaca silenus* species group.

before the most recent glacial maximum (Brandon-Jones, 1996a). The two most recent glacial maxima at about 135,000 and 21-22,000 years ago, are indistinguishable in severity. Instead, they were preceded by a fluctuating, but persistent, temperature decline and succeeded by a rapid temperature increase. Neither of the two most recent glaciations included a cold period preceded by a significantly colder one (Martinson *et al.*, 1987). The more abrupt onset of the earlier glaciation seems the only relevant difference between the two glaciations. At the end of each interglacial, precipitation or temperature seems to have suddenly dropped below a critical threshold for widespread forest maintenance, and this threshold was not re-attained until the beginning of the subsequent interglacial. In the penultimate glaciation, this threshold was broken more abruptly, and the succeeding interstadials offered little remission. The deforestations appear to have occurred quite rapidly as the interglacials ended about 190,000 and 80,000 years ago (Brandon-Jones, 1996a), and the appearance of *Macaca* on Borneo before the most recent glacial maximum, indicates that some reafforestation had occurred by then.

An Indonesian deep-sea core, about 850 km NNW of Australia, yielded grass pollen associated with low sea-level dry glacial periods at 190,000-130,000 years and 38,000-12,000 years ago. Interglacial periods supported woodland/fern vegetation (Kaars, 1991). During the Middle Pleistocene, pine-grassland savannah similar to the open woodlands of Thailand and Luzon (Philippines) occupied areas now characterised by lowland rainforest near Kuala Lumpur, Malaysia (Batchelor, in Morley and Flenley, 1987). Palaeosol development in the intramontane Bandung basin, west Java, indicated an anomalously dry climate at the end of the penultimate glaciation about 135,000 years ago, followed by very warm and humid interglacial conditions from 126,000 to 81,000 years ago (Kaars and Dam, 1995).

At the advent of the last glacial period, about 81,000 years ago, freshwater swamp forest on the Bandung plain was replaced by an open grass-and-sedge-dominated swamp vegetation, indicating a considerably drier climate. A similar climate from 81,000 to 74,000 years ago in the adjacent mountains, is suggested by the strong decline in *Asplenium* ferns. Their resurgence from 74,000 to 47,000 years ago, indicates a return to slightly warmer conditions. Inferred depression of montane vegetation zones and fern scarcity from 47,000 to about 20,000 years ago, suggest distinctly cooler and possibly drier climatic conditions in the Bandung area (Kaars and Dam, 1995). The survival at Niah until about 40,000 years ago of a giant pangolin, otherwise known only from the Middle Pleistocene of Java, conveys some impression of the then prevalent conditions. The tall termite mounds necessary to sustain this extinct species, almost a metre longer than the largest living pangolin, are now characteristic of savannah areas. Its Middle Pleistocene contemporaries, the hippopotamus, antelope, cattle, chital and other deer are all obligate grazers. The terrestrial predators of that era, which include hyenas, three genera of dogs, sabre-toothed 'cats', possibly two 'tigers', a leopard, leopard cat and civets, accord with this impression of a plains community. Data from Niah Cave, the Tabon Cave in the Philippines, and the Ngoum Rock shelter in Vietnam, indicate a cool dry period from 32,000 to 23,000 years ago. The end of this period was even colder, producing a rubble layer in the Ngoum Rock shelter, and probably explaining the scarcity of monkeys and arboreal squirrels at Niah about 19,000 years ago, and the disappearance from this locality almost at sea-level, of two mammal genera, *Hylomys* and *Melogale*, now exclusively montane. This latter drought probably coincided with the most recent glacial maximum when west Javan temperatures fell by 4-7°C. A synchronous dry period occurred in Africa and South America. Since 23,000 years ago the climate moderated, producing a stalagmitic floor in the Tabon Cave (Brandon-Jones, 1996a).

The Javan Quaternary fossil and climate record

Recent clarification, and improved dating, of the Javan fossil mammal record reveals several faunal successions, and has been interpreted to indicate rainforest existence on the island for only 80,000 years. Evidence is absent for mammals on Java before the first major sea level recession at 2.4 Ma. This suggests that before that time, Java may have been largely submerged. The Ci Saat and Trinil faunas of 1.2 and 0.9 Ma respectively, which mark the first appearance there of *Homo erectus*, were preceded by the oldest recognizable fauna, the Upper Pliocene Satir fauna at 2-1.5 Ma. This fauna included *Geochelone atlas*, which persisted until at least 1.2 Ma, after the first immigration of *Stegodon* to Java, implying a more protracted faunal turnover than had been thought. The distribution of this giant tortoise at some time extended from Java to the Siwaliks in north India, and further undermines the concept of rainforest stability. Between the impoverished Satir and the Ci Saat faunal stage, both *Tetralophodon bumiajuensis* and *Hexaprotodon simplex* were replaced by the new immigrants, *Stegodon* and *Hexaprotodon sivalensis*.

A major faunal immigration, the Kedung Brubus fauna, with the greatest abundance of medium to large-sized mammals, indicating relatively open and dry conditions, coincided with a marked sea-level depression at 0.8 Ma. Of its maximum number of 25 species, ten are new records, five of which are probably SE Asian mainland immigrants (*Rhinoceros unicornis*, *Tapirus indicus*, *Manis palaeojavanica*, *Hyaena brevirostris* and *Lutrogale palaeoleptonyx*). Most taxa from the Trinil fauna persist in the Kedung Brubus fauna, and extinctions were negligible. Large bovids dominate both faunas, but with double the number of megaherbivores (eight against four) in the latter fauna. Palynological results from the Sangiran area (with the exception of an anomalous meter section whose abundance of spores and *Podocarpus* tree pollen indicates increased humidity) are poor in tree pollen and spores, but rich in herbaceous plant pollen. Little change, other than a slight impoverishment, is evident in the imprecisely dated Ngandong fauna.

The earliest evidence of both tropical rainforest and *Homo sapiens* derives from the Upper Pleistocene Punung fauna, probably correlated with the warm interglacial from 125,000 years ago onwards. Seventeen Kedung Brubus taxa had disappeared, with ten replacements. These replacements include large numbers of primates such as *Pongo* and *Hylobates*, indicating a major environmental change to humid conditions. The presence of *Homo sapiens*, suggests an age less than 110,000 years. Bergh *et al.* (1996) dated this fauna to 60-80,000 years, based on its similarity with the Jambu and Lida Ajer Sumatran cave faunas, for which aspartic acid racemization dating gave ages of 60-70,000 years and 80,000 years respectively. The invasion probably occurred between 80-110,000 years during the later part of the interglacial. Multiple extinctions with limited replacement, are demonstrated by the Holocene fossil cave faunas, such as Wajak, Sampung and Hoekgrot. *Pongo* is absent from this composite cave fauna, and probably disappeared from Java during the last glacial.

Variations in oxygen isotope levels from benthic and tropical planktonic foraminifera suggest that from about 2.8 Ma to 2.4 Ma ice volumes and sea levels fluctuated, with an overall increase in the former and a decrease in the latter, culminating at about 2.4 Ma in the first major glaciation. Sea level fluctuations then moderated, remaining constant until 0.8 Ma, with a mean of about 70 m below present day level (PDL), and lowest sea levels at about 100 m below PDL. Although insensitive to short-term fluctuations, the one global Pleistocene low sea level event detected by seismostratigraphy commenced at 0.8 Ma when sea levels were apparently exceptionally low for long enough to register on many seismic profiles. Sea level fluctuations then abruptly increased in amplitude, averaging about 90 m below PDL, with sea levels falling as low as 170 m below PDL. This fluctuation mode persisted until recent times (Bergh *et al.*, 1996).

Rainforest dispersal by island hopping

The correlation between glacial landmass emergence and aridity, refutes the notion that rainforest could readily have dispersed across the exposed Sunda and Sahul shelves. Conditions favourable to dispersal returned only when the sea re-attained its interglacial level. This implies that SE Asian rainforest has perhaps permanently been disjunct or insular in distribution, and has dispersed across sea barriers. Extensive areas of grassland or desert on the other hand, would pose insurmountable barriers. The presence of *Presbytis femoralis*, but not *Semnopithecus obscurus*, for example on Singapore, suggests the absence of an overland rainforest connection even between Singapore and peninsular Malaysia, although the dispersal route of *P. femoralis* probably brought it within range of Singapore earlier than did the dispersal route of *S. obscurus*. Such distributional differences tend to confirm that, when exposed, Sundaland was inimical to rainforest dispersal. This would suggest it was predominantly grassland-encompassed desert, perhaps comparable to the dry zone in present-day north Burma.

If marine rainforest dispersal seems implausible, a reliable account exists of a floating island with unusually tall nipa palms being mistaken for a three-masted vessel. The same report tells of a pirate marooned on the bank of a hostile river when his companions were forced to make a hurried embarkation. On seeing a small island floating to the sea, he swam to it and for many days subsisted on its palm fruits. These islands are created by floods undermining the matted roots of riverside nipas (St. John, 1862, pp. 16-17). Natural rafts, sometimes carrying living mammals, have been recorded over a hundred miles off the mouths of tropical rivers, such as the Ganges, the Amazon, the Zaire and the Orinoco (Matthew, 1915, p. 206). Such rafts are likely to have been much more frequent during the post-glacial period of vegetational succession, when primary rainforest gradually reestablished itself at the expense of lower canopy vegetation. Under-storey vegetation, steadily dying off as it became shaded out, would have been increasingly vulnerable to the action of river spates.

Needless to say however, the odds are stacked against successful rafting. Ironically, the better the swimmer the less likely an animal is to cross a strait, because it will have less reluctance about deserting a raft and a greater ability to resist wind and currents. This is probably a major factor in the endemism of the ably swimming proboscis monkey on Borneo, and the absence from Borneo of the tiger which occurs on Sumatra, Java and Bali. To succeed, the raft must offer ample food and protection from the elements, and the animal must be pregnant or accompanied by a member of the opposite sex. These undoubtedly rare coincidences have encouraged the belief that the deep sea channels which remained as barriers when the Sunda and Sahul shelves were exposed were sufficient to have created Wallace's line and other such faunal divides. This ignores the most important consideration that, for successful rafting, landfall must be accompanied by an appropriate climate. Rainforest flora and fauna on a raft will not flourish on arrival unless the raft lodges at a locality with adequate precipitation and temperature.

The deep sea channels were undoubtedly significant in maintaining localised maritime climates when these disappeared elsewhere but, compared to climatological barriers, they were relatively trivial in impeding dispersal. The longevity of the potential dispersal route between Asia and Australia is demonstrated by the presumably Asian origin (other than of those introduced by man) of the New Guinea and Australian rats (Muridae), which are now almost exclusively either endemic species, endemic genera or even endemic subfamilies (Simpson, 1977, p. 115). These successful dispersers are comparatively well able to cope with a range of climates, and perhaps exceptionally adept at survival on the inimical conditions of a raft.

Pre-glacial Bornean primate impoverishment

Although lingering on in Vietnam until about 23,000 years ago (Ha Van Tan, 1985), and formerly occurring in China (Kahlke, 1973) and Java, the main stronghold of the orang-utan appears to be Borneo. This suggests the existence there of both orangs and their rainforest habitat before the first deforestation. Establishing the presence of rainforest on Borneo may seem superfluous but, before this deforestation, the only primate undoubtedly present was the endemic proboscis monkey, *Nasalis larvatus*. Its anatomy, and that of its only close relative, *Nasalis concolor*, on the Mentawai Islands, is that of a predominantly terrestrial monkey, and is very reminiscent of macaques. The genus evidently evolved in forest-woodland (Brandon-Jones, 1996a). Its natural habitat must intermittently have been overwhelmed by mangrove and rainforest. The proboscis monkey remained on Borneo only because its island distribution prevented it from following the climatic and geographic recession of its native vegetation. Even in isolation, this is clear evidence of former extensive areas of open woodland on Borneo. Nevertheless, the fact that the proboscis monkey and the orang-utan have not reclaimed central Sarawak suggests that, at glacial extremes, most of the open woodland disappeared, leaving the two primate species with a very localised distribution in north (and perhaps west) Borneo.

Presbytis and *Hylobates* are represented in the Bornean refugium by the second-wave colonizers, *Presbytis comata* and *Hylobates muelleri*, but not by the initial colonizing ancestors of *P. potenziani* and *H. klossii*. Having established the probability of rainforest there during the first dry period, the absence of these initial colonizing species implies their genera were then absent, otherwise they would be expected to persist in Borneo, cohabiting with their descendent congeners. Such areas of sympatric distribution in Borneo do exist for both sureli species and gibbon species (see Brandon-Jones, 1996b; Mather, 1992). It is possible that the tarsier was also absent and survived the first arid period only on Sulawesi, where its vertical clinging and leaping adaptation would have equipped it well for survival in shrub vegetation, even if rainforest disappeared. Despite Simpson's (1977) reservations, this accords with Groves' (1976) view that Sulawesi might have been a centre of origin for the tarsier. The loris may have survived only in Java, Indochina and southern India. Thus there is the distinct possibility that, until 190,000 years ago at the earlier deforestation, or even until 135,000 years ago at the penultimate glacial maximum, there were only two primate species on Borneo. This does not establish that Bornean rainforest is necessarily of similar age, but it implies that gibbons, a family endemic to Asia, are latecomers to Borneo.

Modern Bornean primate diversity

How then do we account for the present primate diversity on Borneo? Probably the first arrivals after the orang-utan and the proboscis monkey, were *Presbytis comata* and *Hylobates muelleri*, which reached Borneo between the deforestations. It is possible that the loris and the tarsier also invaded during this interval, but more probable that, with the macaques and silvered leaf monkey, they did not arrive until after the second deforestation. The white-fronted sureli, *Presbytis frontata* diverged *in situ* from *P. comata sabana* after the latter event. Its other Bornean congeners, *P. femoralis* (directly) and *P. rubicunda* (indirectly), diverged from *P. comata thomasi* on Sumatra (Brandon-Jones, 1996b). If the south Bornean gibbon is a subspecies of *Hylobates agilis*, it was another Sumatran immigrant to arrive on Borneo since the second deforestation, otherwise it presumably evolved *in situ* from *H. muelleri*. All the extraneous primates evidently rafted to Borneo, because of the unavailability of suitable habitats for dispersal across Sundaland. *P. femoralis* and *P. rubicunda* seem to confirm this by their allopatric distribution in Kalimantan, compared to their sympatric distribution in Sarawak, Borneo. This geographic variation in sympatry is virtually irreconcilable with the concept of prolonged rainforest stability. It is readily explained by inferring recent arrival in separate localities on the west coast (the only other island occupied by *P. rubicunda* is Karimata which it presumably colonized *en route*), followed by parallel eastward dispersal, which was more extensive or rapid in *P. rubicunda*. If rainforest had existed in central Sundaland, the two species would surely have become sympatric much further westward than they do. As predicted by Hershkovitz's (1968) principles of metachromism, pelage colour in both *P. rubicunda* in north Borneo, and *P. melalophos* in south Sumatra, dilutes to a pale colour.

Borneo as a dispersal barrier

It is possible that primate impoverishment on Borneo, before the first deforestation, was simply due to an arid coastal climate which prevented rafted rainforest from establishing itself. This is supported by Stresemann's (1939) conclusion that grassland widespread in SE Asia during the Pliocene persists in south Borneo, and by the apparent entire extirpation and replacement during the Pleistocene of the southwest Bornean freshwater fish fauna (Brandon-Jones, 1996a). By restricting animal and plant colonization of south Borneo the dry zone responsible would have diminished its potential as an intermediate route to Sulawesi. Virtually all the 183 butterfly genera on Sulawesi are Asian, with no special link to Borneo. A set of younger patterns based on the distribution of the 470 species (200 of which are regional endemics) links Sulawesi to the Philippines, Lesser Sunda Islands and especially the Moluccas, in addition to Asia (Vane-Wright, 1991). This circum-Bornean faunal zone is also reflected in Hooijer's (1975) suggestion that Sulawesi, Flores, Timor and intervening small islands had Pleistocene geographic continuity as 'Stegoland'. The proposition of such broad land connections across waters now nearly 3000 metres deep is geologically tenuous and is rendered superfluous by the presence of indistinguishable large Pleistocene stegodonts on Mindanao (in the Philippines) and on Java (Simpson, 1977, p. 113). The possibility of a physical connection of the latter islands with the others is almost negligible. Although Groves (1976) believed that mammal migration between Sulawesi and the Philippines was insignificant, the stegodont dispersal between these islands must have crossed sea barriers. Sulawesi organisms with Indomalayan affinities evidently circumvented Borneo, via Java or the Philippines, many of their ancestors along the route, being eliminated by subsequent glacial drought.

The climatic origin of Wallacea

Simpson (1977, p. 117) concluded that Huxley's line, which approximately coincides with the eastern edge of the Sunda shelf, and Lydekker's line, which roughly corresponds to the western edge of the Sahul shelf, are clear-cut faunal boundaries separated by an unstable zone, now often termed Wallacea. He declined to categorize the intermediate zone as transitional, in order to encourage further research on its biogeography. The possibility that this unstable zone was at least partly created by climatic barriers was appreciated as early as 1845 (Müller, 1846). Lincoln (1975) concluded that Wallace's line was primarily an ecological division, with the Lesser Sunda Islands to the east of the line being drier and smaller with an impoverished but dominantly Oriental avifauna. Smith (1943, p. 140) remarked that the Philippines, Sulawesi and the Lesser Sunda Islands, as compared to Borneo, Java and Sumatra, "are notable for the absence of large mammals, not because they could not have reached them, but because they could not survive upon them if they got there". Pleistocene mammals contradict this. "The former presence of somewhat diverse proboscideans on islands between the Huxley and Lydekker Lines remains a puzzling fact that must be taken into account... ...The large stegodonts, ubiquitous where any Pleistocene mammals are known, surely were not all victims in the first generation of occupants" (Simpson, 1977, p. 113). Perhaps these elephant-like mammals thrived during relatively dry glacial periods of extensive grassland, and died out during excessive aridity when much of it turned to desert. As the climate changed their island distribution would have, as with the proboscis monkey, curtailed their ability to follow their preferred habitat.

Australasian organisms spreading northwestwards would have experienced a similar series of advances and local extinctions on their precarious 'stepping-stone' route to Asia. Relatively few of them, such as the Sulawesi phalangers, have successfully negotiated the climatic adversities (and perhaps superior competition) intermittently intervening to obstruct their passage. Islands along the route have varied (according to their size, topography and geographic situation) in their ability to sustain varying grades of vegetation for varying periods of time. Many species presumably had precursors eliminated from islands behind them on the route, and possibly their descendants from islands ahead of them. Steenis (1935, p. 404) for example, noted that Mount Kinabalu in Sabah, Borneo, is far richer than Timor in Australian temperate plants. Hence the need for caution in inferring dispersal routes from existing distributions. A species' absence can be as informative as its presence. The net result has been an ebb and flow of dispersal correlated with the glacial cycles. Wallace's line and other faunal divides such as Müller's and Weber's, may therefore mark only the approximate midpoint between two or more rainforest glacial refugia in Australasia and SE Asia.

Conclusions

Asian primate distribution indicates that most Asian rainforest was eliminated by glacial drought. Sumatran primate distribution indicates that this occurred at least twice, and that the most recent deforestation was less drastic than its predecessor. The two deforestations appear correlated with the terminations of the two most recent interglacial periods. The Javan fossil record suggests that before these interglacials, conditions generally favoured open country animals. The correlation between landmass emergence and aridity implies that SE Asian rainforest dispersal, successfully negotiated sea barriers. Indications that southwest Borneo had a prolonged arid climate which would have prevented rafted rainforest from establishing itself, provides an explanation for the apparent necessity for most fauna to bypass it on their route to islands further east. The contrast between evident pre-glacial Bornean primate impoverishment and its present primate diversity, provides a model for the effects of climate change on SE Asian island biogeography, and demonstrates how such effects could have restricted faunal and floral interchange between Australasia and SE Asia.

Acknowledgements

I am most grateful to my wife, Chris Brandon-Jones, for the preparation of the maps and to her, Peter Andrews, Jeremy Holloway and Mark Pilkington for constructive criticism of the manuscript.

References

Bergh, G. D. van den, Vos, J. de, Sondaar, P. Y. and Aziz, F. 1996. Pleistocene zoogeographic evolution of Java (Indonesia) and glacio-eustatic sea level fluctuations: a background for the presence of *Homo*. Indo-Pacific Prehistory Association Bulletin 14 (Chiang Mai Papers 1): 7-21.

Biswas, S. and Diengdoh, H. 1978. Notes on stumptailed macaque [*Macaca speciosa* F. Cuvier] and pigtailed macaque [*Macaca nemestrina* (Linn.)] from Meghalaya. Journal of the Bombay Natural History Society 74: 344-345.

Brandon-Jones, D. 1993. The taxonomic affinities of the Mentawai Islands sureli, *Presbytis potenziani* (Bonaparte, 1856) (Mammalia: Primates: Cercopithecidae). Raffles Bulletin of Zoology 41: 331-357.

Brandon-Jones, D. 1995. A revision of the Asian pied leaf monkeys (Mammalia: Primates: Cercopithecidae: superspecies *Semnopithecus auratus*), with a description of a new subspecies. Raffles Bulletin of Zoology 43: 3-43.

Brandon-Jones, D. 1996a. The Asian Colobinae (Mammalia: Cercopithecidae) as indicators of Quaternary climatic change. Biological Journal of the Linnean Society 59: 327-350.

Brandon-Jones, D. 1996b. *Presbytis* species sympatry in Borneo versus allopatry in Sumatra: an interpretation. *In*: Tropical Rainforest Research - Current Issues. Edited by D. S. Edwards, W. E. Booth and S. C. Choy. Kluwer, Dordrecht. Monographiae biologicae 74: 71-76.

Brandon-Jones, D. 1997. The zoogeography of sexual dichromatism in the Bornean grizzled sureli, *Presbytis comata* (Desmarest, 1822). Sarawak Museum Journal 50 (71): 177-200.

Brandon-Jones, D. 1998. The primates of the Mentawai Islands: a conservation imperative. Proceedings of the XVth Congress of the International Primatological Society, 3-8 August 1994, Kuta-Bali, Indonesia in press.

Fooden, J. 1975. Taxonomy and evolution of liontail and pigtail macaques (Primates: Cercopithecidae). Fieldiana: Zoology 67: 1-169.

Geissmann, T. 1995. Gibbon systematics and species identification. International Zoo News 42: 467-501.

Groves, C. P. 1976. The origin of the mammalian fauna of Sulawesi (Celebes). Zeitschrift für Säugetierkunde 41: 201-216.

Ha Van Tan. 1985. The late Pleistocene climate in Southeast Asia: new data from Vietnam. Modern Quaternary Researches in South East Asia 9: 81-86.

Hershkovitz, P. 1968. Metachromism or the principle of evolutionary change in mammalian tegumetary color. Evolution 22: 556-575.

Hooijer, D. A. 1975. Quaternary mammals west and east of Wallace's line. Netherlands Journal of Zoology 25: 46-56.

Kaars, W. A. van der. 1991. Palynology of eastern Indonesian marine piston-cores: A late Quaternary vegetational and climatic record for Australasia. Palaeogeography, Palaeoclimatology, Palaeoecology 85: 239-302.

Kaars, W. A. van der. and Dam, M. A. C. 1995. A 135,000-year record of vegetational and climatic change from the Bandung area, West-Java, Indonesia. Palaeogeography, Palaeoclimatology, Palaeoecology 117: 55-72.

Kahlke, H. D. 1973. A review of the Pleistocene history of the orang-utan (*Pongo* Lacépède 1799). Asian Perspectives 15: 5-14.

Lincoln, G. A. 1975. Bird counts either side of Wallace's line. Journal of Zoology 177: 349-361.

Martinson, D. G., Pisias N. G., Hays J. D., Imbrie J., Moore T. C., Jr. and Shackleton N. J. 1987. Age dating and the orbital theory of the Ice Ages: Development of a high-resolution 0 to 300,000-year chronostratigraphy. Quaternary Research 27: 1-29.

Mather, R. 1992. A field study of hybrid Gibbons in central Kalimantan, Indonesia. Ph.D. Thesis, Cambridge University.

Matthew, W. D. 1915. Climate and evolution. Annals of the New York Academy of Science 24: 171-318.

Morley, R. J. and Flenley, J. R. 1987. Late Cainozoic vegetational and environmental changes in the Malay archipelago. In Biogeographical Evolution of the Malay Archipelago. Edited by T. C. Whitmore. Oxford Monographs on Biogeography 4: 50-59.

Müller, S. 1846. Ueber den Charakter der Thierwelt auf den Inseln des indischen Archipels, ein Beitrag zur zoologischen Geographie. Archiv für Naturgeschichte 12 (1): 109-128.

St. John, S. 1862. Life in the forests of the Far East. Vol. 1. Smith, Elder and Company, London.

Simpson, G. G. 1977. Too many lines: the limits of the Oriental and Australian zoogeographic regions. Proceedings of the American Philosophical Society 121: 107-120.

Smith, M. A. 1943. A discussion of the biogeographic division of the Indo-Australian archipelago, with criticism of the Wallace and Weber lines and of any other dividing lines and with an attempt to obtain uniformity in the names used for the divisions. 3. The divisions as indicated by the Vertebrata. Proceedings of the Linnean Society, London, 154: 138-142.

Steenis, C. G. G. J. van. 1935. On the origin of the Malaysian mountain flora. Part 2. Altitudinal zones, general considerations and renewed statement of the problem. Bulletin du Jardin Botanique de Buitenzorg (3)13: 289-417.

Stresemann, E. 1939. Die Vögel von Celebes. Journal für Ornithologie, Leipzig 87: 299-425.

Thornton, I. W. B. 1996. Krakatau: the destruction and reassembly of an island ecosystem. Harvard University Press, Cambridge, Mass.

Vane-Wright, R. I. 1991. Transcending the Wallace line: do the western edges of the Australian Region and the Australian Plate coincide? Australian Systematic Botany 4: 183-197.

Glossary of terms

This section provides very brief definitions of some of the geological/palaeontological and biological/biogeographical terms used by the authors in this book. The intention of the glossary is to make the book comprehensible to the non-specialist reader rather than to provide a comprehensive discussion of, in some cases, controversial terms.

Geological, including palaeontological, terms

agnostid: type of Lower Palaeozoic trilobite.
allochthon: body of rock tectonically displaced from its place of formation.
andesite: volcanic rock of intermediate composition typical of island arcs.
antiarch: primitive Palaeozoic fish.
APWP: apparent polar wander path. Diagram used in palaeomagnetic work displaying a series of palaeo-poles for a fixed continent to represent the relative motion of the continent with respect to magnetic north. It is of course the continent that moves, but it is simpler to display a series of palaeo-poles on a diagram than plot diagrams showing moving continents with a fixed magnetic pole; APWPs for different continents can then be more easily compared.
backarc basins: small basins floored by oceanic crust formed above subduction zones behind island arcs by poorly understood mechanisms.
basalt: volcanic rock of basic composition typical of oceanic crust.
basement: the underlying or deeper rocks. Typically basement rocks are thought of as the deeper igneous and metamorphic rocks found beneath a sedimentary cover. The term is used to distinguish cover rock sequences from underlying rocks. However, the term is relative, and there is no implied age for the underlying rocks, and can be used to distinguish older sedimentary rocks from younger sedimentary rocks.
bioclastic: character of fragmental material in rocks which is organic debris, typically calcium carbonate shells and skeletons.
biofacies: characteristic assemblage of fossil fauna.
blueschist: rock type formed at high pressures and low temperatures in subduction zone settings, and characterised by the presence of the blue amphibole glaucophane.
calcalkaline: range of igneous rock compositions typical of volcanic arcs, including basalt, andesite and dacite.
carbonates: sedimentary rocks formed of carbonate minerals, principally calcite and dolomite, sometimes with aragonite.
chert: rock formed of fine-grained silica, typically the remains of organisms with siliceous skeletons and commonly found in deep marine environments.
clastic: type of sedimentary rock formed of fragments of rocks and minerals.
conodont: oral apparatus of primitive Palaeozoic craniates (vertebrates).
craton: continental region that has been tectonically stable for a long period, typically for more than several hundred million years.
cyclopygid: type of trilobite.
dacite: volcanic rock of intermediate composition typical of island arcs, especially those underlain by continental crust.
depocentre: site of deposition of sedimentary rocks, in principle the place of the thickest sequence, although the term is often used generally to include the whole area of deposition as a synonym of sedimentary basin.
diamictite: fragmental rock with angular clasts in a mud matrix interpreted as having a glaciomarine origin.
dicynodont: Permian mammal-like reptile.
dikelokephalinid: type of trilobite.
East India letter classification: biostratigraphic scheme for subdivision of Tertiary rocks of SE Asia based on large benthic foraminifera.
eclogite: rock of basic composition which consists essentially of a high density garnet-pyroxene mineralogy indicating metamorphism under high pressures typical of the lower crust or deeper.
Euler pole: the pole of rotation of two tectonic plates on a sphere. Definition of the pole and the angular motion of one plate relative to another fully describes their relative motion.
eustasy: concept of sea level change which affects the whole globe, and is not caused by local tectonics. Causes of eustatic sea level change could include changes in volume of polar ice caps, changes in volume of ocean basins due to displacement of water by sediment, or changes in volume of the mid-ocean ridge system.
extrusive: igneous rocks that are erupted at the surface.

flysch: term of Alpine origin for clastic rocks, typically thick sequences of deep marine sandstones and mudstones, deposited during the early stages of development of a mountain belt and said to be 'syn-orogenic'. These rocks are often deposits of continental slopes formed by turbidity currents.

forearc: region between island arc and trench.

fusulinid: large foraminifera of Carboniferous-Permian age.

glyptomenid: type of brachiopod.

graben: fault-bounded elongate depression characterised by steep and straight bounding faults at the margins with a central subsided block.

granitoid: igneous rocks of granitic composition dominated by quartz and feldspars.

graptolite: order of marine hemichordates of (mainly Lower) Palaeozoic age.

imbricate thrust slices: slices of rocks stacked together by contraction, separated by low angle and sub-parallel thrust faults.

intrusive: igneous rocks intruded within the crust and slowly cooled.

island arc: chain of volcanic islands formed above a subduction zone where oceanic lithosphere is thrust into the mantle.

lithosphere: the outer rigid part of the Earth, including the crust and part of the mantle to depths of about 100 km, forming the tectonic plates.

lowstand: period when eustatic sea level was relatively low.

lyttoniid: group of articulate brachiopods of Permian age.

mafic: igneous material of dark colour. As applied to rocks usually indicates a basic composition with relatively low silica content and is often incorrectly used as a synonym of basic.

magmatism: igneous activity as a result of melting of the crust or mantle.

magnetic anomalies: lineations within the ocean crust formed by igneous activity at linear mid-ocean ridges and alternations in the polarity of the Earth's magnetic field. The anomalies can be mapped and dated and provide the means to trace the motions of plates during the past 200 million years.

marl: calcareous mudrock with more than 25% carbonate.

melange: rock composed of a mixture of blocks in a fine-grained matrix. This mixture may have been formed by sedimentary processes (such as submarine debris flows) or by tectonic mechanisms. Rock of this type are common in active orogenic settings.

molasse: term of Alpine origin for clastic rocks, typically sequences of continental and shallow marine conglomerates and sandstones, deposited late in the development of a mountain belt and often said to be 'post-orogenic'.

nannofossils: fossils of ultramicroscopic size, representing the remains of zooplankton and phytoplankton.

nappe: large thrusted body of rock, typically with basal thrust that is sub-horizontal and has a displacement of several tens of kilometres.

obduction: poorly understood process by which rocks of broadly oceanic character known as ophiolites are thrust onto land.

ophiolite: association of rocks similar to those representative of oceanic crust and mantle but now found on land in orogenic belts. In the ideal ophiolite there are peridotites, gabbros, basalts and pelagic sedimentary rocks in a layered sequence. Some ophiolites may have formed at mid-ocean ridges of major ocean basins but most represent lithospheric fragments from arc-related settings such as backarc basins or forearc regions.

orogeny: process of mountain-building.

orthid: type of brachiopod.

palaeomagnetism: the Earth's magnetic field as recorded in rocks. Palaeomagnetic studies can determine palaeo-latitudes of rocks at the time of their deposition or formation, and can determine rotations since formation. This type of information can contribute to reconstructing the history of plate movements.

palynology: study of microscopic plant material.

palynomorphs: microscopic remains of plant origin, such as pollen grains and spores.

plectambonitoid: type of articulate brachiopod.

pluton: large igneous body intruded into the crust.

rhyolite: volcanic rock of acid composition typical of volcanic arcs formed on continental crust, and commonly erupted explosively.

rifting: process of breaking the crust and lithosphere by extension.

rudist: reef-building bivalve with coral-like appearance of late Mesozoic age.

schist: metamorphic rock with closely spaced planar fabric (schistosity), commonly due to preferred orientation of mica, produced by metamorphic recrystallisation accompanied by directed stress.

siliciclastic: type of sedimentary rock formed of clastic grains of silicate rocks and minerals.

sinolepid: type of Devonian armoured fish.

slab-pull: force exerted by a sinking lithospheric slab at a subduction zone.

splays: strands of a fault, typically in the zone where the fault terminates.

stratigraphy: geological discipline concerned with the description, organisation and classification of stratified rocks, fundamental to our understanding of the history of the Earth.

strike-slip: type of fault or motion in which two block of rocks move past one another with essentially horizontal motion.

strophomenoid: type of articulate brachiopod.

subduction: process by which lithosphere, mainly oceanic, is thrust deep into the mantle at convergent plate boundaries. The principal surface expressions of subduction are the deep oceanic trenches and the volcanic arcs of active margins.

syntaxis: region of abrupt change in orientation of an orogenic belt.

tectonic block: fault-bounded fragment of crust or lithosphere with its own characteristic sequence of strata. Size is not implied, but in many cases the term block as used in regional geology implies microcontinent or island arc-scale fragments.

tectonostratigraphy: study of the stratigraphy of terranes recognising that normal stratigraphic principles need to be applied with caution because of the important tectonic influence on sequences. Relative ages of events within and across terranes can be identified by conventional stratigraphic methods and the sequence of both strata and tectonic events can be displayed on composite diagrams.

terrane: fault-bounded fragment of the crust or lithosphere with its own characteristic stratigraphic sequence. Many mountain belts are now interpreted to be composed of large numbers of terranes which have become fragmented and amalgamated by tectonic processes including plate rifting, subduction, collision and strike-slip faulting.

till: deposit of glacial origin.

trachyandesite: volcanic rock of intermediate composition, but with a more alkaline composition than a normal andesite, common in island arcs.

transform fault: originally defined as type of fault which offsets a mid-ocean ridge but now commonly used for a strike-slip fault which penetrates deep into the lithosphere and forms a plate boundary.

transpression: combination of strike-slip motion and contraction.

transtension: combination of strike-slip motion and extension.

trilobite: Palaeozoic marine arthropod.

turbidites: clastic sedimentary rocks deposited in deep water on or below the continental slopes by currents containing dense mixtures of sediment and water.

ultramafic: material of very dark colour. As applied to rocks normally refers to peridotites or serpentinites, their hydrated equivalents, containing minerals such as olivine and pyroxene, representing mantle material.

unconformity: fundamental discordance in a stratified sequence of rocks representing a break in deposition and time.

yunnanolepid: armoured fish of Late Silurian-Devonian age.

zircon U-Pb ages: absolute ages determined by a method of dating using the mineral zircon (zirconium silicate) which contains radioactive uranium isotopes which decay to lead isotopes.

Biological and biogeographical terms

allopatric: distributions of taxa which are separate, not coincident, overlapping or abutting.

allozymes: enzyme alleles at genetic loci used in electrophoretic analysis of genetic variation between organisms.

anagenesis: transformation in an evolutionary lineage. The transformed states of genes or chemical or morphological characters serve to identify the descendants of the lineage subsequent to the transformation. *See also* apomorphy.

apomorphy: derived (transformed by anagenesis) character or character state. *See also* plesiomorphy.

aril: fleshy, edible surround to a seed.

assumptions 0, 1 and 2: in biogeography, methods of overcoming problems of widespread taxa and redundancy in areas of endemism, given the goal of every area only occurring once in an area cladogram. Assumption 0 treats widespread taxa as monophyletic and allows no manipulation of areas. The other assumptions also allow the area relationships of widespread taxa to be paraphyletic (1) or polyphyletic (2) in order to retrieve information about general area relationships.

autapomorphy: apomorphy that is restricted to a single taxon: that taxon is defined by autapomorphies.

autecology: ecology of a single species.

benthic: aquatic, bottom-living.

bottleneck, genetic: drastic reduction in genetic diversity of an organism by a period of extremely low population, for example, during a colonisation event. *See also* founder effect.

branch-and-bound: algorithm for cladogram construction that starts with a cladogram from a heuristic search (*q.v.*) and then searches for cladograms with topologies of progressively shorter lengths than that of the original, discarding all those that exceed it.

branch swapping: procedure for moving clades (branches) around a cladogram in a search for a more parsimonious solution or topology.

CAFCA: computer program for cladistic analysis.

clade: monophyletic group of organisms.

cladistics: method of classification that groups taxa hierarchically on the basis of homologies (shared apomorphies-synapomorphies) into nested sets, conventionally represented as a cladogram.

cladogenesis: splitting of an evolutionary lineage into discrete daughter lineages.

cladogram (taxon or area): branching diagram indicating hierarchic relationships amongst taxa (or areas) based upon the sharing of apomorphies (or related taxa).

cluster analysis: method of classification that groups items hierarchically into nested sets or non-hierarchically (overlapping clusters that can share items) in terms of overall similarity of their attributes.

coding: in cladistics, conversion of observations on characters and character states into alphanumerical format for cladistic analysis.

COMPONENT: computer method for comparing, and identifying common features (congruence) in, the structure of cladograms where the terminal items of each are the same (areas in area cladograms from different groups of organisms) or related (*e.g.*, parasites and their hosts).

component (of tree): group of taxa (or areas) related by the branching structure in a cladogram.

component analysis: method of identifying the degree of commonality of components (congruence) between trees (*e.g.*, COMPONENT).

congruence (of trees): agreement in tree topology. *See also* component analysis and COMPONENT.

consensus tree (strict, Adams, Nelson): tree (cladogram) produced by a consensus method. Methods of cladistic analysis can yield several trees of different topology but the same minimum length. Consensus methods combine the grouping information in these into a single topology known as the consensus tree.

consistency index (CI): strictly the ensemble consistency index. Measure of the amount of homoplasy (repeated changes in characters) in a data matrix relative to a cladogram derived from it. CI has an upper bound value (no homoplasy) of 1 and a theoretical lower bound of 0 (though this cannot be attained in practice).

contact zone: meeting zone of parapatric (*q.v.*) species.

dendrogram: tree diagram derived in application of a hierarchic method of cluster analysis.

depauperate: biota with fewer taxa than expected (for example, in relation to area, representation of higher groups, etc.).

diploid: organism with a standard pairing of chromosomes. *See also* polyploid.

disjunction: major geographical gap in distribution of an organism that may not necessarily be caused by the absence of suitable habitat.

endemic: found only in the area under consideration.

euphotic zone: stratum near surface of water where sufficient light penetrates to permit photosynthesis.

eurythermal: tolerant of wide variations in temperature.

founder effect: reduction in genetic diversity in an initial colonising population, often followed by genetic drift. *See also* bottleneck.

general area cladogram: cladogram of areas where the topology represents the most parsimonious summary of information in a set of area cladograms for individual taxonomic groups, in some ways a consensus cladogram.

generalised tracks: significantly coincident distribution patterns in panbiogeography.

genetic drift: enhanced, stochastic changes in genetic diversity of small, colonising populations of organisms. *See also* founder effect.

Hennig86: computer program used for cladistic analysis.

heuristic search: method of constructing cladograms that is not guaranteed to find the most parsimonious solution.

homoplasy: any derived character that is not a synapomorphy in relation to a particular tree (cladogram) topology.

ingroup: group of taxa under study in a cladistic analysis. *See also* outgroup.

introgression, genetic: infiltration of genetic material of one species into the genotype of another.

length of cladogram/tree: minimum number of character changes or steps on a cladogram required to account for the data.

lineage: all descendant taxa through time of a common ancestor.

massing centres: concentrations of species within a panbiogeographic track.

megaherbivore: large plant-eating vertebrate.

metapopulations: populations of species occupying discrete patches of suitable habitat and interacting through migration.

monophyly (-letic group): clade defined by synapomorphies; a group that includes all, and only all, of the descendant taxa of a common ancestor.

monotypic: higher taxon consisting of only a single lower taxon (usually a species).

mtDNA: mitochondrial DNA.

node, cladistic: branching point on a cladogram.

node, panbiogeographic: intersection point of two or more generalised tracks.

non-metric multidimensional scaling: method of summarising the distribution of points in multidimensional space in a smaller number of dimensions by minimising disturbance to the rank order of distances between the points.

outgroup: taxon used in cladistic analysis for comparison with group under study (ingroup) to determine character polarisation.

pandemic: taxon distributed universally through the geographical area being studied.

paralogy: (as in paralogy-free subtree analysis) term borrowed from genetics to denote repetition of information in area cladograms.

parapatric: distributions that abut at a contact zone but do not overlap, usually of closely related or sister species.

paraphyly (-letic group): group of taxa in a monophyletic group from which one or more components are excluded.

parsimony: choosing the hypothesis that explains the data most simply. In cladistic analysis this is achieved by minimising the number of character changes inherent in a cladogram topology.

PAUP: A computer program for cladistic analysis.

PeeWee (PIWE): a computer program for cladistic analysis.

phanerogam: seed-plant (conifers and angiosperms).

phenetics: classification of organisms and other items based on overall similarity of their attributes. *See also* cluster analysis.

phylogeny: hypothesis of genealogical relationships of taxa, imposing concepts of ancestry and a time axis on a cladogram.

planktonic: organisms that drift almost passively in bodies of water, usually in the surface layers.

plesiomorphy: ancestral or primitive character state, which may also be an apomorphy of a more inclusive hierarchical level than that under consideration.

polarisation (of characters): determination of the apomorphic and plesiomorphic states of a character, often by outgroup comparison.

polymerase chain reaction (PCR): method of multiplying extracted DNA to facilitate its analysis and comparison.

polyploid: organism with multiples above two (diploid) of the haploid number of chromosomes.

polytomy: node in a tree or cladogram which has three or more distal branches. *See also* resolution.

Q-mode analysis: in a two-way table, classification/comparison of the columns in respect of values in the rows.

R-mode analysis: in a two-way table, classification/comparison of the rows in respect of values in the columns.

redundancy: in trees, refers to repeated information about the relationships of constituent items.

relict: localised remnant of a previously much wider distribution pattern.

resolution (cladistic): extent to which the branching in a tree or cladogram approaches the fully dichotomous.

retention index (RI): strictly the ensemble retention index. For a given cladogram, this measures the amount of similarity in the original data matrix that can be interpreted as synapomorphy, by comparing the actual amount of homoplasy as a fraction of the maximum possible homoplasy. The RI equals 1 for a data set comprising only unique and unreversed synapomorphies (no homoplasy), whereas a value of 0 implies no grouping information at all in the data.

sibling species: closely related species only recently diverged from a common ancestor, probably showing close sister relationship.

sister relationship: shown by two taxa that are more closely related to each other than either is to a third taxon.

subtree: branch of a tree or cladogram.

successive approximation weighting: procedure for *a posteriori* weighting of characters according to their cladistic consistency, for example as indicated by the (rescaled) consistency index for the characters.

sympatry: co-occurrence of taxa in an area.

synapomorphy: apomorphy shared by taxa in a monophyletic group.

synecology: the study of associations or communities of species.

three-item statements (TAS): expression of the relationship between three taxa or areas where two are more closely related to each other than either is to the third.

track: in panbiogeography, the distribution of a taxon, often depicted by lines (representing the shortest distances) linking the localities where it occurs.

ultrametric: distance measures between items being classified form, for any three, an isosceles triangle. In a phylogenetic tree, each terminal taxon would show (if this could be measured precisely) an equal amount of divergence in characters from those of the common ancestor of all the taxa.

vicariance: fragmentation of ancestral species ranges by the appearance of physical (or ecological) barriers.

Index

A

acritarch 28
Alpine-Himalayan belt 99
amphibians 83–89, 135, 136
   temnospondyl 85
Andaman
   Islands 240, 250, 266, 385
   Sea 104, 270, 363
angiosperm 17, 18, 135, 148, 211–229, 244, 255, 279
   origin 211–229
   radiation 212
anoa 136
Antarctica 39, 50, 95, 102, 134, 215, 378
antelope 398
Antidesma 243, 250–251
Aporosa
   biogeography 279–289
   cladistic analysis 279
   distribution 283, 310
   phylogeny 280–283
Apsilochorema
   distribution 91–98
   phylogeny 93–94
areas of endemism 8, 275, 341, 344
   cicadas and Lepidoptera 5, 291–312
   definition 5
   Malesian plants 256
   marine water striders 5, 344
   Moluccan butterflies 317
   Spatholobus 263
Arfak 283, 317, 331
Aru
   basin 114
   Islands 248
Asian plate 217, 225
Assam 94, 218, 239
Australasia 1, 34, 78, 108, 211, 235, 294, 374, 393
Australia 5, 28, 43, 57, 85, 91, 99–123, 134, 172, 200, 212, 237, 244, 267, 285, 291, 317, 330, 342, 362
azooxanthellate corals 168

B

Bali 133, 251, 355, 393
Banda
   arc 10, 114, 279, 291, 328, 384
   Sea 107
Banggai-Sula 27, 140
Baoshan 57–70
bats 16, 136, 292, 385
bears 135
beetles 309, 310, 355
Biak 317, 331
biogeography
   analytical 243–244
   descriptive 243–244
   dispersalist 292
   empirical 243–244
   historical 315

   methodology 3, 262, 291–292, 316–317, 344–346
   molecular 197
   narrative 243–244
   Permian 57–70
   studies 243–244
   vicariant 292
birds 9, 135, 293, 316
   Banda arc 384
   Fiji 365
   Indo-Pacific 361–390
   Maluku 374, 384
   New Caledonia 362–363
   New Guinea 369
   New Zealand 362
   Norfolk Island 365
   Samoa 365
   Sulawesi 136, 375, 384
   Tonga 365
Bird's Head 9, 107, 227, 283, 304, 317, 352, 382
Bismarck
   arc 328
   archipelago 243, 285, 320, 343, 382
   Islands 18, 293, 382
Bonins 12
Borneo 4, 25, 94, 107, 169, 173, 236, 246, 259, 268, 293, 358, 375, 393
   fauna 135
   flora 135
   geological evolution 133–151
   geology 137
   palaeogeography 133
   tectonics 137
brachiopods 27, 28, 43–54, 57–70
   Ordovician 43–54
   Permian 57–70
Brunei 19, 144, 239, 244
Burma 17, 25, 43, 58, 91, 171, 251, 267, 343, 385, 399
   plate 111, 267
Buru 19, 171, 240, 285, 303, 316, 374
Buton 140, 144, 146, 385
butterflies 6, 292, 293, 315, 355. *See also* Lepidoptera
   Seram and Halmahera 315–324

C

caddisfly 17, 91
Cambodia 58, 344
Caribbean 2, 165, 169, 185, 204, 250, 292, 293, 343, 347
Caroline
   arc. *See* South Caroline arc
   Islands 12, 248
   plate 111
   Ridge 108
   Sea 108
Cathaysialand 25, 57, 68, 69, 73, 78, 79
cats 135, 136, 398
cattle 398
Celebes Sea 108, 140, 269
Ceno-Tethys 25, 385
centre of accumulation 166

centre of origin 165, 203, 206, 323, 400
    Paleogene corals 165–192
centre of richness 294
centre of survival 166
cephalopods 51
Chagos Islands 11
Changning-Menglian 37, 58
China 43, 247, 400
    North 25, 27–30, 43, 85
    South 25, 43, 57, 83, 108, 137, 237, 267
chital 398
Chitaura 355–359
Christmas Island
    Indian Ocean 11
    Pacific 200, 203
cicadas 9, 136, 293, 304–305
    Indo-Australian tropics 291–312
Cimmerian 6, 14, 33, 37, 57, 63, 66
civets 398
cladistic
    analysis 3, 243
    biogeography 3, 291, 316, 341
climate 19, 99, 122, 150, 287, 393
    and plate tectonics 211–229
    Cenozoic change 235
    mid and late Tertiary 220
climatic barriers 393
cluster analysis 6, 291, 293, 295, 299
Cocos-Keeling Islands 11–12
conifers 6, 18, 225
conodonts 27
Coral Sea 107, 201, 295, 330, 375
coralline algae 165, 171, 201
corals 27, 341, 352
    algal symbiosis inferred in fossils 169
    Cenozoic links to plate tectonics 172
    cladistic biogeographic analysis 206
    distributional change 183
    Eocene 165, 175–176
    global biogeography 182
    implications of Mesozoic record 190
    Indian Ocean 7
    maintenance factors 182–183
    Mesozoic scleractinian records 171
    Miocene 179
    Neogene 179
    Oligocene 176
    originations 185–189
    Paleogene 165–192
    Paleogene gap in Indo-West Pacific 169
    records 169
    zooxanthellate 168
crabs, coconut 206
crocodilian 85–89
Croizat 3, 316, 361
crustaceans 6

D

deer 136, 398
deforestation
    ages 396–397
diamictites 29, 59, 68

dinosaurs 83–89
dipterocarps 17, 222
dispersal. *See also* rafting
    barrier, Borneo 401
    barriers 393
    island hopping 399
    migration 292
    model 7–9, 206, 292
    mountain plants 211
    path, late Cretaceous and early Tertiary 213–214
    reality of events 255
    timing 198
DNA 198
    hybridization 16
    mitochondrial 199–207, 355
    mutation 198
    sequence data 355–359
dogs 136, 398

E

East India Letter Classification 169
East Philippines arc 108, 317
echinoids 171
Elaeocarpaceae 243–256
Elaeocarpus 243, 253–255
elephants 136
Ephemeroptera 97
Eugeissona 214, 219, 225, 248
Euphorbiaceae 17, 216, 243, 279–289
    geological history 285
Euphorbiaceae-Phyllanthoideae 250–251
Eurasia 25, 39, 57, 83, 84, 99–123, 137, 168, 267
expanding Earth 14, 79, 101
extinction 14, 15, 18, 20, 183, 203, 240, 255, 263, 288, 318, 320, 321, 352, 362, 398, 399, 401
    events 247
    factor 8
    K-T 182, 190
    rates 207
extrusion hypothesis. *See* indentor hypothesis

F

Fanning Island 200
Fiji 13, 93, 204, 223, 243, 301, 322, 343, 361
fish 83, 84, 135, 136, 201, 294, 401
    antiarch 36
    sinolepid 36
    yunnanolepid 36
floras
    Carboniferous 73
    Carboniferous and younger 29
    dispersals from Asia 225–226
    dispersals from Australasia 227
    east of Wallace's Line 223–225
    Eocene 211, 217–219
    Indian Ocean 12
    Jambi 75
    Miocene migrations 222–223
    Permian 73, 74
    Permian, Irian Jaya 78
    Permian, Laos 78

Index

Permian, New Guinea 78
Permian, Papua New Guinea 79
Permian, Thailand 75
Permian, West Malaysia 78
Upper Palaeozoic 73–81
West Malaysia, Sumatra and Thailand 74
foraminifera 64, 70, 87, 165, 171, 176, 399
fossil record
    angiosperms 212–213
    Java mammals 398–399
    Leguminosae 269
    pollen 212
    Spinizonocolpites 246
    value of evidence 14
frogs 134, 309, 385

G

Gag 316
gastropods 27, 51
Gebe 316
gene flow 197
    Pacific patterns 200
genetic change in populations 198
genetic data
    nature and utility 8, 16, 198–200, 357
    SE Asian marine species 200
genetic surveys
    Indian and Pacific oceans 203–204
genetics
    population 198
geological evolution
    Cambrian 34
    Carboniferous 37
    Cenozoic 99–123
    Cretaceous 39
    Devonian 35
    Jurassic 38
    Ordovician 34, 43–54
    Palaeozoic and Mesozoic 25–39
    Permian 37–41, 57–70
    Silurian 35
    Tertiary, Borneo and Sulawesi 133–151
    Triassic 37
Gerromorpha 341
    Indo-Pacific 341–353
giant clam 197, 201
    geographic variation 201–203
gibbons 16, 136, 393, 394
glacial
    cycles 393
    maxima 397
Gondwanaland 4, 17, 25–39, 73, 78, 79, 94, 150, 211, 267, 386
Gramineae 243, 248
    Bambuseae 248–250
grasshoppers 355
    distribution 356
    Sulawesi 355–359
Great Barrier Reef 172, 201
Guangxi 73
Gulf of Thailand 111, 226
gymnosperm 18, 78, 79, 215–229, 333

H

Hainan 30, 52, 236, 267
Halmahera 12, 107, 220, 238, 289, 315, 374
    arc 108, 317, 385
    butterflies 315–324
    geology 317
Hawaii 5, 8, 13, 200, 239, 253, 311
Hawaiian-Emperor seamount chain 108, 183
Heteroptera 327, 341
    aquatic 327
    Indo-Pacific 341–353
Heteroptera distributions 333–339
    Australia 333
    influence of rock type 335
    Melanesian arc 333
    Papuan arc 334
    Solomons arc 336
    Vogelkop 333
Himalaya 33, 57, 61, 94, 211, 237, 250, 267, 279, 387
Huxley's line 393, 401
hybridisation 4, 279, 355
Hydrobiosidae 17, 91–98
hyenas 398

I

indentor hypothesis 9, 102–123
India 25, 66, 214, 239, 248, 267, 394
India-Asia collision 9, 106, 172, 269
Indian
    Ocean 11, 38, 103, 173, 197, 341
    plate 12, 91, 107, 211, 240, 267
Indo-Malayan 168, 352
Indo-Pacific 197, 341
Indo-Pacific gateway 172
Indo-West Pacific 165–192, 168
Indochina 25, 43, 57, 83, 99, 137, 190, 223, 236, 244, 267, 336, 387, 393
Indonesia 13, 30, 43, 75, 100, 150, 173, 225, 240, 244, 285, 315, 342, 388, 394
Indonesian archipelago 133, 169, 171, 190, 205, 289
Indoralian 57, 67
insects
    aquatic 91–98, 327–339. *See also* Heteroptera; Hydrobiosidae
    marine 341–353. *See also* Gerromorpha; Heteroptera
    terrestrial 291–312, 315–324, 355–359. *See also* butterflies; cicadas; grasshoppers; Lepidoptera
Irian Jaya 63, 73, 213, 239, 255, 289, 328
island arc accretion
    New Guinea 327–339
    SE Asia and SW Pacific 107–123
island hopping 133, 149–151, 393
Isthmus of Kra 251, 259, 269
Izu-Bonin-Mariana arc 108

J

Japan 15, 25, 88, 107, 172, 211, 236, 253, 344
Java 11, 107, 133, 169, 176, 217, 236, 247, 267, 282, 293, 355, 375, 393
    Sea 137, 219, 246

K

Kalimantan 120, 133, 217, 236, 385, 400
Kazakhstan 25, 37, 50, 57
Kerala 266
Khorat plateau 83
    vertebrates 83–89
Krakatau 256, 393
Kurosegawa 25, 30

L

Langkawi Island 46–48, 59
Laos 58, 73, 88, 267
Laurasia 25, 75, 84, 91
Laurasian 225
Laurentia 29, 44
Leguminosae 221, 259
lemurs 136
leopard 398
Lepidoptera 11–13, 136, 293, 384
    Gondwanan groups 17–18
    Halmahera and Seram 315–324
    Indo-Australian tropics 291–312
Lesser Sunda Islands 6, 133, 236, 251, 287, 302, 321, 344, 385, 393
Lhasa 25, 57, 83, 91, 269
Lombok 133, 239, 393
lorises 136, 393
Loyalty Rise 107
Luzon 108, 220, 236, 251, 269, 343, 397
    arc 114
Lydekker's line 133, 393

M

macaques 16, 355, 358, 393–402
Madagascar 18, 96, 211, 239, 243, 285, 344
Mahakam delta 145, 222
Makassar Strait 108, 133, 235, 268, 393
Malawa Formation 217
Malay
    basins 222, 226
    peninsula 43, 107, 168, 212, 236, 247, 259, 282, 385, 394
Malaya 57, 267, 285
Malaysia 5, 37, 43, 48, 58, 73, 137, 219, 244, 287, 293, 321, 344, 358, 397
Malesia 229, 235, 243, 247, 279, 285, 341, 343, 344
Malesian plants
    distribution patterns 243–256
Maluku 243, 248, 315, 343, 361. *See also* Moluccas
    geological history 383–384
mammals 19, 134, 214, 385, 398
    glacial and pre-glacial distributions 393–402
Mangkalihat peninsula 142, 217
mangroves 19, 220, 341, 352
Maramuni arc 114
Marianas 12, 108, 173
marine biodiversity 197
marine organisms
    genetic structure 197–207
Marshall Islands 12, 173, 200
marsupials 136, 385
mayflies 97
Mayu 315
Melanesia 327
    geological history 105–123
Melanesian
    arc 9, 108, 279, 302, 328, 378
    archipelago 291
Mentawai
    archipelago 393
    Islands 16, 393
Meratus 259, 268
Mergui/Tenasserim 267
Meso-Tethys 25
metapopulation dynamics 8
Micronesian island arcs 12
Mindanao 237, 247, 289, 293, 334, 383, 401
Mindoro 113, 224, 236, 269
Misool 171
modelling dispersal, speciation and vicariance 7
molecular
    analysis 357–358
    clock 14, 311, 355, 357
    phylogeny 355
moles 136
mollusca 44, 171
Molucca Sea plate 111
Moluccas 9, 150, 223, 286, 293, 315, 355, 401
    geological history 105–123
Mongolia 64, 83
monkeys 16, 393–402
Morotai 333, 374
mosses 235
    disjunctive patterns 236–241
    Malesian 235–241
    Malesian biogeography 235–236
    speciation 236
mountain plants
    dispersal 225–226
Müller's line 402

N

Nanggulan Formation 217
Natuna basins 226
nautiloids 27, 44
Nepal 33, 93, 240, 253, 267
New Britain 108, 361
    arc 331
New Caledonia 107, 173, 176, 228, 236, 244, 301, 333, 361, 378
    Rise 107
New Guinea 9, 94, 168, 236, 243, 279, 293, 315, 327, 352
    geological history 105–123, 328–330
New Guinea arc 111
New Hebrides arc 114
New Zealand 169
    geological history 375–378
    parallel arcs model 19, 292, 379
Nicobar Islands 248, 267
nightjars 361
    evolution 387
Ninetyeast Ridge 173, 184, 218

Index

non-metric multidimensional scaling 6, 292
Norfolk
   basin 111
   Island 292, 321, 362
   Ridge 378
North Fiji basin 13, 115, 381
Northern New Guinea plate 107
Nypa 243, 244–246

O

Obi 308, 315, 383
ocean floor magnetic anomalies 99
oceanic circulation 122
Ontong Java plateau 111, 122, 329
ophiolite
   New Caledonia 108
   New Guinea 339
   Papuan 107
   Sepik 107
   Sulawesi 111, 140
orang-utans 393, 400
Orthoptera 355
   molecular phylogeny 355–359
otters 136

P

Pacific 99, 244
   marine organisms 197–207
   Ocean 12, 103, 173, 203, 344
   plate 103, 168, 173, 317, 328, 352, 378
Palaeo-Tethys 25
palaeogeographic evolution
   Cenozoic 99
   Mesozoic 25–39
   Palaeozoic 25–39
   Sulawesi and Borneo 133–151
   Tertiary 133
palaeogeography
   Australasia 43
   Borneo and Sulawesi
      Eocene 140
      Miocene 144–146
      Oligocene 144
      Pliocene-Recent 147
palaeomagnetic data 1, 29, 99
Palau 12, 344, 389
Palau-Kyushu ridge 111
Palawan 19, 39, 107, 226, 235, 243, 269, 321, 336, 343, 355
palm 17, 243
Palmae 243, 244–247
palynology 211–229
   evidence for Tertiary plant dispersals 211–229
palynomorphs 84, 213
   Eugeissona 248
panbiogeography 3, 291
Pangaea 37, 43, 91, 240
pangolins 136, 398
Papua New Guinea 79, 108, 169, 213, 238, 244, 317, 330
Papuan
   arc 327

   peninsula 286, 309
Papuasia 247, 255, 344
paralogy 5, 294, 341, 350
Parece Vela basin 111
Penang/Kedah 267
Penyu basin 226
Permian biogeography 57–70
Phalacrocoracidae 379–380
phalangers 401
phenetic
   biogeographic methods 5, 291, 292
   classification 243
Philippine Sea plate 103, 168, 269, 381
Philippines 10, 39, 101, 133, 172, 200, 236, 244, 248, 259, 282, 291, 331, 342, 375, 393
Philippines-Halmahera arc 108
phytogeography 73–81, 211–229, 235–241, 243–256, 259–275, 279–289
phytosaurs 84
plant dispersal 211–229
   by birds 255
   plate tectonics and climate 211–229
plants
   distribution
      Malesian, eastern 243
      Malesian, Sundaic 243
   Malesian 243–256
plate motions 100
plate tectonics 1, 99
   and climate 211–229
   Cenozoic 99–123
   Palaeozoic and Mesozoic 25–39
   Paleogene 165
Plecoptera 97
Polynesia 13
population genetics 197
porcupines 136
primates 16, 135, 393–402
   Borneo 393–402
   Borneo impoverishment 400
   distributions 394
   modern Borneo diversity 400
proto-South China Sea 39, 107
Psittacidae 379

Q

Q-mode analysis 3, 293, 294
Qiangtang 25

R

R-mode analysis 3, 291–312
rafting 91, 133, 219, 393, 394, 399
rainforest
   expansion and contraction 393–402
   glacial refugia 402
   refugia 393
rarity 8
reconstruction
   Cambro-Ordovician 34
   Carboniferous 37
   Cretaceous 39

Devonian 35
Eocene 106–107, 107–108
Jurassic 38
Miocene 111–114
Oligo-Miocene 110–111
Oligocene 108
Ordovician 50
Permian 37
Silurian 35
Triassic 37
Red River fault 107
Reed Bank-Dangerous Grounds 25
regional evolution
　Cenozoic 99–123
　Mesozoic 25–39
　Palaeozoic 25–39
reptiles 134, 135, 136
　dicynodont 37
　Mesozoic 83–89
rhinoceroses 135, 136, 398
rifting
　Carboniferous-Permian 32
　Devonian 32
　Triassic-Jurassic 33
rifting and separation
　Gondwanaland terranes 32
rodents 16, 383
Rotuma 13, 301
Ryukyu Islands 113, 244, 309, 342

S

Sabah 19, 120, 135, 179, 227, 238, 251, 270, 402
Sagaing fault 111
Sahul shelf 19, 393, 401
Sakhalin 93
Samoa 13, 301, 343, 361
Sangihe 148, 293
Sapindaceae 214, 310
Sarawak 19, 120, 140, 175, 215, 239, 251, 400
Sarawak/C Borneo 268
scleractinian coral 165
　ecology 168–169
sea level 122, 197
　Eocene-Pliocene 259, 270–274
　Pleistocene 16
sea urchins 203
sea-floor spreading 1
Semitau 25, 144, 259, 268
Sepik basin 330
Sepik-Papuan arc 107
Seram 19, 114, 171, 177, 225, 237, 285, 303, 315, 355, 374
　butterflies 315–324
　geology 317–318
Shan-Thai 43, 57, 83, 239. *See also* Sibumasu
sharks 86
Shikoku basin 111
shrews 16, 135, 136
Siberia 25, 37, 44, 57, 88, 91
Sibumasu 25, 43, 57, 267, 285. *See also* Shan-Thai
　geological history 29–39
　Ordovician biogeography 43–54
　Permian biogeography 57–70

Simao 37, 57
Singapore 244, 267, 343, 399
Solomon
　arc 327
　Islands 9, 108, 244, 289, 293, 378
　Sea 108
South Caroline arc 9, 108
South China Sea 19, 25, 102, 221, 237, 269
South Fiji basin 108, 380
Spatholobus
　biogeography 259, 259–275
　cladistic biogeography 262–263
　history 274
　phylogeny 260
speciation 4, 120, 197, 198, 250, 259, 280, 292, 317, 336, 351, 355
species-area relationship 7
Spinizonocolpites 243, 244
squirrels 135, 398
Sri Lanka 18, 219, 239, 244, 279, 343
starfish 197, 203
stegodonts 149, 401
stoneflies 97
stromatoporoids 27
Sula 114, 307, 315, 344, 375
Sulawesi 5, 30, 96, 99, 137, 165, 225, 236, 243, 259, 285, 291, 315, 334, 343, 355, 361, 393
　fauna 135
　flora 136
　geological evolution 133–151
　geology 137
　palaeogeography 133
　tectonics 137
Sulu 239
　arc 108
Sumatra 16, 172, 236, 247, 267, 288, 309, 393
Sunda shelf 19, 135, 259, 401
Sunda-Java-Sulawesi arcs 108
Sundaic 136, 247, 248, 267
Sundaland 4, 101, 175, 247, 285, 291, 306
Sundanian 16, 211, 279, 291
surelis 393, 394
SW Pacific 91
　geological evolution 99–123

T

Taiwan 107, 226, 237, 251, 344
tapirs 136
Tarim 25, 213
tarsiers 393
Tarutao Island 46
Tasman Sea 107, 378
Tasmania 48, 93, 238, 246, 321
temnospondyls 85
Tengchong 57
terranes 25
　amalgamation and accretion 33
　New Guinea 317, 381
　origins, East and SE Asian 27–30
Thailand 37, 43, 58, 73, 83, 110, 171, 212, 247, 266, 321, 344, 397
　basins 226

Index

Ordovician sequence 46–48
vertebrates 83–89
Three Kings Rise 111
Tibet 29, 64, 83, 213, 240
Tifore 315
tigers 398
Timor 63, 114, 171, 304, 321, 361, 401
Tonga 173, 327, 343, 344, 361
Tonga-Kermadec 108
Triassic 25
Trichoptera 91
trilobites 27, 43–54
Ordovician 44–54
tropical rain forest 19, 211–229, 250, 266, 358
expansion 211
Tukang Besi 114, 140
turtles 84, 86

V

Vanuatu 13, 243, 301, 344, 362
vertebrates
Borneo primates 393–402
Cretaceous 87
Jurassic 85
Mesozoic 83–89
Triassic 84
vicariance
aquatic Heteroptera 338
biogeography, biogeographers 57, 255
cicadas 307
dispersive 293
events, interpreting cladograms 262
grasshoppers 358
importance 4
Lepidoptera 300
Mesozoic plants 17
model for tropical marine organisms 206
modelling 7
Proteaceae 18

rejection of dispersal 291
result of passive allopatric speciation 263
school 292
shrews 16
surface circulation 167
tectonic, Shan-Thai, Tengchong, Baoshan blocks 65
Vietnam 19, 31, 46, 58, 109, 223, 344, 396
Vietnamese basins 226
Visayas 239
Vogelkop 79, 255, 308, 329. *See also* Bird's Head

W

Waigeo 9, 316, 330
Wallace 100, 316, 374
Wallacea 1, 9, 133, 224, 375, 393, 401
climatic origin 401
Wallace's line 116, 133, 211, 222, 343, 393, 401
water bugs 9, 327
New Guinea region 327–339
water striders 341
distribution 342–343
Indo-Pacific 341–353
weasels 136
Weber's line 6, 134, 299, 402
West Philippine basin 108
Woodlark basin 115
Woyla 17, 25, 95, 267

Y

Yapen 317, 331
Yunnan 46, 58, 73, 84, 171, 237, 255, 267
Yunnan/N Thailand 267

Z

Zamboanga 107
zoogeography 91, 330, 393
zooxanthellate coral 165–192